RECRYSTALLIZATION

Edited by **Krzysztof Sztwiertnia**

Recrystallization
Edited by Krzysztof Sztwiertnia

Published by InTech

Janeza Trdine 9, 51000 Rijeka, Croatia

Publishing Process Manager Maja Bozicevic

Technical Editor Teodora Smiljanic

Cover Designer InTech Design Team

Additional hard copies can be obtained from orders@intechopen.com

Recrystallization, Edited by Krzysztof Sztwiertnia
p. cm.
978-953-51-0122-2

For EU product safety concerns:
IN TECH d.o.o., Prolaz Marije Krucifikse Kozulić 3, 51000 Rijeka, Croatia,
info@intechopen.com or visit our website at intechopen.com.

Meet the editor

Krzysztof Sztwiertnia is a professor at the Institute of Metallurgy and Materials Science, Polish Academy of Sciences (IMMS PAS). He graduated at the Academy of Mining and Metallurgy in Krakow, Poland. After graduation, he worked successively in the IMMS PAS, the Institut für Werkstoffe der Technischen Universität Braunschweig, in Germany, and again in the IMMS PAS. His research interests span various areas of crystallographic orientation-related investigations. In particular, he is interested in deformation and recrystallization processes in metals, as well as methods of measurement, description and analysis of texture, and microstructure of polycrystalline materials. Professor K. Sztwiertnia is an author and co-author of over 100 research papers and two monographs.

Contents

Preface

Recrystallization and related phenomena that occur during thermomechanical processing of all types of crystalline materials are areas of intensive research. However, particular subject matters of research differ depending on scientific discipline.

In geology, recrystallization is a process that occurs during natural deformation of rocks and minerals subjected to high temperature and pressure. Grains, atoms or molecules can be packed closer together. Under the influence of these metamorphic processes, new mineral grains can be created in crystalline form. Analysis of the consequences of these processes is often used for quantification in geochronology.

In chemistry and other closely related fields such aspharmacology, recrystallization is often applied as a procedure for purifying compounds.

In metallic materials, recrystallization and related annealing phenomena have been long ago recognized as technologically important and scientifically interesting. Perhaps for this reason, they have been studied most widely. Metallurgical research in this field is mainly driven by requirements of industry.

Significant progress has been made ,expressed in hundreds of publications, reviews and monographs. However there are still considerable gaps in understanding of the recrystallization processes. Lack of a complete explanation can be attributed to high complexities of the phenomenon, which consists of a superposition of the processes of local nucleation and grain growth. These processes depend strongly on the characteristics of the matrix, that is usually complex and heterogeneously deformed. Quantitative characterization of the deformed state and description of grain boundary properties constitute the areas of fundamental importance for the understanding of recrystallization.

Comprehension of the nature of the deformed state as the precursor of recrystallization and the nature of local instabilities in the heterogeneous matrix can be achieved by the techniques of Orientation Imaging Microscopy (OIM), both in scanning and transmission electron microscopes (SEM and TEM). However, the standard techniques of Electron Backscatter Diffraction (EBSD) in SEM - although very useful for testing of advanced recrystallization stages ,proved to be less useful when a high spatial resolution in orientation measurement is required. In order to obtain a better spatial and angular resolution, similar techniques developed for the TEM can be used. The TEM offers spatial resolution an order of magnitude better than these in SEM and it can be used for quantitative nanoscale analysis of the microstructure at the beginning of the process. It is also essential to note the importance of dynamic studies in the SEM and in the TEM that should be capable of providing information about temporal relationships between changes occurring in a material throughout the course of recrystallization. For the above reasons, we have included in one of the chapters an example of an application of OIM/TEM to analyze early recrystallization stages.

The entire book should be seen as a snapshot of the subject at this particular moment in time, as seen by scientists, who work on recrystallization-related issues from wide ranging perspective of scientific disciplines, from geology to metallurgy. The authors wish to emphasize that the progress in the particular field of materials science has been possible today thanks to coordinated action of many research groups that work in materials science, chemistry, physics, geology and other sciences. Thus, it is possible to perform a comprehensive analysis of the scientific problem.The analysis starts from the selection of appropriate techniques and methods of characterization. It is then combined with the development of new tools in diagnostics, and it ends with modeling of phenomena.

The book shows selected results obtained during the last years. Its main topics are recrystallization of metallic materials, recrystallization of minerals and recrystallization in pharmacology.They are grouped in the appropriate sections. Each section is illustrated with problems or applications of the process. For example, chapter 1 in section 1 is focused on recrystallization of Zr-based alloys, and second chapter in section 3 on recrystallization of drugs.

Prof. Krzysztof Sztwiertnia
Institute of Metallurgy and Materials Science, Polish Academy of Sciences, Krakow, Poland

Part 1

Recrystallization of Metallic Materials

1

Development of Texture and Substructure Inhomogeneity by Recrystallization of Rolled Zr-Based Alloys

Yuriy Perlovich and Margarita Isaenkova
National Research Nuclear University "MEPhI"
Russia

1. Introduction

Recrystallization of α-Zr is of the great interest both as a rather complicated scientific phenomenon and as a process of the practical importance for applications in the nuclear industry. Meanwhile in the most known monographs on Zr and Zr-based commercial alloys (Douglass, 1971; Tenckhoff, 1988; Zaymovskiy et al., 1994) the recrystallization of α-Zr is considered on the basis of experimental data, obtained more, than 40 years ago. These data urgently require corrections with taking into account the up-to-day theoretical conceptions and the continuous progress in experimental technique. Zr-based alloys are characterized by $\alpha\leftrightarrow\beta$ phase transformations within the technologically important temperature interval 610°-850°C and by operation of various mechanisms of α-Zr plastic deformation, including slip by basal, prismatic and pyramidal planes as well as twinning in several systems. These features are responsible for very complicated distribution of strain hardening and the corresponding tendency to recrystallization in products from Zr-based alloys. The given chapter makes up some gaps in our knowledge concerning different aspects of recrystallization as applied to α-Zr.

2. Regularities of recrystallization in sheets and tubes of Zr-based alloys with multicomponent rolling textures

The most widespread data on recrystallization regularities in α-Zr pertain to sheets and tubes with the final stable rolling texture of α-Zr (0001)±20°-40° ND-TD $<10\bar{1}0>$ (Douglass, 1971), where ND – normal direction and TD – transverse direction. Meanwhile later the new detailed data were obtained concerning texture development in α-Zr under rolling. In particular, it was established that by rolling of a textureless slab the intermediate texture (0001) ±15°-25° ND-RD $<11\bar{2}L>$ forms, where RD – rolling direction, and keeps its stability up to the deformation degree of 70%, whereupon it converts to the final stable texture (Isaenkova & Perlovich, 1987a, 1987b). Main components of these textures in the order of formation were denoted as T1 and T2. Besides, components (0001)$<10\bar{1}0>$ (T0) and $\{11\bar{2}0\}<10\bar{1}0>$ (T3) are present often in textures of rolled sheets and, especially, tubes. All these texture components form owing to activity of concrete combinations of plastic deformation mechanisms, have their characteristic strain hardening and therefore show different tendencies to recrystallization. In the real case of a multicomponent texture the development of recrystallization must be additionally complicated by interaction between different components in regions of their contact. Indeed, according to (Isaenkova et al., 1988; Perlovich et al., 1989), resulting changes of α-Zr rolling textures in the course of recrystallization can not be reduced to 30°-rotation around basal normals and require for more complex description. In order to investigate this question in more details, the following work was undertaken.

2.1 Materials and methods

Recrystallization was investigated in sheet samples of alloys Zr-2,5%Nb, Zr-2,3%Cr and pure Zr as well as in tube samples of the alloy Zr-2,5%Nb. Sheets were produced by longitudinal or transverse cold rolling up to deformation degrees in the range from 40% to 90% in such a way as to form the following textures: T1, T1+T2, T2, T1+T2+T3. The weak component T0 was present everywhere. Perfection parameters and mutual relationship of different components varied. The channel tube was cold-rolled by 50%-thinning of its wall. All samples were annealed in dynamic vacuum at temperatures 500°-600°C during 1-5 h.

The main used method was X-ray diffractometric texture analysis. Direct pole figures (PF) (0001), $\{11\bar{2}0\}$ and $\{10\bar{1}2\}$ were measured by the standard procedure (Borodkina & Spector, 1981). To reveal PF regions, where texture changes by recrystallization are predominantly localized, diagrams of PFs subtraction (SD) were calculated and constructed. SD involves contours of pole density equal changes, having been drawn by comparison of recrystallization and rolling textures. In addition, PF sections of interest were constructed to follow redistribution of basal and prismatic normals in the course of recrystallization.

2.2 Recrystallization in sheets

Analysis of PFs $\{11\bar{2}0\}$ shows that texture changes by recrystallization can be described as rotation around the motionless basal axis only in the case of the rolling texture consisting largely of the component T2. However, the angle of such rotation varies: e.g. in the case of cold rolling by 60% for pure Zr recrystallized at 500°C this angle is equal to 30°, while for the alloy Zr-2,5%Nb recrystallized at 580°C – only 20°. When the rolling texture consists predominantly of the component T1, recrystallization does not involve lattice rotation around basal normals, - this is confirmed by invariance of PF$\{10\bar{1}2\}$.

Main results concern reorientation of basal normals by α-Zr recrystallization, i.e. changes of PF(0001). Superposition of SD and PF(0001) is shown in Fig. 1 for the sheet alloy Zr-2,5%Nb rolled up to deformation degrees 40, 60 and 80%, corresponding to formation of different textures: T1, T1+T2, T2. The densest cross-hatching indicates zones, where texture are localized predominantly. These zones are situated at slopes of initial texture maxima, increasing scattering of the recrystallization texture. Hence, the model of inhomogeneous strain hardening, proposed in (Perlovich, 1994) for textured BCC-metals, is true for HCP α-phase also. But while taking into account the multicomponent character of observed rolling textures, regularities of recrystallization should be more complicated.

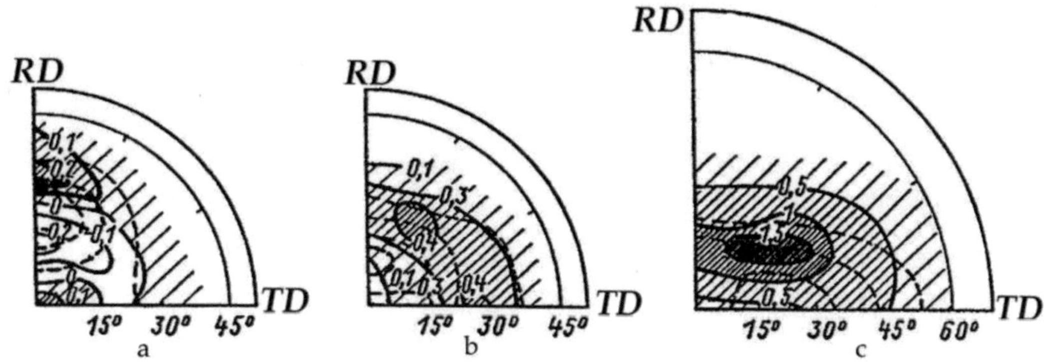

Fig. 1. Superposition of partial PF(0001) (dotted contours) for rolled sheets of the alloy Zr-2,5%Nb by deformation degrees 40% (a), 60% (b), 80% (c) and SD (solid contours), characterizing texture changes by recrystallization. Cross-hatching density reflects an increase of pole density by recrystallization.

For rolling textures T1+T2, where I(T1)>I(T2), it was established that the relative increase of main different components in the recrystallization texture depends on the initial relationship of these components in the rolling texture. In particular:

$$\Delta I(T2)/\Delta I(T1)=f_1[I(T2)/I(T1)], \quad (1)$$

$$\Delta I(T2)/\Delta I(T0)=f_2[I(T2)/I(T0)]. \quad (2)$$

These dependences are drawn in Fig. 2, where linearity of f_1 and nonlinearity of f_2 are evident. The nearer are initial intensities of components T2 and T1, the greater is difference between growth rates of new grains with corresponding orientations. As an area of the

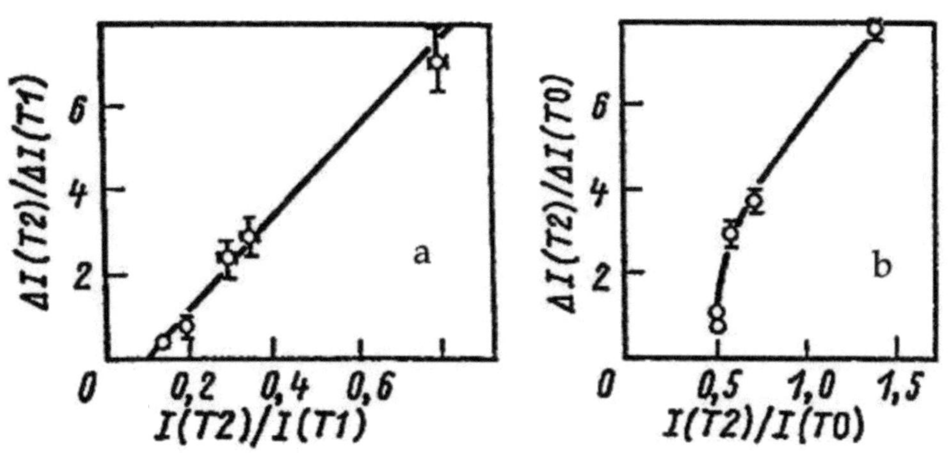

Fig. 2. Relative changing of texture components T2 versus T1 (a) and T2 versus T0 (b) in the recrystallization texture of sheet samples depending on the initial relationship of these components in the rolling texture.

contact surface between deformed grains of these components increases, conditions for growth of grains with the orientation T2 become more favorable. This signifies that T2-grains are growing into T1-grains and absorb them. The weaker dependence connects growths of components T2 and T0 in the recrystallization texture, since the difference between their strain hardening is less than in the case of components T2 and T1. Recrystallization of samples, showing predominance of the component T1, involves essential redistribution of basal normals even to the point of main texture component changing; then, depending on the concrete relationship of components in the rolling texture, T1 can give way to T0 (Fig. 3-a) or T2 (Fig. 3-b). Composition of the alloy influences mainly temperature parameters of the recrystallization process in α-Zr and seems to be of secondary importance for orientation regularities.

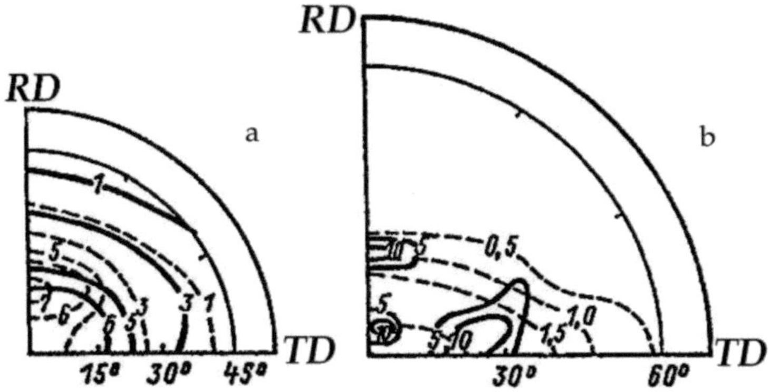

Fig. 3. Partial PF (0001) for sheet alloys Zr-2,5%Nb (a) and Zr-2,3%Cr (b) in rolled (dotted contours) and recrystallized (solid contours) states.

2.3 Recrystallization in tubes

The rolling texture of investigated tubes exhibits predominance of the component $\{11\bar{2}0\}<10\bar{1}0>$ and contains the intensive axial component with the axis $<10\bar{1}0>$ (Fig. 4). Distributions of X-ray reflection (0004) registered intensity in the PF section R-T (radial direction - tangential direction) both for rolled and annealed samples are presented in Fig. 5-a. Redistribution of registered intensity in consequence of tube annealing at 500°C is connected with a general increase of the intensity level by recovery (compare curves 1 and 2 in Fig. 5) and testifies about inhomogeneous recovery in grains with different orientations, resulting from their previous inhomogeneous strain hardening. The most active recovery is localized in grains with basal normals deflected from R-direction by 50°-60° and 90°. Hence, grains with such orientations have the greatest strain hardening and should show the maximal tendency to recrystallization. The curve 2 characterizes the true texture of the rolled tube better, than the curve 1, whose appearance on the inhomogeneous distribution of lattice defects in grains with different orientations.

As the annealing temperature increases, recrystallization process becomes more active and a sharp decrease of pole density near T-direction occurs, testifying about reorientation of basal normals. This effect is especially strong by passing from curve 4 to curve 5 (Fig. 5-a). An increase of the annealing temperature results in shifting of the maximum in the distribution of basal normals to R-direction: after annealing at 550°C is at angular distance of 45° from R-direction, while after annealing at 600°C – at a distance of 60°. The maximum corresponds to grains, which by recrystallization are in favorable conditions for growing, though their initial strain hardening is not maximal. These grains respond to the compromise variant: their strain hardening is so high, that the accumulated energy of lattice distortion ensures their sufficiently quick growth under recrystallization annealing, and at the same time the volume fraction of these grains is sufficiently large for absorption of a significant part of the deformed matrix.

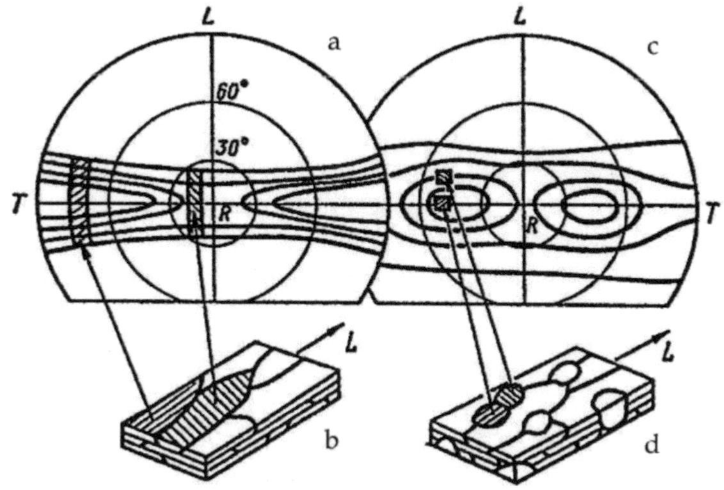

Fig. 4. PF(0001) (a, c) schematic images of microstructure (b, d) for the tube of the alloy Zr-2,5%Nb in rolled (a, b) and annealed (c, d) states. The microstructure (d) corresponds to the initial stage of recrystallization.

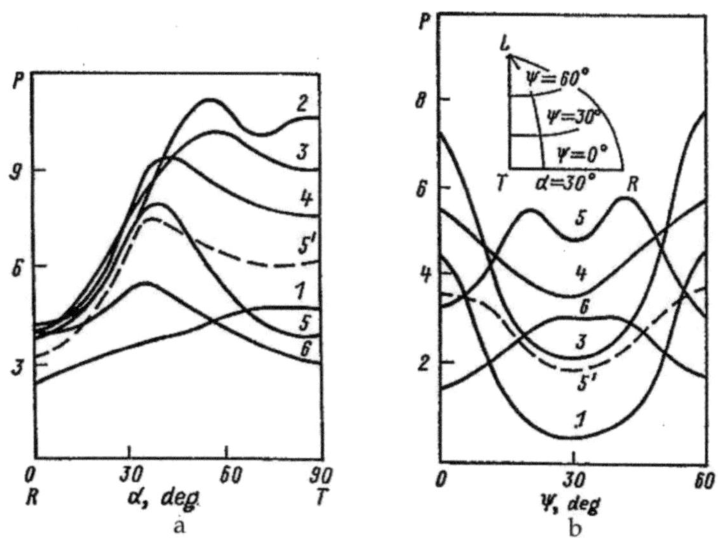

Fig. 5. Intensity distribution in the section R-T of PF(0001) (a) and in the section R-L of PF{11$\bar{2}$0} (b) for tubes of the alloy Zr-2,5%Nb: 1 – initial rolled state; 2-6 – annealing regimes: 2 - 500°C, 3 h; 3 - 530°C, 3 h; 4 - 550°C, 3 h; 5 - 580°C, 3 h; 5′ - 580°C, 1 h; 6 - 600°C, 3h.

As a result of tube annealing the new maxima arise on the meridional section of PF{11$\bar{2}$0} (Fig. 5-b), testifying that, along with redistribution of basal normals in investigated tubes, reorientation of prismatic normals takes place also, leading to development of the typical for α-Zr recrystallization texture (0001) ± α R-T <21$\bar{3}$0 – 11$\bar{2}$0 >. The situation of new maxima in the distribution of prismatic normals corresponds to slopes of initial maxima in the rolling texture, where strain hardening achieves increased values (Perlovich, 1994). The angle of misorientation between deformed and recrystallized grains by the common basal axis depends on the annealing temperature and increases from 20° to 30° when passing from annealing at 580°C to annealing at 600°C (compare curves 5 and 6 in Fig. 5-b). According to the data, presented in Fig. 5-a and 5-b, reorientation of basal normals is somewhat ahead of reorientation of prismatic normals both by time and temperature.

In order to explain the observed development of recrystallization in tubes, the model was proposed, suggesting operation of two different recrystallization mechanisms: (1) growth of new grains with intermediate orientations at high-angle boundaries between regions, having different local textures and originating from different initial grains; (2) growth of new grains with orientations, corresponding to zones of increased strain hardening within rolled initial grains. After rolling up to high deformation degrees, a polycrystal consists of thin plate-like grains, which have their own local textures and contain only low-angle subboundaries. Such structure is shown in Fig. 4-b. Then the first mechanism causes formation of new grains with intermediate orientations of basal normals, as it is shown in Fig. 4-c,d, while the second mechanism results in lattice rotation about basal normals owing to gradual growing of nuclei in zones of increased strain hardening within plate-like grains. Absorption of the deformed matrix by nuclei of new grains, growing along T- and R-directions, is controlled by structure anisotropy of a rolled tube.

Thus, it was shown that by recrystallization of α-Zr along with previously described lattice rotation around basal normals a significant redistribution of these normals takes place both in sheets and tubes. Contrary to the widespread idea, the recrystallization texture of α-Zr varies in a wide range depending on the relationship of main components in the rolling texture.

3. Substructure inhomogeneity of recrystallized sheets from Zr-based alloys

The substructure inhomogeneity is a generally recognized feature of deformed metal materials, controlling the process of their recrystallization. Since the plastic deformation usually is accompanied by arising of the crystallographic texture, developments of the substructure inhomogeneity and the deformation texture prove to be mutually interconnected. It was shown by means of X-ray diffractometric methods (Perlovich et al., 1997; 2000; Perlovich & Isaenkova, 2002), that an actual spectrum of substructure conditions in rolled metals and alloys is extremely wide and that the optimal criterion for systematization of observed substructure inhomogeneities is the grain orientation. The main principle of substructure inhomogeneity is the following: by passing from texture maxima to texture minima, grains (subgrains, blocks etc.) become finer and the lattice distortion increases. A question arises whether the recrystallization removes this inhomogeneity or recrystallized material partially retains it, contrary to the widespread viewpoint concerning its negligible scale.

The above question is very actual as applied to commercial Zr-based alloys in connection with their usage as responsible construction materials in nuclear reactors. The final heat treatment of products of Zr alloys is aimed to attain the stability of their structure and to remove residual stresses of all kinds. The recrystallization is usually believed to satisfy these requirements, since experimental evidences of its efficiency are restricted by those, accessible by standard methods of structure characterization. But standard X-ray methods are selective, i.e. obtained data relate only to grains with some definite orientation, corresponding to their reflecting position by the used measurement geometry. Therefore, these data can be considered as sufficient only under a supposition that recrystallization practically removes the substructure inhomogeneity of products. An aim of the given study is to demonstrate the real substructure inhomogeneity of recrystallized products of Zr-1%Nb and Zr-2.5%Nb alloys using special methods of modern X-ray diffractometry.

3.1 The X-ray method of generalized pole figures

Recent development of X-ray diffractometric technique allowed to elaborate a new method of the fullest description of textured metal materials with taking into account their substructure inhomogeneity. The method involves repeated recording of X-ray line profiles by the geometry of texture measurements, so that, as opposed to the standard description of the substructure by parameters of the X-ray reflection (hkl) from crystallographic planes {hkl} of the single orientation, now the substructure condition of the sample can be characterized by the multitude of line profiles, corresponding to planes {hkl} within grains of different orientations.

The treatment of measured data includes correction for the defocalization effect, approximation of X-ray line profiles with pseudo-Voight functions, calculation of their integral intensity I, physical half-width β and peak position 2θ, construction of distributions $I(\psi,\varphi)$, $\beta(\psi,\varphi)$, $2\theta(\psi,\varphi)$ in the stereographic projection of the sample, where (ψ,φ) – coordinates of reflecting planes {hkl}. These distributions, named Generalized Pole Figures (GPF), characterize substructure conditions along axes <hkl> by all their space orientations. In particular, normalized GPF $I_{hkl}(\psi,\varphi)$ is the usual texture pole figure PF{hkl}, GPF $\beta_{hkl}(\psi,\varphi)$ exhibits the combined effect of coherent block size D_{hkl} and lattice distortion $\Delta d/d_{hkl}$, GPF $2\theta(\psi,\varphi)$ describes the anisotropic elastic deformation $\varepsilon_{hkl}(\psi,\varphi)$ of grains along axes <hkl> due to action of residual microstresses.

The measured GPF $2\theta_{hkl}(\psi,\varphi)$ can be recalculated into GPF $d_{hkl}(\psi,\varphi)$ and further – into GPF $\varepsilon_{hkl}(\psi,\varphi)$, where $\varepsilon_{hkl}(\psi,\varphi) = [d(\psi,\varphi)-d_{av}]/d_{av}$ and d_{av} – the averaged weighted value of interplanar spacing d_{hkl}. Depending on the sign of ε_{hkl}, elastic extension or elastic contraction takes place along axis <hkl> with coordinates (ψ,φ), so that GPF $\varepsilon_{hkl}(\psi,\varphi)$ allows to reconstruct a deformation tensor for grains of main texture components.

Diagrams of the correlation between different GPF are very useful by the analysis of regularities, controlling formation of the inhomogeneous substructure in real metal materials by their technological treatment. These diagrams are constructed in coordinates (β_{hkl}, I_{hkl}) or $(2\theta_{hkl}, I_{hkl})$, so that each their point corresponds to some point (ψ_i,φ_j) in the stereographic projection of the sample and in GPF of reference. When taking into account, that each crystallite of α-Zr, having HCP lattice, has only one axis <0001>, volume fractions of grains with different physical broadening β and peak positions 2θ of X-ray reflections from basal planes (0001), as well as with different values of derivative substructure characteristics along c-axis, can be determined. With this aim for all points (ψ,φ) of PF (0001) values of pole density are recalculated into weight coefficients to be used by the statistical treatment of GPF for the parameter of interest.

3.2 Experimental details and results

Substructure features of recrystallized plates of Zr-1%Nb and Zr-2.5%Nb alloys were studied. Plates were obtained by plain and transverse cold rolling by $\varepsilon \cong 55\%$ of bars, which were cut out from the annealed slab 2 mm in thickness. The direction of cold rolling coincided either with RD of the initial slab or with its TD. All plates were annealed in the evacuated vessel at 580°C during 3h, so that in both alloys the recrystallization of the dominant α-Zr phase took place. The full cycle of X-ray measurements as applied to all plates was carried out twice, that is after rolling and after annealing. The X-ray study was preceded by etching of samples, aimed to remove the surface layer ~40 μm in thickness.

By X-ray studies the texture diffractometer SIEMENS D500/TX with a position sensitive detector was used. The profile of the same X-ray line was registered by each of 1009 successive positions of the sample in the course of texture measurement. For the data treatment both the supplied software and the original programs, elaborated by authors, were applied. All data, obtained for recrystallized samples, are considered in comparison with data for the same samples in the rolled state.

In Fig. 6 incomplete GPF are presented for studied samples both in rolled and recrystallized conditions. All GPF were constructed by measurements of the X-ray line $(0004)_{\alpha\text{-Zr}}$. In Fig. 7 distributions of volume fractions of grains, characterized by different half-widths β of the X-ray line (0004) and by different values of interplanar spacing d_{0001} are constructed for some studied samples. Fig. 8 shows correlation diagrams of GPF $\beta_{0004}(\psi,\varphi)$ and GPF $2\theta_{0004}(\psi,\varphi)$ with PF(0001) for as-rolled (o) and annealed (+) samples.

3.3 Features of obtained distributions

Consideration of obtained data allows to establish the following:

1. As a result of recrystallization, the texture of α-Zr in studied plates changes in accordance with principles, revealed in section 1, so that mutual ratios of main components in the recrystallization texture depend on these ratios in the rolling texture. The most sharp texture changes accompany recrystallization in the Zr-1%Nb plate, obtained by transverse rolling (Fig. 6): the texture with predominance of components (0001)±15°÷20°ND-RD <10.L>, stable by intermediate deformation degrees, transforms into the texture of the central type.
2. Though the significant part of α-Zr crystallites by recrystallization shows a drop of the physical broadening of X-ray line (0004) down to minimal measurable values (Fig. 7), corresponding to the coherent domain size above ~150 nm, the substructure inhomogeneity of all annealed samples is still rather essential and its general character remains the same, i.e. coherent domains becomes smaller and lattice distortions increases by passing from central regions of texture maxima to their periphery. In GPF β(ψ,φ) regions with the most perfect substructure are darkened, so that its perfection increases with the degree of darkening (Fig. 6).

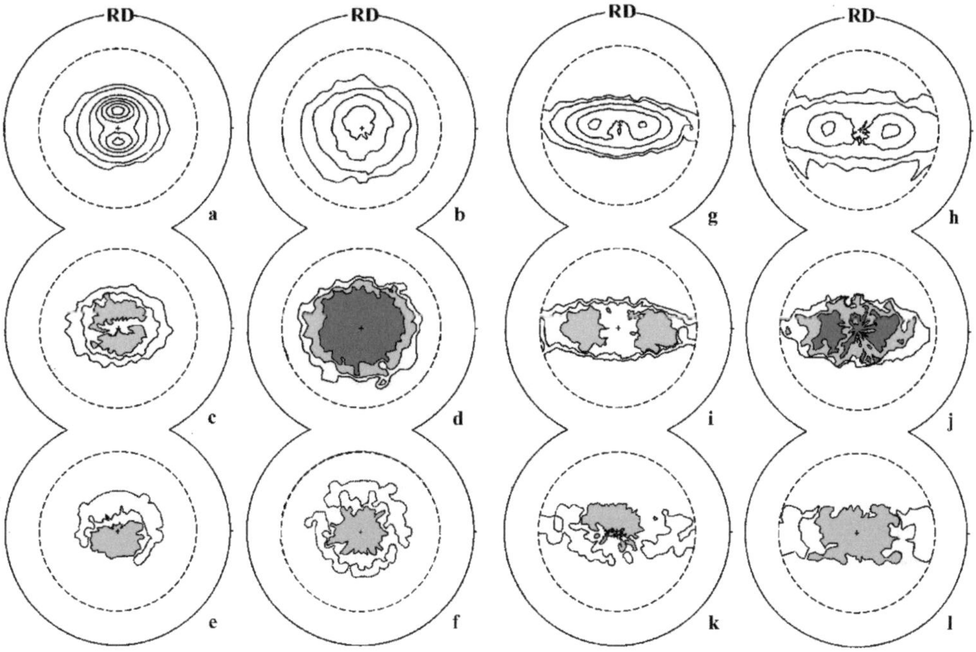

Fig. 6. PF(0001) (a, b, g, h), GPF β_{0004} (c, d, i, j) and GPF ε_c (e, f, k, l) for α-Zr: Zr-1%Nb, transversal rolling, deformed state - (a, c, e), annealed state - (b, d, f); Zr-2.5%Nb, plain rolling, deformed state - (g, i, k), annealed state - (h, g, l). Darkening: GPF β - the dark weakens from β_1= 0,2° to β_2= 0,8°; GPF ε_c – the dark shows regions with ε_c< 0.

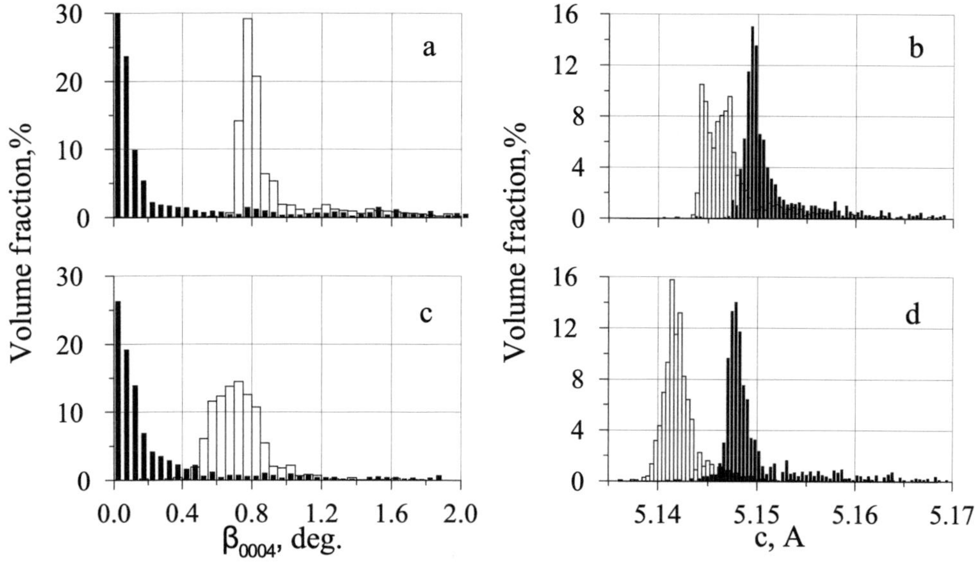

Fig. 7. Volume fractions of α-Zr crystallites with different values of X-ray line broadening β_{0004} (a, c) and lattice parameter "c" (b, d): Zr-1%Nb, transversal rolling – (a, b); Zr-2.5%Nb, plain rolling – (c, d). White columns – deformed state, black columns – annealed state.

3. Contours of equal line broadening in GPF $\beta(\psi,\varphi)$ for recrystallized plates follow contours of equal pole density in their PF(0001), testifying that the substructure inhomogeneity of annealed samples depends on the recrystallization texture and is not connected with their initial rolling texture. Hence, recrystallized plates do not succeed their substructure inhomogeneity to deformed material, but development of this inhomogeneity accompanies formation of the recrystallization texture.
4. Recrystallization results in the decrease of the average elastic microstrain, but nevertheless there are regions in GPF $\varepsilon_{0001}(\psi,\varphi)$ for recrystallized samples, where $\Delta c/c_{av}$ attains rather high values, comparable with those for as-rolled samples. In the orientation space a redistribution of elastic contraction and elastic extension along basal axes takes place in the course of recrystallization (Fig. 6), resulting in the change of the mode of microstress equilibrium (Perlovich et al., 1998): crystallites of rolled α-Zr experience residual contraction and extension along axes <0001>, deflected predominantly in opposite directions from the plane ND-TD, whereas in recrystallized α-Zr the region of elastic contraction surrounds ND and regions of elastic extension are shifted to the plane TD-RD.
5. The increase of c-parameter by recrystallization (Fig. 7) is connected with annealing of lattice defects and with leaving of excessive Nb atoms from the α-Zr solid solution in accordance with the balanced phase diagram (Douglass, 1971). Therefore, the difference in average values of c-parameter between rolled and annealed samples proves to be greater for the alloy with the higher content of Nb (2.5%).
6. In diagrams of correlation between GPF $2\theta_{0004}(\psi,\varphi)$ and PF(0001) (Fig. 8) it is distinctly seen that as a result of recrystallization the distribution of interplanar spacing d_{0001} does not become more homogeneous, than it was in the same sample after rolling; but it acquires the more regular character. Quite evident submission of interplanar spacings in recrystallized samples to usual statistical regularities reflects spontaneity of thermally activated processes, whereas in as-rolled samples these regularities are suppressed by forces, controlling plastic deformation processes.

Obtained experimental evidences of the essential substructure inhomogeneity in recrystallized Zr-based alloys depending on their texture prompt a number of inferences, connecting different aspects of recrystallization:

- The recrystallization texture includes a wide spectrum of grain orientations, so that a significant mutual misorientation of neighboring grains is probable.
- Significant microstrains can arise by meeting of neighboring growing grains and, as a result, a local elastic deformation and an increase of the dislocation density take place within boundary regions.
- Formation of specific dislocation arrangements near grain boundaries are accompanied by some local lattice rotations, so that by passing from the central part of a grain to its periphery the orientation changes, at least, by several degrees.
- Texture minima, where the fraction is localized with the most distorted crystalline lattice and finest coherent blocks, correspond to regions near boundaries of recrystallized grains.

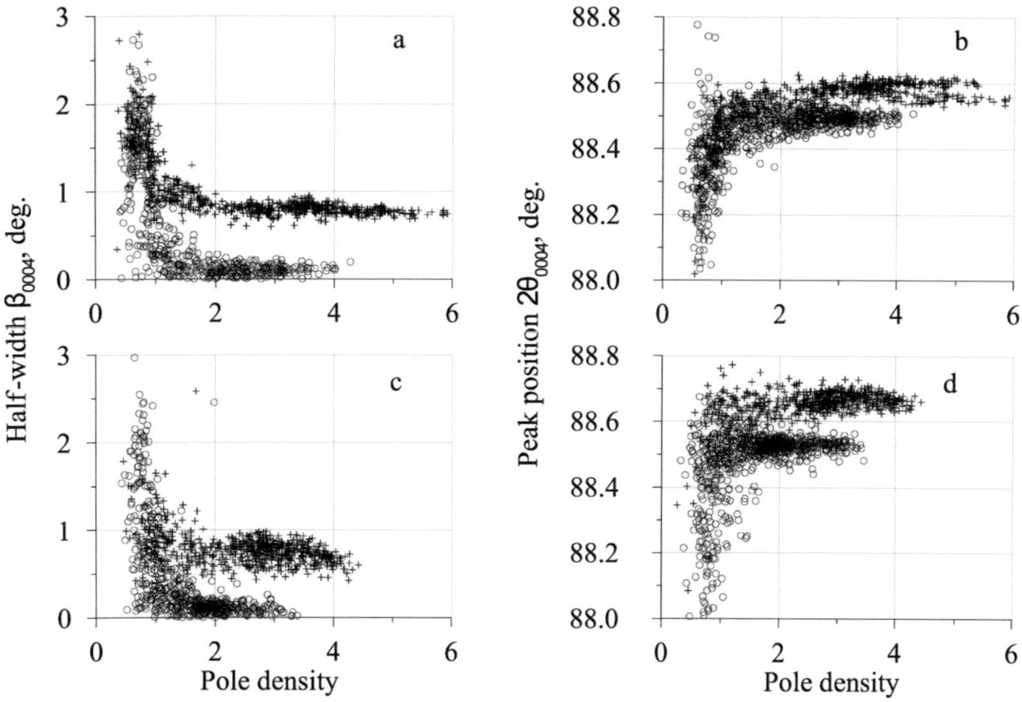

Fig. 8. Correlation diagrams for GPF β_{0004} and PF(0001) - (a, c), GPF $2\theta_{0004}$ and PF(0001) - (b, d): Zr-1%Nb, transversal rolling - (a, b); Zr-2.5%Nb, plain rolling – (c, d). Designation: + - deformed state, o - annealed state.

4. Texture of dynamic recrystallization

The question about an effect of dynamic recrystallization on the texture of rolling at increased temperatures emerges often by study of products from Zr-based alloys. This question is of the general interest and can be answered only by the rather attentive investigation of semi-products, produced at intermediate technological stages. An example of such investigation is presented below.

4.1 Materials, experimental approach and results

Sheets from Zr-2.5%Nb alloy, rolled at 750°C by two deformation routs down to thickness of ~4 mm, were studied. The used routes A and B differ in values of reduction per pass, diminishing in the order A→B, in numbers of successive passes and in presence of intermediate heating. The smaller are reductions per pass, the longer is the rolling procedure and the more significant is cooling of the billet (Perlovich et al., 2006). For compensation of this cooling the intermediate heating is introduced in route B. The alloy Zr-2.5%Nb contains usually two phases - the prevalent low-temperature HCP α-phase and the secondary high-temperature BCC β-phase. The temperature boundaries of (α+β)-region for Zr-2.5%Nb alloy are 610° and 830°C (Douglass, 1971), so that the studied sheets were rolled at the temperature of (α+β)-region, but local temperatures as well as the phase composition of concrete layers under rolling depended on their distance from the surface.

The layer-by-layer study of rolled sheets was carried out by X-ray diffractometric methods and mainly – by texture analysis. Texture analysis of α-Zr included measurement and construction of direct pole figures PF(0001) and PF{11$\bar{2}$0}. Typical PF(0001) and PF{11$\bar{2}$0} for surface, intermediate and central layers of the hot-rolled sheet are presented in Fig. 9. When passing from the surface layer of sheet to the central one, the rolling texture changes essentially. The following characteristic features of the rolling texture are considered: (1) the angular distance γ of texture maxima from normal direction (ND) in PF(0001); (2) the presence of additional texture maxima in PF{11$\bar{2}$0}, arising usually by recrystallization at the angular distance of 30° from maxima of the deformation texture. In PF{11$\bar{2}$0} (Fig. 9) these maxima are denoted by letters R and D, respectively. Layer-by-layer changes of angle γ are shown in figure 10-a; the layer-by-layer inhomogeneity of recrystallization is characterized by changes of ratio P_{recr}/P_{def} in Fig. 10-b, where P_{recr} and P_{def} – intensities of corresponding texture maxima.

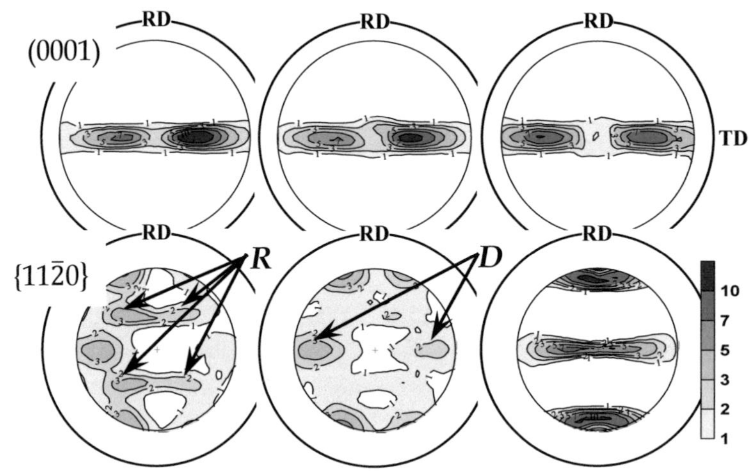

Fig. 9. Typical PF(0001) and PF{11$\bar{2}$0} for surface, intermediate and central layers of the hot-rolled sheet. Angular radius of constructed PF(0001) is equal to 80°, PF{11$\bar{2}$0} – 70°.

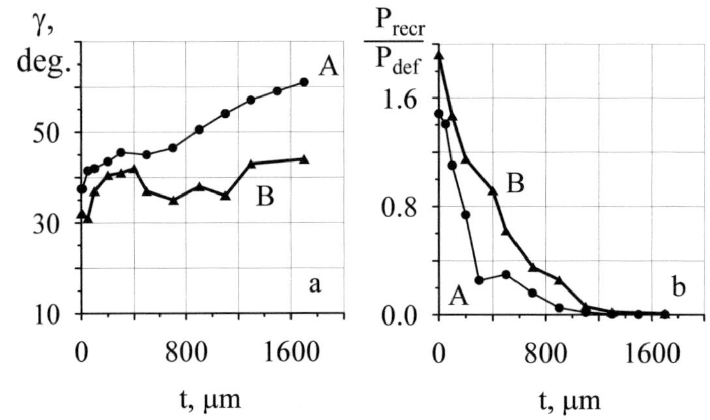

Fig. 10. Layer-by-layer changes of texture characteristics across the thickness of hot-rolled sheets: a – angular distance γ of texture maxima from ND in PF(0001); b - ratio P_{recr}/P_{def}, measured by PF{11$\bar{2}$0}.

4.2 Dynamic recrystallization under rolling

Development of recrystallization in α-Zr under rolling is seen in the surface layer up to 0.8 mm thick (Fig. 10-b). In PF{11$\bar{2}$0} of this layer (Fig. 9) maxima of rolling and recrystallization textures are present simultaneously, testifying that recrystallization of the sheet was only partial. Since texture maxima D and R are lying on different meridians, it is clear, that recrystallization included only grains with basal axes, closest to TD. Fig. 9 demonstrates visually the layer-by-layer inhomogeneity of α-Zr recrystallization in hot-rolled sheets. The angular shift of texture maxima in PF(0001) by passing from surface to central layers (Fig. 10-a) testifies in accordance with results of section 2, that recrystallization involves the reorientation of basal axes along with the known rotation of prismatic axes.

One more noteworthy difference between texture maxima D and R consists in their shape: maxima R are narrow and stretched along parallels of PF, likewise texture maxima in corresponding PF(0001), whereas maxima D are roundish and almost equiaxial. The shape of maxima R is a distinct evidence of dynamic recrystallization: anisotropic development of diffusion processes under hot rolling, including dislocation climb, results in stretching of texture maxima, as it is typical for rolling textures.

And vice versa, when the structure reforms spontaneously, only due to thermal activation, preferred directions in displacements of dislocations and dislocation boundaries are absent, so that texture maxima prove to be equiaxial, as it is usual for textures of static recrystallization. Then the round shape of maxima D in the rolling texture shows, that corresponding grains of α-Zr had time for polygonization in the course of cooling. A volume fraction of recrystallized grains decreases with distance from the surface, since stress relaxation in inner layers occurs by means of α→β→α phase transformations.

By route B the layer of dynamic recrystallization is thicker, than by route A (Fig. 10-b), because this process most probably develops in α-region of Zr-Nb phase diagram and the temperature boundary between α- and (α+β)-regions shifts deep into the sheet by passing from route A to route B in consequence of more intense cooling.

Thus, presented experimental results show, that by rolling of sheets from Zr-2.5%Nb alloy at the temperature 750°C a significant gradient in deformation conditions across the thickness of sheet takes place. The real temperature of concrete layers deviates from the nominal one due to the heat sink to rolls and local heating by deformation. A decrease of the deformation rate promotes development of dynamic recrystallization, suppresses the deformation-induced α→β phase transformation and weakens the unfavorable texture component, formed by rolling in β-phase.

5. Competition between recrystallization and phase transformations by welding of sheets from Zr-2,5%Nb alloy

The usual temperature of recrystallization annealing for α-Zr in cold-rolled products from Zr-based alloys is 580°C, that is close to the lower boundary of the (α+β)-region in the Zr-Nb phase diagram (610°C), where phase transformation (PT) α→β begins. Therefore under conditions of some heat treatments a competition is probable between recrystallization of α-Zr and PT α→β. In particular, such conditions take place by welding of cold-rolled sheets from Zr-based alloys in the thermal influence zone (TIZ) of the welding seam. When taking into account the regular difference between textures of sheets, experienced PT $\alpha \rightarrow \beta$ without and after preliminary recrystallization (Cheadle & Ells, 1966), the inhomogeneity of recrystallization within TIZ of the welding seam was analyzed.

5.1 "Multiplication" of maxima in α-Zr texture by phase transformations $\alpha \rightarrow \beta \rightarrow \alpha$

The preliminary recrystallization of α-Zr influences the texture, which arise in the sheet as a result of PT $\alpha \rightarrow \beta \rightarrow \alpha$ due to realization of the Burgers orientation relationship between α- and β-phases (Douglass, 1971):

$$(0001)_\alpha \parallel \{011\}_\beta,\ \langle 11\bar{2}0\rangle_\alpha \parallel \langle 111\rangle_\beta.$$

Multiplication of initial orientations in consequence of PT $\alpha \rightarrow \beta \rightarrow \alpha$ was analyzed in (Cheadle & Ells, 1966). By absence of variant selection, 35 new orientations of the basal plane arise in addition to the initial one; by mutual coincidence of some orientations their total number decreases down to 24. Since in reality we deal with textured polycrystals of Zr-alloys instead of single crystals, after PT $\alpha \rightarrow \beta \rightarrow \alpha$ the resulting distribution of basal axes in PF(0001) consists of overlapping texture maxima rather than of separate points. Therefore some new orientations of 24 above-mentioned ones, being close to each other, form common maxima. Thus, PT complicates an initial texture, multiplying its maxima in a definite way, though the resulting PF(0001) of the treated sample contains a lesser number of separate maxima, than the Burgers relationship predicts.

5.2 Studied samples and investigation technique

Samples for investigation were cut from sheets of the Zr-2,5%Nb alloy, cold-rolled up to the deformation degree of ~80%. In order to prepare semi-recrystallized samples, after cold rolling they were annealed at 550°C for 1 h. In order to induce PT $\alpha \rightarrow \beta \rightarrow \alpha$, both cold-rolled and annealed samples were subjected to the heat treatment in dynamic vacuum, including heating up to 950°C for 0,25 h, holding at this temperature during 0,5 h and subsequent cooling with an evacuated envelope in air. X-ray texture measurements were carried out by the standard method (Borodkina & Spector, 1981) using the diffractometer DRON-3M and Cu K_α radiation. For construction of complete PF, three mutually perpendicular sections of sheet or tube were investigated to obtain partial PF for their following sewing together (Perlovich & Isaenkova, 2002).

The investigated welded joints were produced by argon-arc welding of cold-rolled sheets of Zr-2,5%Nb alloy, using a non-expendable tungsten electrode with a motion velocity of 60 m/h. The welding direction (WD) was perpendicular to the RD. Texture inhomogeneity near the welding seam, connected with different heating conditions at neighboring regions, was studied layer-by-layer depending on the layer distance from the centre line of the seam.

5.3 Changes in α-Zr texture of cold-rolled and annealed sheets

PF(0001) for cold-rolled sheet is shown in Fig. 11-a, PF(0001) for the same sheet after heat treatment at 950°C – in Fig. 11-b, distributions of pole density along the radius ND-TD – in Fig. 11-c. Analogous PF and distributions for the sheet, which before PT experienced annealing at 550°C, - in Fig. 12-a,b,c. Comparison of PF(0001) in Fig. 11-a and Fig. 12-a shows, that partial recrystallization of the cold-rolled sheet causes extension of the pole density distribution to TD. Black circles in Fig. 11-b indicate ideal positions of texture maxima, arising as a result of PT α→β→α in PF(0001) of the cold-rolled sheet, whereas black circles in Fig. 12-b – calculated positions of PT-induced texture maxima in PF(0001) of the completely recrystallized sheet. In the considered real case of the semi-recrystallized sheet these black circles get only to those texture maxima, which are absent in PF(0001) of the cold-rolled sheet. This feature testifies, that some texture maxima in PF(0001) (Fig. 12-b) are produced by "multiplication" of maxima, belonging to the rolling texture, and some others – to the recrystallization texture.

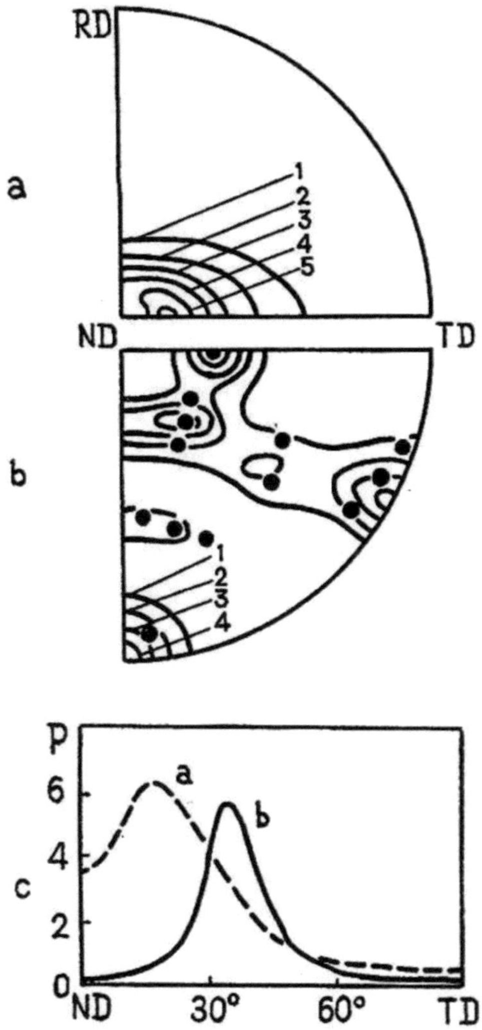

Fig. 11. PT in the cold-rolled sheet of Zr-2,5%Nb alloy, PF(0001): a – cold rolling; b – cold rolling + heat treatment 950°C/0,25 h; c – distributions of pole density along PF radius ND-TD.

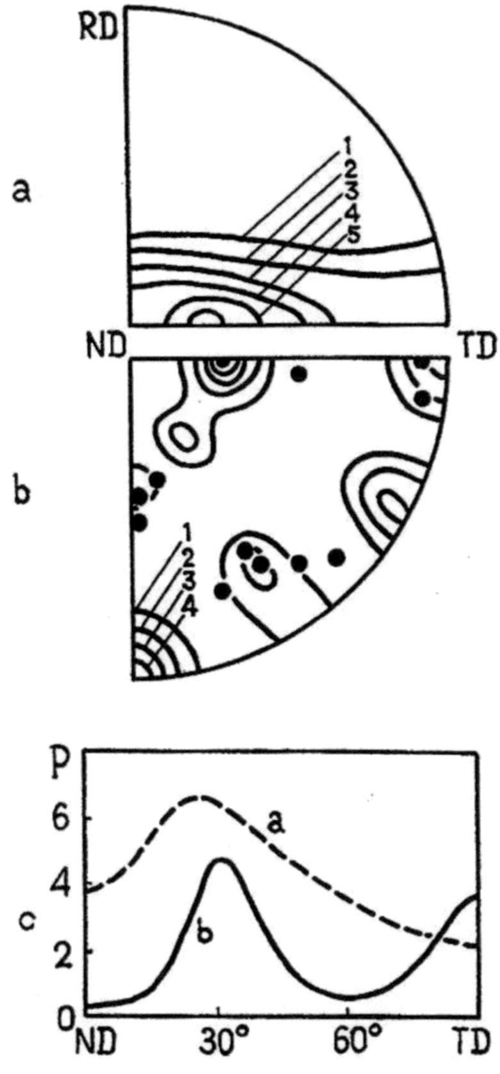

Fig. 12. PT in the semi-recrystallized sheet of Zr-2,5%Nb alloy, PF(0001): a – cold rolling + annealing 550°/1 h.; b - cold rolling + annealing 550°/1 h + heat treatment 950°/0,25 h; c - distributions of pole density along PF radius ND-TD.

In particular, as a result of PT $\alpha\rightarrow\beta\rightarrow\alpha$ in recrystallized grains, a new texture component $\{10\bar{1}0\}<11\bar{2}0>$ arises, having its maximum in PF(0001) at TD. Other maxima of the same set are present at the predicted regions also, confirming additionally the fact of preliminary recrystallization in the course of heating to the β-phase. At the same time, maxima, originating from deformed α-grains, by PT become noticeably weaker, than in the case of PT without preliminary recrystallization (compare Fig. 11-b and 12-b). The real situation in the deformed textured α-phase requires a description, similar to the case of a composite; therefore it would be correct to state a possibility of different outcomes from the competition between recrystallization and PT $\alpha\rightarrow\beta$ depending on deformation degree, grain orientation, heating rate, and so on. Inhomogeneous development of the considered processes, as well as their mutual competition, corresponds apparently to the most general case.

5.4 Recrystallization in the thermal influence zone by welding

Below some observations are presented concerning the competition between recrystallization and PT $\alpha\rightarrow\beta$ in the thermal influence zone (TIZ) with reference to the Zr-2,5%Nb alloy.

Arc welding is accompanied by a local heat treatment of the material in the vicinity of the welding seam. Parameters of a short-time thermal cycle, passing lengthwise TIZ parallel to the welding direction, are different for each longitudinal section of TIZ and depend on its distance from the central line of the seam. Layer-by-layer study of the texture within the TIZ gives an insight into the inhomogeneous structure developed in this zone by welding. In Fig. 13-a a schematic image of a welded joint is drawn; the melting zone is denoted by dense hatching and TIZ – by thin hatching. Three longitudinal sections are shown within the TIZ, and for each section an arrow indicates the corresponding PF(0001), obtained by X-ray diffractometric study just of this section. RD is brought into the centre of these PF in contrast with above-presented PF(0001) in Fig. 11 and 12.

Judging from PF in Fig. 13, different textures have formed in the shown sections of the TIZ depending on the distance from the seam. While in the most remote section the initial distribution of basal axes remains unchanged, the textures of the two closer sections contain new components produced by PT. Pole figures for these sections differ in relationship of PT components, originated from deformed (A) and recrystallized (B) α-grains. A quantitative treatment of obtained experimental data included the calculation of parameters, characterizing the relative contributions of both deformed and recrystallized components in the measured texture. In Fig. 13-b the results of such treatment for 18 successive sections of the TIZ are presented. The upper curve, constructed by PF$\{11\bar{2}0\}$, characterizes the relative fraction of recrystallized grains depending on the distance from the seam, irrespective of whether these grains experienced PT or did not. The following curve, constructed by PF(0001), characterizes the relative fraction of recrystallized grains in volume, covered by PT; it should be noted that in the general case this volume forms only a part of the investigated layer. The lower curve was obtained by subtraction of the (0001)-curve from the $\{11\bar{2}0\}$-curve. It characterizes the fraction, falling on grains, which proved to be recrystallized, but did not experience PT.

The presented curves testify unambiguously that, by sufficient increase of the heating temperature, all grains experience recrystallization prior to PT $\alpha\rightarrow\beta$. Thus, competition

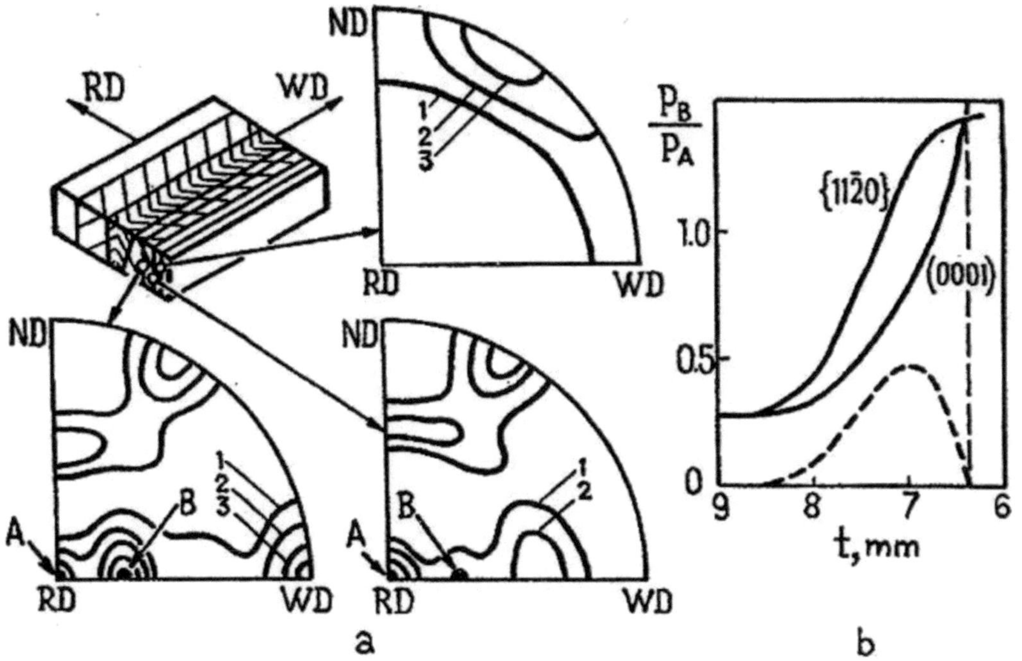

Fig. 13. PT in the thermal influence zone by arc welding: (a) the schematic image of welding seam and PF(0001) for its different section; (b) quantitative treatment of PF (see detailed explanation in the text).

between recrystallization and PT by a high rate of heating results in an absolute predominance of primary recrystallization (variant 1), though at some intermediate regimes of heat treatment two other variants are possible: recrystallization of deformed α-grains without their subsequent PT (variant 2) and PT of deformed α-grains without their preliminary recrystallization (variant 3).

Thus, the most important feature of the processes in the thermal influence zone is their inhomogeneous character, particularly by intermediate regimes of the heat treatment. While in some grains the variant 1 realizes, in other grains the variants 2 or 3 take place. The concrete variant, actual for the given grains, depends in some complicated manner on their orientation.

6. References

Borodkina M.M. & Spector E.N. (1981). *X-ray Analysis of Texture in Metals and Alloys.* Publishing House "Metallurgiya", Moscow, pp. 48-91 (in Russian).

Cheadle B.A. & Ells C.E. (1966) The effect of heat treatment on the texture of fabricated Zr-rich alloys. *Electroch. Techn.*, Vol. 4, No 7-8, pp. 329-336.

Douglass D.L. (1971). *The Metallurgy of Zirconium.* International Atomic Energy Agency, Vienna, pp. 60-76.

Isaenkova M. & Perlovich Yu. (1987a). Kinetics and mechanisms of texture formation in α-Zr by rolling. *Fizika Metallov i Metallovedenie,* Vol. 64, No 1, pp. 107-112 (in Russian).

Isaenkova M. & Perlovich Yu. (1987b). Reorientation of α-Zr crystallites by deformation. *Izvestiya Akademii Nauk SSSR. Metalli,* No 3, pp. 152-155 (in Russian).

Isaenkova M., Kapliy S., Perlovich Yu. & Shmelyova T. (1988). Features of changes of the Zirconium rolling texture by recrystallization. *Atomnaya Energiya,* Vol. 65, No 1, pp. 42-45 (in Russian).

Perlovich Yu., Isaenkova M., Shmelyova T., Nikulina A. & Zavyalov A. (1989). Texture changes in tubes of the alloy Zr-2,5%Nb by recrystallization. *Atomnaya Energiya,* Vol. 67, No 5, pp. 327-331 (in Russian).

Perlovich Yu. (1994). Development of strain hardening inhomogeneity during texture formation under rolling of bcc-metals. In: *Numerical Prediction of Deformation Processes and the Behavior of Real Materials, 15th Riso International Symposium on Material Science,* 5-9 September 1994, S.I. Andersen et al. Eds, Riso National Laboratory, Roskilde, Denmark, pp. 445-450.

Perlovich Yu., Bunge H.J. & Isaenkova M. (1997). Inhomogeneous distribution of residual deformation effects in textured BCC metals. *Textures & Microstructures,* Vol. 29, pp. 241-266.

Perlovich Yu., Bunge H.J., Isaenkova M. & Fesenko V. (1998) The Distribution of Elastic Deformation in Textured Materials as Revealed by Peak Position Figures. *Material Science Forum,* Vol. 273-275, pp. 655-666.

Perlovich Yu., Bunge H.J. & Isaenkova M. (2000) Structure inhomogeneity of rolled textured niobium. *Zeitschrifft fur Metallkunde,* Materials Research and Advanced Techniques, 2000, Vol. 91, No 2, p. 149-159.

Perlovich Yu. & Isaenkova M. (2002) Distribution of c- and a-dislocations in tubes of Zr alloys. *Metallurgical and materials transactions A,* Vol. 33A, No.3, pp. 867-874.

Perlovich Yu., Isaenkova M., Akhtonov S., Filippov V., Kropachev S. & Shtutca M. (2006) Interdependence of plastic deformation and phase transformations in Zr-2.5%Nb alloy under forging by different temperature-rate regimes. *Proceedings of the 9th International Conference on Material Forming ESAFORM 2006,* Glazgow, United Kingdom, April 2006, pp. 439-442.

Tenckhoff E. (1988) Deformation mechanisms, texture and anisotropy in Zirconium and Zircaloy. - ASTM, Special technical publication (STP 966), Philadelphia, 1988. - 77 p.

Zaymovskiy A.S., Nikulina A.V. & Reshetnikov N.G. (1994). *Zirconium Alloys in Nuclear Industry.* Energoatomizdat, Moscow, ISBN 5-283-03767-3, Russia, 256 p. (in Russian).

2

Recrystallization of Dispersion-Strengthened Copper Alloys

Su-Hyeon Kim[1] and Dong Nyung Lee[2]
[1]*Korea Institute of Materials Science,*
[2]*Seoul National University*[2]
Republic of Korea

1. Introduction

1.1 Dispersion-strengthened copper alloys

Pure copper exhibits high electrical and thermal conductivities, but it has low strength at room temperature as well as at elevated temperatures. Dispersion-strengthened (DS) copper alloy exhibits a high strength without sacrificing its inherent high conductivities, and maintains excellent thermal and mechanical stability at elevated temperatures by retaining its microstructures (Nadkarni, 1984). These unique characteristics are mainly attributed to the presence of uniformly dispersed thermally stable particles, which are typically oxides. Unlike precipitation-hardened copper alloys, which lose their strength by heating above the initial aging temperatures, the non-metallic oxide particles in oxide DS copper alloys, such as alumina, silica, and beryllia, neither coarsen nor go into solution, effectively preventing recrystallization and consequent softening of the alloys. Alumina DS copper alloys are not recrystallized even after exposure to temperatures approaching the melting point of copper (Preston & Grant, 1961). This is due to the pinning effect of the nano-sized alumina particles on the movement of the boundaries and dislocations. A unique combination of high strengths and high conductivities at elevated temperatures makes alumina DS copper alloys good candidates for high temperature electric materials (e.g., electrodes, lead wires, and connectors) (Nadkarni, 1984) as well as potential components in nuclear energy applications (Sumino et al., 2009).

Alumina DS copper alloys can be recrystallized when boron is added (Kim & Lee, 2001, 2002). Boron is often intentionally added as an oxygen scavenger during fabrication of the alloys (Gallagher et al., 1992). Long term annealing of boron-added alumina DS copper alloys results in an unexpected transformation from fine γ-Al_2O_3 to coarse $9Al_2O_3$-$2B_2O_3$ with a concurrent recrystallization of the matrix to form a large and elongated grain structure (Kim & Lee, 2002). Whereas Ni-based DS alloys are used in a coarse-grained condition to increase high-temperature creep resistance (Gessinger, 1976; Stephens & Nix, 1985), key applications of alumina DS copper alloys require them to be in a fully work-hardened state. Consequently, a large decrease in room temperature strength due to recrystallization is not desirable. Therefore, an understanding of the recrystallization behaviour of DS copper alloys is important from both practical and theoretical perspectives.

1.2 Recrystallization of particle-containing alloys

The presence of dispersed particles critically affects the plastic deformation and recrystallization behaviour of the matrix. The presence of particles accelerates or retards recrystallization of the matrix, depending on the interparticle spacing, size, mechanical properties, and thermal stability of the particles (Humphreys & Hatherly, 1995). Closely spaced fine particles exert a pinning effect on the movement of boundaries (Zener drag) resulting in retardation or even complete suppression of recrystallization. However, alloys with widely spaced particles larger than ~1 μm show accelerated recrystallization. Non-deformable large particles can introduce deformation zones around the particles during deformation, providing favourable nucleation sites for recrystallization (particle stimulated nucleation, PSN). Under certain conditions, particle-containing alloys transform from a deformed structure to a recrystallized grain structure in the absence of conventional discontinuous recrystallization accompanying a long-range motion of the boundaries. During low-temperature annealing, small particles give rise to boundary pinning, and subsequent coarsening of the particles at high temperatures may allow the subgrains to grow, forming recrystallized grain structures. This phenomenon is sometimes known as continuous recrystallization.

1.3 Purpose of the study

While several studies exist on the fabrication methods, mechanical properties, and deformation behaviour of alumina DS copper alloys, there is a lack of understanding of their recrystallization behaviour. This study examines the recrystallization behaviour of boron-added alumina DS copper strips rolled under different conditions. Particular attention is given to several anomalous phenomena, such as unique recrystallized grain structures and textures, as well as the dependency of recrystallization characteristics on prior rolling conditions. The results of several microscopy studies to elucidate microstructural evolution during rolling and annealing are presented, and the effects of dispersed particles on recrystallization are examined.

2. Research methods

2.1 Materials

The material used in this study was commercially available alumina DS copper alloy strips, Glidcop Al25, produced by SCM Metal Products. This material contains 0.25wt% Al in the form of Al_2O_3 particles as well as 0.02wt% B used for oxygen scavenging. The thickness of the as-received strips was 840 μm. The chemical composition of the as-received strips was measured by inductively coupled plasma (ICP) analysis and given in Table 1.

Al	B	P	Fe	S	As	Mn	Cu
0.275	0.023	0.0001	0.0001	0.0001	0.0034	0.0001	Balance

Table 1. Chemical composition of the as-received strips measured by ICP (wt%)

2.2 Rolling and annealing

The as-received strips were rolled under lubrication using a two-high rolling mill whose roll diameter was 126 mm to make two different specimens, as listed in Table 2. The cold-rolled strips were further rolled to reduce their thickness by 25% with one pass at room temperature. The thickness of the hot-rolled strips was reduced by 27% with one pass after heating the strips at 813 K for 10 minutes. Isothermal annealing of the as-received and rolled strips was carried out in a salt bath. After the heating, the strips were quenched in water.

Specimen	Number of passes	Total reduction	Rolling temperature	Lubrication
Cold-rolled strip	1	25%	Ambient	Yes
Hot-rolled strip	1	27%	813 K	Yes

Table 2. Rolling conditions of the as-received strips

2.3 Microstructure and texture analysis

The microstructures of the strips were investigated by optical microscopy and transmission electron microscopy (TEM) in the transverse direction (TD) and the normal direction (ND). The specimens were cut from the strips, mechanically polished, and chemically etched in $FeCl_3$ solution prior to optical microscopy. For the TEM study, the specimens were electrically polished in a nitric acid solution to make a thin foil using a twin-jet electro-polisher, while the dispersed alumina particles were extracted from the material using a carbon replica method.

The macroscopic textures of the strips were determined by measuring (111), (200), and (220) pole figures with an X-ray diffraction goniometer in the back reflection mode with Co Kα radiation. The specimens were mechanically polished parallel to the rolling plane and chemically etched in a nitric acid solution. Three-dimensional orientation distribution functions (ODFs), complete pole figures, and orientation densities were calculated from the measured pole figures using the WIMV program (Matthies et al., 1987). The orientations of individual crystallites were calculated from the Kikuchi patterns obtained by TEM (Young, et al. 1973) in the TD section of the specimens. Misorientations between adjacent crystallites were calculated using 24 symmetry operations (Randle, 1993).

2.4 Analysis of the mechanical properties

Tensile tests of specimens with a gauge length of 30 mm along the rolling direction (RD) were carried out at room temperature at a crosshead speed of 1 mm/min. The micro-Vickers hardness of the specimens was measured under a load of 25 g for 10 s.

3. Results

3.1 Characterization of the as-received strips

Figure 1 shows the microstructures of the as-received strips observed under an optical microscope. The material exhibited a highly deformed microstructure consisting of fine band-like substructures aligned nearly parallel to the RD. Figure 2 shows the longitudinal section TEM microstructure observed in the surface and centre regions of the as-received strips. The average band thicknesses of the surface and the centre regions were 0.127 and 0.129 μm, respectively. Additional band boundary characteristics measured on the centre region are given in Table 3. The grain structure of the as-received strips was characterized by a fine band-like grain structure with a high-angle boundary character.

The mechanical properties of the as-received strips are given in Table 4. The high strengths and hardness indicate that the strips were heavily deformed.

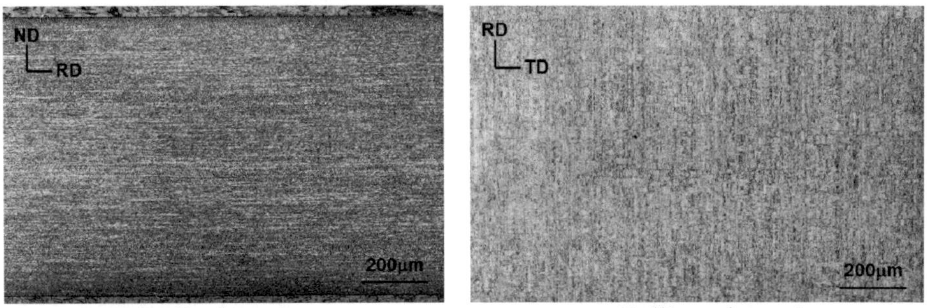

Longitudinal section Rolling plane view

Fig. 1. Optical micrographs of the as-received strips

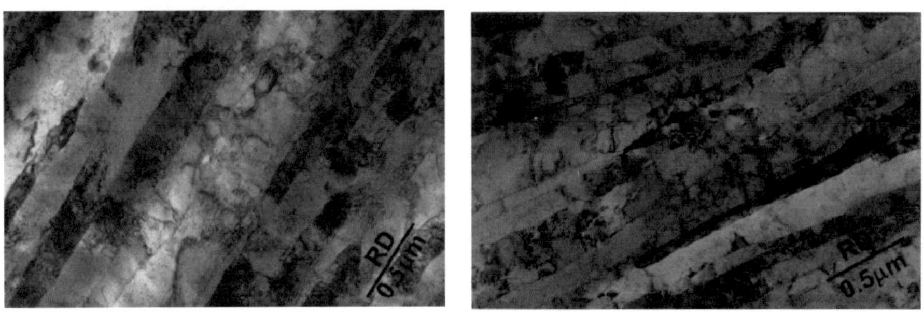

Surface region Centre region

Fig. 2. Longitudinal section TEM micrographs of the as-received strips

Specimen	Average band thickness (μm)	Average boundary misorientation (deg)	High angle boundary fraction (misorientation ≥ 15 deg)
As-Received	0.129	30.6	0.52

Table 3. Band boundary characteristics of the as-received strips

Specimen	Tensile strength (MPa)	Yield strength (MPa)	Elongation (%)	Hardness (Hv)
As-Received	553	515	14	169

Table 4. Mechanical properties of the as-received strips

Figure 3 shows the texture evolution of the as-received strips. The texture was characterized by the β-fibre, running from the copper orientation {112}<111> over the S orientation {123}<634> to the brass orientation {011}<211> in the Euler orientation space. The well-developed β-fibre texture indicated that the received strip was in a heavily rolled state, which is consistent with the microstructure evolution shown in Figures 1 and 2. Figure 4 shows the orientation densities along the β-fibre of the surface and the centre regions. The orientation densities of the brass and the S components were higher than the copper component, which is unlike plane-strain rolled pure copper sheets where the copper component is dominant (Hirsch & Lücke, 1988). Also noteworthy is the fact that the density of the brass component was lower than that of the S component in the surface region, while the brass and the S components were almost equally dominant in the centre region.

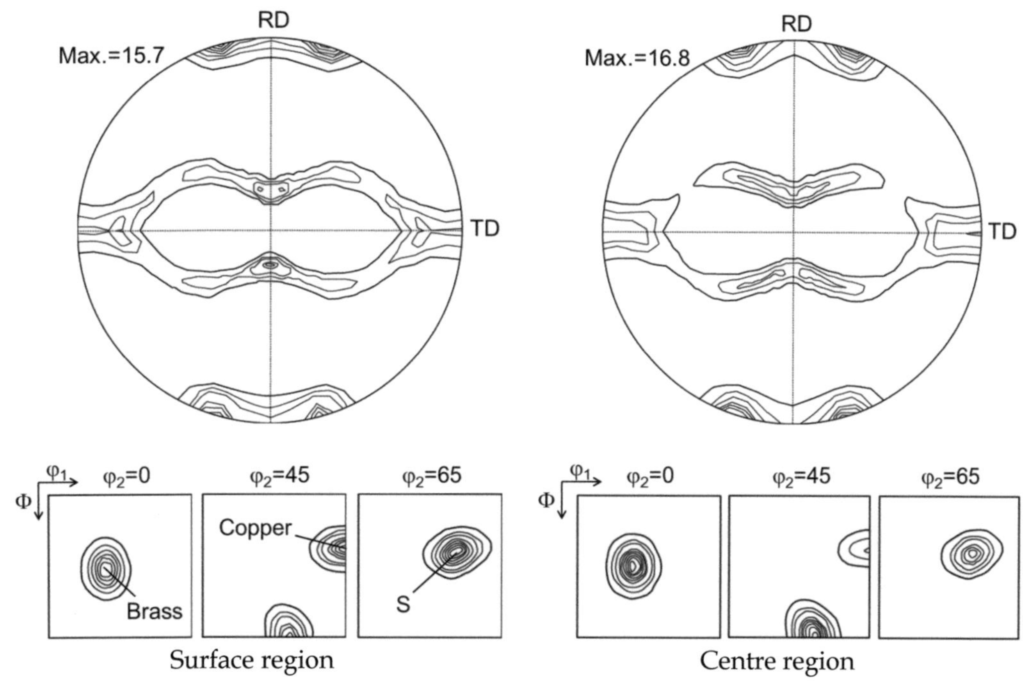

Fig. 3. (111) pole figures and ODFs of the surface and the centre layers of the as-received strips (Kim & Lee, 2002)

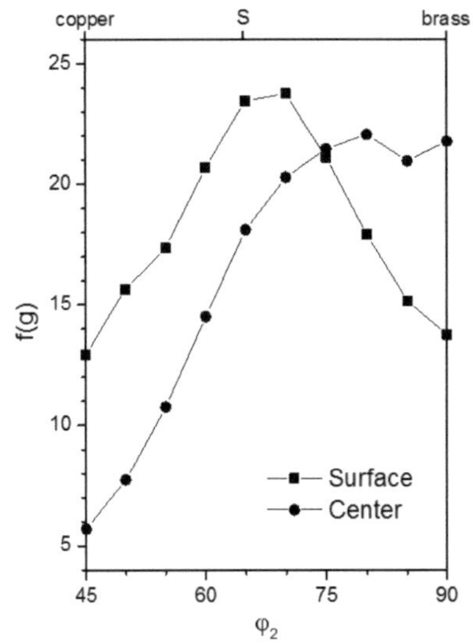

Fig. 4. Orientation densities along the β-fibre of the surface and the centre regions of as-received strips

Figure 5 shows optical microstructures of the as-received strips annealed at 1123 K for 1 hr. Recrystallization occurred in the centre region while no recrystallization took place in the surface region. The plate-like morphology of the recrystallized grains and the ragged shape of the grain boundaries are similar to other extruded or rolled dispersion-strengthened alloys after recrystallization (Klug et al., 1996; Chou, 1997). The TEM microstructures of a recrystallized grain (Figure 6) show dispersed particles aligned parallel to the rolling direction in the recrystallized regions. The micro-Vickers hardness values of the centre and the surface regions were 136 and 168, respectively. This result indirectly indicates that the centre region was recrystallized but the surface region was not. Figure 7 shows the TEM plane view observation of a recrystallized grain in the centre region. A large plate-like recrystallized grain was identified. The textures of the annealed strips are visible in Figure 8. The surface region retained the β-fibre texture, which was similar to the texture in the rolled state. The centre region exhibited a strong texture component, which could be approximated by {112}<312>. The texture of the centre region originated from the recrystallized grains.

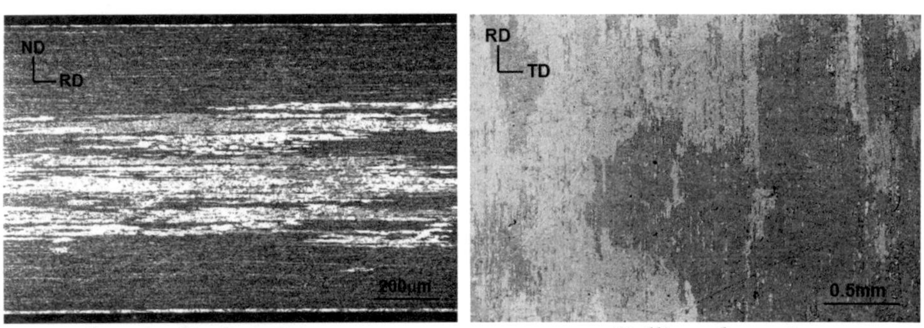

Longitudinal section Rolling plane view

Fig. 5. Optical micrographs of the as-received strips annealed at 1123 K for 1 hr

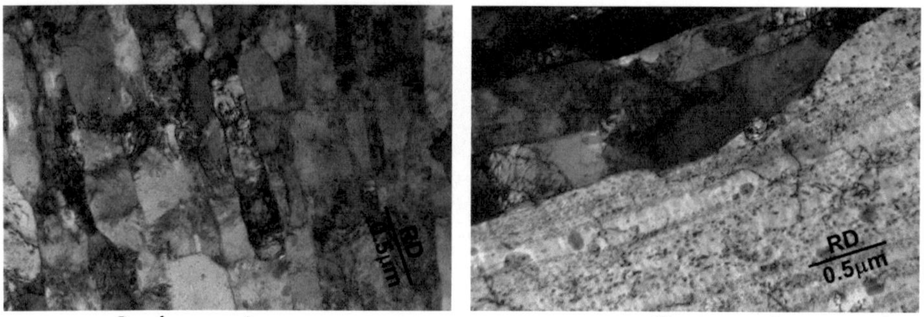

Surface region Centre region

Fig. 6. Longitudinal section TEM micrographs of the as-received strips annealed at 1123 K for 1 hr

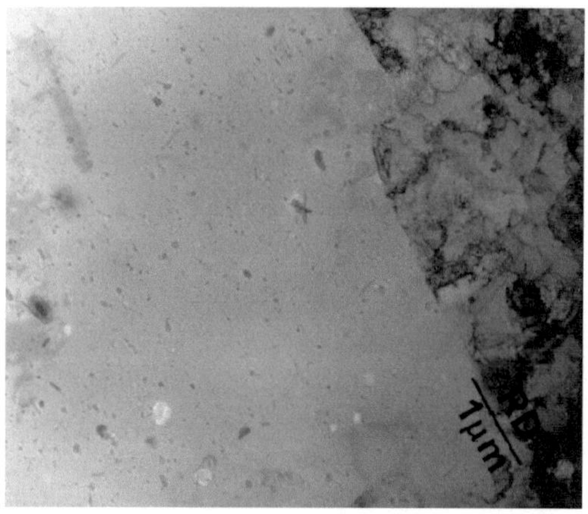

Fig. 7. Rolling plane view TEM micrographs of the centre region of the as-received strips annealed at 1123 K for 1 hr

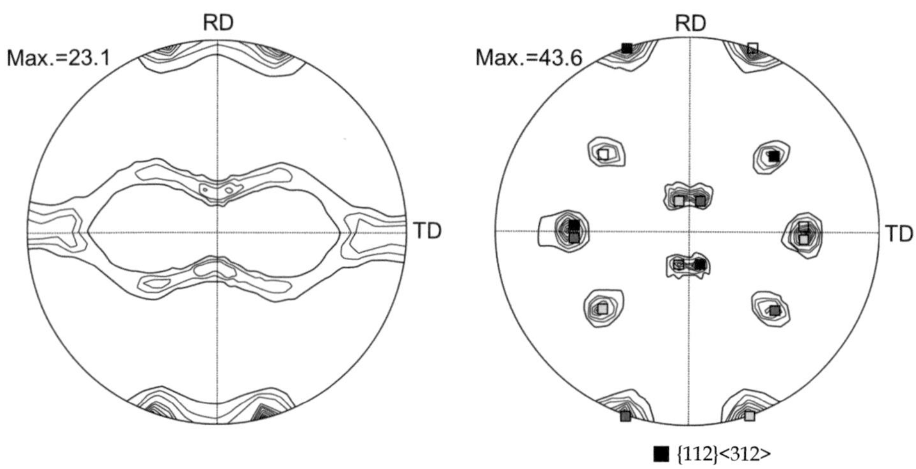

Surface region Centre region

Fig. 8. (111) pole figures of the as-received strips annealed at 1123 K for 1 hr (Kim & Lee, 2002)

3.2 Properties of the rolled strips

Figures 9 and 10 show optical and TEM microstructures of the cold-rolled and hot-rolled strips. Band-like structures aligned parallel to the RD were observed that were similar to those of the as-received strips. No dynamically recrystallized grains were found in the hot-rolled strips. Table 5 details the band structure characteristics of the cold-rolled and hot-rolled strips measured in the centre regions of each strip. By comparing the band structure characteristic of the as-received strips given in Table 3, the thickness of the band was decreased by cold rolling and increased by hot rolling. Cold rolling also increased the high-angle boundary fraction.

Table 6 shows the mechanical properties of the rolled strips. The cold-rolled strip showed higher strengths and hardness than the hot-rolled strip. By comparing the properties to those of the as-received strips, it can be seen that both cold rolling and hot rolling increased the strengths and hardness of the strips while decreasing their elongation.

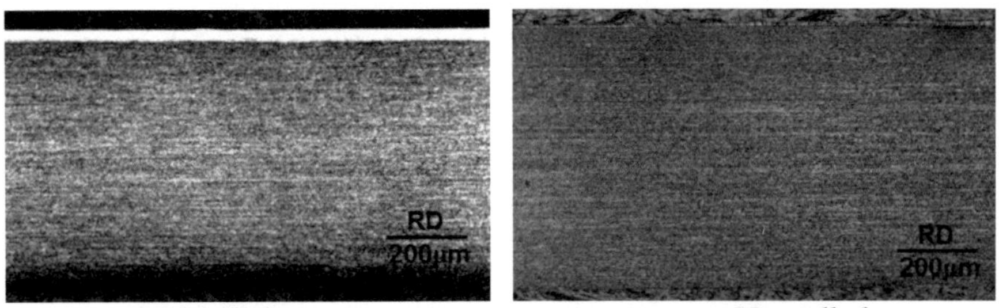

Cold rolled Hot rolled

Fig. 9. Longitudinal section optical micrographs of the rolled strips (Kim & Lee, 2002)

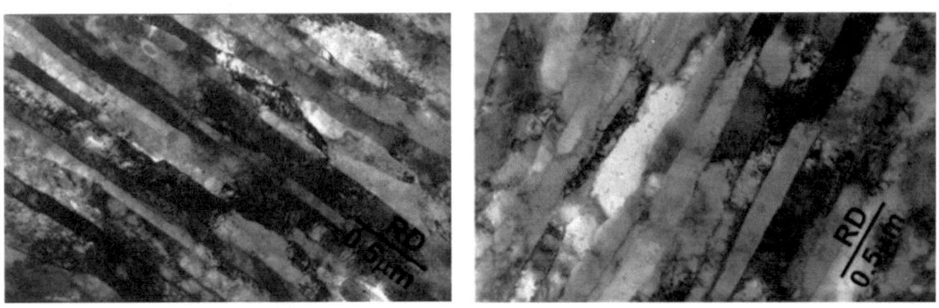

Cold rolled Hot rolled

Fig. 10. Longitudinal section TEM micrographs of the centre region of the rolled strips (Kim & Lee, 2002)

Specimen	Average band thickness (μm)	Average boundary misorientation (deg)	High angle boundary fraction (misorientation ≥ 15 deg)
Cold rolled	0.116	27.9	0.60
Hot rolled	0.141	24.5	0.51

Table 5. Band structure characteristics of the cold-rolled and hot-rolled strips

Specimen	Tensile strength (MPa)	Yield strength (MPa)	Elongation (%)	Hardness (Hv)
Cold rolled	605	579	5	184
Hot rolled	580	553	5.5	179

Table 6. Mechanical properties and hardness of the cold-rolled and hot-rolled strips

The textures of the rolled strips were similar to those of the as-received strips. Figure 11 shows the orientation densities along the β-fibre of the surface and the centre regions of the rolled strips. The textures of the rolled strips were characterized by the strong β-fibre.

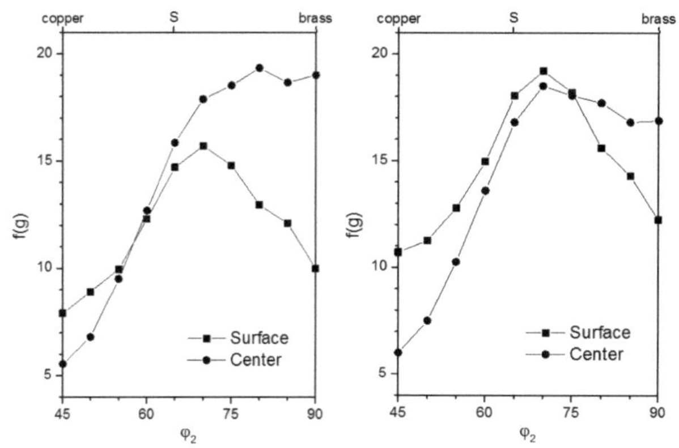

Fig. 11. Orientation densities along the β-fibre of the surface and centre regions of the rolled strips

Figure 12 shows the optical microstructures of the rolled strips annealed at 1123 K for 1 hr. TEM micrographs of the centre region are given in Figure 13. Similar to the as-received strips, the cold-rolled strips exhibited recrystallization in the centre region while the hot-rolled strips did not show recrystallization since no recrystallized grains were observed throughout the examined area. However, substantial band growth appeared on the hot-rolled strip. Table 7 gives the band structure characteristics of the hot-rolled and annealed strips. By comparing the results in Tables 5 and 7, it appears that annealing increased the band thickness and high-angle boundary fraction of the hot-rolled strip.

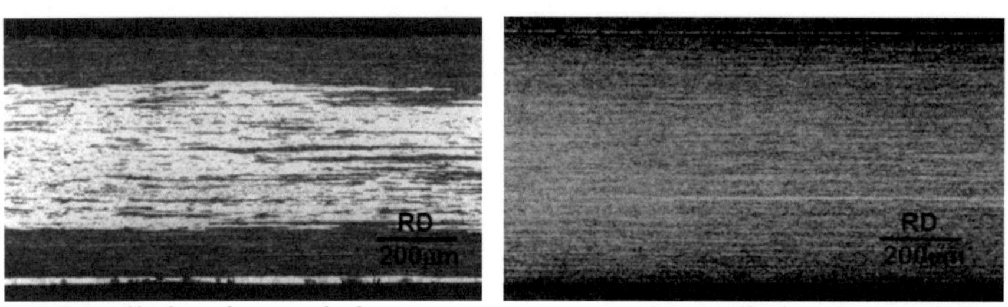

Fig. 12. Longitudinal section optical micrographs of the rolled strips annealed at 1123 K for 1 hr (Kim & Lee, 2002)

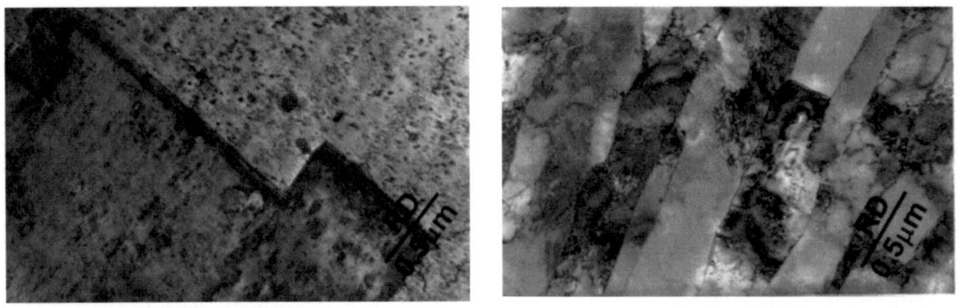

Cold rolled and annealed Hot rolled and annealed

Fig. 13. Longitudinal section TEM micrographs of the centre region of the rolled strips annealed at 1123 K for 1 hr (Kim & Lee, 2002)

Specimen	Average band thickness (μm)	Average boundary misorientation (deg)	High angle boundary fraction (misorientation ≥ 15 deg)
Hot rolled and annealed	0.270	36.0	0.81

Table 7. Band structure characteristics of the hot-rolled strip annealed at 1123 K for 1 hr

Figure 14 shows (111) pole figures of the cold-rolled strip annealed at 1123 K for 1 hr. The texture of the surface region was characterized by the β-fibre, and the recrystallization texture in the centre region was indexed by {112}<312>. The texture of the hot-rolled strip after annealing is shown in Figure 15. Both the surface and the centre regions retained most of the β-fibre rolling texture.

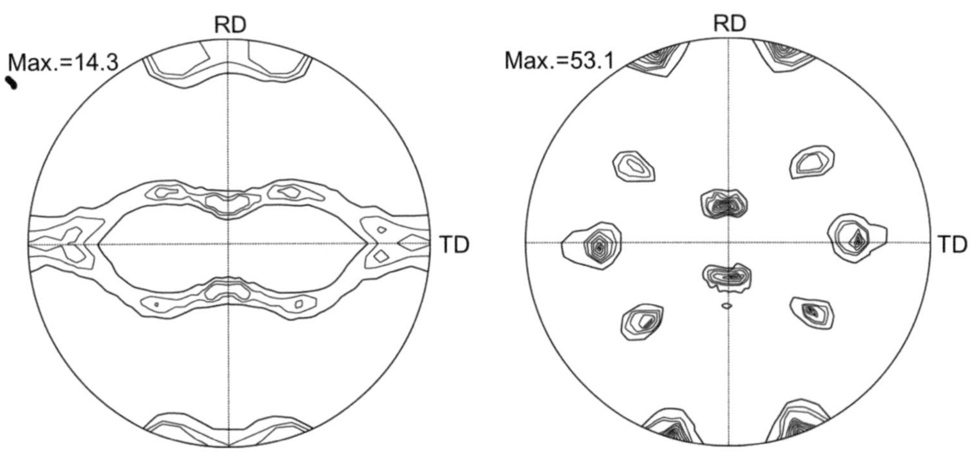

Surface Centre region

Fig. 14. (111) pole figures of the cold-rolled strip annealed at 1123 K for 1 hr (Kim & Lee, 2002)

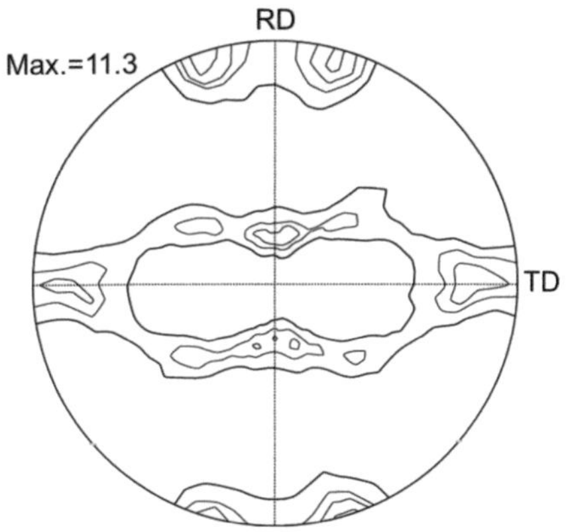

Fig. 15. (111) pole figures of the hot-rolled strip annealed at 1123 K for 1 hr (Kim & Lee, 2002)

4. Discussion

Earlier studies (Preston & Grant, 1961; Nadkarni, 1984) have shown that alumina DS copper alloys resist recrystallization up to their melting points due to the presence of thermally stable alumina particles. The present study showed that alumina DS copper alloys recrystallized after moderate-temperature annealing when boron was added. This is attributed to a reduction in the particle-pinning effect caused by the transformation of particles from fine alumina to coarse aluminium boron oxide. Additionally, large particles already present in the deformed state can introduce deformation zones that act as nucleation sites for recrystallization. Alumina DS copper alloys are fabricated by internal oxidation of Cu–Al alloy powders, consolidations of the powders into fully dense shapes, and further cold rolling to final shapes. The internal oxidation involves the mixing and heating of the alloy powders with oxidants like Cu_2O. Frequently, residual oxygen, or unconverted Cu_2O, may react with hydrogen introduced during alloy processing. This produces a large internal pressure of water vapour and results in blister formations. The material used in this study, Glidcop, is made oxygen-free by intentionally adding boron as an oxygen scavenger. Figure 16 shows a coarse particle observed in the as-received strips, which was identified as $9Al_2O_3$–$2B_2O_3$ by indexing its diffraction patterns. Phase transformation of the particles is expected during the fabrication of a strip since it is subjected to a heating process.

Fig. 16. Identification of an existing large particle observed in the as-received strip (Kim & Lee, 2002)

The remainder of the discussion explores several anomalous phenomena observed during the annealing of the alumina DS copper alloy.

4.1 Unique recrystallized microstructure

The recrystallized microstructure of the boron-added alumina DS copper alloy strip was characterized by the following features:

- Recrystallization only in the centre region
- Plate-like morphology of recrystallized grains
- Very large recrystallized grains

Figures 5 and 12 show that recrystallization occurred only in the centre region of the strips. In order to observe how the microstructure evolved, both the as-received strips and the cold-rolled strips were quickly annealed. Figure 17 shows optical micrographs of the as-received strip annealed at 923 K for 10 s and 15 min. Recrystallized grains emerged along lines originating exclusively from the centre region. Detailed TEM observations revealed that large bands were present in the deformed state in the centre region and appeared to promote recrystallization. Figure 18 shows a large band found in the centre region of the as-received strip, along with its orientation. Figure 19 reveals that similar bands were present in the cold-rolled strips. The orientations of the large bands in the as-received and cold-rolled strips included cube, RD-rotated cube, copper, and ND-rotated copper. Among them, the ND-rotated copper orientation was similar to the recrystallization texture {112}<312> observed in the annealed strips. These pre-existing large bands survived the early stages of annealing as shown in Figures 20 and 21.

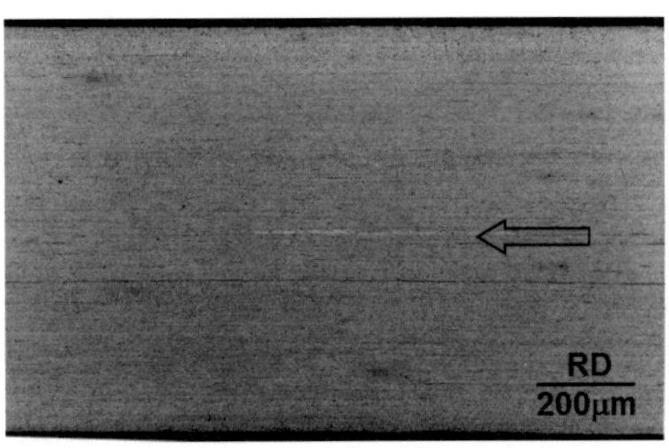

Annealed for 10 s

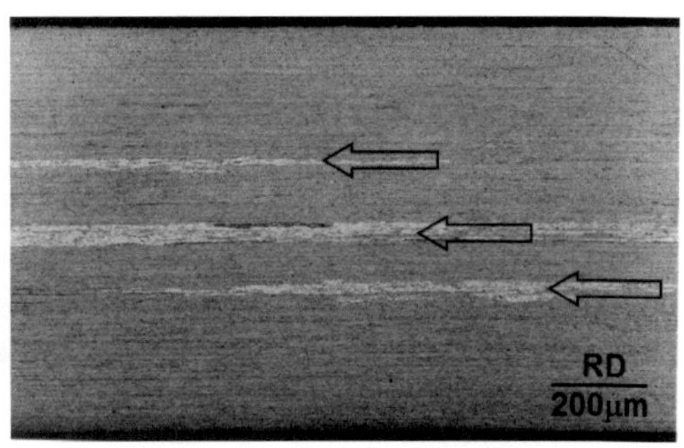

Annealed for 15 min

Fig. 17. Longitudinal section optical micrographs of the strips annealed at 923 K

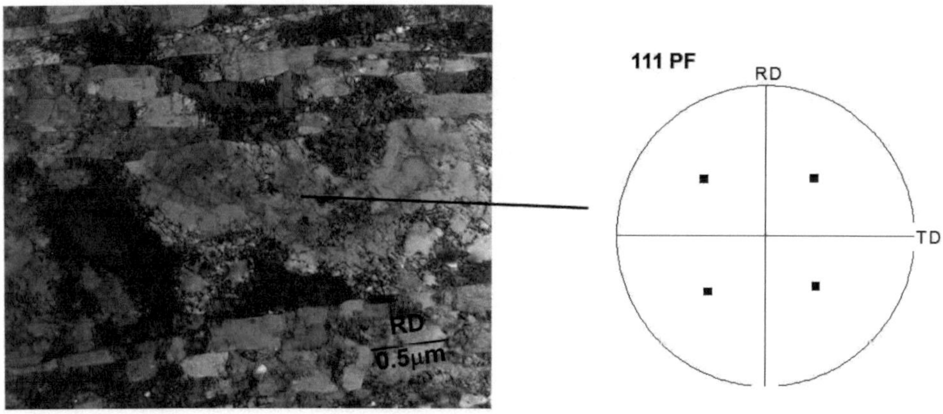

Fig. 18. Longitudinal section TEM micrograph showing a large band and its orientation in the centre region of the as-received strip (Kim & Lee, 2002)

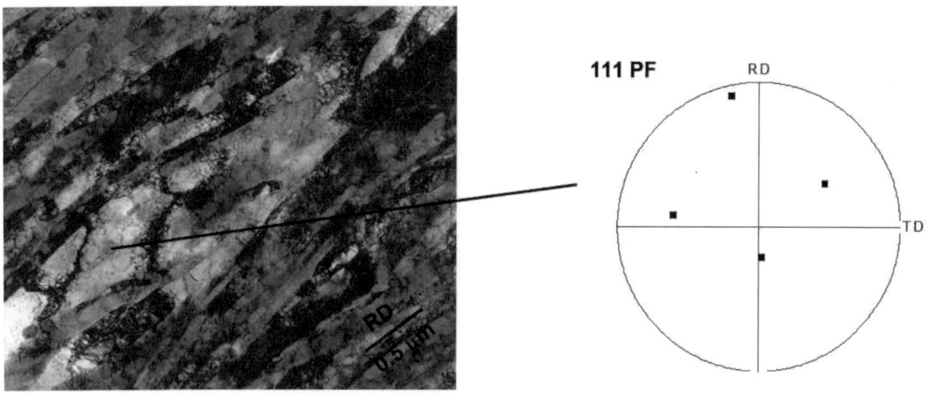

Fig. 19. Longitudinal section TEM micrograph showing a large band and its orientation in the centre region of the cold-rolled strip (Kim & Lee, 2002)

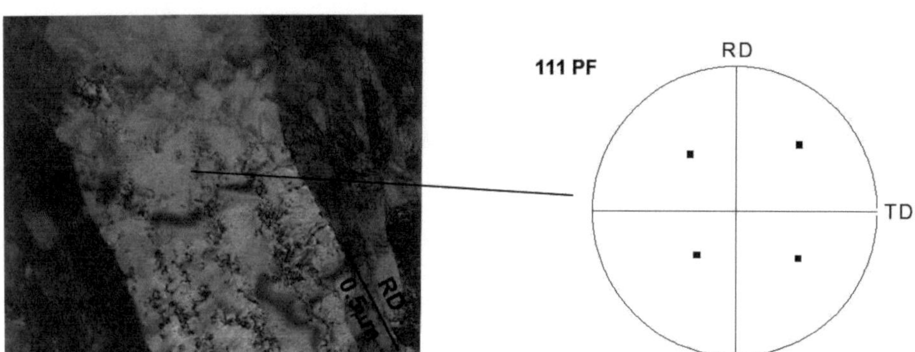

Fig. 20. Longitudinal section TEM micrograph showing a large band and its orientation in the centre region of the as-received strip annealed at 1123 K for 1 s

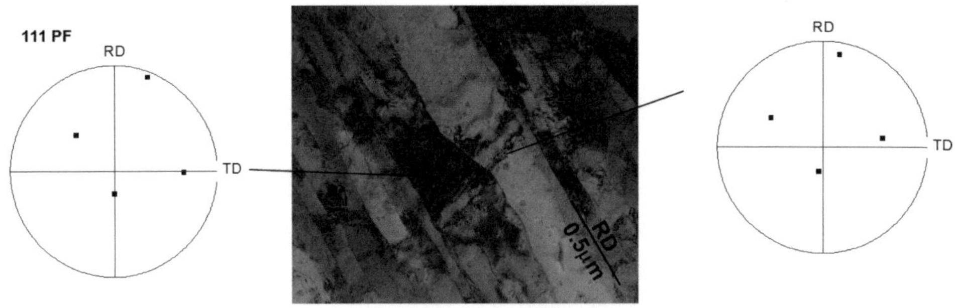

Fig. 21. Longitudinal section TEM micrograph showing a large band and its orientation in the centre region of the cold-rolled strip annealed at 1123 K for 1 s

Recrystallization can be divided into two consecutive processes: nucleation and growth. A nucleus must be some minimum size in order for further growth to occur, or else it will shrink and vanish. Subgrain coalescence is a requisite process to form these critical-sized nuclei. Band coalescence in the present material is unlikely when the band boundaries exhibit high-angle characteristics and their movement is hindered by the presence of dispersed particles. A more likely explanation is that the pre-existing large bands provide favourable nucleation sites for recrystallization. It appears that the large bands present in the centre region are the preferred recrystallization nucleation sites. It is possible that the large bands originated from large grains formed during the manufacturing process as similar grains have been observed in extruded alumina DS copper alloys in previous studies (Afshar & A. Simchi, 2008; H. Simchi & A. Simchi, 2009).

The plate-like morphology of the recrystallized grains in the alumina DS copper alloys can be related to the distribution of the dispersed particles. When recrystallized grains grow, the moving boundaries are pinned by the particles. The pinning pressure of the particles on the boundary movement is given by Equation 1 (Humphreys & Hatherly, 1995):

$$P_Z = 3F_V \gamma_b / d \quad (1)$$

where:

- F_V is the volume fraction of the particles
- d is the particle size
- γ_b is the boundary energy

If particles are randomly distributed, the pinning pressure will be directionally isotropic. On the other hand, if the distribution of the particles is anisotropic, there will be an anisotropic pinning pressure on the boundaries. Figure 5 shows that the particles in the as-received strips were aligned along the rolling direction. The pinning pressure parallel to the rolling plane should be lower than that along the thickness direction. Therefore, the plate-like recrystallized grain shape can be mainly attributed to the planar distribution of the particles. The directional distribution of the particles might be driven by the rolling of the strip. A plate-like morphology of the recrystallized grains is often reported in the recrystallization behaviour of other dispersion-strengthened alloys (Klug et al., 1996; Chou, 1997), although other dispersion-strengthened alloys show equiaxed recrystallized grain structures (Miodownik et al., 1994; Miodownik et al., 1995).

Another unique recrystallization characteristic of the alumina DS copper alloy is that the recrystallized grains are very large. Early researchers (Singer & Gessinger, 1982; Mino et al., 1987; Kusunoki et al., 1990) reported that the very large recrystallized grains found in dispersion-strengthened alloys are formed through secondary recrystallization. They concluded that primary recrystallization occurred immediately before secondary recrystallization, or during plastic deformation - dynamic recrystallization. Later studies (Klug et al., 1996) suggested that primary recrystallization was responsible for the formation of large grains because microstructural changes are driven by stored energy acquired from plastic deformation. While plastically deformed alumina DS copper alloy possesses a sufficient driving force for recrystallization, a barrier to recrystallization exists due to the particle pinning effect. Microstructural inhomogeneity, such as large bands, provides preferential nucleation sites, and a large nucleus at a large band can grow with a size advantage over the surrounding matrix. Therefore, the emergence of very large recrystallized grains is a result of preferential nucleation at pre-existing large bands. The annealing behaviour of alumina DS copper alloy might be regarded as secondary recrystallization since very large recrystallized grains are formed when they overcome the particle-pinning pressure. However, the microstructure of the alumina DS copper alloy suggests that the driving force for recrystallization is stored energy by plastic deformation. Thus, while the evolution of the annealed alumina DS copper alloy microstructure appears to be due to secondary recrystallization, the mechanism that forms the very large recrystallized grains is due to primary recrystallization.

4.2 Unique recrystallization texture

The recrystallization texture of the annealed alumina DS copper alloy can be approximated by {112}<312>. To our knowledge, this texture has not been reported for other copper alloys. The recrystallization texture is determined by the orientations of the new grains and their growth rates. The present study discussed the role of these two factors and how they determine the unique recrystallization texture of the alumina DS copper alloy.

4.2.1 Selective nucleation

As discussed previously, pre-existing large bands provided favourable nucleation sites for recrystallization. Pre-existing large particles could introduce particle deformation zones that act as nucleation sites. Figure 22 shows the recrystallizing grains formed around the particles and their orientations observed in the as-received strips after rapid annealing. The orientation of grain A was similar to that of the deformed matrix, and multiple twinning could cause grains B and C to generate different orientations. It is known that PSN usually gives rise to weak recrystallization textures (Humphreys & Hatherly, 1995). Band coalescence is unlikely but possible when the pinning of the boundary movement is relaxed. Figure 23 shows that the band growth took place by coalescence of similarly oriented bands. Various orientations could be generated from new grains resembling the matrix orientations through PSN and band coalescence, as well as by subsequent twinning. Since no specific grain orientations dominated as the new grains evolved, the well-developed strong recrystallization texture {112}<312> could not be caused by new grain evolution.

4.2.2 Selective growth

According to the theory of selective growth, the recrystallization texture is determined by the relative growth rates of the boundaries. The velocity of the moving boundary (V) is a function of the boundary mobility (M) and the driving pressure (P), given by:

$$V=MP \tag{2}$$

P can be expressed as follows:

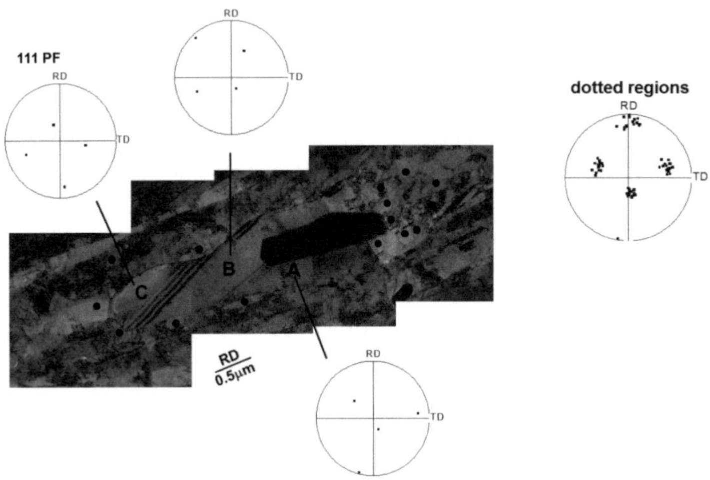

Fig. 22. Longitudinal section TEM micrograph showing individual grains around a particle and their orientations in the as-received strips annealed at 923 K for 10 s

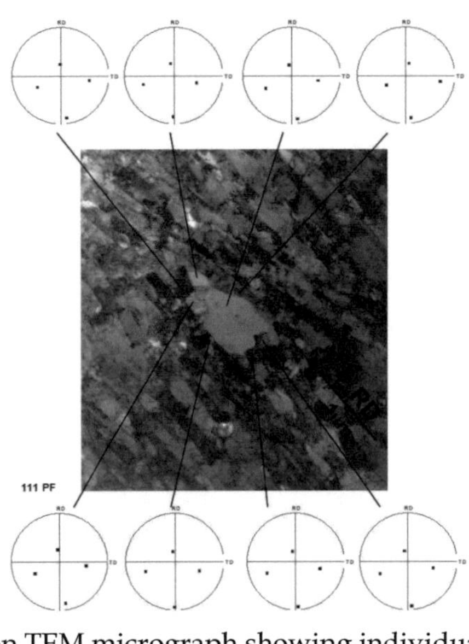

Fig. 23. Longitudinal section TEM micrograph showing individual grains and their orientations in the cold-rolled strips annealed at 1123 K for 3 s (Kim & Lee, 2002)

$$P=P_D - P_C = P_D - 2\gamma_b / R \tag{3}$$

where:

P_D is the stored energy,
P_C is the opposing pressure from the boundary curvature
γ_b is the boundary energy
R is the radius of the grain.

In particle-strengthened alloys, the Zener pinning pressure (P_Z) arises from the particles, and P can be expressed as follows (Humphreys & Hatherly, 1995):

$$P=P_D - P_C - P_Z = P_D - 2\gamma_b / R - 3F_V \gamma_b / d \tag{4}$$

where:

F_V is the volume fraction of the particles
d is the particle size.

Recrystallizing grains will grow only when P is positive. P increases with increasing grain size and decreasing boundary energy. The low-angle boundaries and twin boundaries have a lower boundary energy than the high-angle boundaries. Based on Equation 4, only large grains with low-angle boundaries or twin boundaries can overcome the pinning pressure. High-angle boundaries can be stagnant, even though they have higher mobility than low-angle boundaries. The recrystallization texture {112}<312> is defined as ND-rotated copper, which is occasionally found in large bands in the deformed state, as shown in Figures 19 and 21. Recrystallizing grains with {112}<312> orientations have a chance to face the surrounding deformed matrix with low-angle boundaries because {112}<312> orientations deviate slightly from the deformation texture. Furthermore, {112}<312> orientations have a twinning relationship between the two equivalent orientations among them. Figure 24 shows the orientations of two adjacent recrystallized grains observed in the cold-rolled and annealed strips. The boundary shape and orientation relationship indicated that the grain boundary of the two adjacent recrystallized grains was a twin boundary. Therefore, the unique recrystallization texture was determined by the preferential growth of large recrystallizing grains with low-angle boundaries or twin boundaries, even though those boundaries had low mobility.

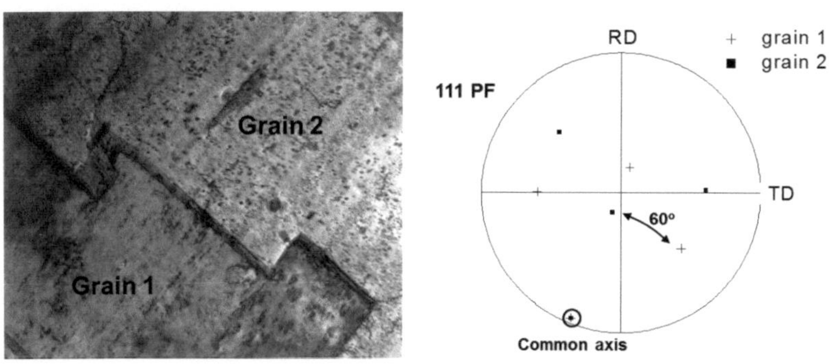

Fig. 24. Longitudinal section TEM micrograph and (111) pole figure showing two adjacent recrystallized grains of the cold-rolled strips annealed at 1123 K for 1 hr

4.3 Dependency of recrystallization on prior rolling conditions

As described in Section 3.2, the response to annealing of the alumina DS copper alloy is influenced by prior rolling conditions. The annealing behaviour of the cold-rolled strip is characterized by recrystallization, whereas recovery by band growth occurs in the hot-rolled strip. Similar results have been reported for other dispersion-strengthened alloys (Petrovic & Ebert, 1972; Singer & Gessinger, 1982). As recovery and recrystallization are competitive processes, dynamic recovery during hot rolling could reduce the potential energy in the alloy. This argument would also apply when comparing results between the cold-rolled and hot-rolled strips; since the hardness of the hot-rolled strip is lower, the recovery process during hot-rolling is governed by normal band growth. In our tests, after annealing, a continuous band growth occurred in the hot-rolled strip, which became a coarse band structure with high-angle boundary characteristics (see Table 7).

Subjecting the as-received strips to hot rolling gave rise to band growth and increased the hardness. Plastic deformation during hot rolling could increase the dislocation density, increasing the hardness. Therefore, a reduction in potential energy may not occur during hot rolling. It is not clear at this time why the hot-rolled strips became resistant to recrystallization. One explanation would be the homogeneity of the microstructural evolution. Microstructural inhomogeneity often occurs during plastic deformation, and these regions are frequently sites of initial recrystallization. The deformation becomes more homogeneous as the deformation temperature increases (Humphreys & Hatherly, 1995). A reduction in microstructural inhomogeneity during hot rolling could be responsible for the suppression of discontinuous recrystallization.

An alternative explanation is based on the assumption that coarse particles are sheared into finer particles during hot rolling (Kim & Lee, 2002). The shear strength of the particle might decrease with increasing temperature. Particle shearing could result in a decrease in interparticle spacing, which in turn could give rise to the higher hardness and the corresponding difficulty in recrystallization.

5. Conclusions

The recrystallization behaviour of boron-added alumina DS copper alloy strips was studied. The results may be summarized as follows.

Recrystallization occurred only in the centre region of the strips. Pre-existing large bands provided a favourable nucleation site for very large recrystallized grains.

The morphology of the recrystallized grains was plate-like due to the planar alignment of the dispersed particles.

The recrystallization texture was indexed to {112}<312>. Preferential growth of the large recrystallizing grains against the particle pinning appeared to determine this unique recrystallization texture.

The hot-rolled strip underwent recovery accompanied by continuous band growth, but without recrystallization.

6. References

Afshar, A. & Simchi, A. (2008). Abnormal Grain Growth in Alumina Dispersion-Strengthened Copper Produced by An Internal Oxidation Process. *Scripta Materialia*, Vol.58, No.11, (June 2008), pp.966-969, ISSN 1359-6462

Chou, T.S. (1997). Recrystallization Behavior and Grain Structure in Mechanically Alloyed Oxide Dispersion Strengthened MA956 Steel. *Materials Science and Engineering A*, Vol.223, No.1-2, (February 1997), pp. 78-90, ISSN 0921-5093

Gallagher, D.E. ; Hoyt, E.W. & Kirby, R.E. (1988). Surface Segregation of Boron in Dispersion-Strengthened Copper. *Journal of Materials Science*, Vol.27, No.21 (November 1992), pp. 5926-5930, ISSN 0022-2461

Gessinger, G.H. (1976). Mechanical Alloying of IN-738. *Metallurgial Transactions A*, Vol.7, No.8, (August 1976), pp. 1203-1209, ISSN 0360-2133

Hirsch, J. & Lücke, K. (1988). Mechanism of Deformation and Development of Rolling Textures in Polycrystalline FCC Metals-I. Description of Rolling Texture Development in Homogeneous CuZn Alloys. *Acta Metallurgica*, Vol.36, No.11, (November 1988), pp. 2863-2882, ISSN 1359-6454

Humphreys, F.J. & Hatherly, M. (1995). *Recrystallization and Related Annealing Phenomena*, Pergamon, ISBN 0-08-041884-8, Oxford, United Kingdom

Kim, S.-H. & Lee, D.N. (2001). Recrystallization of Alumina Dispersion Strengthened Copper Strips. *Materials Science and Engineering A*, Vol.313, No.1-2, (August 2001), pp. 24-33, ISSN 0921-5093

Kim, S.-H. & Lee, D.N. (2002). Annealing Behavior of Alumina Dispersion-Strengthened Copper Strips Rolled Under Different Conditions. *Metallurgical and Materials Transactions A*, Vol.33, No.6 (June 2002), pp. 1605-1616, ISSN 1073-5623

Klug, R.C. ; Krauss, G. & Matlock, D.K. (1996). Recrystallization in Oxide-Dispersion Strengthened Mechanically Alloyed Sheet Steel. *Metallurgical and Materials Transactions A*, Vol.27, No.7, (July 1996), pp. 1945-1960, ISSN 1073-5623

Kusunoki, K. ; Sumino, K. ; Kawasaki, Y. & Yamazaki, M. (1990). Effects of the Amount of γ' and Oxide Content on the Secondary Recrystallization Temperature of Nickel-Base Superalloys. *Metallurgical Transations A*, Vol.21, No.2, (February 1990), pp.547-555, ISSN 0360-2133

Matthies, S. ; Vinel, G.W. & Helming, K. (1987). *Standard Distribution in Texture Analysis*, Akademie-Verlag, ISBN 3-05-500249-0, Berlin, Germany

Mino, K. ; Nakagawa, Y.G. & Ohtomo, A. (1987). Abnormal Grain Growth Behavior of An Oxide Dispersion Strengthened Superalloy. *Metallurgical Transactions A*, Vol.18, No.6, (June 1987), pp. 777-784, ISSN 0360-2133

Miodownik, M.A. ; Martin. J.W. & Little. E.A. (1994). Secondary Recrystallization of Two Oxide Dispersion Strengthened Ferritic Superalloys : MA956 and MA 957. *Materials Science and Technology*, Vol.10, No.2, (February 1994), pp. 102-109, ISSN 0267-0836

Miodownik, M.A. ; Humphreys, A.O. & Martin, J.W. (1995). Growth of Secondary Recrystallized Grains during Zone Annealing of Oxide Dispersion Strengthened Alloys. *Materials Science and Technology*, Vol.11., No.5, (May 1995), pp. 450-454, ISSN 0267-0836

Nadkarni, A. (1984). Dispersion Strengthened Copper Properties and Applications, In : *High Conductivity Copper and Aluminum Alloys*, E. Ling, P.W. Taubenblat, (Ed.), 77-101, TMS-ASME, Warrendale, PA

Petrovic, J.J. & Ebert, L.J. (1972). Electron Microscopy Examination of Primary Recrystallization in TD-Nickel. *Metallurgical Transactions*, Vol.3, No.5, (May 1972), pp.1123-1129, ISSN 0360-2133

Petrovic, J.J. & Ebert, L.J. (1972). Abnormal Grain Growth in TD-Nickel. *Metallurgical Transactions*, Vol.3, No.5, (May 1972), pp.1131-1136, ISSN 0360-2133

Preston, O. & Grant, N.J. (1961). Dispersion Strengthening of Copper by Internal Oxidation. *Transactions of the Metallurgical Society of AIME*, Vol.221, (February 1961), pp.164-173

Randle, V. (1993). *The Measurement of Grain Boundary Geometry*, Institute of Physics Publishing, ISBN 0-7503-0235-6, London, United Kingdom

Simchi, H. & Simchi, A. (2009). Tensile and Fatigue Fracture of Nanometric Alumina Reinforced Copper with Bimodal Grain Size Distribution. *Materials Science and Engineering A*, Vol.507, No.1-2, (May 2009), pp. 200-206, ISSN 0921-5093

Singer, R.F. & Gessinger, G.H. (1982). The Influence of Hot Working on the Subsequent Recyrstallization of a Dispersion Strengthened Superalloy-MA 6000. *Metallurgical Transactions A*, Vol.13, No.8, (August 1982), pp. 1463-1470, ISSN 0360-2133

Stephens, J.J. & Nix, W.D. (1985). The Effect of Grain Morphology on Longitudinal Creep Properties of INCONEL MA 754 at Elevated Temperature. *Metallurgical Transactions A*, Vol.16, No.7, (July 1985), pp. 1307-1324, ISSN 0360-2133

Sumino, Y. ; Watanabe, H. & Yoshida, N. (2009). The Microstructural Evolution of Precipitate Strengthened Copper Alloys by Varing Temperature Irradiation. *Journal of Nuclear Materials*, Vol.386-388, No.C, (April 2009), pp. 654-657, ISSN 0022-3115

Young, C.T. ; Steele, J.H. & Lytton, J.L. (1973). Characterization of Bicrystals Using Kikuchi Patterns, *Machine Design*, Vol.4, No.9, (September 1973), pp. 2081-2089, ISSN 0024-9114

3

Application of Orientation Mapping in TEM and SEM for Study of Microstructural Evolution During Annealing – Example: Aluminum Alloy with Bimodal Particle Distribution

K. Sztwiertnia, M. Bieda and A. Kornewa
Polish Academy of Sciences, Institute of Metallurgy and Materials Science, Krakow, Poland

1. Introduction

There are still considerable gaps in the understanding of the recrystallization processes of metallic materials, which reduce the possibility of controlling their course and introducing technological modifications aimed at obtaining desirable properties. The lack of a complete explanation can be attributed to the high complexity of the phenomenon, which consists of a superposition of the local nucleation and grain growth processes. These processes depend strongly on the characteristics of the matrix, which is typically complex and heterogeneously deformed. The phenomenology of the process and its energetic causes are known because they were examined long ago, e.g., (Humphreys & Hatherly, 2002). On the other hand, the relevant physical mechanisms that control the nucleation and growth of new grains are not entirely clear. This uncertainty exists, among other reasons, because the origin of the crystallographic orientations of the nuclei is usually not known.

1.1 Crystallographic orientation and orientation characteristics of materials

The crystallographic orientation is a feature of a material that is defined at any point of the sample at which the ordering of the crystal lattice is not disturbed (or not significantly disturbed). It can be generally said that almost all of the basic quantities that characterize a polycrystalline material and its properties have a direct or complex relationship to the orientation (g_i), which is a function of the coordinates x_i, y_i, z_i of a point in the sample. The $g(x,y)$ function described in a plane of the sample defines the orientation topography (commonly called the orientation map). According to the definition of the term, the orientation at any point (x_i, y_i) in the sample is given by a rotation that brings the local sample reference system with its origin at the point (x_i, y_i) into coincidence with the crystal reference system. The orientation is described unambiguously by three parameters, which can be expressed in different ways. Usually, for the convenience of calculation, the Euler angles φ_1, Φ, φ_2 are applied. In some cases, the parameters of the rotation axis (θ, ψ) and the rotation angle ω are used because this scheme is easy to visualize. If the orientation is described by a greater number of parameters, then the parameters depend on each other.

Such is the situation in the case of crystallographic indices {hkl}<uvw>, commonly used in practice. Extensive analysis of orientation problems can be found in (Morawiec, 2004). If an orientation map is obtained, the grains or subgrains may be reconstructed by identifying areas whose pixels have orientations within a specified range. The knowledge of the orientation topography enables the identification of grain and subgrain boundaries, as well as other microstructural inhomogeneities, by selection of misorientations between neighboring measuring points. This approach enables stereological analysis with regard to the crystallographic orientation. Based on the orientation topography, the orientation characteristics of the microstructure can be determined (Pospiech et al., 1993). The set of orientation characteristics comprises the "principal distributions", which are texture functions determined by the whole set of measurements and "partial distributions", in which only part of measurements are needed. The most important of the principal distributions is the well-known orientation distribution function (ODF), which describes the crystallographic texture of a material. The ODF is defined by the density of the global orientation distribution of the grains (taking into account their volume fraction). Another "principal distribution" is the orientation difference distribution function (ODDF). The ODDF contains all possible misorientations between the measured orientations. To investigate local textures (or microtextures) in selected areas of inhomogeneities or in the environment of preferred orientations, partial distribution functions are applied. The most important of these functions is the misorientation distribution function (MODF), which describes the distribution of misorientation between the nearest neighbor grains. The other partial orientation distributions are statistical quantities related to orientation and misorientation, which may be related to the properties of a material, its anisotropy or specific stages through which material passes (Pospiech et al., 1993).

Many of the essential properties of polycrystalline materials and their anisotropy depend directly or indirectly on the topographical arrangement of orientations. Using the orientation topography, the material properties that depend on the character and distribution of the grain boundaries can be described, for example, segregation or corrosion. The knowledge of the orientation topography is of basic importance for the understanding of many processes that occur in the material, such as deformation, recrystallization, phase transformation or diffusion.

It is, therefore, not surprising that the characterization of the microstructure based on the sets of measured orientations has advanced as a well-established technique, known as Orientation Microscopy (OM). The main concept behind this technique is the automatic collection and indexing of many electron diffraction patterns that are correlated with sample coordinates. The development of new generations of computer-controlled electron microscopes has improved their spatial resolution and increased the rate at which large sets of Electron Back Scattered Diffraction (EBSD) patterns can be collected and processed in a Scanning Electron Microscope (SEM), e.g., (Dingley, 1984; Wright & Adams, 1992; Adams & Dingley, 1994; Schwartz et al., 2009). Systems created orientation topographies using EBSD in SEM are now very common, and commercially available versions of this technology are essentially fully automated, e.g., (HKL, 2007; TSL, 2007).

1.2 Orientation imaging microscopy for recrystallization study

Obviously, the analysis and modeling of the recrystallization of a deformed metallic material requires description of the microstructure evolution during annealing that is as complete as possible. Such a description may be based on the orientation topographies obtained by OM techniques in systematic measurements of a sample that undergoes a specific deformation and annealing process, e.g., (Zaefferer et al., 2001; Sztwiertnia, 2008).

To study the recrystallization (particularly its early stages), a high spatial resolution in the orientation measurement is required. Unfortunately, the spatial resolution that can be achieved by EBSD/SEM measurements is relatively low. It falls approximately one order of magnitude behind the spatial resolution in conventional SEM imaging, and still further behind when compared to the spatial resolution of a transmission electron microscope (TEM). The measurement is limited in this way because the inherent resolution of EBSD is governed not by the diameter of the beam spot at the point of impact on the surface, but primarily by the excitation volume. This quantity is the fraction of the interaction volume of the primary electrons within the sample from which the pattern-forming electrons are back diffracted and leave the crystal, without further scattering. This volume is strongly dependent on the type of electron gun and the material being investigated. Tilting of the sample during the EBSD measurement further degrades the spatial resolution and produces resolution along the beam direction on the sample surface that is approximately three times worse than the resolution along the direction perpendicular to the beam, e.g., (Schwartz et al., 2009). For these reasons, the best achievable spatial resolution in EBSD/SEM special cases is in the order of approximately 30 nm; however, the practical limit is approximately 100 nm. This limitation restricts the utility of EBSD/SEM for the investigation of very fine-grained and deformed microstructures, as in the case of the early stages of recrystallization. To obtain better spatial and angular resolution, similar systems have been developed for TEM, e.g., (Haessner et al., 1983; Dingley, 2006; Morawiec, 1999; Morawiec et al., 2002; Rauch & Dupuy, 2005). Despite some restrictions, such as the currently unsolved problems of image analysis, difficulties in measurement automation and sample preparation, the OIM technique applied to TEM offers spatial resolution better than 10 nm and can be used for quantitative analysis of structures at the nanoscale. Such a system, built at the Institute of Metallurgy and Materials Science (Morawiec et al., 2002; Sztwiertnia et al., 2006; Bieda-Niemiec, 2007), was used for a study of the recrystallization of 6013 aluminum alloy (Sztwiertnia et al., 2007; Bieda et al., 2010).

The 6013 aluminum alloy was chosen as a prototype material that represents a group of commercial alloys with a bimodal second phase particle distribution. The second-phase particles are used to control the strengthening, grain size and texture of the alloy. Such alloys can be interesting for examination of the role of the second phase particles in the recrystallization process, e.g., (Humphreys & Hatherly, 2002; Sztwiertnia et al., 2005; Ardakani & Humphreys, 1994).

To elucidate the mechanisms of the alloy microstructure transformation during annealing, *in situ* TEM experiments and combined calorimetric–microscopic investigations of bulk samples were carried out. The *in situ* experiments were necessary to provide information about the temporal relationships between changes that occur in the metal at the beginning of recrystallization. Dynamic studies using SEM or TEM should be capable of providing the required information. Because of its already mentioned limitations, the EBSD/SEM measurements of localized strain in the deformed polycrystal, which are particularly interesting as nucleation sites, give rather poor information about the orientation. For the *in situ* studies of such regions, measurements using convergent beam electron diffraction

(CBED) or microdiffraction in a TEM are more suitable, although the proximity of the free surface in the thin foils can be a complicating factor during annealing. The first *in-situ* TEM observations, obtained by Bailey in 1960 and Hu in 1963, indicated differences in the recrystallization processes that can occur during the annealing of bulk samples and thin foils. As a consequence, many researchers have been skeptical about the results of such experiments up to now. However, the results obtained later by other authors e.g., (Roberts & Lehtinen, 1972; Hutchinson & Ray, 1973; Sztwiertnia & Haessner, 1994) allow the determination of the experimental conditions, which ensures that the changes directly observed in an annealed foil are at least similar to those occurring in a bulk sample. In general, recrystallization is easiest in orthogonal sections from a rolled sheet in which the grain boundaries, extending from top to bottom of the foil, present the most favorable distribution of driving potential for migration. Grooving grains are not strongly inhibited in the thin foil regions and frequently extend nearly to the edge of the foil (Hutchinson & Ray, 1973). Nevertheless, because of the thermal grooving, the recrystallization front always stops in foil thinner than a certain critical value, which approximately depends on the fineness of its microstructure. One can increase the usable foil thickness by increasing the accelerating voltage. The impact of the sample thickness on grain boundary movement explained in greater detail by Roberts & Lethinen, 1972.

In highly deformed 6013 aluminum alloy, the critical thickness for foils cut from planes perpendicular to the sheet is so low that it allows *in situ* observation of nucleation events (and to some extend the growth of the nuclei) to be carried out in a conventional TEM operated at 200 kV (Sztwiertnia et al., 2005; Sztwiertnia et al., 2007; Bieda et al., 2010). To examine the significance of the *in situ* experiments, the thin foils annealed in the TEM were compared to thin foils prepared from bulk samples heated in a calorimeter to obtain a specified recrystallization stage. The comparison shows that the processes occurring in both types of foils were at least qualitatively the same.

2. Example: Recrystallization of aluminum alloy with bimodal particle distribution

2.1 Material and investigation methodology

The changes of the microstructure during annealing were examined in the case of the polycrystalline aluminum alloy 6013 (Table 1), which was previously identified as the prototype of materials with a bimodal precipitate distribution.

Mg	Si	Cu	Mn	Fe	others	Al
1.15	1.0	1.1	0.3	0.5	0.15	remainder

Table 1. 6013 aluminum alloy chemical composition (% wt.).

Samples for testing were supersaturated, then aged and reversibly cold-rolled up to the 75 and 90 % of the 10 mm value. The deformed samples were examined by means of non-isothermal annealing in a differential calorimeter (DC). It was found that the spectrum of released stored energy contained several peaks. Next, a new series of samples were heated in the calorimeter to the selected temperatures, rapidly cooled, and then analyzed in the TEM. The tests in the TEM were complemented by SEM examinations. Because of the poor quality of the orientation topographies measured by the standard EBSD/FEG/SEM techniques, these measurements were used only to detect the presence of newly recrystallized grains (that is, only those diffraction patterns were taken into account that had a high image quality and showed local areas with low dislocation densities). The high degree of deformation also resulted in low quality TEM diffraction patterns. However, it was still possible to measure enough single orientations in the TEM to construct orientation topographies for all of the microstructural elements of the cold-rolled material. The investigations also included a comparison of the deformed alloy microstructure with that of the pure metal and the investigation of the recrystallization process dynamic. The latter investigation consisted of the *in situ* measurements in the TEM. This measurement was necessary to obtain information about the time sequence of changes occurring in the material at the beginning of the recrystallization. In the investigated material, the sequence of events that occurs in the deformation zones is of particular interest because such areas undergo intense nucleation.

2.2 Deformation state

First, the deformation microstructure of the alloy was compared with that of the pure metal. The microstructure of commercially available pure aluminum (3N Al), reversibly cold rolled to 90%, has been chosen for comparison. Both microstructures are built of elongated in the rolling direction (RD) and lie nearly parallel to the sheet plane grains and subgrains, as shown in Fig. 1 b and 2 a. Some significant differences became evident when the measured orientation topographies were compared. In the alloy matrix deformed to 75 %, the distances between the high angle grain boundaries (HAGB) in the normal direction (ND) to the sheet plane were typically smaller than 1 µm, as shown in Fig. 2 c. In the pure metal, the thickness of similarly oriented layers often exceeded 10 µm, even for the deformation of 90 %, as shown in Fig. 1 a, c. These layers were composed of parallel bands or clusters of subgrains that were strongly elongated in the RD. In some of them, relatively small but accumulating disorientation angles[1] occurred (Fig. 1c). Because of this accumulation, large disorientation angles (up to ~ 20 °) between the first and the last subgrain in the band occurred frequently. Orientation changes of this type are characteristic for transition bands (Dillamore et al., 1972). The alternation of orientation patterns was also recognized. The high frequency of the low angle grain boundaries (LAGB) inside the band indicates a well-developed subgrain structure. With no further analysis of the pure metal deformation microstructure, we can only conclude that it consists of thick deformation and transition bands, and the density of HAGBs in the ND is low.

The crystallographic orientation characteristics of the alloy were quite different from those of the pure metal. The matrix consisted of well-developed HAGBs. The distances between them along ND were much smaller than those in the pure metal and never exceed a few hundreds of nanometers. LAGBs in the elongated thin matrix grains were less ordered and created a less-expanded subgrain structure. In the laminar microstructure of the alloy, small (<<1 µm) and large (1÷3 µm) precipitate particles of the second phase were scattered. Around the large particles, zones of localized strain were identified. These deformation zones consisted of

[1]A given misorientation can be described by a rotation axis and an angle of rotation. For a material with crystal symmetry, there is more than one angle of rotation. The angle with the absolute value smallest of all possible angles of rotation is called the disorientation angle.

ultrafine (50 - 200 nm) grains and more or less bent microbands of the matrix (Fig. 3 b, 4 b). The orientation image of the alloy clearly suggests that microstructural evolution occurs by grain subdivision at a very small scale compared to the original grain size, which was approximately 100 μm in this case. This result is presented in Fig. 2 c and 3 d, showing large orientation variations over a region as small as a few micrometers. Such rapid changes in orientation may demonstrate that volumes characterized by a combination of slip systems can be very small.

Fig. 1. As-deformed microstucture of 90% cold-rolled aluminum, the TEM bright field image (b) and orientation topographies (a and c). On the map (c) the color change indicates a deviation from the initial orientation (dark blue) to the disoriented orientation (light blue); black lines (a) - low angle grain boundaries (>1 °), red lines (c) - the boundaries with a disorientation angle > 5 °; the points where the diffraction pattern has not been solved are shown in white.

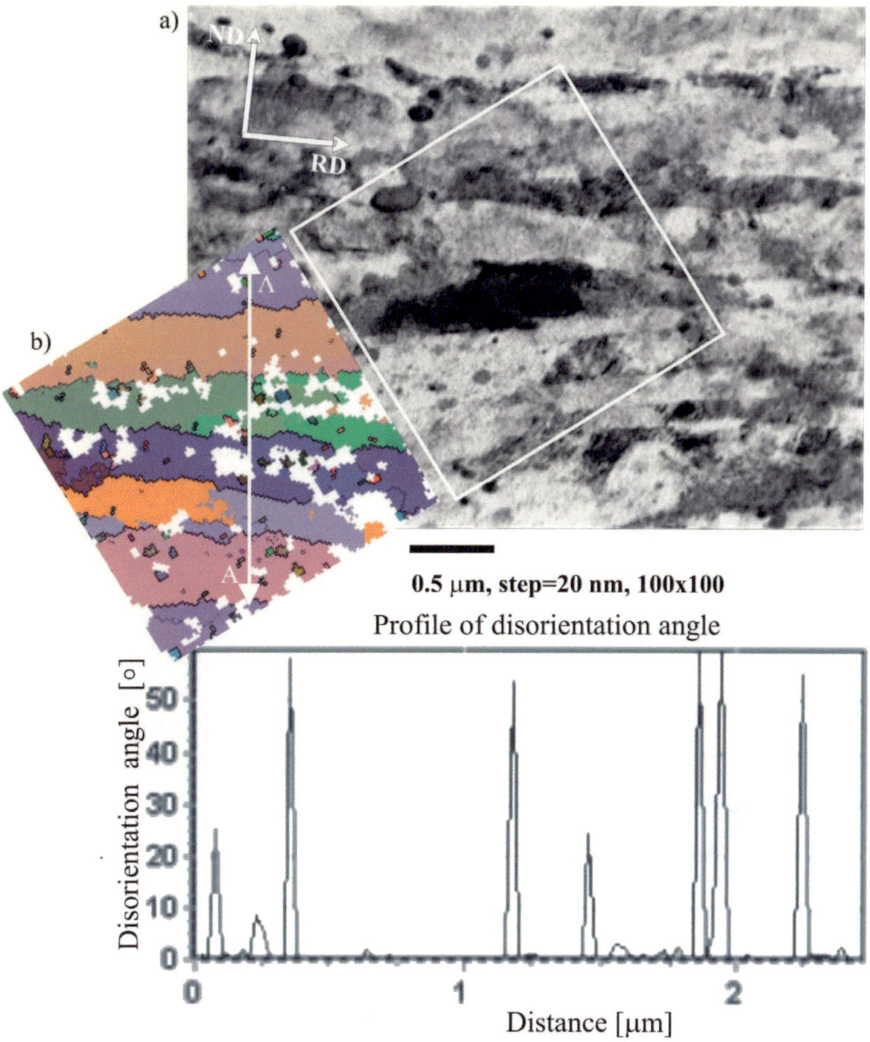

Fig. 2. a) As-deformed microstucture of 75% cold-rolled 6013 alloy, longitudinal section, TEM. b) Orientation topography of the matrix area, black lines show high angle grain boundaries; white regions are not indexed. c) Disorientation angle profile along A-A line (b), (Sztwiertnia et al., 2007).

The global crystallographic textures of both materials were similar and corresponded to the well-known rolling texture of FCC metals with high stacking fault energies (such as pure aluminum). Such a texture is characterized by the concentration of components along two orientation fibers. The main one, called the α fiber, runs diagonally through the orientation space containing the preferred texture components S {123}<634> and Cu {112}<111>. The other one, called the β fiber, includes Goss {011}<100> and Bs {011}<211> components (Sztwiertnia et al., 2005).

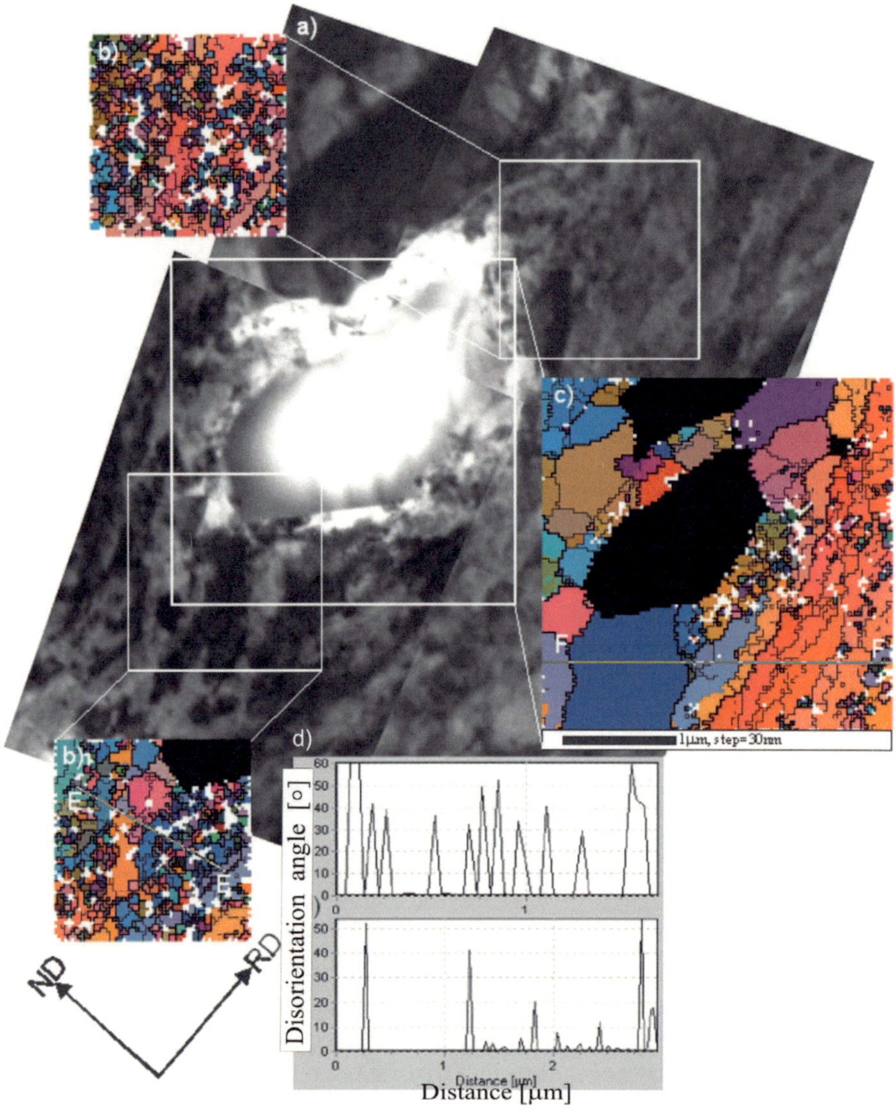

Fig. 3. a) As-deformed microstucture of 75% cold-rolled 6013 alloy, the deformation zone around the large particle, longitudinal section, TEM. Orientation topographies in areas of the deformation zone before (b) and after (c) heating *in situ* in TEM; particles of the second phase are shown in black, white regions are not indexed; thick lines indicate high angle grain boundaries, thin lines indicate low angles grain boundaries. d) Examples of disorientation angle profiles along E-E and F-F (Sztwiertnia et al., 2007).

2.3 Microstructure changes during annealing

The cold-rolled alloy was tested using the non-isothermal annealing method in a differential calorimeter. On the basis of these tests, as well as the microscopic analysis of the microstructures of the appropriately annealed samples, it was possible to state that the two separate peaks of the stored energy release correspond to the two stages of the recrystallization process (Fig. 6 a).

The *in situ* tests in the TEM allowed the determination of the sequence of events occurring at the beginning of recrystallization in the deformation zones around the large particles and in the matrix beyond those areas. First, the deformation microstructure was carefully examined. In the deformation zones around the large second phase particles, small grains and distorted fragments of microbands were identified. The small grains were approximately 50-200 nm in size. The strong orientation changes, greater than 15°, either identify HAGBs lying at distances lower than 200 nm or they are an effect of strong grain bending, as shown in Fig. 3 b, 4 b. To precisely distinguish between these two phenomena, small-step orientation maps were generated. The broad distribution of orientations in the deformation zones tended to group in the range of the deformation components after rotation around the transverse (TD) or the normal direction (ND) to the sheet plane (Sztwiertnia et al., 2005). This type of rotations suggests that at least a part of the strong orientation changes may be a result of the accumulation of small disorientations along a bent grain. In the matrix outside of the DZs, the HAGBs lie roughly parallel to the sheet plane at distances of 0.5 - 1 µm along the ND (Fig. 2 b).

After the sample was annealed in the microscope, the same areas as in the deformation state were investigated. Figures 3 c and 4 c show examples of DZ orientation maps after *in situ* annealing. Nuclei and new grains appeared in the vicinity of the deformation zone. The orientations of crystallites in the deformed state commonly lay in the area of a particular new grain or nucleus (Fig. 4 d, e). For each orientation of a new grain, at least one similarly oriented fragment of the deformed matrix was found. Some of the nuclei were growing within the zone defined by the migration of HAGBs, a result that was confirmed by a significant reduction of their density in those zones after annealing to the temperature of the first peak (Fig. 3 c, d and 4 b, c). The shape of some new grains suggests that they could have been formed as a result of the local recovery of strongly bent fragments of matrix microbands, as shown in Fig 4 c.

Partial misorientation distribution functions (PMDF) were calculated between the grains in the deformed state and the new grains that appeared at the same location (Fig. 5). PMDFs for both the 75 and 90 % deformed materials show a random distribution of misorientations. This distribution suggests that there was no special orientation relationship describing favored growth.

In the deformation zones at very early stages of the recrystallization, broad-spectrum HAGBs appear to have been active and mobile. At that time, no migration of HAGBs in the matrix areas outside the zones was observed.

To examine the significance of the *in situ* experiments, the microstructures of TEM-annealed thin foils were compared with the microstructures of thin foils prepared from bulk samples heated in a calorimeter to obtain a specified recrystallization stage. The comparison shows that the processes occurring in the thin foils and in the bulk samples were, at least qualitatively, the same. The temperature range of the first recrystallization peak was found to produce nucleation, which is accompanied by some limited enlargement of new grains in the sheet plane.

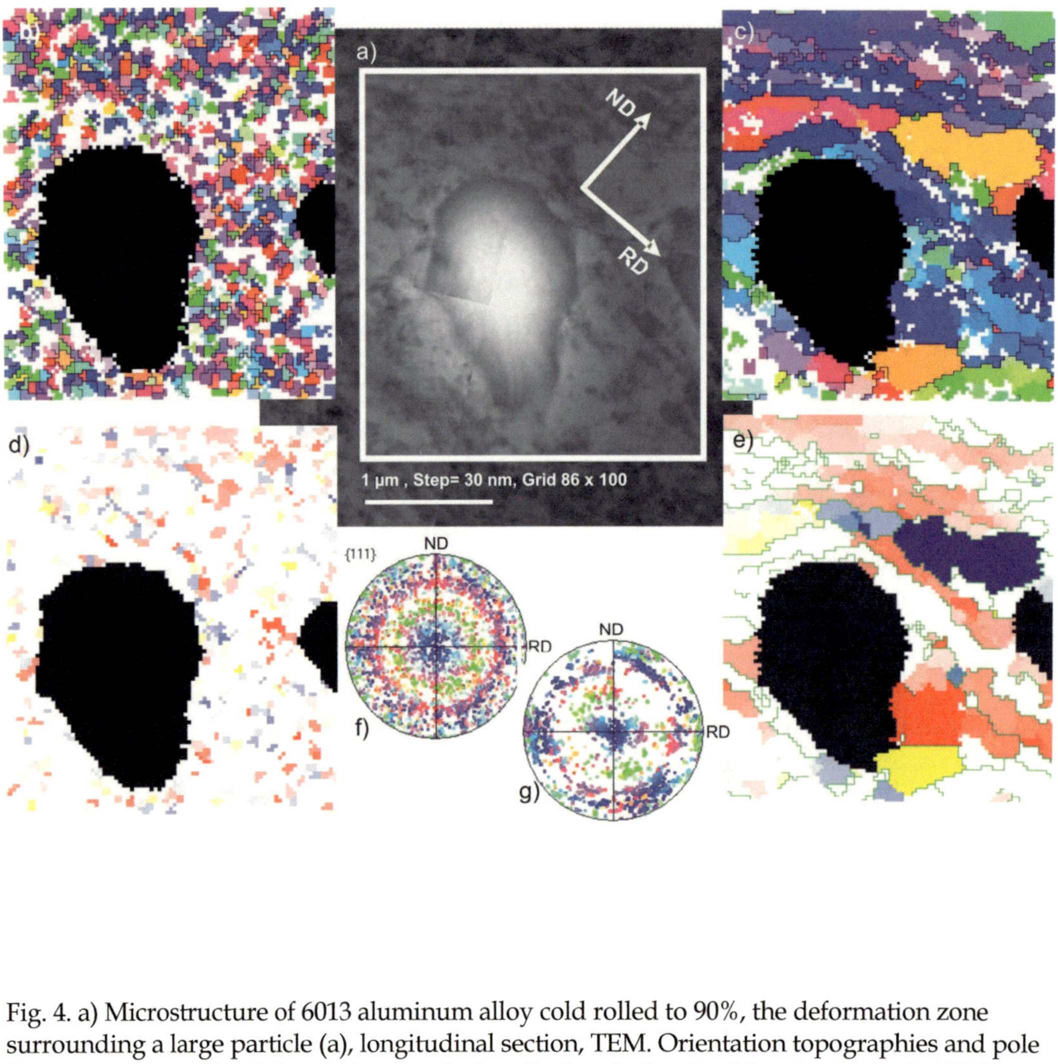

Fig. 4. a) Microstructure of 6013 aluminum alloy cold rolled to 90%, the deformation zone surrounding a large particle (a), longitudinal section, TEM. Orientation topographies and pole figures in areas of the deformation zone before (b, f) and after (c, g) heating *in situ* in TEM; particles of the second phase are shown in black, white regions are not indexed; thick lines indicate high angle grain boundaries, thin lines indicate low angle grain boundaries. Areas of similar orientations (blue, red, yellow) before and after annealing (d, e) (Bieda et al., 2010).

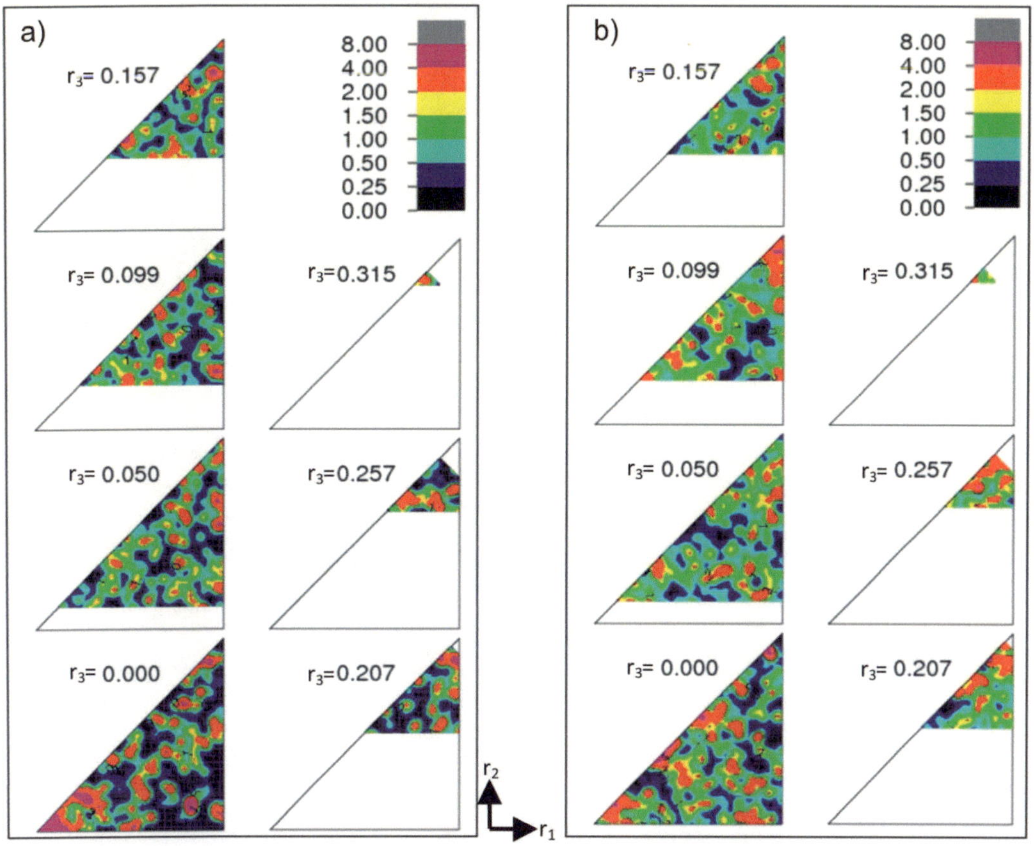

Fig. 5. Partial misorientation distribution functions showing the orientation relationships between the crystallites from the deformation zones (before annealing) and the new grains appearing in their positions after annealing (a) 75% cold-rolled 6013 aluminum alloy, (b) 90% cold-rolled 6013 aluminum alloy; Rodrigues' representation r_1, r_2, r_3, cross-section r_3=const., asymmetric domain (O, O) (Bieda et al., 2010).

By way of example, Fig. 6 c shows the orientation topography in the 75 % cold-rolled sample that was heated to 330 °C in a calorimeter. The appearance of new grains with sizes up to a few micrometers can be observed around the large particle. These new grains, similar to those in the *in situ* experiment, were formed as a result of nucleation and the consequent growth of nuclei inside the zone. At the examined temperature, the arrangement of LAGBs can also be observed in the elongated grains of the matrix outside the zone. The annihilation of these boundaries takes place at higher temperatures, near the end of the first peak.

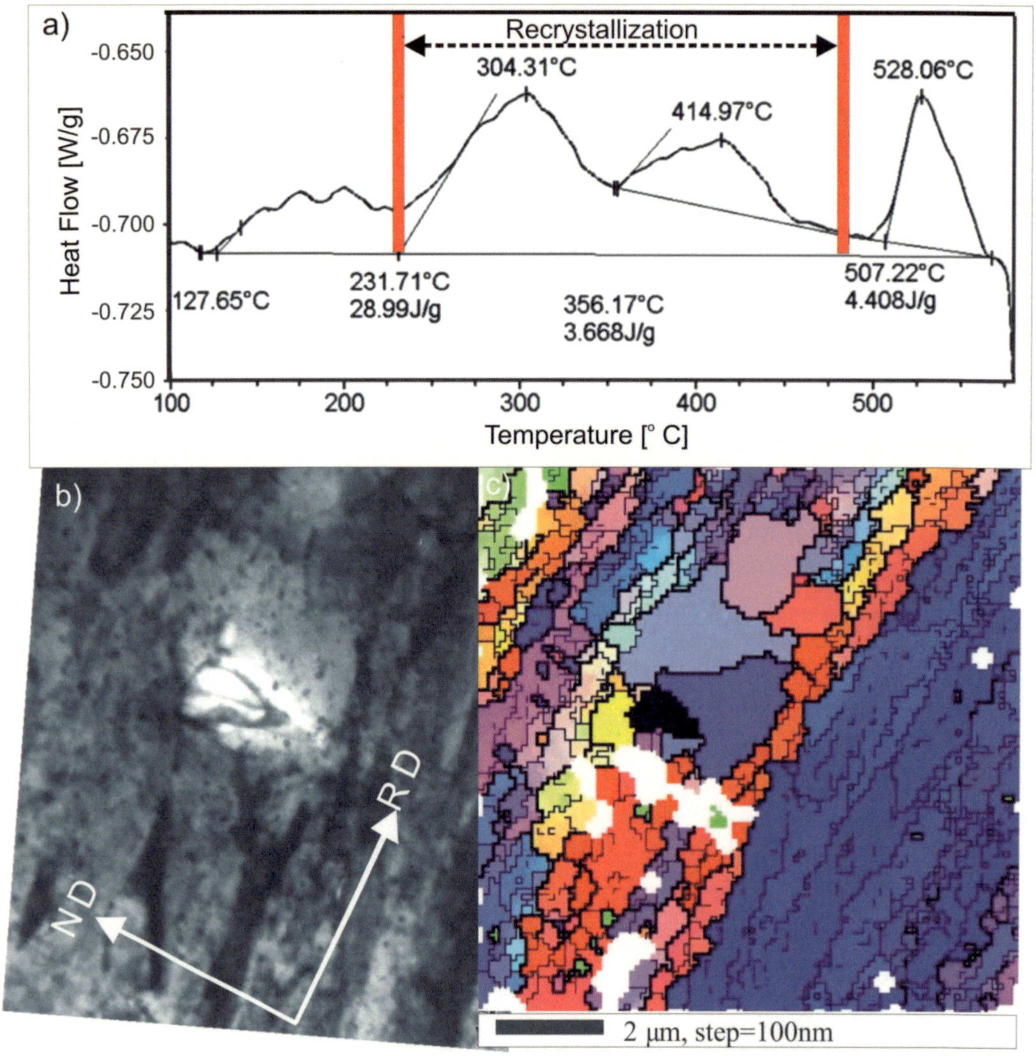

Fig. 6. a) Power differences, representing the release of stored energy from 75 % cold-rolled 6013 aluminum alloy as a function of annealing temperature. b, c) As-deformed microstructure and orientation topography of 6013 aluminum alloy 75 % cold-rolled and subsequently heated in the calorimeter to 330 °C: particles of the second phase are shown in black, white regions are not indexed; thick lines indicate high angle grain boundaries, thin lines indicate low angle grain boundaries; TEM, (Sztwiertnia et al., 2007).

The phenomenon of the annihilation of LAGBs and the growth of new grains in the sheet plane (mainly, parallel to the RD) was also observed in the SEM results (Fig. 7). The SEM measurements of the orientation topographies were made on samples heated to appropriate temperatures from the range of the first peak (Fig. 5 a). These observations fully confirm the TEM results and show that after the recrystallization of the deformation zones, local recovery processes (coalescence of subgrains) take place in the matrix. These local recovery processes lead to the annihilation of the LAGBs between chains of subgrains lying parallel to the sheet plane and, consequently, to the production of long grains with a low density of lattice defects. The growth of the elongated grains in the ND occurs rarely or not at all within the temperature range of the first peak. The migration of HAGBs in the matrix becomes the main process within the temperature range of the second peak. Heating of the sample to the temperature of the end of this peak leads to the complete discontinuous recrystallization of the material (Fig. 7 c). The recrystallized microstructure is dominated by elongated grains (up to 30 μm in length along the RD). It also includes groups of smaller grains, which are often almost equiaxial. After the discontinuous recrystallization, the grains are a few times thicker than they are in the state observed at the end of the first stage.

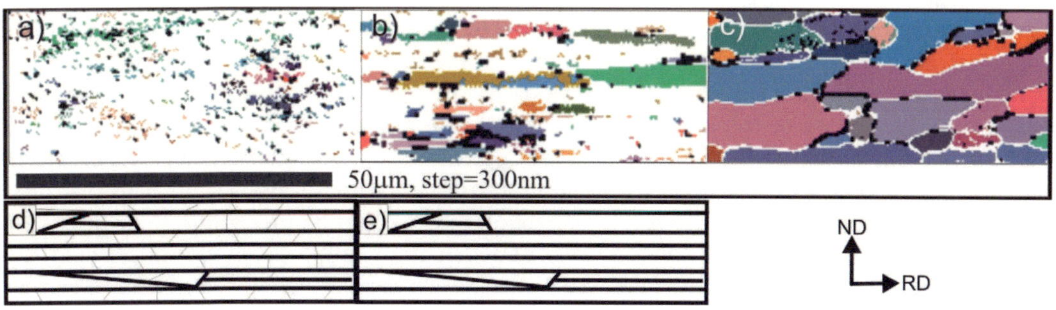

Fig. 7. Orientation topographies of recrystallized grains in 6013 aluminum alloy 75% cold rolled and subsequently heated in the calorimeter to: (a) 330 °C, (b) 350 °C and (c) 480 °C, regions of unsolved diffraction (approximately corresponding to the deformed areas) are shown in white, EBSD/SEM/FEG. d) and e) Schematic representation of the deformed microstructure before and after annealing to the temperatures from the end of the first recrystallization peak, Fig. 2; thick lines indicate high angle grain boundaries, thin lines indicate low angle grain boundaries.

3. Summary

Orientation topography provides basic local and global information about microstructures by allowing the identification and description of occurring regularities. The OM technique in TEM is a useful tool for the quantitative and qualitative characterization of fine crystalline and deformed microstructures in polycrystalline materials. It is possible to obtain information about grain distribution, misorientation between grains, material phases, the local orientation distribution function, and the misorientation distribution function. Replacement of the SEM measurements by TEM measurements improves the spatial resolution to a few nanometers. Both SEM and TEM can be used for complementary analysis of crystalline materials at the "micro" and "nano" scale, respectively. Together with *in situ* studies, orientation mapping in TEM can provide additional information about the behavior of a material during annealing, particularly in zones of greater deformation. The example of orientation characteristics presented in this chapter illustrates only some aspects of the applicability of these techniques. Orientation mapping in a transmission electron microscope was successfully applied to the study of microstructural changes during the initial stage of recrystallization in an aluminum alloy with a bimodal second-phase particle distribution. The images of the microstructure in the representative areas of a sample of deformed aluminum alloy 6013, described by measurements of orientation topography, shows greatly advanced grain fragmentation.

In situ investigations in TEM, calorimetric measurements, and orientation mapping in TEM and SEM demonstrate that the recrystallization of the tested material can be considered to occur as a number of partly overlapping processes that proceed in two stages. These stages correspond to the two separate stored energy release peaks. In the initial stage, the deformation zones around large second phase particles act as sites for particle-stimulated nucleation. This nucleation is accompanied by the growth of nuclei. However, the migration of high angle grain boundaries only occurs in deformation zones. At the same range of temperatures, some enlargement of new grains in the matrix (outside of the deformation zones) was also observed. The formation of grains elongated primarily in the direction parallel to the rolling direction may be correlated to the processes of local recovery, which is triggered in the deformation zones. Grain elongation then continues to develop along the bands of the deformed matrix in the directions of low orientation gradients. The elongated grains appear because of the annihilation of low angle grain boundaries between chains of subgrains lying in layers parallel to the sheet plane. As a consequence, new grains often have a plate-like character, with their shorter axis parallel to the sheet plane normal direction. Their lengths along the rolling direction may exceed 50 μm, while their thickness corresponds approximately to the distance between high angle grain boundaries in the normal direction outside the deformation zones; this dimension was not observed to exceed a few micrometers. In the second stage, high angle grain boundaries were observed to migrate in the direction of the high orientation gradient. This migration, mostly in the normal direction, was limited to "free areas" of the deformed matrix between bands of new grain formation in the initial stage of recrystallization.

4. References

Adams B.L., Dingley D.J (1994), *Orientation Imaging Microscopy: New Possibilities for Microstructural Investigations using Automated BKD Analysis,* Mater. Sci. Forum, 157-62 31.

Ardakani M.G., Humphreys F.J. (1994), *The annealing behavior of deformed particle-containing aluminum single crystals.* Acta Metal. Mater., 42, 763.

Bailey J.E. (1960), *Electron microscope observations on the annealing process occurring in cold-worked silver,* Phil. Mag., 5, 833.

Bieda M.,.Sztwiertnia K, Korneva A., Czeppe T., Orlicki R. (2010), *Orientation mapping study on the inhomogeneous microstructure evolution during annealing of 6013 aluminum alloy,* Solid State Phenom, 16, 13.

Bieda-Niemiec M. (2007), *Opracowanie systemu do automatycznego pomiaru map orientacji w transmisyjnym mikroskopie elektronowym do analizy mikrostruktury drobnoziarnistych materiałów metalicznych,* Thesis, Kraków IMIM PAN, in polish.

Dillamore I.L., Morris P.L., Smith C.J.F., Hutchinson W.B. (1972), *Transition Bands and Recrystallization in Metals,* Proc. Roy. Soc., 329A, 405.

Dingley D.J. (1984), *On-line determination of crystal orientation and texture determination in SEM,* Proc. Roy. Microsc. Soc., 19, 74.

Dingley D.J. (2006), *Orientation Imaging Microscopy for the Transmission Electron Microscope,* Mikrochim. Acta, 155, 19.

Haessner, F., Pospiech, J. and Sztwiertnia, K. (1983), *Spatial arrangement of orientations in rolled copper,* Mat. Sci. Eng.1, 1.

HKL (2007) http://www.oxford-instruments.com

Hu H. (1963), *Electron Microscopy and Strength of Crystals,* Interscience, London, 564.

Humphreys F.J. and Hatherly M. (2002), *Recrystallization and Related Annealing Phenomena,* Pergamon Press, Oxford.

Hutchinson W.B., Ray R.K. (1973), *On the feasibility of in situ observations of recrystallization in the high voltage microscope,* Phil. Mag., 28, 953.

Morawiec A. (1999), *Automatic orientation determination from Kikuchi patterns,* J. Appl. Cryst. 32, 788.

Morawiec A. (2004), *Orientations and Rotations. Computations in Crystallographic Textures.* Berlin, Heidelberg, New York: Springer-Verlag.

Morawiec A., Fundenberger J.J., Bouzy E., Lecomte J.S. (2002), *EP-a program for determination of crystallite orientations from TEM Kikuchi and CBED diffraction patterns"* J. Appl. Cryst., 35, 287.

Pospiech J., Lücke K. and Sztwiertnia K. (1993), *Orientation Distribution and Orientation Correlation Functions for Description of Microstructures,* Acta Metall. Mater. , 41, 305.

Rauch E.F., Dupuy L. (2005), *Rapid spot diffraction patterns identification through template matching,* Arch. Metall. Mater., 50, 87.

Roberts W., Lehtinen B. (1972), *On the feasibility of in situ observations of recrystallization in the high voltage electron microscope,* Phil. Mag., 26, 1153.

Schwartz A.J., Kumar M.,. Adams B.L, Field D.P. (2009), *Electron Backscatter Diffraction in Materials Science,* ISBN 978-0-387 88135-2, Springer.

Sztwiertnia K. (2008), *On recrystallization texture formation in polycrystalline fcc alloys with low stacking fault energies,* Int. J. Mater. Res., 99, 178.

Sztwiertnia K., Bieda M., Korneva A., Sawina G. (2007), *Inhomogeneous microstructural evolution during the annealing of 6013 aluminium alloy,* Inżynieria Materiałowa, Vol. 3 XXVIII, 476.

Sztwiertnia K., Bieda M., Sawina G. (2006), *Determination of crystallite orientations using TEM. Examples of measurements,* Arch. Metall. Mater. , 51, 55.

Sztwiertnia K., Haessner F. (1994), *In situ observations of the initial stage of recrystallization of highly rolled phosphous copper,* Mater. Sci. Forum, 157-162, 1069.

Sztwiertnia K., Morgiel J., Bouzy E. (2005), *Deformation Zones and their Behaviour during Annealing in 6013 Aluminim Alloy,* Arch. Metall. Mater. 50, 119.

TSL (2007) http://www.edax.com

Wright S.I., Adams B.L. (1992), *Automatic analysis of electron backscatter diffraction patterns*, Met. Trans. A 23, 759.

Zaefferer S, Baudin T, Penelle R (2001), *A study on the formation mechanisms of the cube recrystallization texture in cold rolled Fe-36% Ni*, Acta Mater. , 49, 1105.

4

Crystal Growth: Substructure and Recrystallization

Vadim Glebovsky
Institute of Solid State Physics, the Russian Academy of Sciences
Russia

1. Introduction

Single crystalline refractory transition metals (molybdenum, tungsten, niobium, and tantalum) exhibit a unique combination of properties, namely, high strength, plasticity, Young modulus, wear resistance, number, and low coefficient of linear expansion as well as high radiation resistance which is what makes single crystals of these metals of high purity the most suitable materials to be widely employed in science and engineering. Single crystals of high purity tungsten are well suited to production of the deflectors of the charged particles beams in the linear accelerators, the colliders and, also, to be successfully used as the target-converters for the sources of the positron beams. The attributes of single crystals of molybdenum alloys, compared to their polycrystalline counterparts, include more stable microstructures, lower creep rates, better compatibility with nuclear fuels and lower diffusion penetrability (Liu & Zee, 1996). On the other side, studies of the X-ray wave field in crystals or so called effects of the dynamic scattering theory are of high interest although the first observations of X-ray anomalous transmission are made more than fifty years ago. The necessary high degree of structural perfection is achieved for a limited number of crystals, such as silicon, germanium, and related families; almost no observations of the dynamic effects have been made in metals. Studies of X-ray anomalous transmission in the transition metals are of considerable interest, particularly in tungsten which has a simple structure and a high absorption coefficient.

A method of electron-beam floating zone melting (EBFZM) is widely used to grow single crystals of high-purity refractory transition metals for years (Pfann, 1966; Shah, 1980). The growth of the perfect single crystals of the refractory metals presents difficulties because of the low defect formation energy and the stringent constraints on the level of the temperature gradients. The single crystals of molybdenum and tungsten, grown from the melt by this method, tend to have the specific substructure, characterized by the high dislocation density, reaching 10^5-10^7 cm^{-2}. The main part of these dislocations is collected in the walls, forming the dislocation substructure of the three orders of magnitude. On the one hand, the substructure is due to polygonization of dislocations arising during the growth by one of the known mechanisms (Bolling & Finestein, 1972, Kittel, 1996, Nes & Most, 1966). On the other hand, there is inevitable inheritance of the seed crystal substructure in the growing single crystal, which consists in the fact that the favorably oriented low-angle boundaries grow up into a crystal. Considerable efforts have therefore been made to improve the structural quality of the tungsten single crystals (Cortenraad *et al.*, 2001a). Modern methods of preparing the tungsten single crystals can produce the specimens having the dislocation density of about 10^5 cm^{-2}. A chemical composition, a growth rate, a number of passes by the liquid zone, geometry of the crystals and some other parameters of the growth in varying degrees affect the substructure, but in any case, the substructure of the crystals of molybdenum and tungsten grown from the melt is far imperfect (Glebovsky *et al.*, 1988; Glebovsky & Semenov, 1993-1994, 1995, 1999)

However, the single crystals, free of the specific substructure, can be grown by the secondary recrystallization process, which consists of the plastic deformation procedure and the high-temperature annealing procedure. The plastic deformation procedure of monocrystalline specimens can be produced by rolling in the vacuum rolling machines. The high temperature annealing procedure can be performed with the help of the anneal devices inside the rolling machines or in the EBFZM set-ups. The studies of structural perfection of the single crystals grown from the melt and by recrystallization are made by using the methods of X-ray rocking curves and angular scanning topography. To monitor the subgrain substructure of the tungsten single crystals, the X-ray anomalous transmission method has been employed as well. The optimal recrystallization process involves the deformation of single crystals with the [111] growth axis by rolling along the (112) plane. The 6-12% deformation is found to be optimal to get the polycrystals with the large grains of high perfection. The vacuum conditions are most suitable for vacuum rolling to avoid oxidation of the crystal surfaces during deformation at high temperatures. As a result, the single crystals of molybdenum and tungsten have the substructure which is characterized by both the record-low dislocation density and the small mosaic (Glebovsky & Semenov, 1999). For comparison, the tungsten single crystals, grown from the melt, contain the subgrains of the first order, elongated along the growth axis with the misorientation angles of 8-10′ of an arc. The crystallographically perfect tungsten single crystals, obtained by recrystallization, do not contain the subgrains of the first and second orders at all, and the maximum misorientation angles of the subgrains of the third order are less than 1′ of an arc. The structural changes in the perfect single crystals as a result of the thermal stresses, when they have been used as the seed crystals for growing the single crystals from the melt, have been studied.

It is well known that ideal growth techniques and technologies do not exist. All methods and technologies have their own advantages and disadvantages, so the main tasks of researchers consist in developing the advantages and in reducing negative effects of the disadvantages. The EBFZM method has its unique advantages which open wide prospects in the production of the high-purity refractory metals. It would be a mistake if the prospects will not be realized because of the complexities associated with the structural features of the single crystals grown from the melt.

The idea of the chapter is to show the most reliable ways of improving the structural quality of single crystals of the high-purity refractory metals. The recrystallization example for the tungsten single crystals shows how perspective and reliable are these ways in obtaining the structurally perfect single crystals.

2. Brief comments on the electron beam float-zone melting and growing single crystals

Zone-melting techniques (particularly, the EBFZM method) are useful for metals which are very reactive in the liquid state at high temperatures, so they cannot be processed at any contact with other materials (Pfann, 1966). Quoting Pfann, discovered the zone melting technique: "I regard the conception and development of zone melting as an exiting scientific advance. And I cannot help being saddened to hear it occasionally referred to as simply a technical innovation that was mysteriously evoked by the need for transistor grade germanium and silicon. I regard zone melting as elegant both in its simplicity and its surprising complexity." By far the EBFZM method, which is the crucibleless zone melting technique, is characterized by simplicity and complexity, but it is still the best one for melting refractory metals and their alloys. There are some well-known advantages of the method: small volume of the melt; the high efficient EB guns; the well-defined thermal gradients; no contamination from the crucible materials - because instead of them the surface tension of liquid metals works; the non-contaminating way of heating - the electron beams; effective purification which can be achieved due to evaporation of impurities in a vacuum. In parallel, there are some disadvantages of the EBFZM method: it can only be used in a vacuum; limitation of sizes of crystals due to surface overheating and thus decreasing the surface tension of the liquid zone; the high axial temperature gradients in a solid; the high thermal stresses and the high density of the unremovable dislocations; limitation of geometry and mass of the crystals produced by the EBFZM method. These or other properties of the method are marked as the merits or shortcomings, it would be wrong to perceive clearly. Thus, one of the major advantages of the method is absence of refractory crucibles and holding the liquid zone by the surface tension. However, high sensitivity of the surface tension to the surface-active impurities and the temperature gradients converts the recognized advantage into a serious drawback, which prevents growing the single crystals of large diameters, because mass of the liquid zone is too large so that surface tension forces are able to hold it. A similar comment can be done concerning the temperature distribution and the temperature gradients. Certainly, the well-defined thermal gradients are the advantage of the method, but their high values lead to formation of the specific substructure in single crystals, which creates great problems for both the physics research and industrial application. The electronic heating is the really controlled non-contaminating way of heating, but it may only be used in a vacuum, which is also a kind of contradiction when discussing the advantages and the disadvantages of the method.

3. Features of the single crystals growth

The single crystals of the high-purity refractory metals are widely used in modern material science and technology (Alonzo *et al.*, 1995; Calverly *et al.*, 1957; Glebovsky *et al.*, 1998; Hay *et al.*, 1968; Liu & Zee, 1996; Moest *et al.*, 1998). This necessitates both studying purification processes and developing advanced techniques of growing single crystals of high-purity refractory metals with modern electron beam (EB) guns (M. Cole *et al.*, 1968; Glebovsky *et al.*, 1986). Crucibleless techniques with electronic heating are extremely important for melting, studying and preparing refractory metals because of their high chemical reactivity in the liquid state. Early zone refining theories such as progressive freezing, zone refining, zone crystal growing, nonideal separation and optimization are studied and discussed elsewhere (Pfann, 1966; Shah & Wills, 1975). Equipment and technologies such as types of the electron guns used, the drive mechanisms, heating and cooling, floating-zone melting, and stirring are also discussed in detail elsewhere (Shah, 1980). In the EBFZM method the liquid zone is held in place between two vertical collinear solid rods by its surface tension (Fig. 1). Single crystals of high-purity refractory metals can be grown exclusively by EBFZM because of their extremely high melting temperatures and chemical reactivity (Calverly *et al.*, 1957; Hay *et al.*, 1968; Alonzo *et al.*, 1995; Moest *et al.*, 1998; Glebovsky *et al.*, 1998). This necessitates both studying purification processes and developing advanced methods of growing single crystals metals using modern electron beam guns (M. Cole *et al.*, 1968; Glebovsky *et al.*, 1986). The main purpose in this field is to study the real structure of single crystals as a function of the technological parameters of the EBFZM method (Langer, 1980; Riedle *et al.*, 1994, 1996).

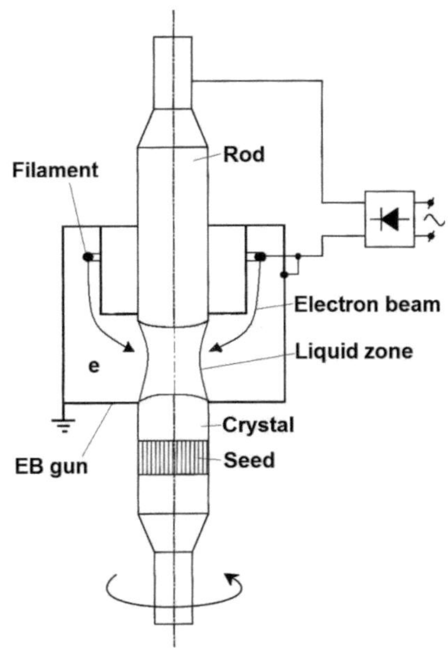

Fig. 1. Thermal zone of the EBFZM.

For effective melting and growing, the original EB guns have been elaborated on because the EB guns is the most important element of the EBFZM set-ups (Glebovsky *et al.*, 1986; Shah, 1980). In Fig. 2 the EB gun is shown, which consists of a cathode, an anode, and the focusing electrodes. The main features of the EB gun are: (a) the rod (crystal) serves as the anode, (2) the focusing electrodes made of a water-cooled copper, which makes the EB gun geometrically solid even at very high temperatures in the liquid zone, (3) the focusing electrodes form a stable circular electron beam field and focus it on the liquid zone, (4) the EB gun produces the well-defined thermal gradients on the crystal under the crystallization front. The cathode is made of a circular tungsten filament of 55 mm in dia. An arrangement of the focusing electrodes makes it possible to vary the electron-beam field from a diffuse pattern to sharp one. The advantage of the EB gun is its effectiveness at the refining and growing procedures during service of about 200 hours, compared to the known EB guns which can be used for no longer than 20-30 min. Thus, the original EB gun can be used continuously, both for refining of refractory metals and growing the single crystals at the growth rates of up to 50 mm/min, diameters up to 35 mm and lengths up to 1100 mm (Glebovsky *et al.*, 1986). The growth of single crystals is usually accompanied by purifying liquid metals to high purity. It is demonstrated by preparation of the high-purity refractory metals with the residual impurities at the level of detection of the modern analytical techniques (Alonzo *et al.*, 1995; Bdikin *et al.*, 1999; Bozhko *et al.*, 2008; Chaika *et al.*, 2009; Brunner & Glebovsky, 2000a, 2000b; Cortenraad *et al.* 2001a, 2001b, 2001c, 2001d; Ermolov *et al.*, 1999, 2002; Glebovsky *et al.*, 1998; Markin *et al.*, 2006, 2010; Moest *et al.*, 1998; Shipilevsky & Glebovsky, 1989). The most problematic metals in growing the single crystals of the refractory metals (molybdenum, tungsten, niobium, and tantalum) are two - molybdenum and tungsten. Therefore, the focus of this chapter is devoted to just these two metals, although almost all the results can be easily applied to other two metals - niobium and tantalum.

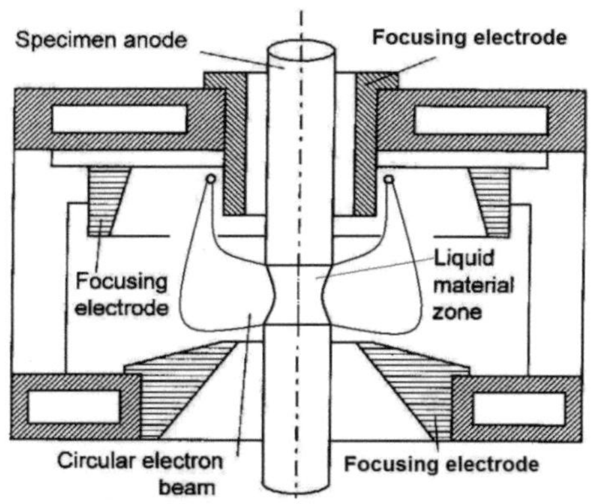

Fig. 2. Circular EB gun with focusing electrodes made of water-cooled copper.

4. Substructure of the molybdenum and tungsten single crystals

Single crystals of refractory metals with the relatively simple *bcc*-lattice grown by EBFZM have the specific dislocation substructure with the size of subgrains, which can be divided into three orders of magnitude. Table 1 shows approximate parameters of the substructures. The chemical composition, especially the content of the interstitial impurities, the growth rate, the number of the liquid zone passes, geometry of the crystal and other parameters significantly affect structural perfection of crystals.

There are several mechanisms of appearance of both the dislocations and substructure of single crystals at growing from the melt (Hurle, 1977; Reid, 1966): under influence of thermal stresses during growing and cooling of single crystals, due to the impurity concentration gradients in the solid phase, due to supersaturating of the lattice with vacancies, inheritance of the substructure of the seed crystal into the growing crystal. In reality, apparently, several mechanisms can operate simultaneously, or some of them will dominate. Obviously, in single crystals of the sufficiently pure refractory metals, effect of impurities on the substructure is unimportant (Akita *et al.*, 1973). However, despite of a large number of studies in this area still remain unclearness related with influence of some factors in formation of the substructure.

Order of substructure	Average size of subgrains	Misorientation angles between subgrains
First order	1 mm < d < 8 mm	30′ < θ < 4⁰
Second order	50 μm < d < 1 mm	30″ < θ < 30′
Third order	0 < d < 50 μm	0 < θ < 30″

Table 1. Estimated parameters of the crystalline substructure.

In growing the single crystals of molybdenum and tungsten by EBFZM one of the main monitored parameters is the growth rate (or the rate of liquid zone traveling, or the rate of the EB gun displacement). The growth rate is essential both at the crystallization stage, and at the post-crystallization annealing stage, which begins at an interface between the liquid zone and single crystal. At high temperatures, dislocations, regardless of nature of their origin, have very high mobility, so that there polygonization of the dislocation substructure has taken place. Apparently, along with increased dislocation mobility the stresses at the interface between the solid and liquid phases also contribute to formation of the polygonized structure. The typical substructure of the tungsten single crystal with the growth axis [001], revealed at the (010) plane parallel to the growth axis, is shown in Fig. 3. It is clearly seen that the boundaries of the subgrains of the first order extend for the considerable distances along the growth axis. They represent the walls or the dislocation network formed by potential for the *bcc*-lattice dislocations with the Burgers vectors b_1 = a/2 [111] and b_2 = a [100].

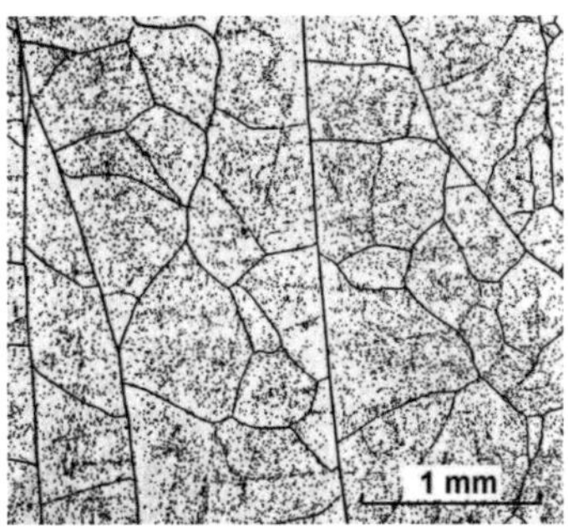

Fig. 3. Substructure of the tungsten single crystal with the vertical growth axis [001] grown at the rate of 2 mm/min. Electrolytic etching in 25% solution of NH_4OH.

The impurities and doping have significant effect on the substructure of single crystals grown from the melt. When microalloying occurs at the definite growth rate, the flat crystallization front is quite stable, and concentration supercooling does not develop. Changes in the dislocation substructure are associated with increase of the dislocation density inside subgrains, decrease of the average size of subgrains and increase of the misorientation angles between subgrains. In the case of doping in significant concentrations, the growth of single crystals becomes impossible at any growth rate. Figure 4 shows the longitudinal and transverse cross-sections of the polycrystalline Mo-2%W alloy, grown from the melt by EBFZM. This polycrystalline ingot contains the large grains elongated along the growth axis. Figure 5 shows the microstructure of the polycrystalline molybdenum ingot of low purity, also grown from the melt by EBFZM. The resulting structure is distinctive: the single-crystalline core surrounded with the polycrystalline periphery.

Fig. 4. Microstructure of the polycrystalline Mo-2%W alloy, grown from the melt by EBFZM (longitudinal and transverse cross-sections).

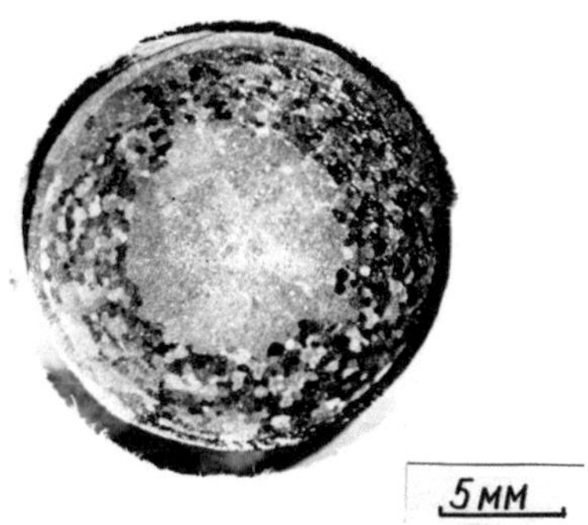

Fig. 5. Microstructure of the polycrystalline molybdenum ingot of low purity, grown from the melt.

To study effect of the growth rate on the substructure, the single crystals of molybdenum and tungsten, having the growth axes along the [001] axis, are grown. The growth rate varied from 0.2 to 50 mm/min, the rotation rate of the growing crystal - from 0 to 100 revolutions per minute. The range of the growth rates under study actually covers all growth rates, implemented in the EBFZM method: the lower limit depends on intense metal evaporation, and the upper limit – by possibility of full melting and stability of the liquid zone. The single crystals are grown in three passes of the liquid zone: the first pass at the growth rate of 6 mm/min, the second pass at the growth rate of 2 mm/min, the third pass - on the seed crystal at the rate from the above range of the rates.

4.1 The substructure of the molybdenum single crystals

The substructures of the molybdenum single crystals vary seriously depending on the growth rate (Glebovsky *et al.*, 1988; Glebovsky & Semenov, 1994, 1995; Glover *et al.*, 1970, Liu & Zee, 1996). At high growth rates the substructure is characterized by the high dislocation density and the more stressed state. The latter is confirmed by zone annealing the molybdenum single crystal of 8 mm in dia at the temperature close to the melting temperature and at the traveling rate of the EB-gun of 0.5 mm/min. At the low growth rates the dislocations have enough time to be polygonized, because single crystals stay at elevated temperatures for a longer time. Studies of the molybdenum single crystal grown at the growth rate of 6 mm/min show that after zone annealing the crystal has been polygonized to greater extent.

The single crystals grown at the growth rate of 0.5 mm/min have the specific developed substructure with an average size of subgrains of the second order of about 100 μm and the dislocation density, calculated from the etch pits, of 3×10^5 cm^{-2}. The substructure of the single crystal grown at 6 mm/min is characterized by the individual etch pits and the lack of the polygonized boundaries. The dislocation density is higher - of 1×10^6 cm^{-2}. The specific substructures of the molybdenum single crystals grown at different growth rates are shown at Fig. 6, 7, and 8. Most clearly the dependence of the misorientation angles of the subgrains on the growth rate is detected by the divergent X-ray beam patterns (Fig. 9). The misorientation angles of the subgrains of the second order in accordance with the value of discontinuities on the line (400) is 50′ of an arc for the growth rate of 0.5 mm/min, and 20′ of an arc - for the growth rate of 6 mm/min.

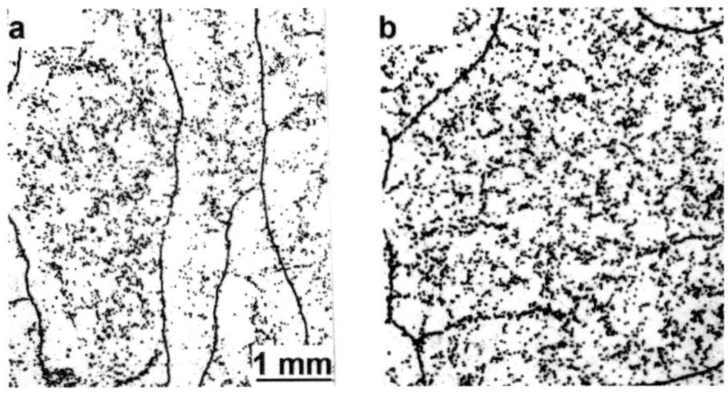

Fig. 6. Specific substructures of longitudinal (a) and transverse (b) cross-sections of the molybdenum single crystal, grown at 2 mm/min.

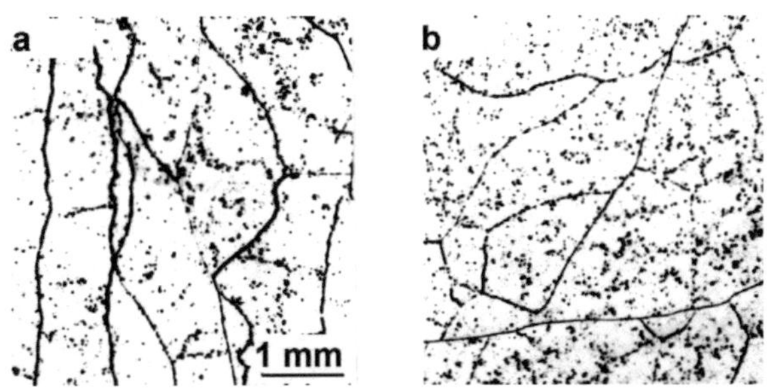

Fig. 7. Specific substructures of longitudinal (a) and transverse (b) cross-sections of the molybdenum single crystal, grown at 10 mm/min

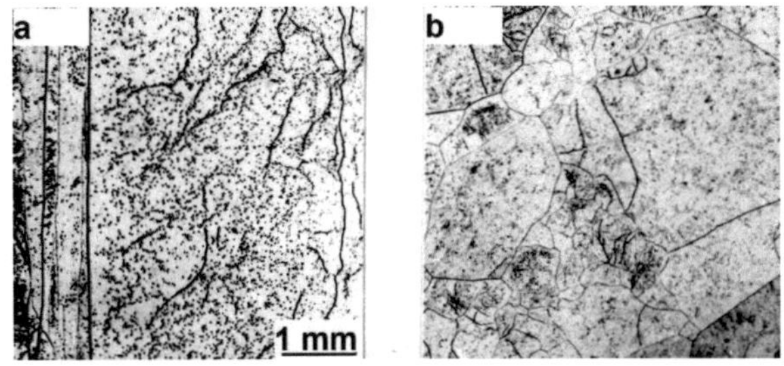

Fig. 8. Specific substructures of longitudinal (a) and transverse (b) cross-sections of the molybdenum single crystal, grown at 40 mm/min.

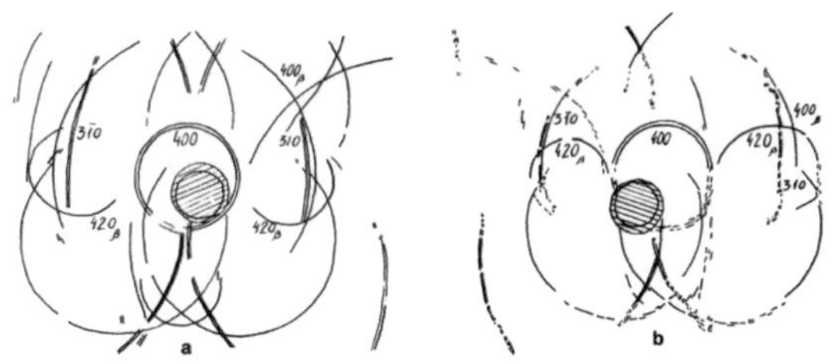

Fig. 9. Divergent X-ray beam patterns of the molybdenum single crystals, grown at rates: a- 0.5 mm/min, b- 6 mm/min.

Another feature of the substructure of the single crystals of molybdenum is radial heterogeneity, detectable at all growth rates. In the central part of the single crystals the boundaries are almost absent; however, at the periphery is observed intense polygonization. Radial inhomogeneity of the single crystals of 8 mm in dia is clearly visible on the divergent X-ray-beams patterns, where the value of shifts at edges of the line (310) is bigger than in the central part: at the edges – 50′ of an arc, in the center – 20′ of an arc. Dependence of nature of the substructure of the single crystals on radial inhomogeneity as well as on the growth rate can be explained by dislocation motion in the non-uniform temperature field of the single crystals.

4.2 The substructure of the tungsten single crystals

The typical substructures of the tungsten single crystals with the growth axis [100] and 10-12 mm in dia, grown at growth rates from 0.5 mm/min up to 40 mm/min, are shown in Fig. 10, 11, and 12. At the growth rates of 0.5-4 mm/min on the transverse cross-sections in the (001) plane, the distribution of the subgrains of the second order by size is close to normal. On the longitudinal cross-sections in the plane (100), pronounced elongated subgrains along the growth axis are found. At 40 mm/min, the polygonization process is not completed: the dislocation density inside the subgrains increases. At two different sites of the tungsten single crystal grown at the growth rates of 2 mm/min and 40 mm/min, the dislocation density increases by a half of an order of magnitude - from $8x10^4$ cm^{-2} to $2x10^5$ cm^{-2}. There is also decrease of the average size of the subgrains, and the misorientation angles are increased to $3\text{-}4^0$ of an arc. Apparently, such influence of the growth rate on the substructure of the tungsten single crystals is common for all refractory transition metals.

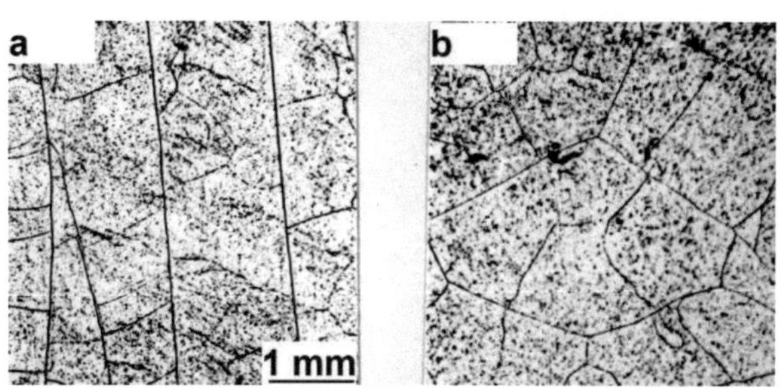

Fig. 10. Specific substructures of longitudinal (a) and transverse (b) cross-sections of the tungsten single crystal grown at 0.5 mm/min.

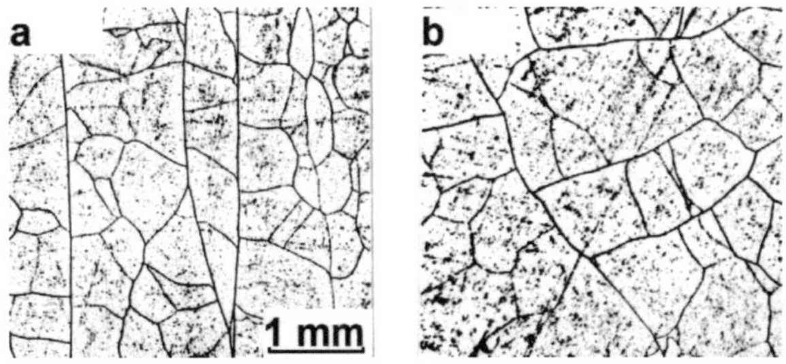

Fig. 11. Specific substructures of longitudinal (a) and transverse (b) cross-sections of the tungsten single crystal, grown at 2 mm/min.

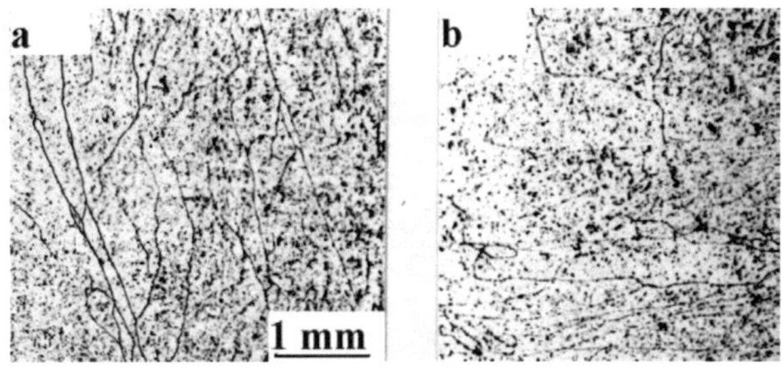

Fig. 12. Specific substructures of longitudinal (a) and transverse (b) cross-sections of the tungsten single crystal, grown at 40 mm/min.

Note significant irregularity of the substructure of the tungsten single crystals in the radial direction: the central part and the periphery are notably different, similar to differences observed in the molybdenum single crystals. The metallographic studies of the substructure of the single crystals of molybdenum and tungsten show that between them there is fundamental similarity. The rate of rotation, and the focusing and power fluctuations of the electron beam, leading to emergence of the constrictions and other defects on the surface of the single crystals, generally have no noticeable effect on the substructure of the single crystals. The fact that the temperature field is non-uniform in the radial direction has been confirmed by existence of the curvilinear crystallization front in vicinity of the seed during zone melting (Fig. 13). In turn, heterogeneity of the temperature field leads to non-uniform mechanical stresses which have different effects on the dislocation motion rate.

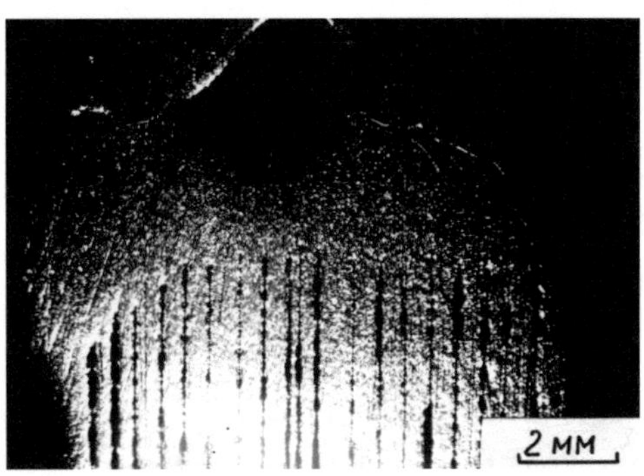

Fig. 13. Macrostructure of the longitudinal thin section of the tungsten single crystal of 11 mm in dia, showing the curvilinear crystallization front in vicinity of the seed.

Although unevenness of the substructure of single crystals creates some problems for physical studies and practical applications (Markin *et al.*, 2006, 2010; Mundy *et al.*, 1978), there are recent results on successful use of the as-grown single crystals for the manufacture of the STM tips (Bozhko *et al.*, 2008; Chaika *et al.*, 2009). It is demonstrated the main advantage of the single crystalline W[001] STM tips: sharpness, stability, and the predictable atomic structure. With these tips a set of the complimentary atomically resolved images of the complicated Si(557)5x5 stepped surface reconstruction is reproducibly received and revealed its atomic structure. The example of instability of the W[001] tip illustrates how the known tip axis orientation and the apex atom jump lengths may allow one to predict the atomic structure of the real single crystalline tip that can be of high importance for correct interpretation of the ultimately high resolution STM data. Nevertheless, presence of the specific substructures in single crystals, which prevents expansion of the practical application of single crystals, while insoluble problem for the researchers, that stimulates search for ways of obtaining the perfect single crystals of these metals.

5. Effect of thermal stresses on the substructure of the single crystals

The EBFZM method is characterized by presence of the high axial temperature gradients, especially near the solidification front in both the solid and liquid phases. Since the temperature gradient is a nonlinear function of the distance from the solidification front, *i.e.*, $d^2T/dz^2 \neq 0$, this leads to thermal stresses in growing single crystals that can cause multiplication of the dislocations. For a cylindrical crystal with dissipated radiating crystallization heat, the axial temperature gradients can be estimated from the known formulas relating the temperature, the black-body coefficients, the Stefan-Boltzmann constant, the crystal diameter, and thermal conductivity (Table 2).

Metal	The distance from the liquid zone Z, cm						
	0	0,2	0,4	0,6	0,8	1,0	1,2
Molybdenum	-618	-557	-505	-461	-422	-389	-359
Tungsten	-1453	-1207	-1021	-877	-763	-671	-596

Table 2. Axial temperature gradients (K/cm) in the solid state for single crystals of molybdenum and tungsten

The temperature along the axis of single crystals has been measured by optical micropyrometry. The holes of 1 mm in dia and 7-8 mm in depth, located along the axis of the single crystals, serve as a black-body model. Due to the high axial temperature gradients, the black-body model is appeared to be essentially non-isothermal. This increases the temperature measurement error up to $\pm 100^0$, but it is still possible to obtain the reproducible temperature profiles for all metals studied. Figure 14 shows the temperature distribution along the axis of the cylindrical tungsten single crystal of 15 mm in dia. It is seen that near the crystallization front the temperature along the crystal falls particularly sharply to 2000K at 40 mm from the front. The control of the temperature profile along the axis of the liquid zone is performed by measuring an electron current. Owing to the design of the EB gun, the temperature profile of the liquid zone can be effectively changed from diffuse one to sharp, which gives additional opportunity to manage the crystal growth.

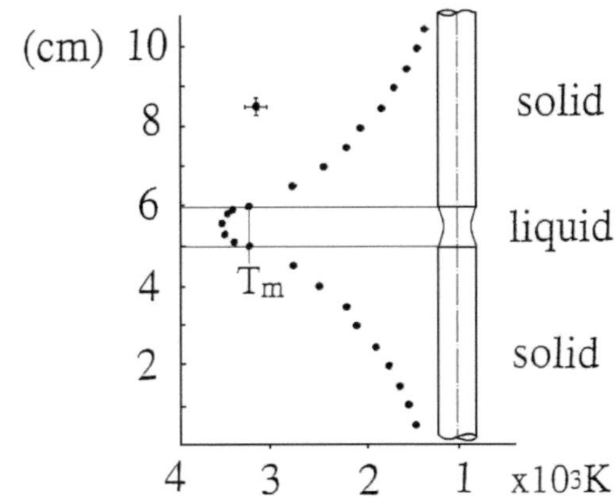

Fig. 14. Temperature distribution along the axis of the cylindrical tungsten single crystal of 15 mm in dia.

For the case of the ax-symmetric temperature distribution along the cylindrical single crystal, resulting thermal stresses are mainly determined by the axial temperature gradients. At pre-melting temperatures, an elastic limit of metals is practically zero and thermal stresses are completely removed by the dislocations, *i.e.*, there is plastic deformation. Since at the growth of crystals from the melt the crystallization heat should released, then inevitably there is the temperature gradient in the solid phase, which leads to the certain density of unremovable dislocations (M. Cole, *et al.*, 1961; Buckley-Golder & Hurphreys, 1979; Esterling, 1980; Nes & Most, 1966; Otani, 1984;). The estimate of this dislocation density can be performed by:

$$\rho \geq \alpha \; gradT \; / \; b, \tag{1}$$

Here, α - a linear coefficient of thermal expansion, K^{-1}; b - the Burgers vector, cm; *gradT* - the temperature gradient at the crystallization front, K/mm. For tungsten single crystals, such evaluation reveals that in order to have the dislocation density of ρ = 10^4 cm^{-2}, the temperature gradient must be less than 50 K/cm. Since the real temperature gradients are usually higher for an order of magnitude, in the melt-grown tungsten single crystals the dislocation density is typically 10^5-10^6 cm^{-2} or even more. The dislocation density in the boundaries is usually for an order of magnitude higher than in the bulk of the subgrains. The estimate by means of the formula (1) with the temperature gradient of 100 K/mm gives the dislocation density of $\rho \geq 5x10^5$ cm^{-2} (Table 3).

Dislocation density ρ, cm^{-2}	Temperature gradients, K/cм	
	Molybdenum	Tungsten
10^6	2900	3600
10^5	290	360
10^4	29	36

Table 3. Numerical estimates of temperature gradients at the crystallization front for single crystals of molybdenum and tungsten of 16 and 11 mm in dia, respectively.

Note that dislocations in single crystals are formed under action of thermal stresses in the growth process, and in the cooling process as well. Depending on the cooling rate, the number of the imposed dislocations in the growing crystal can be even higher than the number of the dislocations appeared during the growth (Nes & Most, 1966). The result is in decreasing subgrains sizes, as observed in the thermal shock at welding up both the seed crystal and initial rod together or cutting the crystal by the electron beam. In addition to the axial temperature gradients, quantitative estimates for the stationary stage of the crystal growth have shown, that to form the substructure, the cooling rate is important as well, which is realized in the growing and cooling processes when the crystal growth process is over. For example, in aluminum single crystals obtained by zone melting, the residual dislocation density is of $\rho \sim 10^{?}$ cm^{-2} when cooled to room temperature at ~ 10^{-3} K/s (about a week), while at the increasing cooling rates on an order of magnitude, the residual dislocation density increases to $\rho \sim 10^{4}$ cm^{-2}.

In the case of growing the single crystals of molybdenum and tungsten by EBFZM, one can get the estimate from below for the maximum cooling rate of the crystal growth process. The maximum of this magnitude occurs in the solid phase just below the crystallization front. Multiplying the axial temperature gradient on the growth rate, one obtains the value of the maximum cooling rate of the crystal during the growth:

$$\frac{dt}{dz} \times \frac{dz}{dt} = \frac{dT}{dt}. \qquad (2)$$

Taking the data of the temperature gradients (Table 3) and the growth rate of 2 mm/min, the most frequently used in practice, one can obtain the cooling rate for molybdenum and tungsten, respectively, 2K/s and 5 K/s, *i.e.,* the cooling rates are very high, taking into account the corresponding values for aluminum (10^{-3} K/s). It should be noted that significant reducing the crystallization rate in the EBFZM method is impossible. The fact that molybdenum and tungsten have very high vapor pressure at $T > T_m$ and at the growth rate of 0.5 mm/min, the metal losses by evaporation can reach 30% of initial mass.

However, even if the crystallization rate tends to zero, desired reduction of the dislocation density still can not be achieved. The reason is that the crystallization rate never coincides with the growth rate (or the rate of the EB gun displacement). In fact, the crystallization rate affects significantly the hydrodynamic processes developing in the liquid zone (Kobayashi, 1970; Kobayashi & Wilcox, 1982; Murphy, 1987; Surek & Chalmers, 1975). These processes give rise to oscillations in the growth rate; moreover, the instantaneous crystallization rate in these moments of time can be significantly greater than the cooling rate, as shown by estimates for molybdenum and tungsten. Presence of such oscillations of both the temperature and growth rate is shown elsewhere (Mullins & Sekerka, 1964; Wilcox & Fuller, 1965). The frequency of these oscillations is close to the inverse of the thermal time constant of the melt-crystal system:

$$f \approx \frac{a}{S} \qquad (3)$$

Here, f – a frequency of oscillation, Hz; a – a thermal diffusivity, cm^2/s; S – a cross-sectional area of the crystal, cm^2. Oscillations of the crystallization front in presence of impurities in a crystal lead to substantial change in the distribution coefficient and, consequently, the so-called transverse striations in the crystals observed by autoradiography.

Considerable interest represents an estimate of the cooling rate from T_m to room temperature. The simplest case can be considered, when the one-dimensional quasilinear heat conductivity equation with the constant coefficients is numerically solved. The process of heat propagation in a homogeneous rod can generally be described by the equation:

$$\rho' C_p \frac{\partial T}{\partial \tau} = \frac{\partial \left(K \frac{\partial T}{\partial X} \right)}{\partial x} + f(x,t) \tag{4}$$

where T (x, t) - a temperature at the point X of the rod at the moment of time t; C_P - heat capacity per unit mass at constant pressure; ρ_m - metal density; K - thermal conductivity; f - density of the heat sources (sinks); X - a coordinate along the rod length L. If one assumes that K, C_P, and ρ' are constant, the equation (4) can be rewritten as:

$$\frac{\partial T}{\partial t} = a^2 \frac{\partial^2 T}{\partial X^2} + f(X,t) \tag{5}$$

Thus, it is necessary to find the continuous solution at $T = T$ (x, t) of the equation (5) for

$$\overline{D} = \{0 \le X \le L; 0 \le t \le t^1\}$$

if

$$\begin{aligned} T(X,0) &= T_0(X); 0 \le X \le L \\ T(0,t) &= T_1(t); 0 \le t \le t_1 \\ T(L,t) &= T_2(t); 0 \le t \le t_2 \end{aligned}$$

For a uniform rod with the diameter d, cooling due to radiation and thermal conductivity after switching off the electron beam one obtains instead of the equation (5):

$$\frac{\partial T}{\partial t} = a^2 \frac{\partial^2 T}{\partial X^2} - \frac{4\varepsilon T^4}{d\rho' C_p} \tag{6}$$

In the task the most interesting is the value of $\partial T/\partial t$ with the limitations $0 \le X \le L$; and $t=0$. This value is calculated numerically, and it provides the cooling rate at the upper limit. For these calculations, the following numerical values of the physical parameters are used. For molybdenum: K = 0.909 $Wcm^{-1}K^{-1}$, C_P = 0.235 $Jg^{-1}K^{-1}$, ρ_{Mo} = 10.2 gcm^{-3}, d = 1.6 cm. For tungsten: K = 0.945 $Wcm^{-1}K^{-1}$, C_P = 0.172 $Jg^{-1}K^{-1}$, ρ_W = 19.2 gcm^{-3}, d = 1.1 cm. The cooling rate for molybdenum is 2×10^4 K/s, and for tungsten - 5×10^4 K/s. Although these values exceed the cooling rates at the stationary stage of the crystal growth, they can not significantly impact on deterioration of the substructure, because in 5-10 seconds for the most part of the crystal they become comparable with the cooling rates at the stationary phase. The dislocation density can increase only a few in the surface layer near the end of the crystal. Therefore, when the crystals of refractory metals are grown from the melt, it is absolutely impractical to cool slowly, from scientific or technological points of view.

6. Recrystallization of single crystals

Because of relatively low perfectness of single crystals of some semiconductors and refractory metals, grown by EBFZM, there are some studies made to improve the substructure of single crystals. It is well known that the growth of the semiconductor alloys crystals is one of old problems in physics and practice of the crystal growth. The studies on growing such crystals include the casting-recrystallizing-annealing procedures and require careful balancing of the pseudo binary melt stoichiometry, which inherently is a quite difficult process. Increase in recrystallization efficiency can be achieved by adjusting the casting conditions and the suitable thermal gradient during recrystallization (Yadava *et al.*, 1985). This means, that the mechanism of the crystal growth is a combination of both the chemical potential gradient and temperature gradient of zone melting processes, so the growing processes of semiconductor alloys crystals are complicated at their practical realization.

In comparison with the growth of semiconductor crystals, the growth of such simple metals like molybdenum and tungsten seems to be a relatively non-problematic task. However, the main obstacles are both the high melting temperatures and high temperature gradients along single crystals when growing from the melt. Because of these obstacles the growth of single crystals of refractory metals becomes the very complicated task which attracts attention of many scientists for a half a century. Numerous studies concerning this problem have shown that these metals, in spite of their crystallographic simplicity, require special studies, knowledge, and equipment.

Using collected experimental information, the single crystals of molybdenum and tungsten, free of the specific substructure, are produced by recrystallization including plastic deformation and high-temperature annealing (Bdikin *et al.*, 1999; Katoh *et al.*, 1991). Plastic deformation of the crystals has been done by rolling in the vacuum rolling machine or in the standard mills in air. High temperature annealing is produced with the help of the heating devices located inside the rolling machine or in the set-up for electron-beam zone melting and growing single crystals. The comparison of structural perfection of the single crystals grown from the melt and grown by recrystallization is done by both X-ray rocking curves and angular scanning topography.

As a rule, the recrystallized single crystals of molybdenum and tungsten have the substructures characterized by both the record-low dislocation density and small-angle mosaic. It should be noted that the lower dislocation density in the single crystals can only be achieved by recrystallization. As mentioned already before, the optimal procedure involves deformation of the single crystals with the <111> growth axis by rolling along the (112) plane (Bozhko *et al.*, 2008). To monitor both the substructure and perfection of the tungsten single crystals, the anomalous X-rays transmission has been employed as well.

6.1 Experiments on recrystallization of the tungsten single crystals

The tungsten single crystals under study are grown by the EBFZM method (Glebovsky *et al.*, 1986). The single crystals have the different crystallographic growth axes. The high-purity tungsten powders of chemical purity 99.99% are used as a starting material. The as-grown single crystals are 11-22 mm in dia and 100 mm in length. The structural studies involve the X-ray diffraction microscopy methods, namely, angular scanning topography and rocking curves (Aristov *et al.*, 1974; Bozhko *et al.*, 2008; Brunner & Glebovsky, 2000a, 2000b; Ermolov *et al.*, 1999, 2002; Riedle *et al.*, 1994, 1996).

Several deformation systems are studied, which differ by the crystallographic parameters and strain, because two parameters such as the growth axis and the deformation direction are most important to get the strained single crystals before high-temperature annealing. Crystallographic systems tested are [100]/(010), [100]/(011), [110]/(110), [110]/(111)], [111]/(110) and [111]/(112) where [growth axis]/(rolling plane). The 6-12% deformation by rolling of the two-cant bars with fixed crystallography is found to be optimal to get the large grains. The vacuum conditions are most suitable for the crystal deformation by rolling because they enable one to avoid oxidation of the crystal surfaces during deformation at high temperatures.

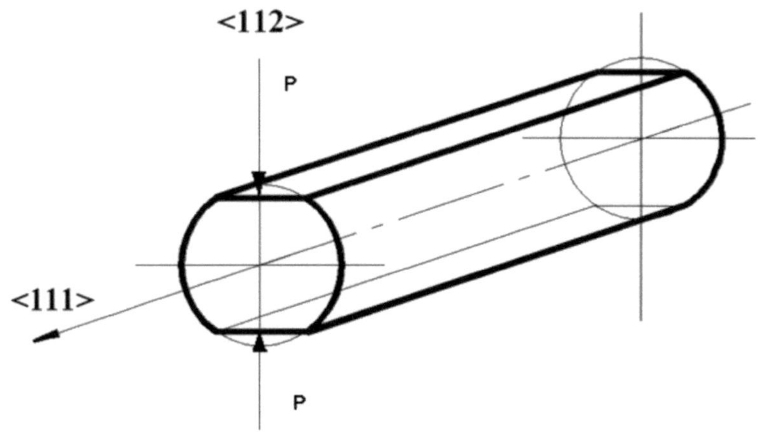

Fig. 15. Two-cants bar with fixed crystallography for rolling. The crystallographic parameters are for the system [111]/(112) where [111] is the growth axis and (112) is the rolling plane (P).

Sample	Growth axis	Rolling plane	Strain, %	Size of grains
1	[100]	(010)	13.4	Grains, 8 mm
2	[100]	(011)	11.6	Grains, 8 mm
3	[100]	(010)	12.3	Grains, 8 mm
4	[100]	(011)	11.6	Grains, 8 mm
5	[110]	(110)	7.9	Grains, 15 mm
6	[110]	(111)	7.5	Grains, 15 mm
7	[111]	(110)	7.5	Grains, 15 mm
8	[111]	(112)	6.6	Grains, 25 mm

Table 4. Parameters of the samples subjected to vacuum rolling followed by high-temperature annealing.

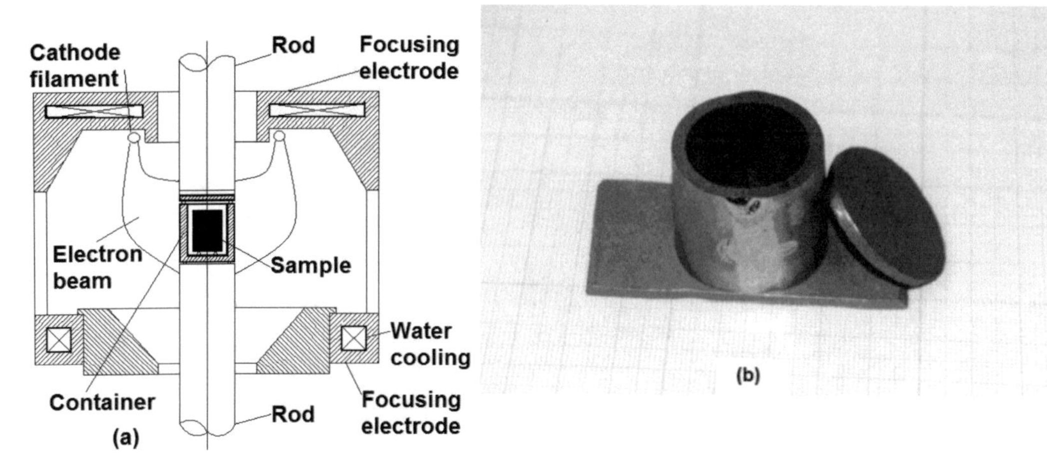

Fig. 16. Electron-beam annealing of a single crystal in a container, a-EB gun, b-container.

The billets to be plastically deformed are produced from the cylindrical tungsten single crystals and have a form of the two-cant bar so that crystallography has been fixed before rolling (Fig. 15). Plastic deformation is performed in one pass at the temperature 900ºC in the vacuum rolling machine or, in a number of cases, in the standard rolling mill in air. In Table 4 are listed the main data for the recrystallized single crystals. The hypercritically strained tungsten single crystals are annealed at 2500ºC in the EBFZM set-up by the defocused electron beam (Fig. 16, a). In order to obtain the more uniform temperature field in the specimens, the tungsten container is used. Its height is 40 mm, the diameter 30 mm, the wall thickness 3 mm (Fig. 16, b). The tungsten specimen of the maximal length 30 mm and the diameter 20 mm is installed inside the container on the tungsten holders. The container has been arranged coaxially with the EB gun equipped with a circular cathode.

6.2 Recrystallized tungsten single crystals in comparison with ones grown from the melt

In order to elucidate influence of recrystallization on the real substructure of tungsten single crystals, the topograms of angular scanning of the specimens, cut from the melt-grown single crystal before deformation, are taken. In Fig. 17, the topograms show the mosaic substructures of two as-grown tungsten single crystals grown in identical conditions. The size of the subgrains is about 1-2 mm, the misorientation angles between subgrains are about 50″ of an arc. On the angular scanning topograms are well seen the small-angle boundaries, their misorientation angles make up tens of the angular minutes, for some single crystals they can exceed 1° of an arc. Single crystals of such structural quality are not suitable for producing, *i.e.*, deflectors and targets using in the runs concerned with channeling of high-energy particles beams. It is necessary to decrease substantially the dislocation density in such crystals and to deplete them from the small-angle boundaries. In contrast to the melt-grown single crystals, necessary structural perfection can only be achieved by the recrystallization processes consisting of plastic deformation and high-temperature annealing.

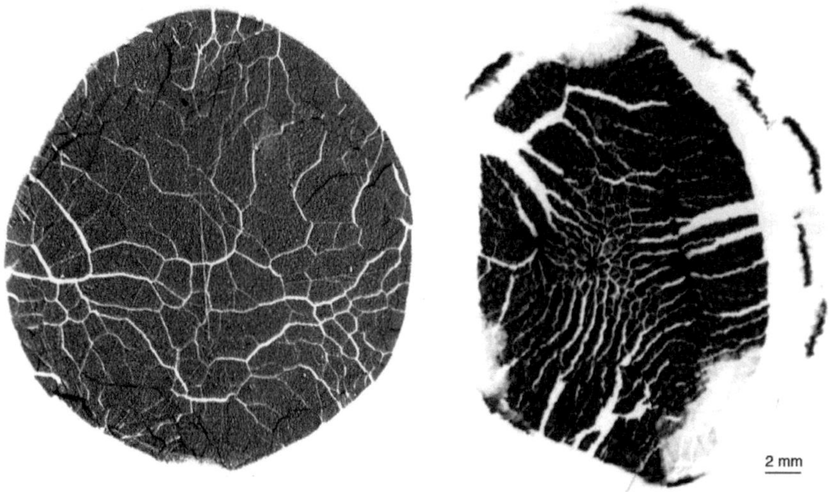

Fig. 17. Angular scanning topograms of two as-grown tungsten single crystals, (110) reflection, CuK_{α}.

The rocking curves are recorded in the dispersion-free system using the perfect silicon single crystal as a monochromator (Kittel, 1996). To measure the rocking curves, the angle between the X-ray source and the detector is fixed under the first order Bragg diffraction angle. A crystal is rotated with respect to the incoming X-ray beam, and an intensity of the diffracted beam is measured as a function of the angle. The X-ray diffraction rocking curves are measured in three different sites on a specimen surface. The sites differ in a rotation angle of 45^0 in order to exclude texture effects. A half-width (FWHM) of the rocking curves is ~1^0 of an arc. Different subgrains show up as individual peaks on the rocking curves. The angular scanned topograms are obtained in the Θ-2Θ scanning regime and used to study the position, dimensions, and misorientation angles of subgrains. This X-ray diffraction technique is based on a principle of the Bragg reflection (Bdikin *et al.*, 1999; Cortenraad *et al.*, 2001a, 2001b, 2001c, 2001d).

In Table 1 are listed main data for those single crystals. The final recrystallized specimen is the polycrystal incorporating the distinct large grains (Fig. 18). After the recrystallization procedure the large grains do not contain the small-angle boundaries and have the relatively low dislocation density. The optimal procedure, involving deformation of the single crystal with the <111> growth axis by rolling along the (112) plane, is the most suitable for recrystallization. The ~6% deformation is found to be optimal for these crystallographic parameters to get few single nuclei and then large grains to be grown. Higher deformation leads to too many nuclei and as a result - to smaller grains grown. The vacuum conditions are most suitable for rolling because they enable one to avoid oxidation of the crystal surfaces during deformation at high temperatures. By this technique the high purity single crystals of the low dislocation density and free of the small angle boundaries are produced. In several cases the rolled specimens are annealed outside the container. That results in formation of the surface damaged layer of up to 500 μm thick. The middle part of such specimens is virtually free from the small-angle boundaries and has the perfect structure although there are many small subgrains at the periphery.

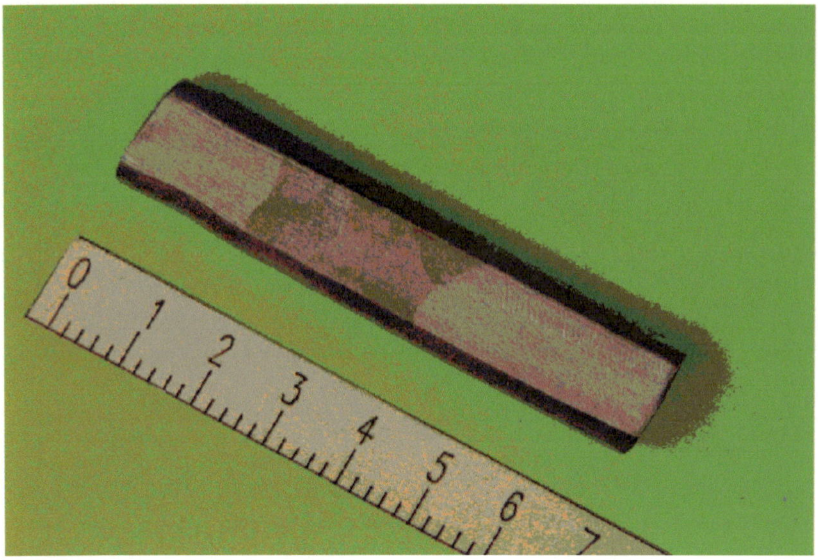

Fig. 18. Recrystallized tungsten polycrystal incorporating three distinct large grains.

Using a tungsten container for high-temperature annealing enables both avoiding damage of the specimen surface layer and decrease of the temperature gradient. The specimens with record structural perfection are obtained by using this technique. The angular scanning topogram and corresponding rocking curve, taken from the specimen annealed in the container are shown in Fig. 19. The angular scanning topogram shows the perfect surface of the specimen, without any boundary or other defects (compare with the topogram for as-grown specimens in Fig. 17). The rocking curve has only one sharp peak with the width at a half height (FWHM) of ~50″ of an arc. Correspondingly, the dislocation density is about $5x10^4$ cm^{-2} for this specimen.

Strong changes of the substructure of the perfect tungsten single crystal has been found when the latter has been used as a seed crystal for growing a new tungsten crystal from the melt by EBFZM (Glebovsky & Semenov, 1999). Before growing, the dislocation density of the seed crystal is of about 10^4 cm^{-1}, and it does not contain any subgrains. After growing from the melt, the substructure of the seed single crystal indicates significant deterioration - subgrains of 500 μm are appeared, and they are elongated along the growth axis with the misorientation angle of 8-10′ of an arc (Fig. 20). At the growth of single crystals of refractory metals from the melt, the main contribution to formation of the substructure is made by the dislocations arising under influence of the thermal stresses during the growth and cooling of the growing crystal. During growing from the melt, dislocations may arise due to thermal stresses in the solid and under action of the impurity concentration gradients due to lattice oversaturation with vacancies. Subgrains are formed by fresh dislocations and walls in spite of the fact that the seed crystal does not contain any subgrains before growing.

(a)

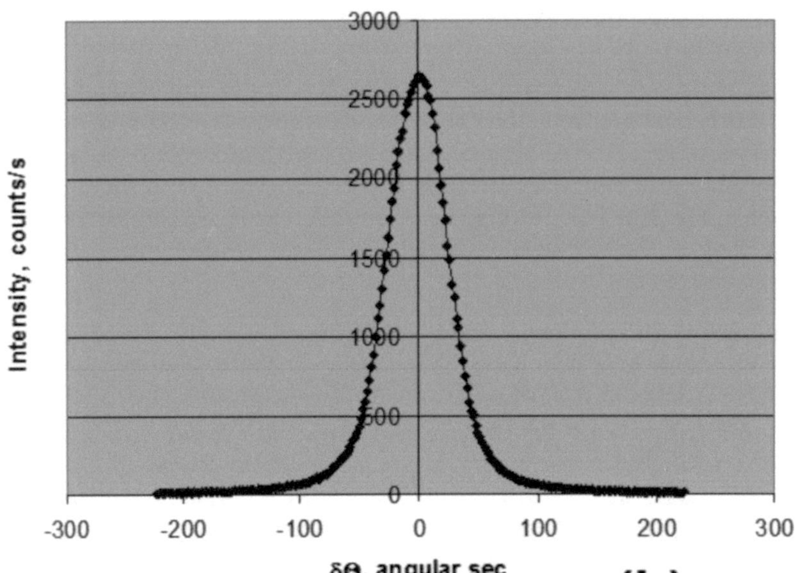

(b)

Fig. 19. Angular scanning topogram (a) and rocking curve (b) of the tungsten single crystal, annealed in the container; the surface plane (110).

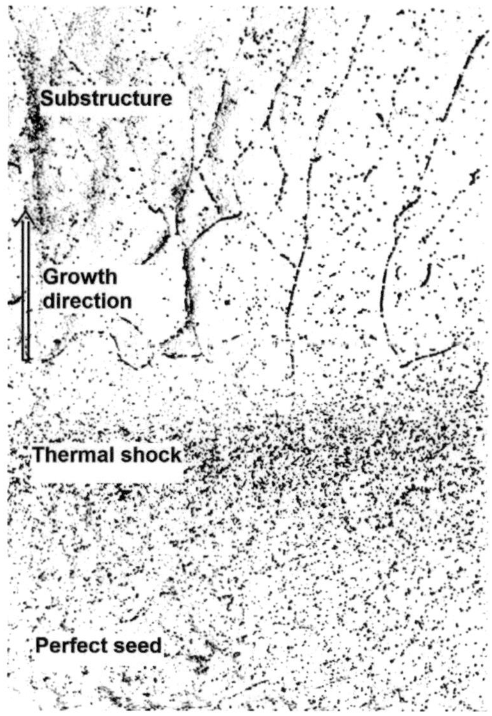

Fig. 20. Appearance of the fresh dislocation walls in the tungsten single crystal during the growth from the melt.

6.3 Anomalous X-ray transmission in the tungsten single crystals

A diverging X-ray beam is the beam where the angles of incidence of the X-rays on the specimen differ for different parts of the specimen (Bdikin *et al.*, 1999). Thus, for the given wavelength the diffraction conditions are satisfied along a certain line on the surface of the specimen. The classical system is used to record anomalous X-ray transmission (Borrmann effect). A fine-focus tube with a focus measuring 50x50 μm^2 is used as a source of the diverging X-ray beam. The distance between the specimen and the photographic film is 10 cm and that between the source and the specimen is 18 cm (Fig. 21). An asymmetric extraction geometry (the distance between the source and the specimen is not equal to that between the specimen and the photographic film) is used to eliminate focusing of the diffracted rays over the radiation spectrum. In the case of dynamic diffraction (perfect crystal) at points, where the diffraction conditions are satisfied, amplification of intensity is observed in the transmitted beam. In kinematic approximation (crystal containing defects), intensity is suppressed. The fact that transmitted radiation is recorded, and radiation intensity is the same in the diffracted and forward transmitted beams, together with observation of amplification of characteristic $CuK_{\alpha1,2}$ radiation intensity in these directions indicate that diffraction is of dynamic nature. Observation of splitting of these beams into a $CuK_{\alpha1,2}$ doublet (Fig.22) and a shift as the specimen rotates indicate that these beams are of diffraction origin and are not the topological characteristic of the specimen. Diffractometering of the waves which passed in the directions R and T also confirms that intensities of the diffracted and transmitted beams are the same.

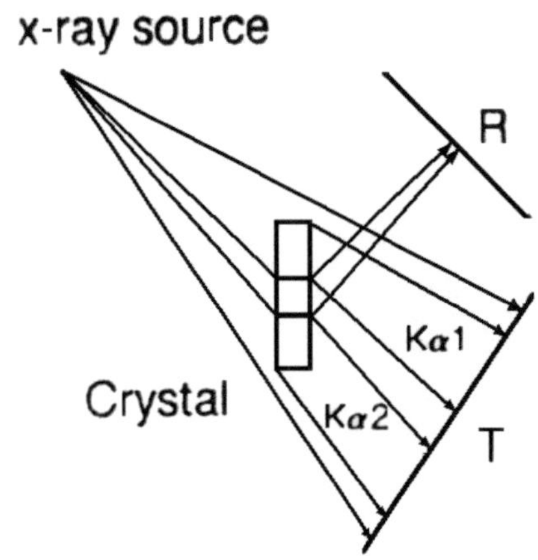

Fig. 21. Scheme for recording the anomalous transmission of X-rays. R-reflection, T-transmission.

Fig. 22. Anomalous X-rays transmission in a perfect W single crystal; lines are from transmitted beams.

Calculations using the kinematic theory show that for Cu-radiation and the specimen thickness of 0.3 mm ($\mu t \sim 98$) absorption is 10^{43}. With the available radiation sources (no more than 10^7-10^8 pulse/s) it is impossible to obtain recordable transmission intensity. For Mo-radiation, absorption is lower than that for Cu-radiation. The linear absorption coefficient of the 0.3 mm thick tungsten single crystal is calculated to be $\mu t = 53.4$ on the Mo-radiation characteristics.

If the absorption coefficient is conceived as $\sigma = \mu_i t + y_i$, where μ_i is a coefficient of interference absorption and y_i is responsible for renormalization of the atomic amplitude of scattering, then for the perfect tungsten single crystal can be obtained $\mu_i = 369.2$, $y_i = -6.63$. The measurements of the absorption coefficient of a diffracted Mo-radiation yield $\sigma = 7.3$. Under assumption that y_i is independent of the defects concentration one obtains $\mu_i = 464.3$ cm^{-1}, *i.e.*, in the real specimen the interference absorption coefficient is by 25% greater than it has to be in the perfect crystal.

The effects associated with dynamic X-ray propagation can only be observed in crystals for which the distance between the dislocations does not exceed an extinction length L. The calculations for (110) CuK_α reflection for the defect-free tungsten crystal give $L = 1.7$ μm, *i.e.*, the critical dislocation density is $N_d = L^{-2} = 3.5\text{x}10^7$ cm^{-2}. From the view point of the specific features of X-ray scattering, the crystal lattice defects fall into two classes. The first class involves localized defects, *e.g.* vacancies, which, virtually, do not deform the reflective atomic planes and, therefore, do not change the rocking curve width. Occurrence of such defects in the crystal only results in decrease of intensity of transmitted radiation, *i.e.*, the growth of the absorption coefficient. The second class involves, for example, dislocations, presence of which gives rise to distortion of the atomic planes and violation of the crystal lattice period. These defects lead to increase of both the angular divergences of the diffracted beam (rocking curve width) and interference absorption coefficient. In the defect-free crystal the waves propagating in the crystal have the finite natural angular width $\delta = \lambda/L$, where λ is the characteristic radiation wavelength.

In order to determine the type of the defects which predominate in the perfect tungsten single crystals and their density, the rocking curves for the (110) reflection in the Bragg and Laue geometries are recorded, and measured the interference damping factor μ_i. In the Laue geometry the width of the rocking curve is 32″ of an arc whereas in the Bragg geometry the width of the rocking curve is 72″ of an arc. The calculated values for these widths in the defect-free crystal are 5.5″ of an arc and 48.6″ of an arc in the Laue and Bragg geometries, respectively. Hence, broadening of the rocking curves in the real crystal compared with the defect-free crystal is around 25″ of an arc for both rocking curves. The difference between the measured values of the rocking curve width and the interference absorption coefficient, and the calculated values for the defect-free crystal can be used to determine the dislocation density N_d in the crystals under study. Estimates of the dislocation density using the width of the rocking curve give $N_d = 2\text{x}10^5$ cm^{-2}, and using the interference absorption coefficient – $N_d = 4\text{x}10^5$ cm^{-2}. These data show a good agreement with the dislocation density determined using etch pits, $N_d = 5\text{x}10^5$ cm^{-2}. Thus, the dislocation density in the real crystal is substantially lower than the critical density. Note that the dislocation density determined from the interference absorption coefficient is higher than that obtained from the rocking curve width. This is evidently attributable to presence of the defects in the crystal (such as the point defects) which increase absorption but make no contribution to the rocking curve width.

The technique used to study the Borrmann effect allows to obtain the angular scanned transmission and reflection topograms for the tungsten crystals, which reveal subgrains larger than 1-2 mm. Comparison of the transmission and reflection topograms suggests that the subgrain substructure exhibits similar reflection but the transmission topograms have a higher image contrast as a result of dynamic narrowing of the diffracted beam for the thick crystal (the smaller width of the rocking curve in the Laue geometry). The dislocation density determined by the interference absorption coefficient is larger than that determined by the rocking curve width. This is likely due to occurrence of the localized defects in the crystal, for example, the vacancies which virtually do not deform the reflective atomic planes. It is worth mentioning that the reflection topograms include the weak misorientations at the edges which on the transmittance topograms are not manifested since defectness is higher at the crystal edges than in the center. That excludes the effect of anomalous transmission.

The image of the mosaic subgrains is absent on the transmission topograms. This is because the concentration of the defects in the subgrains exceeds the critical concentration $N_d > L^{-2}$, and the regime of dynamic diffraction is not realized in them. Note that in the diffraction direction in the Bragg geometry, the subgrain formation and other structural features, leading to change in the direction of the diffracted beam, are clearly seen on images. This can be attributed to the fact that the distances between the dislocations in such boundaries are less than L and, accordingly, the small-angle boundaries are not transparent from a standpoint of dynamic diffraction that gives rise to the shadowing effect on the topogram.

The obtained results suggest the conclusion about the character of influence of recrystallization on the real structure of the tungsten single crystals. Importantly that effect of anomalous X-ray transmission manifests itself in the subgrains of ~1 mm in size in the as-grown single crystals as well. After recrystallization the sizes of the subgrains having the perfect structure appears to be much larger. In accordance with the dynamic theory of X-ray scattering, the calculated value of FWHM of the rocking curve for the perfect tungsten single crystal is 48" of an arc. Unfortunately, it is quite difficult to estimate the dislocation density in a limit of small broadening of the rocking curve. Thus, it is possible only to declare - the dislocation density in the tungsten single crystals is very low, about 10^4 cm^{-2}.

7. Conclusions

The dislocation substructure of single crystals of molybdenum and tungsten is characterized by significant similarity and remains virtually unchanged at the growth rates of 0.5-5 mm/min. Significant changes in the substructure, reflected in increasing fragmentation of subgrains and the misorientation angles of up to 3-4^0 of an arc, take place when the growth rates increase to 10-20 mm/min and above. After sudden increase of the growth rates from 2 mm/min to 40 mm/min, the dislocation density inside subgrains increases on an order of magnitude, reaching 5x10^6 cm^{-2}.

Due to the high temperature gradients near the crystallization front and the high cooling rates, the growth of the perfect single crystals of molybdenum and tungsten from the melt is impossible. Even using the perfect seed crystal, free from the small-angle boundaries with the misorientation angles 3' of an arc, the method does not allow growing crystals of satisfactory structural quality.

The numerical estimates show that the cooling rates of single crystals can reach 10^4 K/s. This results in small increase in the dislocation density in the thin surface layer during cooling, which is quite acceptable. Therefore, to cool single crystals slowly after growing by EBFZM is impractical from all points of view.

The conditions of the growth of single crystals with the ultimately low dislocation density and the small subgrains spread are revealed. The relatively low dislocation density and the lack of the small-angle boundaries are achieved by recrystallization. The optimal procedure involves the 6% deformation of the single crystalline specimen with the <111> growth axis by rolling in vacuum along the <112> plane and *in-situ* high-temperature annealing.

To monitor the subgrain substructure of the tungsten single crystals, the anomalous X-ray transmission method is effective. The Borrmann effect is observed in the recrystallized perfect tungsten single crystals. The dislocation density determined by diffraction data is close to that determined by etch pits ($\sim 5x10^{-4}$ cm^{-2}). This opens potentialities for controlling the dislocation density by the X-ray diffraction techniques. The perfect single crystals can be employed as the novel crystalline deflectors to monitor the beams of relativistic charged particles and for other applications. The recrystallization example shows how perspective and reliable is this way in obtaining structurally perfect single crystals of tungsten and other refractory metals as well.

8. Acknowledgement

The author has a great pleasure to express sincere acknowledgment to my colleagues and friends Valery Semenov, Sergey Ermolov, Eugene Stinov, Sergey Markin, Boris Shipilevsky, and Sergey Bozhko from the Institute of Solid State Physics, Chernogolovka, Russia, for favorable attitude, cooperation in science and life. The author is very grateful to Wolfgang Gust from the Max-Planck Institute fuer Metallforshung, Stuttgart, Germany, to Hidde Brongersma from the Technical University of Eindhoven, Eindhoven, The Netherlands, and to Wayne King from the Lawrence Livermore National Laboratories, Livermore, USA, for fruitful discussions and friendliness for many years. The skillful technical assistance of Victor Lomeyko from the Institute of Solid State Physics, Chernogolovka, Russia, is greatly acknowledged.

9. References

Akita, H., Sampare, D.S., & Flore, N.F. (1973). Substructure control by solidification control in copper crystals. *Metallurgical Transactions*, Vol. 4, pp.1593-1597, ISSN 1073-5623.

Alonzo, V., Berthier, F., Glebovsky, V.G., Priester, L., & Semenov, V.N. (1995). Study of microstructure of a Ni(100) single crystal grown by electron beam floating zone melting. *Journal of Crystal Growth*, Vol.156, pp.480-486, ISSN 0022-0248.

Aristov, V.V., Shmytko, I.M., & Shulakov, E.V. (1974). Application of the X-ray divergent-beam technique for the determination of the angles between crystal blocks. *Journal of Applied Crystallography*, Vol.7, No.4, pp.409-413, ISSN 1600-5767.

Bdikin, I.K., Bozhko, S.I., Semenov, V.N., Smirnova, I.A., Glebovsky, V.G., Ermolov, S.N., & Shekhtman, V.S. (1999). Observation of anomalous transmission of X-rays in tungsten single crystals. *Technical Physics Letters*, Vol.25, No.12, (December 1999), pp.933-935, ISSN 1063-7850.

Bolling, C.F., & Finestein, D. (1972). On vacancy condensation and the origin of dislocations in growth from the melt. *Philosophical Magazine*, Vol.25, No.45, pp.45-66, ISSN 1478-6435.

Bozhko, S.I., Glebovsky, V.G., Semenov, V.N., & Smirnova, I.A. (2008). Study on the growth of tungsten single crystals of high structural quality. *Journal of Crystal Growth*, Vol.311, No.1, pp.1-6, ISSN 0022-0248.

Brunner, D., & Glebovsky, V.G. (2000). The plastic properties of high-purity W single crystals. *Materials Letters*, Vol.42, (February 2000), pp. 290-296, ISSN 0167-577X.

Brunner, D., & Glebovsky, V.G. (2000). Analysis of flow-stress measurements of high-purity tungsten single crystals. *Materials Letters*, Vol.44, (June 2000), pp. 144-152, ISSN 0167-577X.

Buckley-Golder, J.M., & Hurphreys, C.J. (1979). Theoretical investigation of temperature distribution during Chochralski crystal growth. *Philosophical Magazine*, Vol.39, No.1, pp.41-57, ISSN 1478-6435.

Calverly, A., Davies, M. & Lever, R. (1957). The floating-zone melting of refractory metals by electron bombardment. *Journal of Scientific Instruments*, Vol.34, No.4, pp.142-144, ISSN 0950-7671.

Chaika, A.N., Semenov, V.N. Glebovsky, V.G., & Bozhko, S.I. (2009). Scanning tunneling microscopy with single crystalline W[001] tips: High resolution studies of Si(557)5x5 surface. *Applied Physics Letters*, Vol.95, No.17, 173107, ISSN 0003-6951.

Cole, M., Fisher, D.S., & Bucklow, J.A. (1961). Improved electron beam device for zone melting. *British Journal of Applied Physics*, Vol.12, No.10, pp.577-578, ISSN 0508-3443.

Cole, G.S. (1971). Inhomogeneities and their control via solidification. *Metallurgical Transactions*, Vol. 2, pp.357-370, ISSN 1073-5623.

Cortenraad, R.; Ermolov, S.N.; Semenov, V.N.; Denier van der Gon, A.W.; Glebovsky, V.G.; Bozhko, S.I., & Brongersma, H.H. (2001). Growth, characterization and surface cleaning procedures for high-purity tungsten single crystals. *Journal of Crystal Growth*, Vol.222, pp. 154–162, ISSN 0022-0248

Cortenraad, R., Ermolov, S.N., Moest, B., Denier van der Gon, A.W., Glebovsky, V.G. & Brongersma, H.H. (2001). Crystal face dependence of low energy ion scattering signals. *Nuclear Instruments & Methods in Physical Research, B*, Vol.174, pp.173-180, ISSN 0168-9002.

Cortenraad, R., Ermolov, S.N., Denier van der Gon, A.W., Glebovsky, V.G., Brongersma, H.H., Manenschijn, A., Gartner, G., & Belozerov E.V. (2001). Cleaning procedures for single crystal tungsten substrates. *Inorganic Materials*, Vol.37, No.7, pp.673-677, ISSN 0020-1685.

Cortenraad, R., Ermolov, S.N., Semenov, V.N., Denier van der Gon, A.W., Glebovsky, V.G., Bozhko, S.I., Stinov, E.D. & Brongersma, H.H. (2001). Electron-beam growing and purification of W crystals. *Vacuum*, Vol.62, pp.181-188, ISSN 0042-207X.

Ermolov, S.N., Cortenraad, R., Semenov, V.N., Denier van der Gon, A.W., Boghko, S.I., Brongersma, H.H., & Glebovsky, V.G. (1999). Growth and characterization of monocrystalline tungsten substrates. *Vacuum*, Vol.53, pp.83-86, ISSN 0042-207X.

Ermolov, S.N., Glebovsky, V.G., Cortenraad, R., Moest, B., Stinov, E.D., Denier van der Gon, A.W., & Brongersma, H.H. (2002). Low-energy ion scattering by various crystallographic planes of tungsten single crystals. *Physics of Metals & Metallography*, Vol.93, pp.443–449, ISSN 0031-918X.

Esterling, D.M., (1980) Dislocation dissociation in some *bcc*-metals. *Acta Metallurgica*, V.28, pp.1287-1294, ISSN 0956-7151.

Glebovsky, V.G., Lomeyko, V.V., & Semenov, V.N. (1986). Set-up for electron-beam zone melting of refractory materials. *Journal of Less-Common Metals*, Vol.117, pp.385-389, ISSN 0022-5088.

Glebovsky, V.G., Semenov, V.N., & Lomeyko, V.V. (1988) Influence of the crystallization conditions on the structural perfection of molybdenum and tungsten. *Journal of Crystal Growth*, Vol.87, No.1, pp.142-150, ISSN 0022-0248.

Glebovsky, V.G., & Semenov, V.N. (1993-1994), Electron-beam floating zone melting of refractory metals and alloys: art and science. *International Journal of Refractory Metals & Hard Materials*, Vol.12, pp.295-301, ISSN 0263-4368.

Glebovsky, V.G., & Semenov, V.N. (1995). Growing single crystals of high-purity refractory metals by electron-beam zone melting. *High-Temperature Materials & Processes*, Vol.14, pp.121–130, ISSN 0334-6455.

Glebovsky, V.G., Sidorov, N.S., Stinov, E.D., & Gnesin, B.A. (1998). Electron-beam floating zone growing of high-purity cobalt crystals. *Materials Letters*, Vol. 36, August, pp.308-314, ISSN 0167-577X.

Glebovsky, V.G., & Semenov, V.N. (1999). The perfection of tungsten single crystals grown from the melt and solid state. *Vacuum*, Vol.53, pp.71–74, ISSN 0042-207X.

Glover, A.H., Wilcox, B.A., & Hirth, J.P., (1970). Dislocation substructure induced by creep in molybdenum single crystals. *Acta Metallurgica*, V.18, pp.381-397, ISSN 0956-7151.

Hay, D.R., Scogerboe, R.K., & Scala, E. (1968). Electron beam zone purification and analyses of tungsten. *Journal of Less-Common Metals*, Vol.15, No.2, pp.121-127, ISSN 0022-5088.

Hurle, D.T. (1977). Control of diameter in Czochralski and related crystal growth techniques. *Journal of Crystal Growth*, Vol.42, pp.473-482, ISSN 0022-0248.

Katoh, M., Iida, S., Sugita, Y., & Okamoto, K. (1991). X-ray characterization of tungsten single crystals grown by secondary recrystallization method. *Journal of Crystal Growth*, Vol. 112, pp.368-372, ISSN 0022-0248.

Kittel, C. (1996). *Introduction to Solid State Physics* (7th edition), Wiley, New York, USA.

Kobayashi, N. (1977). Power required to form a floating zone and the zone shape. *Journal of Crystal Growth*, Vol.43, pp.417-424, ISSN 0022-0248.

Kobayshi, N., & Wilcox, W.C. (1982). Computational studies of convection in a cylindrical floating zone. *Journal of Crystal Growth*, Vol.59, pp.616-624, ISSN 0022-0248.

Langer, J.S. (1980) Instabilities and pattern formation in crystal growth. *Revue of Modern Physics*, Vol.52, No.1, pp.1-28, ISSN 0034-6861.

Liu, J., & Zee, H.R. (1996). Growth of molybdenum-based alloy single crystals using electron beam zone melting. *Journal of Crystal Growth*, Vol.163, pp.259-265, ISSN 0022-0248.

Markin, S.N., Ermolov, S.N., Sasaki, M., van Welzenis, R., Stinov, E.D., Glebovsky, V.G., & Brongersma, H.H. (2006). Scattering of low-energy ions from the surface of a W(211) single crystal. *Physics of Metals & Metallography*, Vol.102, No.3, pp.274-278, ISSN 0031-918X.

Markin, S.N., Ermolov, S.N., Sasaki, M., van Welzenis, R.G., Glebovsky, V.G., & Brongersma, H.H. (2010). On a peculiarity of low-energy ion scattering from well-ordered *bcc* W(211) surface. *Nuclear Instruments and Methods in Physics Research B.*, Vol.B268, pp.2433-2436, ISSN 0168-9002.

Moest, B., Glebovsky, V.G., Brongersma, H.H., Bergmans, R.H., Denier van der Gon, A.W., & Semenov, V.N. (1998). Study of Pd single crystals grown by crucibleless zone melting. *Journal of Crystal Growth*, Vol.192, pp.410-416, ISSN 0022-0248.

Mullins, W.W., & Sekerka R.F. (1964). Stability of a planar interface during solidification of a dilute binary alloy. *Journal of Applied Physics*, Vol.35, No.2, pp.444-451, ISSN 0021-8979.

Mundy, J.N., Rotman, S.J., Lam, N.Q., Hoff, H.A., & Nowicki, L.J. (1978). Self-diffusion in tungsten. *Physical Revue*, V.B12, No.12, pp.6566-6575, ISSN 1098-0121.

Murphy, Y.A. (1987). Numerical simulation of flow, heat and mass transfer in a floating zone at a high rotational Reynolds numbers. *Journal of Crystal Growth*, Vol.83, No.1, pp.23-34, ISSN 0022-0248.

Nes, E., & Most, W. (1966). Dislocation densities in slow cooled aluminum single crystals. *Philosophical Magazine*, Vol.13, No.124, pp.855-859, ISSN 1478-6435.

Otani, S., Tanaka, T., & Ishizawa, Y. (1984). Temperature distribution in crystal rods with high melting points prepared by a floating zone technique. *Journal of Crystal Growth*, Vol.66, No.2, pp.419-425, ISSN 0022-0248.

Pfann, W.G. (1966). *Zone Melting* (2nd edition), Wiley, New-York, USA.

Reid, C.N. (1966). Dislocation widths in anisotropic *bcc*-crystals. *Acta Metallurgica*, V.14, pp.13-16, ISSN 0956-7151.

Riedle, J., Gumbsh, P., Fishmeister, Glebovsky, V.G., & Semenov, V.N. (1996). Cleavage fracture and the brittle-to-ductile transition of tungsten single crystals. *Fracture - Instability Dynamics, Scaling, and Ductile/Brittle Behavior*, R.L. Blumberg, J.J. Mecholsky, A.E. Carlsson & E.R. Fuller, eds., Vol.409, pp.23–28, MRS.

Riedle, J., Gumbsh, P., Fishmeister, Glebovsky, V.G., & Semenov, V.N. (1994). Fracture studies of tungsten single crystals. *Materials Letters*, Vol.20, (August 1994), pp.311-317, ISSN 0167-577X.

Shah, J.S., & Wills, H.H. (1975). Zone melting and applied techniques. In: *Crystal Growth*, B.R. Pamplin, ed., Pergamon Press, London, New York, p.194.

Shah, J.S. (1980). Zone refining and its applications (2nd edition). In: *Crystal Growth*, B.R. Pamplin, ed., pp.301-355, Pergamon Press, Oxford.

Surek, T., & Chalmers, B. (1975). The direction of the surface of crystal in contact with its melt. *Journal of Crystal Growth*, Vol.29, No.1, pp.1-11, ISSN 0022-0248.

Shipilevsky, B.M., & Glebovsky, V.G. (1989). Competition of bulk and surface processes in the kinetics of hydrogen and nitrogen evolution from metals into vacuum. *Surface Science*, Vol.216, pp.509–527, ISSN 0039-6028.

Wilcox, W.R., & Fuller, L.D. (1965). Turbulent free convection in Czochralski crystal growth. *Journal of Applied Physics*, Vol.36, pp.2201-2205, ISSN 0021-8979.

Yadava, R.D.S., Sinha, S., Sharma, B.B., & Warrier, R. (1985). Grain growth mechanism during recrystallization of a tellurium rich $Hg_{1-x}Cd_xTe$ cast. *Journal of Crystal Growth*, Vol.73, No.2, pp.343-349, ISSN 0022-0248.

5

Recrystallization Behavior During Warm Compression of Martensite Steels

Pingguang Xu[1] and Yo Tomota[2]
[1]*Japan Atomic Energy Agency, Tokai, Ibaraki,*
[2]*Ibaraki University, Hitachi, Ibaraki,*
Japan

1. Introduction

The application of high strength-toughness-ductility structural steels is beneficial to reduce the body weight of automotives and to improve the usage efficiency of energy without any potential damage of safe and security of human beings. Grain refinement is an important fundamental research field for the development of such low alloy structural steels. The conventional thermo-mechanically controlled process (TMCP) of ferrite or ferrite-pearlite steels including severe deformation at a lower temperature of ferrite transformation and rapid cooling is usually utilized to refine the grain size down to about 5 microns. For ferrite/pearlite steels, the grain refinement through dynamic recrystallization was observed to take place at a true strain of 1.2 at 873K, and the fully recrystallized ferrite/cementite microstructure may be realized at a strain of 2.0 (Torizuka, 2005). The final grain size is dependent on the Zener-Hollomon parameter, Z, which is given by

$$Z = \dot{\varepsilon} \cdot \exp(\frac{Q}{RT}) \tag{1}$$

where $\dot{\varepsilon}$, Q, R and T refer to the strain rate, the activation energy, the gas constant, and the absolute temperature, respectively. Because the severe deformation in uni-directional rolling cannot meet the requirement on Z-value for full recrystallization, the multi-directional groove rooling process has been developed recently. Unfortunately, the groove rolling process is unsuitable to the commercial production of wide steel plates.

In various microstructure types, the martensite was expected to have a low critical strain requirement for grain refinement because the high density dislocations, the supersaturated solute carbon and the ultrafine laths are helpful to raise the stored energy. The advantage of martensite was firstly claimed (Miller, 1972) as an initial microstructure to obtain ultrafine ferrite-austenite microstructures through cold-rolling and annealing of Ni(-Mn)-C martensite steels. The cold rolling and annealing of lath-martensite was confirmed (Ameyama, 1988) much effective to make ultrafine grained microstructures of low carbon steels. Ueji et al (2002) have claimed that the formation of ultrafine grained microstructure by cold rolling followed by annealing is closely related to the fine substructures and the high density dislocations of martensite. However, the cold rolling of martensite steel requires much higher loading capacity of mills and the microcracks may occur in the surface layers and/or the side edges of martensite steel plates.

Hayashi *et al* (1999, 2002) reported that the multi-pass warm groove-rolling of low-carbon martensite could produce an ultrafine ferrite-cementite microstructure less than 1μm. Tomota *et al* proposed to realize the grain refinement by warm compression or warm rolling of martensite steel plates through the dynamic recrystallization and confirmed that the dynamic recrystallization occurs at a low critical strain during warm compression for martensite of a low-carbon SM490 steel (Bao, 2005a). This technique has been successfully employed to get ultrafine grained ferrite-cementite microstructure (Li, 2008) and ultrafine grained ferrite-austenite microstructure (Xu, 2008a). Furuhara *et al* (2007) found that an initial microstructure of high carbon martensite is preferable to reduce the critical strain for full dynamic recrystallization, showing that the cementite particles can act as hard particles to promote the dynamic recrystallization during the martensite warm deformation. Though the carbon enriched retained austenite usually transforms to martensite to improve ductility, its detailed role during warm compression was necessary to be clarified. Therefore, the neutron diffraction was applied to *in situ* investigate the precipitation of austenite and the recrystallization of ferrite in 17Ni-0.2C martensite steel during warm compression. The existence of carbon enriched austenite particles during warm compression was found to further promote the dynamic recrystallization of ferrite from lath martensite by playing a role of the hard second phase (Xu, 2008b).

In this chapter, the research progress in dynamic recrystallization and grain refinement during warm compression of martensite steels was reviewed systematically while the advantages of *in situ* neutron diffraction were emphasized as a powerful beam technique suitable for clarifying the microstructure evolution during various thermo-mechanically controlled processes.

2. Dynamic recrystallization during warm compression in low alloy martensite steel

The dynamic recrystallization is one of the effective ways to refine microstructure to improve strength-toughness balance. Torizuka *et al* (2005) carried out a systematic study on warm compression and warm rolling for a conventional low-carbon ferrite/pearlite steel, realized a submicron ferrite-cementite microstructure through continuous dynamic recrystallization during heavy deformation up to a true strain of 3~4 and successfully produced the ultrafined grained high strength wires, rods and even steel plates with 300mm in width. However, it is difficult to apply such severe deformation with a high Z-value to commercial production of wider plates or sheets.

The average grain size of dynamically recrystallized microstructure was found to be dependent on Z-value and independent on the initial micrsotructure type such as ferrite or ferrite-pearlite or martensite (Ohmori, 2002; Ohmori, 2004; Bao, 2005a; Bao, 2005b). Though the initial grain size and the compressive strain in cases of ferrite or ferrite-pearlite initial microstructure have no direct effect on the average grain size of final ferrite microstructure, it is not clear about the effects of initial austenite grain size, strain and pre-tempering treatment on the dynamic recrystallization and the final grain size during the warm compression of martensite. Li *et al* (2008) investigated the effects of initial microstructure, deformation strain and pre-tempering on dynamic recrystallization of ferrite from lath martensite by using a commercial low-carbon steel, JIS/SM490 (mass%: 0.16C-1.43Mn-

0.41Si-0.014P-0.004S-0.01Cu-0.027Al-0.028N). The transformation temperatures measured by dilatometry at a heating and cooling speed of 5K/s were 1010K for Ac1, 1133K for Ac3 and 895K for Ar1, respectively (Bao, 2005a). Cylindrical compressive specimens with ϕ4mm×6mm were spark cut from the 15×15mm martensite steel bars obtained by water quenching after solid solution treatment at 1273K for 3.6ks. The specimens were heated up to 873K or 923K, held there for 1s, compressed by true strain $\varepsilon = 0.3$, 0.55 or 0.7 at 1.7×10^{-3} s^{-1}, and then quenched into water. Some specimens were pre-tempered at 873K or 923K for 3.6ks before the warm compression. The deformed specimens were cut along the longitudinal direction and the central area of the sectioned plane was observed with a field emission scanning electronic microscope (FE-SEM).

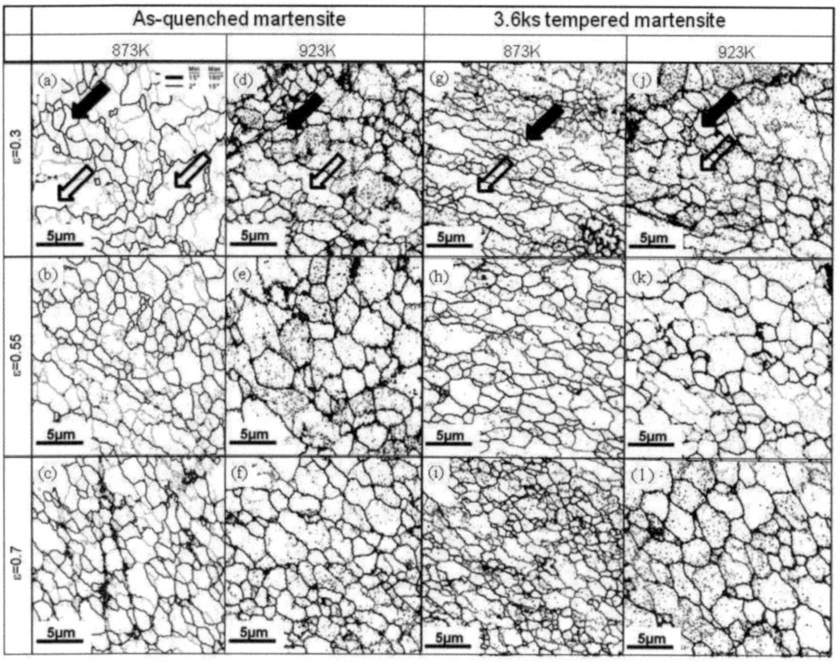

Fig. 1. Boundary misorientation mappings for the specimens deformed at 1.7×10^{-3} s^{-1} where the initial martensite was quenched from 1273K.

Fig.1 summarized the boundary misorientation mappings for the specimens warm compressed at 1.7×10^{-3} s^{-1}, where the solid arrows marked the recrystallized ferrite grains surrounded only by high angle grain boundaries (here called as Type I grains, according to Torizuka's definition (2005)) and the open arrows marked the recrystallized ferrite surrounded partially by high angle boundaries (here called as Type II grains). The Type I grains can be observed after the true strain 0.3 in all cases, revealing much lower critical strain for the initiation of dynamic recrystallization of low carbon martensite microstructure. This is surprizingly low because the critical strain is higher than 1.0 for the conventional ferrite-pearlite initial microstructure (Torizuka, 2005). The Type II grains equaxilize gradually and its size becomes almost equal to that of Type I grains with increasing of strain. It is also found that the grain size of Type I grains at a lower temperature (873K) deformation is smaller than that at a higher temperature (923K), revealing that the grain size is dependent on the Z-value. Moreover, compared with (c) and (i), it can be found that the 3.6ks prior tempering at 873K leads to the increase of Type I grains and smaller Type II grains, suggesting the prior tempering promote the dyanmic recrystallization. However, compared with (f) and (l), the 3.6ks prior tempering at 923K does not acclerate the grain refinement process. The observation of initial micorstructure before warm compression shows that the pretempering at 873K enables the dispersive precipitation of cementite particles and the dislocationes in martensite do not disappear so much, while the pretempering at 923K reduces the dislocation density and increases the size of cementite particles (Li, 2008). Consequently, the acceleration of dynamic recrystallization after the lower temperature pre-tempering (873K) is mostly related to the dispersive precipitation of cementite particles, while the delay of dynamic recrystallization after the higher tempatrue pretempering (923K) is mostly related to the decrease in dislocation density.

3. Dynamic recrystallization and austenite precipitation during warm compression in high nickel martensite steels

While the submicron ultrafine grained ferrite/cementite microstructure shows the limited ductility, the submicron ultrafine grained austenite/ferrite duplex microstructure possesses up to 1000MPa tensile strength and 25% uniform elongation (Tomota, 2008). Because the austenite can evidently improve the tensile ductility through the transformation induced plasticity (TRIP) effect, it leads to higher possibility for industrial applications. As mentioned in Section 1, warm compression or rolling of martensite microstructure with initial fine substructures and the high density dislocations of martensite can be utilized to accelerate the recrystallization and refine grains at a low critical strain (Bao, 2005a and 2005b). It was suspected that the competition of austenite precipitation and ferrite recrystallization played an important role to obtain ultrafine grained structures (Enomoto, 1977). However, the effects of austenite amount and its thermomechanical stability related to carbon concentration on the dynamic recrystallization of ferrite and the formation of ultrafine grain microsructure have not been clarified (Maki, 2001).

Xu et al investigated the effects of austenite precipitation and the carbon enrichment in austenite on the dynamic recrystallization during warm compression by using 18Ni and 17Ni-0.2C (mass%) martensite steels (Xu, 2008a) where 843K and 773K were chosen as the pre-tempering temperature according to the phase diagrams of Fe-Ni and Fe-Ni-C alloys, respectively. Cylindrical samples with 6.5 mm in length and 4mm in diameter for compression tests were prepared by spark cutting and surface grinding. The samples were heated up to the deformation temperature at a heating rate of 5K/s and then deformed at 8.3×10^{-4}/s followed by water quenching. The longitudinally sectioned specimens were electrochemically polished to avoid the stress-induced martensite transformation of austenite.

3.1 SEM microstructure observation and EBSD microstructure characterization

Fig.2 showed the different pre-tempered 17Ni-0.2C microstructures before and after warm compression at 773K. Different from the non-tempered martensite (Fig. 2(a)), small particles can be found in the tempered martensite matrix (marked by circle in Fig. 2(b)), and these particles are coarser (marked by circle in Fig.2(c)) after long-time pre-tempering. In the specimen warm compressed without any pre-tempering, ultrafine equiaxed grains (marked by white arrow) predominate the SEM microstructure while some elongated grains (marked by gray arrow) related to recovery can be also found. In contrast, the warm deformed specimen with 3.6ks pre-tempering shows a fully equiaxed microstructure, and the ferrite grains are much finer. However, the warm compression after 36ks pre-tempering results in a partially equiaxed microstructure, revealing the long-time pre-tempering retards the ferrite recrystallization evidently and a larger compressive strain is necessary to achieve full recrystallization.

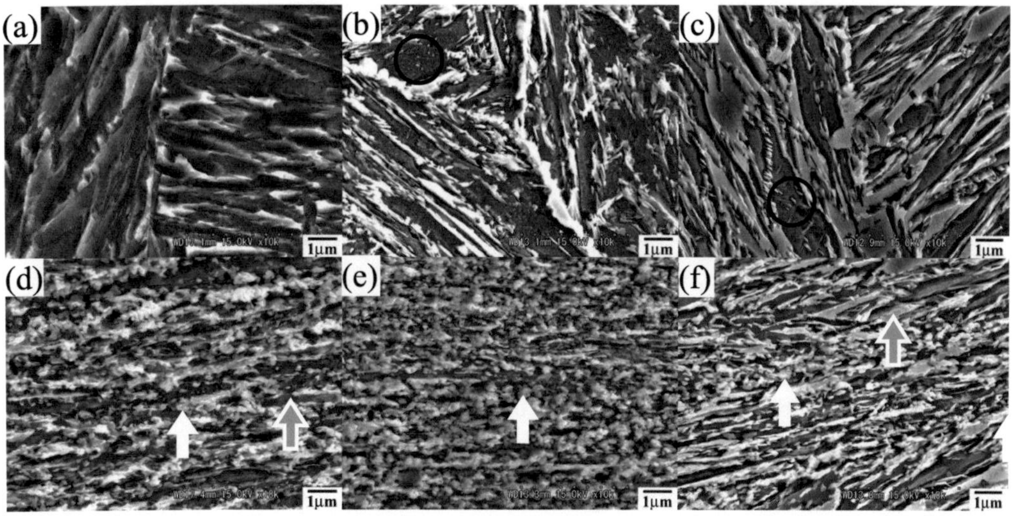

Fig. 2. SEM microstructures of 17Ni-0.2C steels before (a, b and c) and after (d, e and f) warm compression at 773K, ε=0.6, 8.3×10^{-4}/s. (a, d) non-tempered; (b, e) 3.6ks tempered; (c, f) 36.0ks tempered. The compression axis is along the vertical direction. (Xu, 2008a)

Fig.3 showed the electron back-scattering diffraction (EBSD) maps of 18Ni steel microstructures before and after warm deformation, where the compression axis is along the vertical direction. The volume fraction of the retained austenite amount in the non-tempered microstructure is less than 1%, and increases hardly after the 3.6ks pre-tempering.

In the ε=0.6 warm compressed specimen without pre-tempering, the equiaxed ferrite grains are dominative although many grains are still with small misorientation; in case of 3.6ks pre-tempering, the equiaxed ferrite grains are much few and the average grain size is a little coarser, suggesting that the decrease in dislocation density in the initial microstructure by pre-tempering brings clearly negative influence on the dynamic recrystallization.

After a larger strain warm compression (ε=0.8), the fully equiaxed ferrite grains can be obtained in both cases and their average grain sizes are approximately equal. Although blocky austenite grains are not observed in the microstructures after warm compression, the local regions with a confidence index of less than 0.2 (marked by circles) are related to the metastable austenite at high temperature, partially transformed to martensite /austenite islands during rapid cooling to room temperature for the sample preparation. Considering that such conventional microstrure quenching(freezing) process is not so ideable, the *in situ* microstructure evaluation involving neutron diffraction (see Section 4) is much important for investigation on the high temperature microstructure evolution.

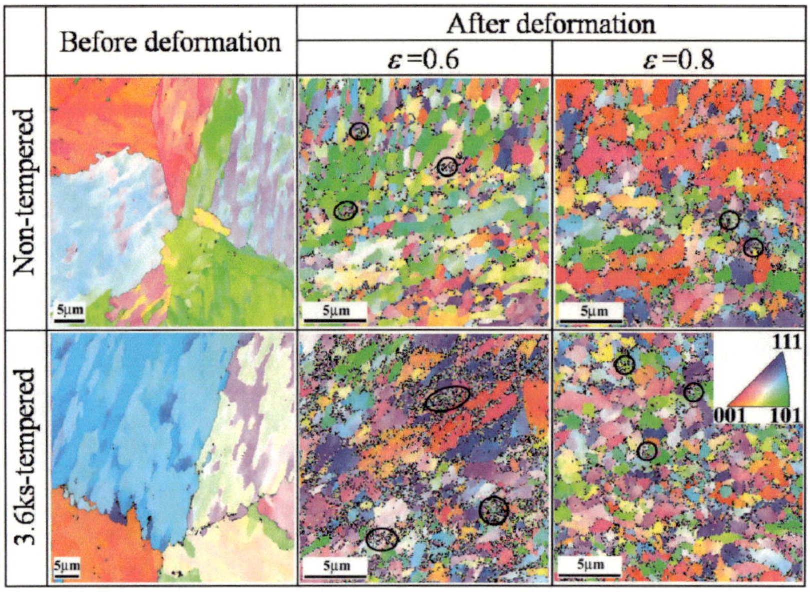

Fig. 3. Crystallographic orientation characteristics of 18Ni martensite steel before and after warm compression at 843K, 8.3×10^{-4}/s. (Xu, 2008a).

Fig.4 gives the EBSD maps of 17Ni-0.2C microstructures before and after warm compression. The austenite volume fraction is plotted in Fig.5. For the non-tempered specimen, the austenite amount increases much after deformation (from 9.5% to 17%) and its average grain size of austenite decreases evidently because of the precipitation of ultrafine grained austenite during warm deformation. After warm deformation, the equiaxed ferrite grains in the local regions with bulky austenite (marked by circle) are with large misorientation but those in the regions with no bulky austenite (marked by reactangle box) are with small misorientation, showing that the ferrite dynamic recrystallization is related to the existence of bulky austenite grains.

The 3.6ks pre-tempering leads to an clear increase of austenite amount. After the ε=0.6 warm compression, the ferrite-austenite duplex microstructure becomes fully equiaxed and the volume fraction of austenite increases from 18% before deformation to 36.8% after deformation and the average size of austenite grains decreases from 1.53μm before deformation to 0.94μm after deformation, which also proves that the warm deformation promotes the precipitation of ultrafine grained austenite. Meanwhile, the ferrite grains are refined to 0.85μm after deformation, smaller than that of 3.6ks pre-tempered 18Ni steel after warm compression at a strain of 0.8. It reveals that the higher Z-value (17Ni-0.2C, 773K, $Z=1.21\times10^{14}$ s^{-1}; 18Ni, 843K, $Z=4.55\times10^{12}$ s^{-1}) and the high carbon content are helpful to refine the fully recrystallized microstructure.

When the pre-tempering time prolongs to 36.0ks, the austenite amount approaches 33.8% after the isothermal tempering treatment. After the ε=0.6 warm deformation, the ferrite grains are elongated and coarse, showing that the dynamic recrystallization is delayed evidently. Meanwhile, the austenite amount measured at room temperature decreases abruptly to 13.8%. According to the phase diagram, the equilibrium microstructure at

773K consists of ferrite, austenite and cementite. The austenite amount after long-time pre-tempering in Fig.5 increases evidently due to the austenite precipitation at an elevated temperature below the martensite reverse transformation starting temperature, which is consistent with the previous reports for a Fe-Ni martensite alloy (Enomoto, 1977) and for a 18Ni-12Co-4Mo maraging steel (Moriyama, 2001). The decreasing austenite amount after warm compression is related to the weakening of austenite thermal stability caused by the carbon depletion in austenite after cementite precipitation. Since the cementite particles are mostly in nanometer scales, TEM observations are necessary.

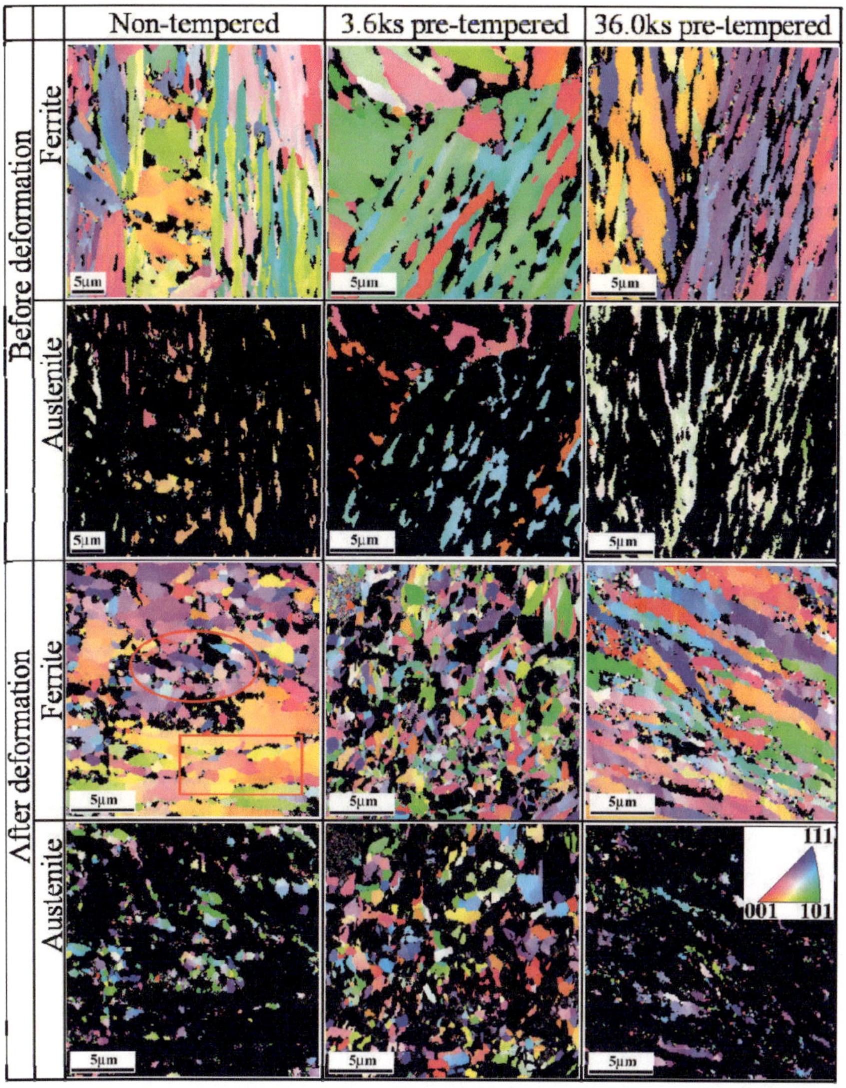

Fig. 4. Crystallographic orientation characteristics of different pre-tempered 17Ni-0.2C martensite steel specimens before and after warm compression at 773K, 8.3×10^{-4}/s. The compression axis is along the vertical direction (Xu, 2008a).

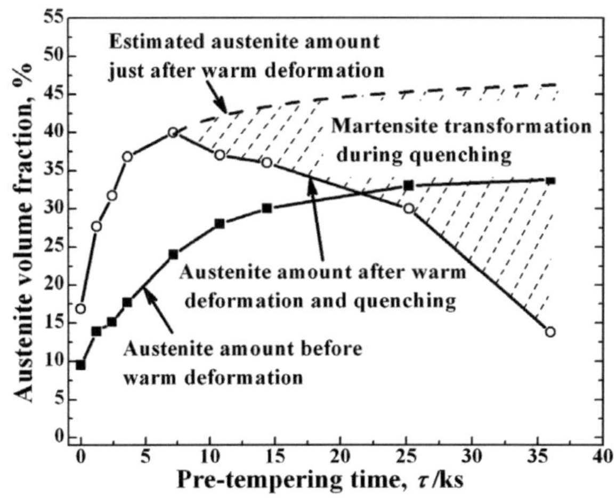

Fig. 5. Change in the austenite amount of 17Ni-0.2C steel by pre-tempering and warm deformation at 773K according to EBSD phase mapping statistics (Xu, 2008a).

3.2 TEM microstructure observation and EDX composition analysis

Fig.6 shows the bright field images for the non-tempered (a) and pre-tempered (b, c) microstructures of 17Ni-0.2C steel after warm compression. In the case of no pre-tempered condition, large ferrite grains near the austenite particles can be found with high density dislocations and some new grains with no intragranular dislocations form in these regions. It is found in Fig. 6(a) that the high angle grain boundaries surrounded an individual ferrite grain. In the warm compressed specimen after 3.6ks pre-tempering, the ultrafine equiaxed ferrite and austenite grains can be found frequently (Fig. 6(b)), suggesting that the warm deformation promotes the austenite precipitation and the ferrite dynamic recrystallization. The misorientation analysis confirms that the grain boundaries between these grains and their neighbors are of high angle boundaries, exhibiting that the dynamic recrystallization of ferrite can be confirmed really to take place during the warm deformation.

In case of the 36ks tempering, the elongated martensite laths within low dislocation density are clearly observed, which is related to the recovery of martensite. It is also found that the cementite precipitates preferably around the carbon-enriched austenite grains located at the martensite lath boundaries rather than within the martensite laths. The warm deformation accelerates the carbon diffusion and promotes the dynamic precipitation of spherical cementite particles within the martensite laths (Fig.6(c)).

For a pre-tempering time longer than 7.2ks, the carbon partitioning from martensite to austenite is gradually replaced by the cementite precipitation. Because of the precipitation of cementite particles and the increase of austenite amount, the average carbon concentration in austenite must decrease, finally lower than that in the 3.6ks tempered and deformed specimen. As a result, the thermal stability of austenite decreases, some austenite grains transform to martensite upon cooling after deformation and the fresh nanometer-scale martensite particles form around the cementite and austenite particle (Fig.6(c)). Hence the austenite amount after rapid cooling to room temperature becomes smaller (see Fig.4). The low density dislocations and no carbon-enriched austenite lead to the delayed dynamic recrystallization in the 36ks pre-tempered and warm compressed specimen when compared with the 3.6ks pre-tempered and warm compressed specimen.

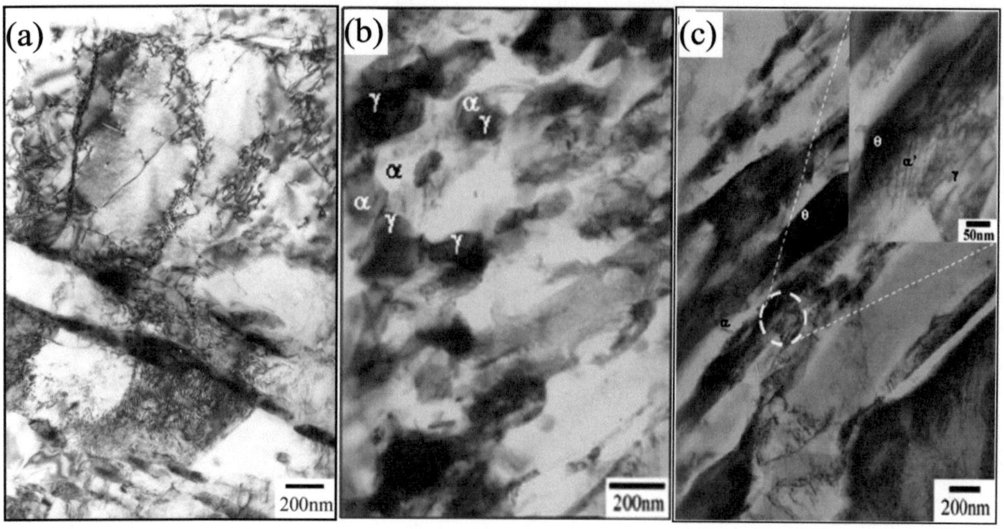

Fig. 6. Transmission electron microscopic microstructures of different pre-tempered 17Ni-0.2C steel after warm compression at 773K, 8.3x10^{-4} /s: (a) non-tempered; (b) 3.6ks pre-tempered; (c) 36ks pre-tempered.

According to the Fe-Ni and Fe-Ni-0.2C equilibrium phase diagrams, the nickel-rich austenite should appear at 843K in 18Ni and the nickel-rich and carbon-rich austenite and cementite should appear at 773K in 17Ni-0.2C. The EDX composition analysis proves that the metastable austenite (or martensite) is nickel-enriched, for example, the nickel content is about 20-26mass% (austenite) versus 15-16mass% (ferrite) in 17Ni-0.2C. Though the EDX analysis can not provide a reliable carbon concentration in constituent phases, the austenite in the 3.6ks pre-tempered 17Ni-0.2C after warm compression should be carbon-eniriched because (a) no cementite particles are found in such microstructure, (b) the stability of austenite phase is much higher than that in 18Ni, (c) the diffusion of carbon is more rapid than that of nickel and (d) the interstitial solubility of carbon is much higher in austenite than in ferrite. Considering that the 3.6ks pre-tempering decreases the dislocation density of martensite substructures, the accelerate dyamic recrystallization is mainly related to the formation of a large amount of carbon-enriched austenite.

3.3 Effect of carbon-enriched austenite on dynamic ferrite recrystallization

Generally, the retained austenite at room temperture is softer than the corresponding martensite. If there is no *in situ* quantitative micromechanical data of two duplex microstructure at elevated temperature, it will be much difficult to clarify why the existence of carbon-enriched austenite promotes the dynamic ferrite recrystallization during warm compression. Fortunately, the *in situ* neutron diffraction study about 3.6ks pre-tempered 17Ni-0.2C steel (Xu, 2008b) showed that the heterogeneous deformation occurs during warm compression at 773K and the ferrite matrix is subject to larger plastic deformation than the carbon-enriched austenite, i.e. the former is softer than the latter. Moreover, it was found that before and after approaching the critital strain ε=0.13, the ferrite (hkl) reflections showed clear difference among their relative integrated intensities because of the occurrence of dynamic ferrite recrystallization. In addition, the splitting of austenite reflection peaks and its disappearance during warm compression obtained by a neutron diffraction study (Xu, 2008b) help us to understand the change in carbon concentration of austenite during the warm compression. Consequently, the influencing mechanism of carbon-enriched austenite on the heterogeneous deformation and the ferrite-austenite duplex microstructure evolution can be summarized as shown in Fig.7.

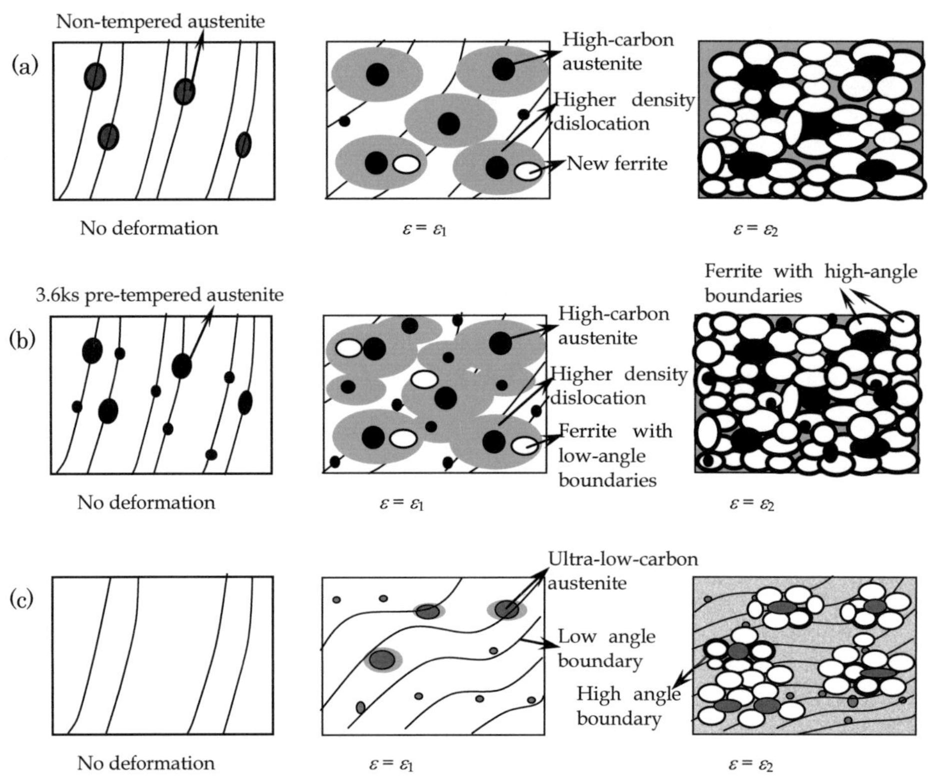

Fig. 7. Schematic illustration for microstructure evolution of martensite steels with different carbon contents during warm deformation, where the compression axis is along the vertical direction and $\varepsilon_1 < \varepsilon_2$: (a) 17Ni-0.2C, non-tempered, at 773K; (b) 17Ni-0.2C, 3.6ks pre-tempered, at 773K; (c) 18Ni, at 843K. (Xu, 2008a)

For the carbon-added high-nickel steel, the amount of carbon-enriched austenite increases due to the proper pre-tempering and the austenite particles becomes harder than the pre-tempered martensite (or recovered ferrite matrix) during warm compression. Hence the larger local plastic flow takes place in the regions around austenite particles. When the local plastic strain increases to a critical value ($\varepsilon = \varepsilon_1$), the recrystallized ferrite grains preferably nucleate in these local regions near the austenite particles according to the related recrystallization model (Doherty, 1997). When the true strain increases to another critical value ($\varepsilon = \varepsilon_2$), all the ferrite grains are equiaxed and fully recrystallized (Fig. 7(b)).

In case of no pre-tempering, the austenite amount is relatively less than that in case of 3.6ks pre-tempering and the carbon content in austenite is lower than that in the latter. Although the warm compression promotes the austenite precipitation and the carbon enrichment in austenite, the plastic deformation partitioning between austenite particles and recovered ferrite matrix is a little weaker. Consequently, at the true strains $\varepsilon=\varepsilon_2$, the recrystallized ferrite amount is correspondingly less than that in the case of 3.6ks pre-tempering (Fig. 7(a)).

When the pre-tempering time is extended to more than 14.4ks, the precipitation of cementite particles leads to relative carbon depletion in austenite, and the plastic deformation partitioning between the austenite and the recovered ferrite matrix is no longer apparent. Hence the dynamic recrystallization of ferrite is retarded markedly (Maki, 2001). The coarse austenite grains formed during long-time pre-tempering are metastable because of low carbon concentration, easy to deform during warm compression, and then transform to martensite (or ferrite matrix) upon rapid cooling to room temperature.

For the ultra-low-carbon high-nickel steel, the plastic deformation partitioning is relatively weak since the carbon-enriched austenite is impossible to form and martensite is easy to recover. Moreover, the dislocation density in the martensite matrix is lower than that in carbon-added high-nickel steel and the block is also coarser. As a result, a larger compressive strain is needed for the full recrystallization (Fig. 7(c)).

As a summary, the following conclusions should be mentioned. The carbon addition is beneficial to reduce the critical strain for full recrystallization of high-nickel martensite steels during warm compression. The increment of carbon-enriched austenite volume fraction accelerates the dynamic recrystallization of ferrite through plastic deformation partitioning in the 17Ni-0.2C steel. Proper pre-tempering promotes the precipitation of the carbon- and nickel-enriched austenite, and then promotes the dynamic recrystallization in the 17Ni-0.2C steel. On the other hand, long-time tempering leads to the carbon depletion in austenite so as to delay the dynamic recrystallization. In the 18Ni steel, a larger warm compression strain is required for full recrystallization during warm compression, mainly because of no carbon-enriched austenite.

4. Dynamic microstructure evolution in high nickel martensitic steel during warm compression studied by *in situ* neutron diffraction

As mentioned in above sections, a large strain and a high Zener-Hollomon value may be unnecessary to obtain ultrafine grained microstructures if the martensite initial microstructure is employed during annealing after cold working (Miller, 1972; Ameyama, 1988) or during warm deformation (Hayashi, 1999, Bao, 2005a). The microstructural observation of warm compressed specimens with different amounts of pre-existing austenite has shown that a dynamically recrystallized microstructure can be obtained in a 17Ni-0.2C martensite steel at a small strain ($\varepsilon \leq 0.6$). In order to make clear the microstructure evoluton and to investigate the effect of the austenite particles on dynamic recrystallization, the *in situ* Time-Of-Flight (TOF) neutron diffraction for the as-quenched 17Ni-0.2C steel during warm compression was carried out using the ENGIN-X neutron diffractometer at ISIS, Rutherford Appleton Laboratory.

Since the warm deformation at a temperature just below the austenite transformation starting temperature accelerated the austenite transformation significantly, a little low temperature (773K) was selected in order to investigate the dynamic recrystallization of ferrite. Here, 3.6ks pre-tempering treatment at 773K was carried out to obtain about 18vol% austenite in the initial microstructure. Cylinder specimens with 8mm diameter and 20mm length were prepared by spark cutting and surface grinding.

A 100kN hydraulic loading rig attached with a radiant furnace with a control error of ±1K was employed to realize the thermomechanically controlled process (TMCP). The specimen was heated up to 773K and neutron diffraction spectrum during the isothermal holding was collected repeatedly with 1min acquisition time to investigate the effect of the tempering process on the microstructure evolution. After 600s isothermal holding at 773K, the specimen was compressed at a strain rate of 8.3×10^{-4} s^{-1} and the neutron diffraction spectra were acquired at 773K. In order to decrease the experimental error, each neutron spectrum was summed with the consecutive one for the purpose of applying the Rietveld refinement.

Because it was difficult to distinguish the diffraction peaks of ferrite and martensite using the relative weak neutron spectra, the tempered/deformed martensite and the recrystallized ferrite during warm compression are here simply designated as the ferrite matrix. The austenite volume fraction was determined by the Rietveld refinement using the General Structure Analysis System (GSAS) software package (Larson, 2004), taking all diffraction peaks measured into consideration. The texture indexes (Bunge, 1982) of warm deformed austenite and ferrite were evaluated by the spherical harmonic preferential orientation fitting with an assumption of cylindrical sample symmetry, and the series was truncated at a maximum expansion order of l_{max}=8. Single peak fitting with the third TOF profile function (Larson, 2004) was employed to obtain the integrated intensities and the lattice spacings of (*hkl*) peaks. The preferred orientations and the lattice strains of ferrite and austenite were also analyzed by these integrated intensities and lattice spacings, respectively.

4.1 Microstructure evolution during isothermal holding

Fig.8 gives the axial neutron spectrum acquired in 600s at room temperature and the corresponding Rietveld refinement result where the small residual error reveals that the profile fitting quality is good enough. The austenite amount of the 3.6ks pre-tempered 17Ni-0.2C martensite steel is about 21.0±0.3%, a little higher than that measured by EBSD technique for an electrochemically polished sample (about 18%), which is possibly related to the weak stability of austenite, *i.e.* martensite transformation near surface (Chen, 2006). Actually, the neutron diffraction, as an important materials characterization technique different from the X-ray and electronic diffraction, enables to measure a large sample with a large gauge volume and to get bulk average information with high statistics and to evaluate the microstructure evolution under various environments, especially for multiphase materials.

Fig.9 shows the change in austenite mass fraction during isothermal holding at 773K. The austenite amount increases slowly and results in about 1.5% increment after 600sec isothermal holding. This precipitation speed is consistent with that estimated by the microstructure observation where the austenite amount increases from 9.5% before tempering to about 18% after 60min tempering at 773K (Xu, 2008a). These results also reveal that the neutron diffraction technique is suitable to evaluate the austenite precipitation during isothermal holding.

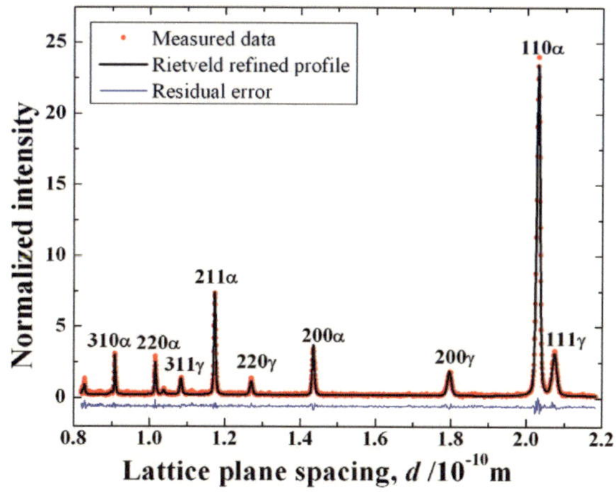

Fig. 8. Neutron diffraction spectrum of 3.6ks pre-tempered 17Ni-0.C martensite steel obtained at room temperature and its Rietveld refinement result taking all diffraction peaks measured into consideration. (Xu, 2008b)

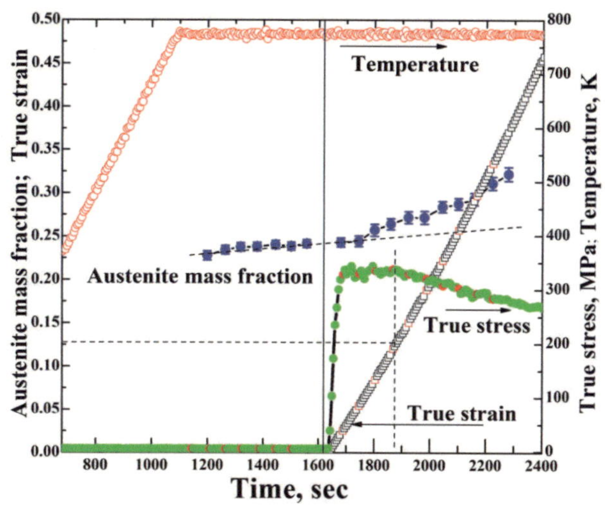

Fig. 9. Change in austenite fraction during 773K isothermal holding and warm compression.

4.2 Change in austenite diffraction spectra during warm compression

Fig.9 gives the changes in austenite fraction and true strain monitored with a high-temperature extensometer during warm compression as a function of time, showing that the austenite amount increases gradually with increasing compressive strain, *i.e.* from 24% at ε=0.0 to 32% at ε=0.33. Compared with the isothermal holding test, it is clear that the isothermal compression accelerates the austenite precipitation. In addition, the flow stress approaches the maximum value at a true strain about ε=0.13, and then gradually decreases.

Accordingly, the stress-strain curve is divided into two parts: (1) Region I, work hardening; (2) Region II, work softening.

Fig.10 compares the (111) diffraction peaks of austenite obtained in the axial direction at different loading steps during the warm compression at 873K. Though the austenite (111) peak intensity at ε=0.05 has almost equal to that at ε=0.0, the (111) lattice plane spacing at ε=0.05 (d_{111} =0.208428±0.000013 nm) is smaller than that at ε=0.0 (d_{111}^0 =0.208826±0.000007 nm). In other words, the lattice strain of austenite, $\varepsilon_{111} = (d_{111} - d_{111}^0) / d_{111}^0$, is about -1906×10^{-6}, which is comparable with the elastic strain of the specimen estimated by $\varepsilon = \sigma / E$ =-1781×10^{-6} where the external stress σ is about 330MPa. Here the high-temperature Young's modulus *E* is taken as 185GPa (Sawada, 2005). That is to say, the (*hkl*) peak shifts of austenite related to the lattice strain are mostly due to the external stress.

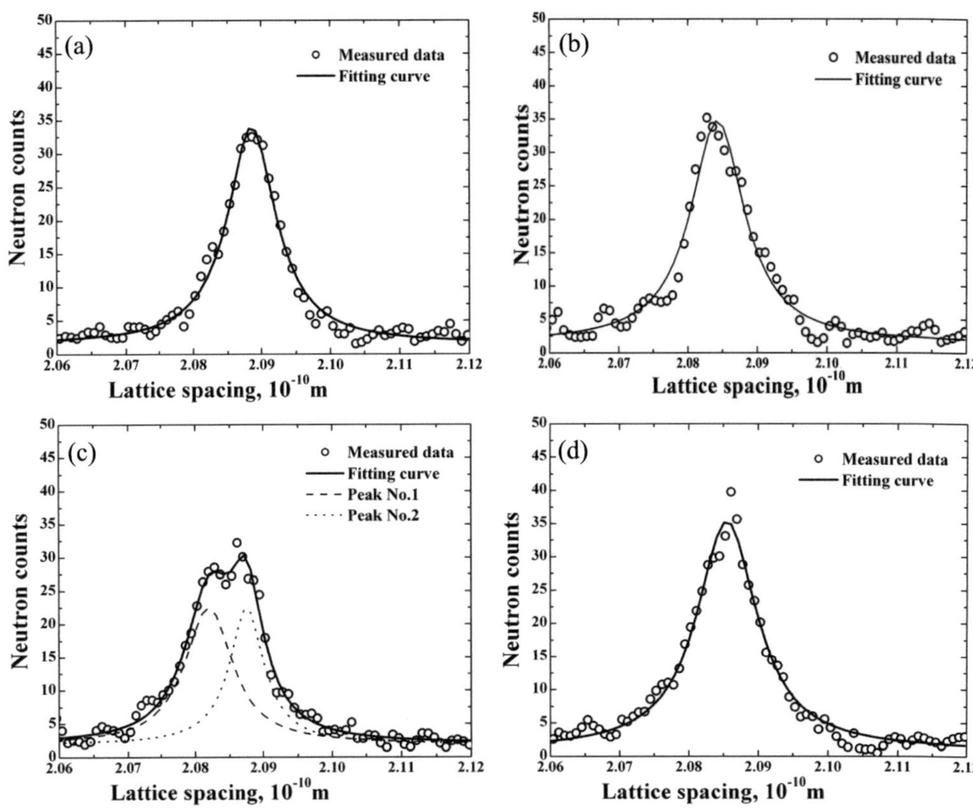

Fig. 10. Austenite (111) peaks and fitting results: (a) non-splitting, ε=0.0; (b) peak shift, ε=0.05; (c) peak splitting at ε=0.15; (d) disappearance of peak splitting at ε=0.33.

As the true strain increases to 0.15, the external stress increases only a little bit (see Fig.9), so that the amount of further peak shift is very limited. After this stage, the decreasing external stress with increasing true strain leads to a smaller peak shift in the opposite sense. On the other hand, the austenite (111) peak at ε=0.15 splits into two peaks with lower peak intensities (see Fig.10c). However, this peak splitting gradually disappears by ε=0.33, by which stage there is again a single peak with a higher intensity. Considering the peak intensities of such neutron spectra were not strong, the third TOF profile function (Larson, 2004) was employed to fit the austenite (111) peaks. The splitting of austenite (111) peak suggests that newly precipitated austenite possesses lower carbon and nickel contents from those of the pre-existing austenite. Then, the chemical composition of austenite gradually becomes uniform with increasing compressive strain.

4.3 Change in ferrite diffraction spectra during warm compression

Fig.11 compares the (110) diffraction peaks of ferrite obtained in the axial direction at different loading steps at 773K. The (110) lattice plane spacing at ε=0.05 (d_{110}=0.203425±0.000007 nm) is smaller than that at ε=0.0 (d_{111}^{0}=0.203780±0.000007 nm). In other words, the lattice strain of ferrite, $\varepsilon_{110} = (d_{110} - d_{110}^{0}) / d_{110}^{0}$, is about -1742×10^{-6}, lower than that of austenite ε_{111}. The *hkl*-specific elastic moduli E_{hkl}^{K}, calculated by the Kröner model, are 247.9 GPa for the austenite along the [111] direction and 225.5 GPa for the ferrite along the [110] direction (Hutchings, 2005), respectively. Therefore, the austenite is subject to a higher average phase stress than the ferrite matrix at the beginning of warm compression, *i.e.* the pre-existing austenite is harder than the ferrite matrix. In comparison with diffraction peaks obtained at ε=0.0, all (*hkl*) ferrite peaks show clear peak shifts at ε=0.05 due to the external stress about 330MPa. The ferrite (110) peak intensity decreases substantially and the ferrite (211) peak intensity increases a little. When the strain increases to ε=0.15, the ferrite (110) peak intensity decreases slowly but the ferrite (211) and (200) peak intensities decrease significantly.

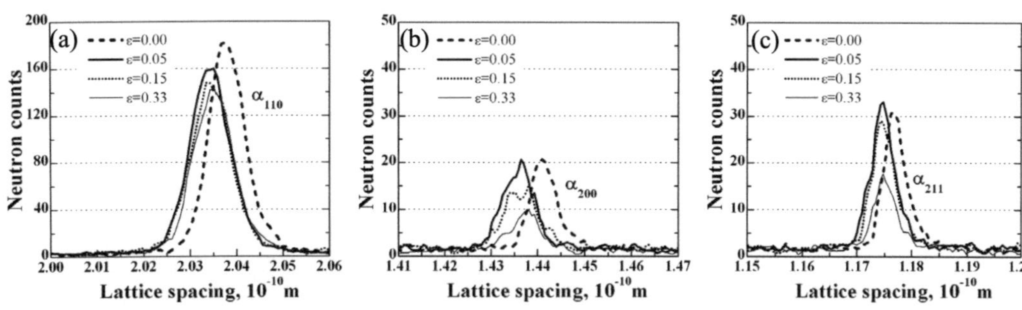

Fig. 11. Comparison of ferrite axial diffraction peaks of the 17Ni-0.2C steel at different compressive strains: (a) ferrite (110) peak; (b) ferrite (200) peak; (c) ferrite (211) peak. (Xu, 2008b)

Fig.12 shows the change in integrated intensities of different diffraction peaks with increasing of the compressive strain, obtained by single peak fitting of neutron spectra measured both in the axial and the radial directions of cylinder specimen. Before the compressive strain increases to 0.13, *i.e.* before the true stress reaches the maximum value, the ferrite (110) integrated intensity obtained in the axial direction decreases rapidly but that obtained in the radial direction decreases slowly. When the compressive strain is beyond ε=0.13, the decrease in ferrite (110) intensity in the axial direction becomes much slower while the decrease in the radial direction accelerates markedly. For the ferrite (200) and (211) peaks, the integrated intensities obtained in the axial direction increase slowly before ε=0.13 and then decrease evidently beyond ε=0.13, but such data obtained in the radial direction show little change during the warm compression. Based on these changes in ferrite crystallographic orientation, the strain corresponding to the maximum true stress can be regarded as the onset strain for dynamic recrystallization of ferrite from martensite.

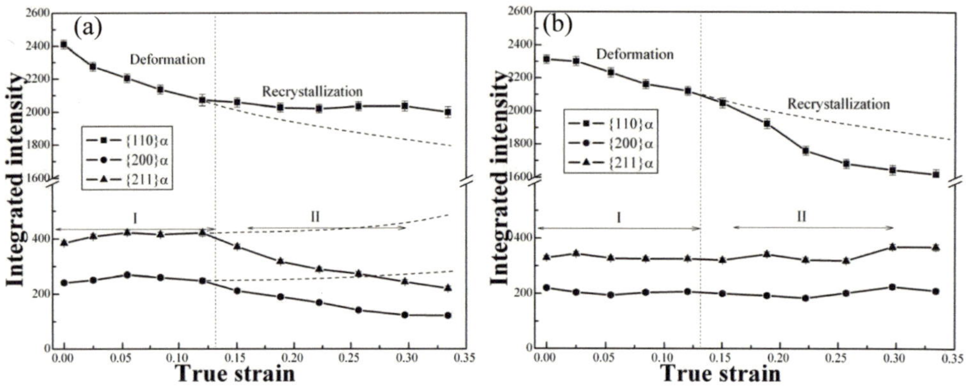

Fig. 12. Change in integrated intensity of ferrite in 17Ni-0.2C steel during warm compression: (a) obtained from the axial neutron spectra; (b) obtained from the radial neutron spectra. Regions I and II correspond to the ferrite deformation and the ferrite dynamic recrystallization, respectively. (Xu, 2008b)

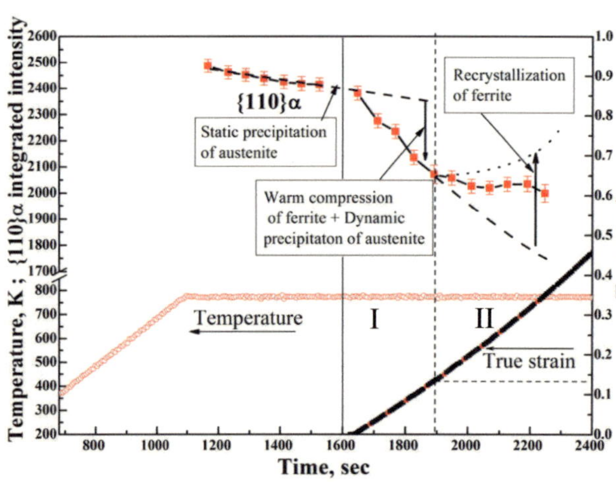

Fig. 13. Illustration for the microstructure evolution during isothermal holding and warm compression, based on the change in integrated intensity of ferrite (110) peak obtained in the axial direction.

Fig.13 illustrated the microstructure evolution based on the change in integrated intensity of ferrite (110) peak acquired in the axial direction during the isothermal holding and warm compession of 17Ni-0.2C martensite steel with 3.6ks prior tempering treatment. During the isothermal holding, the ferrite (110) integrated intensity decreases slowly due to the static precipitation of austenite. During the warm compression in the Region I, the warm compression of ferrite and the dyanmic precipitation of austenite accelerate the reduction of the ferrite (110) integrated intesity and form a clear deviation from the change trend of ferrite (110) integrated intensity in during the isothermal holding; when the strain surpasses the critical strain 0.13, the compessed ferrite grains begins to recrystalize, which leads to an evident increase of ferrite (110) integraed intensity as marked by the dotted line if there is no other microstructure evolution. The co-existence of the ferrite recrystallization as marked by the dotted line and the ferrite deformation as marked by the dashed line (which is partially related to austenite precipitation) means that the so-called ferrite dyamic recrystallization really takes place. In other words, the microstructure evolution in the Region II is involved in the dyanamic ferrite recrystallization and the dynamic austenite precipitation, and their compitition effect is believed to result into the slow decrease in the ferrite (110) integrated intensity.

4.4 Effect of the existence of austenite grains on dynamic recrystallization of ferrite

Following the dynamic recrystallization, the flow stress decreases gradually to a stable stress. Since the external loading and the composition change in constituent phases both affect the lattice plane spacing, it is difficult to compare the lattice compressive strains directly during warm compression beyond ε=0.05. Considering that the plastic deformation changes the texture as described above, the texture indexes of the two constituent phases can be used to make an indirect comparison between the plastic strains of austenite and ferrite.

According to Bunge's definition (Bunge, 1982), the texture index J is a parameter to characterize the sharpness of the texture by the integral of the square of the texture function $f(g)$, without considering the details of the crystallographic orientation distribution:

$$J = \oint [f(g)]^2 dg = \sum_{l,\mu,\nu} \frac{1}{2l+1} \left| C_l^{\mu\nu} \right|^2 \quad (2)$$

where l, μ and ν are the series expansion orders and $C_l^{\mu\nu}$ is the corresponding expansion coefficient. For a random texture, J=1.0 and for an ideal texture of single orientation, $J \rightarrow \infty$. Generally, heavier plastic deformation leads to sharper texture. In addition, the TOF neutron spectrum measurement covers a wide range of lattice plane spacings which is partially equivalent to measuring one reflection over a wide range of sample orientations. Therefore, the texture indexes calculated from the TOF neutron spectra by the GSAS software package (Larson, 2004) can be employed to compare the plastic strains of austenite and ferrite.

As shown in Fig.14, the texture indexes of the ferrite phase are evidently larger than those of austenite in both the axial and radial directions while the difference between the two directions is related to the details of the crystallographic orientation distribution. This suggests that the plastic deformation occurs preferentially in the ferrite matrix during warm compression.

The different peak shifts in Section 4.2 & Section 4.3 reveal that the austenite grains are harder than the ferrite grains. The texture index of the ferrite matrix is larger than that of austenite, suggesting that the plastic deformation occurs preferentially in the ferrite matrix. According to the related recrystallization literature (Doherty, 1997), the recrystallized ferrite grains nucleate preferentially in the heterogeneously deformed regions near large hard particles. Therefore, the existence of carbon-enriched austenite is able to accelerate the dynamic recrystallization in the ferrite matrix.

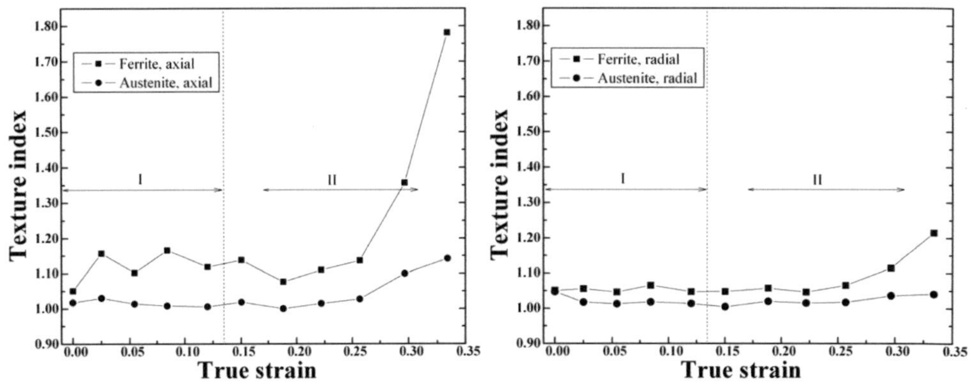

Fig. 14. Change in texture indexes of austenite and ferrite during warm compression: (a) obtained from the axial direction; (b) obtained from the radial direction. (Xu, 2008b)

The TOF neutron diffraction experiment helped us to understand the austenite precipitation and dynamic recrystallization behaviors of the 17Ni-0.2C martensite steel during isothermal warm compression after isothermal holding at 773K, and the main results were as follows: (1) the warm deformation at 773K accelerated the austenite precipitation in 17Ni-0.2C martensite steel. Splitting of the austenite (111) peak was found to occur and then disappear during warm compression. The splitting is ascribed to the different carbon concentrations in the dynamically precipitated austenite and the pre-existing austenite, and the disappearance of splitting is related to carbon homogenization at larger strain. (2) The austenite in the 17Ni-0.2C martensite steel is harder than the ferrite matrix at 773K. Heterogeneous deformation occurs preferentially in the ferrite matrix, leading to the acceleration of dynamic recrystallization.

5. Summary and future works

The low-carbon martensite steel and the high-nickel martensite steels have been warm compressed with and without prior tempering treatment to realize the microstructure refinement based on dynamic recrystallization at lower Zener-Hollomon parameter Z. Because of the lower critical strain for fully recrystallization compared with the warm deformation of ferrite or pearlite/ferrite, the warm deformatioin of martensite may be employed to the production of future ultrafine grained multiphase steels. The carbon addition promotes the formation of hard second phase particles such as cementite and austenite, which accelerates the ferrite recrystallization through the formation of local high strain regions near the hard particles during warm deformation.

Neutron diffraction has been applied as a powerful tool to investigate the microstructure evolutions of bulk materials during tensile/compressive deformation, heating/cooling and under other specific environmental conditions (te Velthuis, 1998; Tomota, 2005; Xu, 2006a; Xu, 2006b). Recently, TOF (*hkl*) multiple reflection spectra obtained by neutron diffraction have been analyzed to evaluate the crystallographic textures during forward and reverse diffusional phase transformations (Wenk, 2007) and the preferred orientations of ferrite and austenite in a 0.2C-2Mn steel before and after hot compression (Xu, 2009). The *in situ* microstructure and texture evolution during thermomechical controlled process will be studied further in order to well optimize the multiphase microstructure (Xu, 2012).

6. Acknowledgments

The authors thank Dr. Y. Adachi at National Institute for Materials Science, Japan for his support on the TEM microstructure observation. They also appreciate Dr. E.C. Oliver at ISIS Facility, Rutherford Appleton Laboratory, United Kingdom for his support on the neutron diffraction.

7. References

Ameyama, K.; Matsumura, N. & Tokizane, M. (1988). Ultrafine Austenite Grains Obtained by Thermomechanical Processing in Low and Medium Carbon Steels. *Journal of the Japan Society for Heat Treatment*, Vol.28, no.4, pp.233-240, ISSN 0288-0490.

Bao, Y.Z.; Adachi, Y.; Toomine, Y.; Suzuki, T.; Xu, P.G. & Tomota, Y. (2005a). Dynamic Recrystallization Behavior in Martensite in 18Ni, 17Ni-0.2C and SM490 Steels. *Tetsu-to-Hagané* (Journal of the Iron and Steel Institute of Japan), Vol.91, pp. 602-608, ISSN 0021-1575.

Bao, Y.Z.; Adachi, Y.; Toomine, Y.; Xu, P.G.; Suzuki, T. & Tomota, Y. (2005b). Dynamic Recrystallization by Rapid Heating Followed by Compression for a 17Ni-0.2C Martensite Steel. *Scripta Materialia*, Vol.53, pp. 1471-1476, ISSN 1359-6462.

Bunge, H.J. (1982) *Texture Analysis in Materials Science*. Butterworth & Co., ISBN 0-408-10642-5, London, pp.88-98.

Chen, S.C.; Tomota, Y.; Shiota, Y.; Toomine, Y. & Kamiyama, T. (2006). Measurements of Volume Fraction and Carbon Concentration of the Retained Austenite by Neutron Diffraction. *Tetsu-to-Hagané* (Journal of the Iron and Steel Institute of Japan), Vol.92, pp.557-561, ISSN 0021-1575.

Doherty, R.D.; Hughes, D.A.; Humphreys, F.J.; Jonas, J.J.; Juul Jensen, D.; Kassner, M.E.; King, W.E.; McNelley, T.R.; McQueen H.J. & Rollett, A.D. (1997). Current issues in recrystallization: a review. *Materials Science and Engineering A*, Vol.238, pp.219-274, ISSN 0921-5093.

Dong, H. & Sun, X.J. (2006). Deformation induced ferrite transformation in low carbon steels. *Current Opinion in Solid State and Materials Science*, Vol.9, pp.269-276, ISSN 1359-0286.

Enomoto, M. & Furubayashi, E. (1977). A Crystallographic Study of Austenite Formation from Fe-Ni Martensite during Heating in Alpha-Gamma Region. *Transactions of the Japan Institute of Metals*, Vol.18, pp.817-824, ISSN 0021-4434.

Furuhara, T.; Yamaguchi, T.; Furimoto, S. & Maki, T. (2007). Formation of Ferrrite+Cementite microduplex structure by warm deformation in high carbon steels. *Material Science Forum*, Vol.539-543, pp.155-160, ISSN 0255-5476.

Glovre, G. & Selllers, C.M. (1973). Recovery and recrystallization during high temperature deformation of α-iron. *Metallurgical Transactions*, Vol.4, pp. 765-775, ISSN 0026-086X.

Hayashi, T.; Torizuka, S.; Mitsui, T.; Tsuzaki, K. & Nagai, K. (1999). Creation of low-carbon steel bars with fully fine ferrite grain structure through warm grooved rolling. *CAMP-ISIJ* (Current Advances in Materials and Processes), Vol.12, pp.385-388, ISSN 1882-8922.

Hayashi, T. & Nagai, K. (2002). Inprovement of strength-ductility balance for low carbon ultrafine-grained steel through strain hardening design. *Transactions of the Japan Society of Mechanical Engineers. A*, Vol.68, pp.1553-1558, ISSN 0387-5008.

Hutchings, M.T.; Withers, P.J.; Holden, T.M. & Lorentzen, T. (2005). *Introduction to the Characterization of Residual Stress by Neutron Diffraction*, ISBN 0-415-31000-8, Taylor & Francis, New York, pp.230-238.

Larson, A.C. & Von Dreele, R.B. (2004). General Structure Analysis System (GSAS), *Los Alamos National Laboratory Report*. LAUR 86-748, pp.147-148.

Li, J.H.; Xu, P.G.; Tomota, Y. & Adachi, Y. (2008) Dynamic Recrystallization Behavior in a Low-carbon Martensite Steel by Warm Compression. *ISIJ International*, Vol.48, no.7, pp.1008-1013, ISSN 1485-1664.

Maki, T.; Okaguchi, S. & Tamura, I. (1982). Dynamic recrystallization in ferritic stainless steel. Strength of metals and alloys (ICSMA 6) : Proceedings of the 6th International Conference, pp.529-534., ISBN 0080293255. Melbourne, Australia, 16-20 August 1982.

Maki, T.; Furuhara, T.; & Tsuzaki, K. (2001). Microstructure Development by Thermomechanical Processing in Duplex Stainless Steel. *ISIJ International*, Vol..41. pp. 571-579. ISSN 1485-1664.

Miller, R.L. (1972). Ultrafine-grained microstructures and mechanical properties of alloy steels. *Metallurgical Transactions*, Vol.3, pp.905-912, ISSN 0026-086X.

Moriyama, M.; Takaki, S. & Kawagoishi, N. (2001). Influence of Reversion Austenite on Fatigue Property of 350 ksi Grade 18Ni Maraging Steel. *Journal of the Japan Society for Heat Treatment*, Vol.41. pp.266-271, ISSN 0288-0490.

Najafi-Zadeh, A.; Jonas, J.J. & Yue, S. (1992). Grain refinement by dynamic recrystallization during the simulated warm-rolling of interstitial free steels. *Metallurgical Transactions A*, Vol.23, pp.2607-2617. ISSN 0360-2133.

Ohmori, A. ; Torizuka, S. ; Nagai, K. ; Yamada K. & Kogo, Y. (2002). Evolution of ultrafine-grained structure through large strain-high Z deformation in a low carbon steel. *Tetsu-to-Hagané* (Journal of the Iron and Steel Institute of Japan), Vol.88, no. 12, pp.857-864, ISSN 0021-1575.

Ohmori, A.; Torizuka, S.; Nagai, K.; Yamada, K. & Kogo, Y. (2004). Effect of Deformation Temperature and Strain Rate on Evolution of Ultrafine Grained Structure through Single-Pass Large-Strain Warm Deformation in a Low Carbon Steel. *Materials Transactions*, Vol.45, pp.2224-2231, ISSN 1345-9678.

Poorganji, B.; Miyamoto, G.; Maki, T. & Furuhara, T. (2008). Formation of ultrafine grained ferrite by warm deformation of lath martensite in low-alloy steels with different carbon content. *Scripta Materialia*, Vol.59, pp.279-281, ISSN 1359-6462.

Reed, R.C. & Root, J.H. (1998). Determination of the Temperature Dependence of the Lattice Parameters of Cementite by Neutron Diffraction. *Scripta Materialia*, Vol.38, pp.95-99. ISSN 1359-6462.

Sawada, K.; Ohba, T.; Kushima, H. and Kimura, K. (2005). Effect of microstructure on elastic property at high temperatures in ferritic heat resistant steels. *Materials Science and Engineering A*, Vol.394, pp.36-42, ISSN 0921-5093.

te Velthuis, S.G.E.; Root, J.H.; Sietsma, J.; Rekveldt, M.T. & van der Zwaag, S. (1998). The ferrite and austenite lattice parameters of Fe-Co and Fe-Cu binary alloys as a function of temperature. *Acta Materialia*. Vol.46, pp.5223-5228, ISSN 1359-6454.

Tomota, Y.; Suzuki, T.; Kanie, A.; Shiota, Y.; Uno, M.; Moriai, A.; Minakawa, N. & Morii, Y. (2005). In situ neutron diffraction of heavily drawn steel wires with ultra-high strength under tensile loading. *Acta Materialia*. Vol.53, pp.463-467, ISSN 1359-6454.

Tomota, Y.; Narui, A. & Tsuchida, N. (2008). Tensile behavior of fine-grained steels, *ISIJ International*, Vol.48, pp.1107-1113, ISSN 1485-1664.

Torizuka, S. (2005) Production of ultrafine-grained steel bar and plate by high Z-large strain deformation in ferrite region. *Ferrum* (Bulletin of The Iron and Steel Institute of Japan), Vol.10, no.3, pp.188-195, ISSN 1341-688X.

Tsuji, N.; Matsubara, Y.; Saito, Y. & Maki, T. (1998). Occurance of Dynamic Recrystallization in ferritic Iron. *Journal of the Japan Institute of Metals*, Vol.62, pp.967-976. ISSN 0021-4876.

Tsuji, N.; Okuno, S.; Koizumi, Y. & Minamino, Y. (2004). Toughness of Ultrafine Grained Ferritic Steels Fabricated by ARB and Annealing Process. *Materials Transactions*, Vol. 45, no.7, pp.2272-2281, ISSN 1345-9678.

Ueji, R.; Tsuji, N.; Minamino, Y.; & Koizumi, Y. (2002). Ultragrain refinement of plain low carbon steel by cold-rolling and annealing of martensite. *Acta Materialia*, Vol.50, pp.4177-4189, ISSN 1359-6454.

Wenk, H.R.; Huensche, I. & Kestens, L. (2007). In-Situ Observation of Texture Changes during Phase Transformations in Ultra-Low-Carbon Steel. *Metallurgical and Materials Transactions A*, Vol.38, pp.261-267, ISSN 1073-5623.

Xu, P.G.; Tomota, Y.; Lukas, P.; Muransky, O. & Adachi, Y. (2006a). Austenite-to-ferrite transformation in low alloy steels during thermomechanically controlled process studied by in situ neutron diffraction. *Materials Science and Engineering A*, Vol.435, pp.46-53, ISSN 0921-5093.

Xu, P.G. and Tomota, Y. (2006b). Progress in materials characterization technique based on in situ neutron diffraction. *Acta Metallurgica Sinica*, Vol.42, pp.681-688, ISSN 0412-1961.

Xu, P.G.; Li, J.H.; Tomota, Y. and Adachi, Y. (2007). Effect of Carbon Addition on Ultrafine Grained Microstructure Formation by Warm Compression for Fe-18Ni Alloys, *Materials Science Forum*, Vol.558-559, pp.601-606, ISSN 0255-5476.

Xu, P.G.; Li, J.H.; Tomota, Y. & Adachi, Y. (2008a). Effects of Volume Fraction and Carbon Concentration of Austenite on Formation of Ultrafine Grained Ferrite/Austenite Duplex Microstructure by Warm Compression, *ISIJ International*, Vol.48, no.11, pp.1609-1617, ISSN 1485-1664.

Xu, P.G.; Tomota, Y. & Oliver, E.C. (2008b). Dynamic Recrystallization and Dynamic Precipitation Behaviors of a 17Ni-0.2C Martensite Steel Studied by *In Situ* Neutron Diffraction, *ISIJ International*, Vol.48, no.11, pp.1618-1625. ISSN 1485-1664.

Xu, P.G.; Tomota,Y.; Suzuki, T.; Yonemura, M. & Oliver, E.C. (2009). In Situ TOF Neutron Diffraction for Isothermal Ferrite Transformation during Thermomechanically Controlled Process of Low Alloy Steel. *Netsu Shori* (Journal of the Japan Society for Heat Treatment), Vol.49. special issue, pp.470-473. ISSN 0288-0490.

Xu, P.G.; Tomota, Y.; Vogel, S.C.; Suzuki, T.; Yonemura, M. & Kamiyama, T. (2012) Transformation Strain and Texture Evolution during Diffusional Phase Transformation of Low Alloy Steels Studied by Neutron Diffraction, *Reviews on Advanced Materials Science*. Vol.33. (in press). ISSN 1605-8127.

6

The Deformability and Microstructural Aspects of Recrystallization Process in Hot-Deformed Fe-Ni Superalloy

Kazimierz J. Ducki
Silesian University of Technology
Poland

1. Introduction

The behaviour of metals and alloys during hot plastic working has a complex nature and it varies with the changing of such process parameters as (Zhou et al., 1994): deformation, strain rate and temperature. The high-temperature plastic deformation is coupled with dynamic recovery and recrystallization processes which influencing the structure and properties of alloys. One of crucial issues is finding the relationship between the hot plastic deformation process parameters, microstructure and properties. Since the sixties of the last century, theoretical and experimental investigations have been carried out to find those interdependence for steels and nickel alloys.

In the recent years, the constitutive equations describing hot plastic deformation processes have started to take into account the so-called internal variables determining the material condition. These variables include substructural parameters such as (Hansen, 1998; Sellars, 1998): grain size, grain shape, recrystalized volume fraction, dislocation density, subgrain size, subgrain misorientation angle and stacking fault energy (SFE). Determination of the above-mentioned parameters of a deformed material structure description requires the application of analytical methods primarily based on quantitative metallography and transmission electron microscopy (TEM). Taking those substructure parameters into consideration in calculations should enable the correct modelling of structural phenomena during hot plastic deformation and enhance the technological processes control for the purpose of obtaining the assumed structures of required properties (McQeen et al., 2002).

The Fe-Ni superalloys precipitation hardened by intermetallic phase of γ' - $Ni_3(Al,Ti)$ type are one of the groups of construction materials intended for operation in cryogenic and elevated temperatures. These alloys are difficult to deform and are characterized by high values of yield stress at a high temperature. High deformation resistance of Fe-Ni alloys is caused by a complex phase composition, high activation energy of the hot plastic deformation process and a low rate of dynamic recrystallization. When choosing the conditions for hot plastic working of Fe-Ni alloys, the following factors should be considered (Bywater et al., 1976; Kohno et al., 1981; Ducki et al., 2006): the matrix grain size, plastic deformation parameters and the course of the recrystallization process. The grain size is of particular importance. Grain refining leads to an increased rate of recovery and dynamic recrystallization and to a smaller diameter of recrystallized grains. This is important, for the grain refinement in Fe-Ni superalloys has an advantageous influence on increasing their yield point and fatigue strength (Koul et al., 1994; Härkegård et al., 1998).

In the presented study, research has been undertaken on the influence of the initial microstructure of austenite and the parameters of hot plastic working on deformability, grain and subgrain size, and dislocation density in a high-temperature creep resisting Fe-Ni alloy. It is assumed that the results obtained will be used for optimizing hot plastic working processes and forecasting the microstructure and functional properties of products made of Fe-Ni superalloys.

2. Material and methodology

The examinations were performed on rolled bars, 16 mm in diameter, of an austenitic A-286 type alloy. The chemical composition is given in Tab. 1.

Content of an element [wt. %]															
C	Si	Mn	P	S	Cr	Ni	Mo	V	W	Ti	Al	Co	B	N	Fe
0.05	0.56	1.5	0.026	0.016	14.3	24.5	1.35	0.42	0.10	1.88	0.16	0.08	0.007	0.0062	55.3

Table 1. Chemical composition of the investigated Fe-Ni superalloy

In order to model the conditions of alloy heating prior to plastic processing, the investigations were carried out on samples after initial soaking at high temperatures. Sections of rolled bars, which the samples for investigations were made of, were subjected to two variants of preheating, i.e. 1100°C/2h and 1150°C/2h with subsequent cooling in water. Heat treatment of this type corresponds to the soaking parameters of the investigated superalloy before hot plastic processing (Kohno et al., 1981).

The research on the alloy deformability was performed in a hot torsion test on a Setaram torsion plastometer 7 MNG. The plastometric tests were performed every 50°C in a temperature range of 900÷1150°C, with a constant holding time of 10 minutes at the defined temperature. Solid cylindrical specimens (Ø 6.0 × 50 mm) were twisted at a rotational speed of 50 and 500 rpm, which corresponds to the strain rate of 0.1 and 1.0 s^{-1}, respectively. To freeze the structure, the specimens after deformation until failure were directly rapid cooled in water (Fig. 1).

The data obtained in the plastometric torsion test were entered in an Excel spreadsheet in the form of columns containing the recorded values. Processing of the measured data by means of filtration, cutting, shrinking and planishing was conducted using the Matlab 6 program. A correction of the torque moment, due to diversified rotational speed values and increase of the sample temperature during torsion, was calculated by the method of joint action of speed and temperature from the following relations (Hadasik, 2005):

$$M''' = M + \Delta M''' \tag{1}$$

$$\Delta M''' = M(N, \dot{N}, T) - M(N, \dot{N}_r, T + \Delta T) \tag{2}$$

$$\Delta M''' = A \cdot N^B \cdot \exp(C \cdot N) \cdot \dot{N}^{D+\frac{E}{T}} \cdot \exp\left(\frac{F}{T}\right) - A \cdot N^B \exp(C \cdot N) \cdot \dot{N}^{D+\frac{E}{T+\Delta T}} \cdot \exp\left(\frac{F}{T+\Delta T}\right) \quad (3)$$

where: M – recorded torque moment [Nm]; M''' – corrected torque moment value [Nm]; $\Delta M'''$ – torque moment correction taking account of a joint action of speed and temperature [Nm]; N – number of sample torsion, $\dot{N}$ – given torsion speed [rpm]; $\dot{N}_r$ – recorded rotational speed [rpm]; T – deformation temperature [°C]; ΔT – temperature increment during torsion [°C]; A, B, C, D, E, F – material constans.

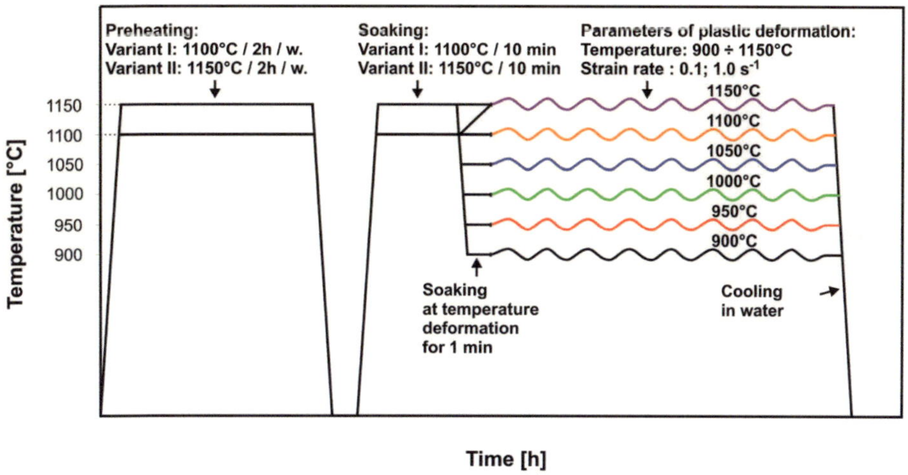

Fig. 1. Scheme of plastometric examination of the Fe-Ni alloy

The corrected data constituted a basis for the determination of equivalent deformation ε as a function of the number of the sample's rotations during torsion (Hadasik, 2005):

$$\varepsilon = \frac{2}{\sqrt{3}} \times \operatorname{arcsin} h\left(\frac{\pi \bar{R} N}{L}\right) \quad (4)$$

where: $\bar{R}$ equivalent radius corresponding to 2/3 of the outer radius R of the sample, L – measured sample length.

Yield stress σ_p was determined according to relation (5) taking account of the corrected torque moment M''', sample radius R, parameters m, p and axial force F_o (Hadasik, 2005):

$$\sigma_p = \left[\left(\frac{\sqrt{3} \cdot M'''}{2\pi R^3}\right)^2 \times (3+p+m)^2 + \left(\frac{F}{\pi R^2}\right)\right]^{0,5} \quad (5)$$

where: p – parameter reflecting stress sensitivity to deformation size; m – parameter reflecting stress sensitivity to deformation rate.

On the flow curves determined, the following parameters characterising plastic properties of the alloy in the torsion test were defined:

- σ_{pp} – maximum yield stress on the flow curve;
- ε_p – deformation corresponding to the maximum yield stress;
- σ_f – stress at which the sample is subject to failure;
- ε_f – deformation at which the sample is subject to failure, the so-called threshold deformation.

Relations between the yield stress and alloy deformation, and the deformation conditions were described using the Zener-Hollomon parameter Z (Zener et al., 1944):

$$Z = \dot{\varepsilon} \times \exp\left(\frac{Q}{RT}\right) = A \times \left[\sinh\left(\alpha\sigma_{pp}\right)\right]^n \tag{6}$$

where: $\dot{\varepsilon}$ – strain rate, Q – activation energy of the hot plastic deformation process, R – molar gas constant, T – temperature, and A, α, n – constants depending on grade of the investigated material.

The activation energy of the hot plastic deformation process Q was determined in accordance with the procedure specified in the work by (Schindler et al., 1998). The solution algorithm consisted in transforming equation (6) to the following form:

$$\dot{\varepsilon} = A \times \exp\left(\frac{-Q}{RT}\right)\left[\sinh(\alpha\sigma_{pp})\right]^n \tag{7}$$

Further procedure was based on solving equation (7) by a graphic method with the application of a regression analysis.

Structural inspections were performed on longitudinal microsections taken from the plastically deformed samples until failure after so-called "freezing" (Fig. 2). The specimens were etched using a reagent: 54 cm³ HF, 8 cm³ HNO_3 and 38 cm³ distilled H_2O. Due to the deformation inhomogeneity, microscopic observation was conducted in a representative region located at a distance of ca. 0,65÷0,75 of the specimen radius.

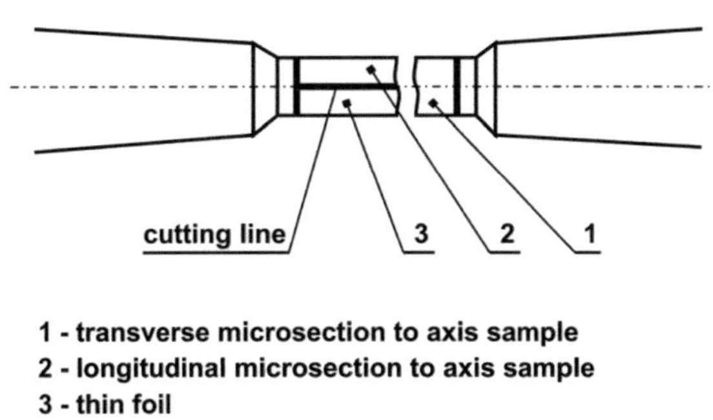

Fig. 2. Scheme of material cutting for metallographic microsections and thin foils from plastometric samples

A quantitative analysis of the investigated structures was carried out by means of a computer program MET-ILO v. 3.0 (Szala, 1997). For the analyzed microstructures, in accordance with the methodology presented in paper (Cwajna et al., 1993), the following stereological parameters were determined:

- average area of grain plane section $\bar{A}$ [μm²]:

$$\overline{A} = \frac{1}{N_A} \quad (8)$$

where: N_A – average grains number on area unit [μm⁻²];

- variability coefficient of the grain plane section area $\nu(A)$:

$$\nu(A) = \frac{S(A)}{\bar{A}} \times 100\,[\%] \quad (9)$$

where: $S(A)$ – empirical standard deviation of grain section area;

- volume fraction of dynamically recrystallized grains in the structure V_V [%];
- grain elongation coefficient δ (Feret coefficient):

$$\delta = \frac{F_x}{F_y} \quad (10)$$

where: $F_{x,\,y}$ – Feret diameters in x and y axes direction;

- classical, dimensionless shape coefficient ξ:

$$\xi = \frac{4\pi\overline{A}}{P^2} \quad (11)$$

where: P – perimeter of grain plane section.

The examination of the substructure was carried out by means of a JEM-100B Joel transmission electron microscope. Direct measurements on the TEM micrographs allowed the calculation of the structural parameters: the average subgrain area $\bar{A}$, and the mean dislocation density ρ. The mean subgrain areas were determined by a planimetric method making use of a semi-automatic image analyser MOP AMO 3 type. The measurements were conducted on the TEM images. The analysed microsections of thin foils involved measurements of about 150 subgrains for each sample. The mean dislocation density was calculated by use of a method based on counting the inter-section points of a network superimposed over the micrograph with dislocation lines. The dislocation density ρ as calculated for the thin foils according to the relation (Klaar et al., 1992):

$$\rho = \frac{x \cdot (n_1 / l_1 + n_2 / l_2)}{t} \quad [\mathrm{m}^{-2}] \quad (12)$$

where: x – a coefficient which defines the fraction of invisible dislocations with Burgers vectors *a/2<111>* for the A1 structure: $x = 2$ for image of dislocations observed in (111) reflex, $x = 1.5$ for image of dislocations observed in (200) reflex, $x = 1.5$ for image of dislocations observed in (220) reflex; $l_{1(2)}$ – the total length of the horizontal (vertical) lattice lines; $n_{1(2)}$ – the number of intersections of the horizontal (vertical) lattice lines with dislocations; t – the thickness of the foil.

The thickness of the foil in the investigated areas can be approximately calculated following the formula (Head et al., 1973):

$$t = n \bullet \zeta_{hkl} \tag{13}$$

where: n - a number of extinction lines; ζ_{hkl} - a value of extinction.

The values of ζ_{hkl} given by (Head et al., 1973) must be considered as a rough estimation of the actual value of extinction for the investigated material.

3. Results and discusion

3.1 Initial microstructure of the alloy

The application of two variants of initial solution heat treatment simulating the heating conditions allowed diversifying significantly the initial microstructure of the Fe-Ni alloy before hot plastic deformation. The alloy in its initial state differed primarily in the average grain size. After solution heat treatment under the conditions of 1100°C/2h/w., in the alloy microstructure presence was found of twinned austenite with medium-size grain ($\bar{A}$ = 2120 μm^2) with a small amount (ca. 0.3 wt.%) of insoluble particles (Fig. 3a). The increasing of the solution heat treatment temperature to 1150°C at an analogous soaking time resulted in an increase of the austenite grain ($\bar{A}$ = 6296 μm^2) and a reduction in the quantity and size of undissolved primary particles (Fig. 3b). For the analyzed variants of solution heat treatment, the microstructure of the samples was characterized by equiaxial grains, as evidenced by the elongation coefficient δ approximate to 1 and the dimensionless shape factor ξ of approximately 0.6.

Presence of titanium compounds, such as TiC carbide, $TiN_{0.3}$ nitride, $TiC_{0.3}N_{0.7}$ carbonitride, $Ti_4C_2S_2$ carbosulfide and Lavesa Ni_2Si phase and boride MoB was disclosed in the phase composition of the undissolved particles (Ducki, 2010).

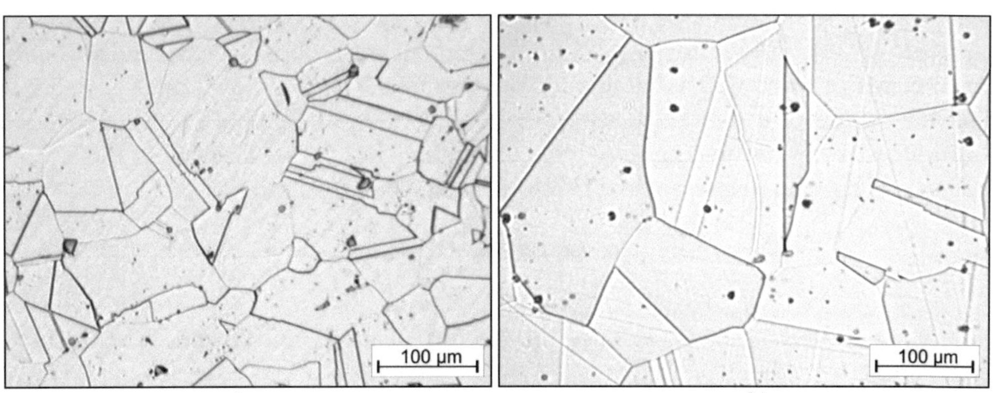

a) b)

Fig. 3. Diversified microstructure of alloy after solution heat treatment at: a) 1100°C/2 h/w., $\bar{A}$ = 2120 μm^2; b) 1150°C/2 h/w., $\bar{A}$ = 6296 μm^2

3.2 Deformability of the alloy

Knowledge of the phenomena occurring during hot deformation of materials enables selecting the correct conditions for plastic working and shaping their material characteristics. The results of the plastometric investigations, in the form of the calculated alloy flow curves at temperatures of 900÷1150°C for two options of initial soaking are shown in Fig. 4 and 5. The curves obtained for the option of initial soaking at 1100°C/2h and strain rate 0.1 s^{-1} have a shape characteristic of a material in which dynamic recovery and recrystallization take place (Fig. 4). High deformation values were obtained for the alloy in a wide range of torsion temperatures, i.e. 950÷1100°C. An increase of strain rate to 1.0 s^{-1} results in a significant increase of yield stress values and a distinct decrease of the alloy deformability at all temperatures analysed. This phenomenon can be explained by a higher alloy consolidation rate as well as too slow removal of the reinforcement as a result of dynamic recovery and recrystallization.

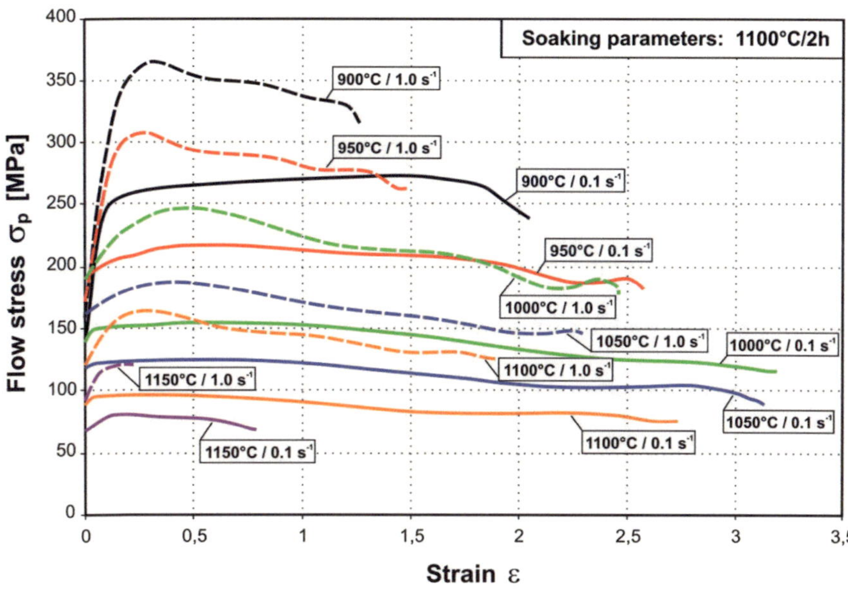

Fig. 4. The effect of deformation temperature on the flow stress of Fe-Ni alloy after initial soaking at 1100°C/2 h. Strain rate: 0.1 s^{-1} and 1.0 s^{-1}

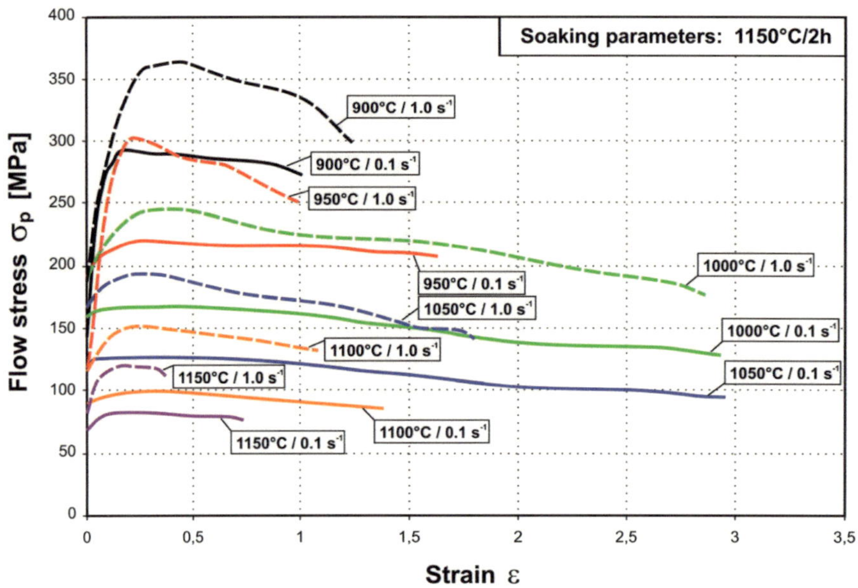

Fig. 5. The effect of deformation temperature on the flow stress of Fe-Ni alloy after initial soaking at 1150°C/2h. Strain rate: 0.1 s^{-1} and 1.0 s^{-1}

An increase of the initial soaking temperature to 1150°C/2h significantly reduces the alloy deformability for the two strain rates, both at low and high deformation temperatures (Fig. 5). In this case, fairly high deformation values were obtained for the alloy in a narrow range of torsion temperatures, i.e. 1000÷1050°C. Such behavior of the alloy may be explained by a larger growth of austenite grains at this soaking temperature and, consequently, lower recovery and dynamic recrystallization rates.

The values determined for the maximum yield stress σ_{pp}, maximum deformation ε_p, stress until failure σ_f and threshold deformation ε_f depending on the temperature and strain rate are presented in Figs. 6-9. For the option of initial soaking at 1100°C/2h and torsion speed of 0.1 s^{-1}, the alloy under discussion shows a contiuous drop of σ_{pp} from values 277 MPa at a temperature of 900°C to the value of 81 MPa at 1150°C (Fig. 6). The threshold deformation ε_f rises initially together with the torsion temperature, reaching the maximum of (3.19/3.14) at 1000÷1050°C, and then falls (Fig. 8). An increase of the strain rate to 1.0 s^{-1} results in an increase of σ_{pp} to maximum values of 367 MPa at the temperature of 900°C (Fig. 6) and a decrease of the threshold deformation to the maximum of 2.47/2.34 at 1000÷1050°C (Fig. 8).

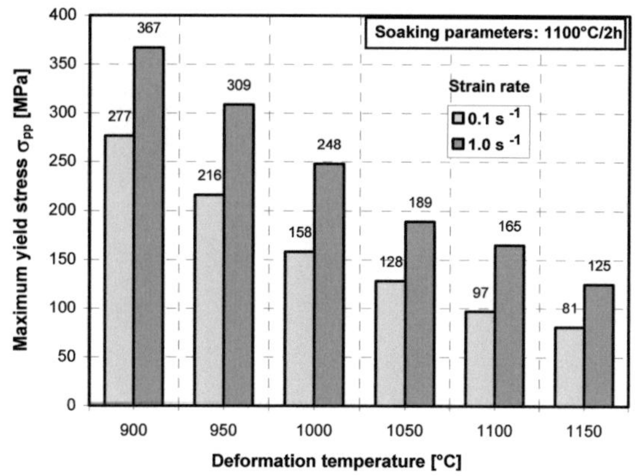

Fig. 6. The effect of deformation conditions on maximum flow stress. Initial alloy soaking: 1100°C/2 h

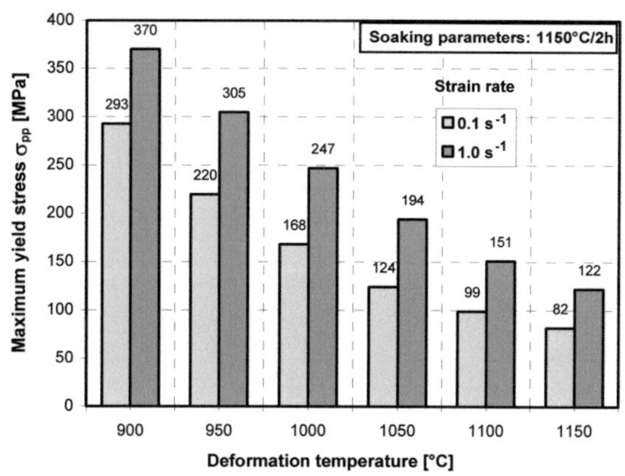

Fig. 7. The effect of deformation conditions on maximum flow stress. Initial alloy soaking: 1150°C/2 h

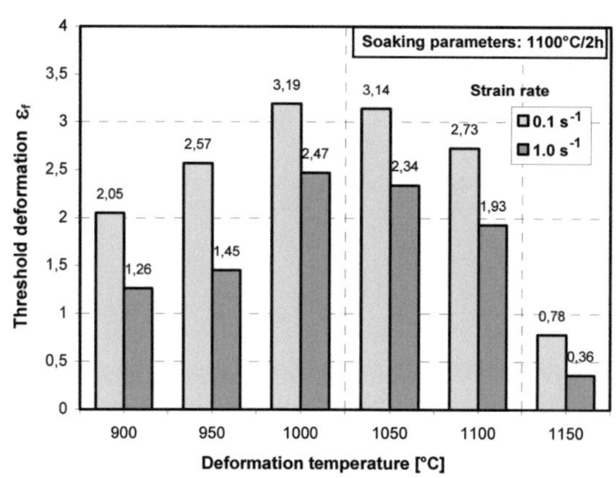

Fig. 8. The effect of deformation conditions on threshold deformation of the alloy. Initial soaking: 1100°C/2 h

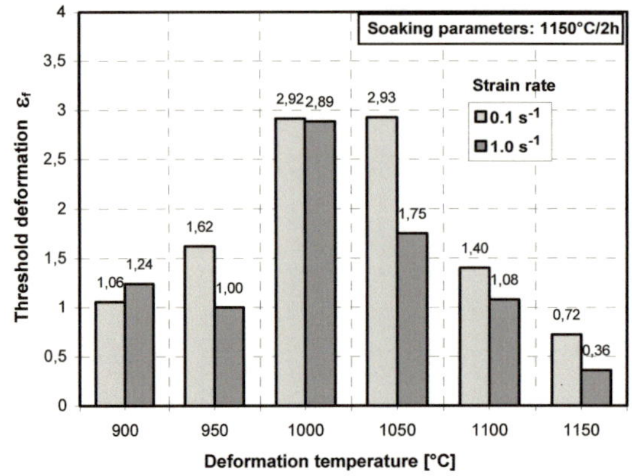

Fig. 9. The effect of deformation conditions on threshold deformation of the alloy. Initial soaking: 1150°C/2 h

An increase of the alloy initial soaking temperature to 1150°C/2h at a strain rate of 0.1 s^{-1} results in a slight increase of σ_{pp} to maximum values of 293 MPa at 900°C (Fig. 7) and decrease of ε_f to the maximum of 2.92÷2.93 in the range of 1000÷1050°C (Fig. 9). An increase of the torsion speed to 1.0 s^{-1} results in further increase of the σ_{pp} value to maximum values of 370 MPa at 900°C (Fig. 7), and decrease of ε_f to the maximum values of 2.89/1.75 at the temperature of 1000÷1050°C (Fig. 9).

The activation energy of the hot plastic deformation process Q was calculated by the means of a computer programme ENERGY 3.0 (Schindler et al., 1998). The activation energy, Q, necessary to initiate dynamic recrystallization in the Fe-Ni alloy, was determined on the basis of the linear dependencies presented in Fig. 10.

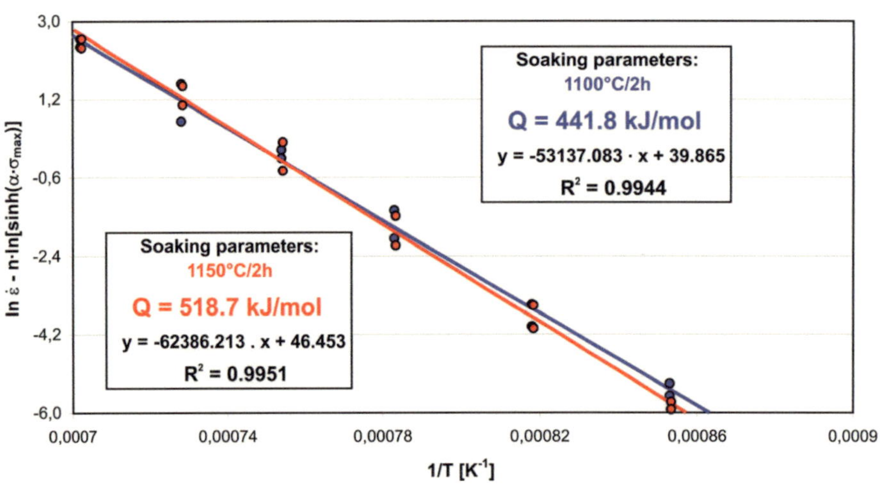

Fig. 10. The plot for determination of the activation energy for hot plastic deformation of the Fe-Ni alloy. Initial alloy soaking: 1100°C/2 h and 1150°C/2 h

The activation energy, Q, of hot plastic deformation for the Fe-Ni alloy depends on the temperature of initial soaking and equals as follows:

- $Q = 441.8$ [kJ/mol] – for initial alloy soaking 1100°C/2 h;
- $Q = 518.7$ [kJ/mol] – for initial alloy soaking 1150°C/2 h.

The higher value of the activation energy, Q, of hot plastic deformation for the alloy after initial soaking at 1150°C/2h can be justified by higher values of the maximum yield stress σ_{pp}, a larger growth of the initial austenite grain and a higher degree of matrix saturation with alloying elements.

The dependencies between maximum yield stress σ_{pp} and Zener Hollomon Z parameter are presented in Fig. 11. For both options of initial soaking, a power dependence ($R^2 = 0{,}98$) of the alloy yield stress was obtained as a function of the Z parameter. So determined function dependencies between the maximum yield stress σ_{pp} and the Z parameter had a form of power function:

- for the alloy after initial soaking 1100°C/2 h:

$$\sigma_{pp} = 0.43 \times Z^{0.151} \text{ [MPa]} \tag{14}$$

- for the alloy after initial soaking 1150°C/2 h:

$$\sigma_{pp} = 0.34 \times Z^{0.133} \text{ [MPa]} \tag{15}$$

Higher values of the Z parameter for the alloy after initial soaking at 1150°C/2h result from higher values of the plastic deformation activation energy Q.

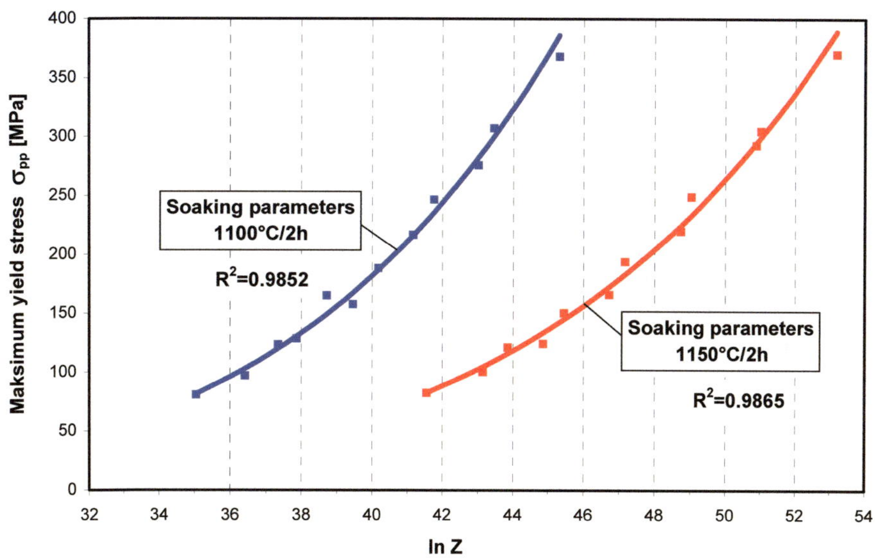

Fig. 11. Dependence of the maximum flow stress σ_{pp} on the Zener-Hollomon parameter Z. Initial alloy soaking: 1100°C/2 h and 1150°C/2 h

3.3 Microstructure of hot-deformed alloy

The recovery and dynamic recrystallization which occur in the Fe-Ni alloy during hot plastic deformation affect the size of the austenite grain. The results of investigations of the alloy microstructure after initial soaking at 1100°C/2h and deformation in a temperature range 900÷1150°C and a strain rate of 0.1 and 1.0 s^{-1} are presented in Figs. 12a-d.

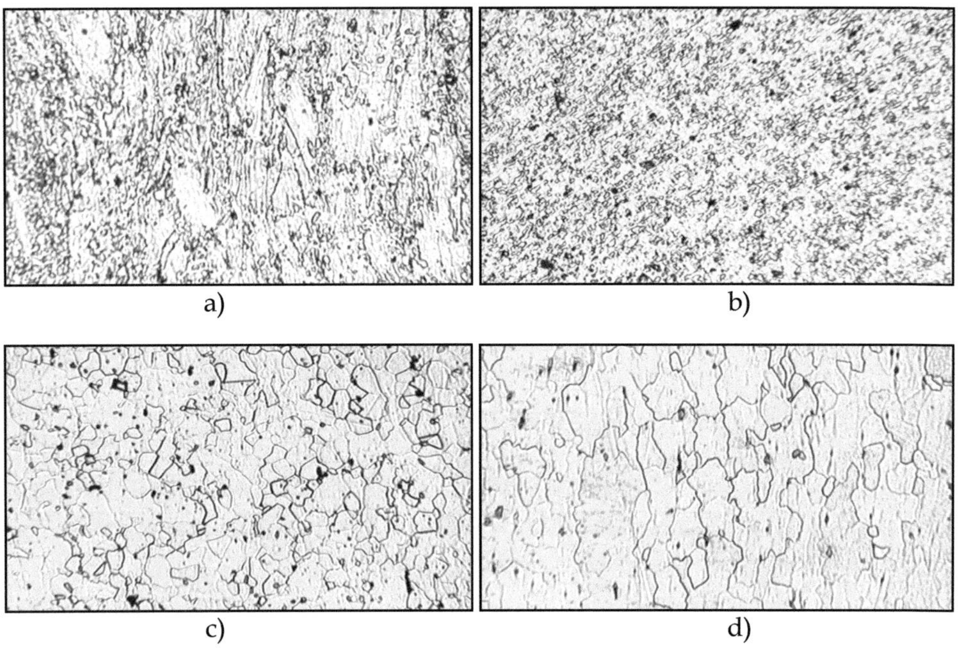

Fig. 12. The microstructure of the alloy after plastic deformation at: a) 900°C/1.0 s^{-1}, b) 950°C/0.1 s^{-1}, c) 1100°C/1.0 s^{-1}, d) 1150°C/0.1 s^{-1}. Initial soaking: 1100°C/2h

After deformation at 900°C for both of the torsion rates applied, the alloy structure is not completely recrystallized, which is indicated by the presence of primary elongated grains and fine recrystallized grains (Fig. 12a). At a torsion temperature in the range of 950÷1100°C, the alloy structure consisted of dynamically recrystallized grains (Fig. 12b and 12c). With an increasing deformation temperature, a gradual growth of the recrystallized grains was observed. The recrystallized grains were characterized by a deformed grain boundary line, which indicates an extensive cumulated deformation in the specimens. At the highest torsion temperature, 1150°C, some elongated grains of dynamically recrystallized austenite are observed in the alloy structure (Fig. 12d).

An increase of the initial soaking temperature of the alloy to 1150°C/2h results in increasing the initial austenite grain size and decreasing the kinetics of dynamic recovery and recrystallization for both of the analyzed strain rates (Figs. 13a-d). After deformation at 900°C/0.1 s^{-1} and 900÷950°C/1.0 s^{-1}, the alloy microstructure was not completely recrystallized and it was composed of deformed primary grains and dynamically recrystallized grains of small sizes (Fig. 13a and 13b). The new, fine recrystallized grains nucleated at primary grain boundaries, creating the so-called "necklace". Within the deformation temperature range from 950÷1000 to 1100°C, the alloy structure is fine-grained and completely dynamically recrystallized (Fig. 13c). The highest deformation temperature of 1150°C induced deformation localization, as evidenced by the elongated dynamically recrystallized grains of varying sizes (Fig. 13d).

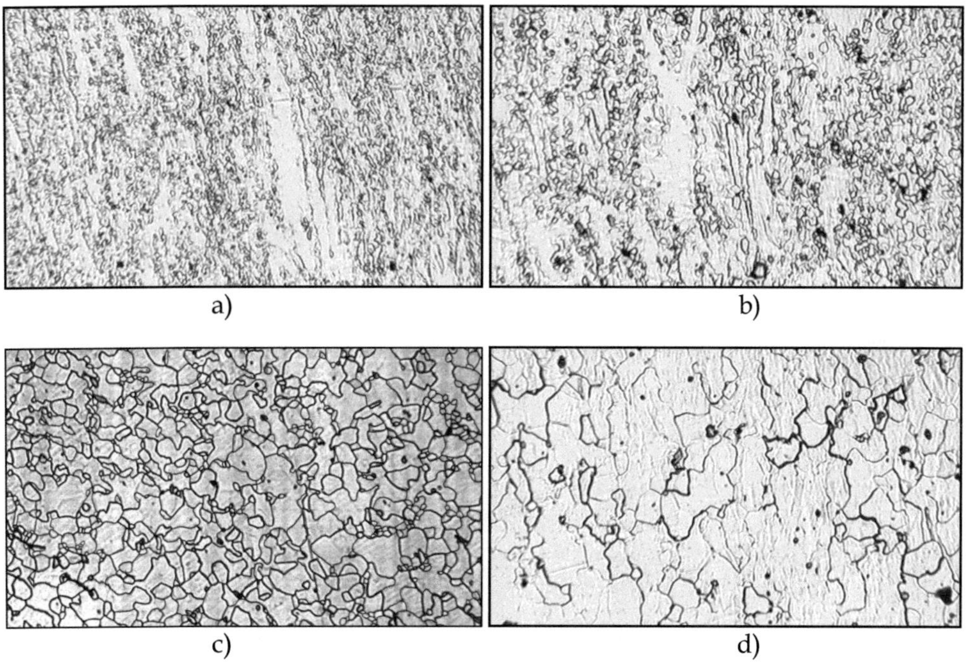

Fig. 13. The microstructure of the alloy after plastic deformation at: a) 900°C/1.0 s^{-1}, b) 950°C/0.1 s^{-1}, c) 1050°C/1.0 s^{-1}, d) 1150°C/0.1 s^{-1}. Initial soaking: 1150°C/2h

The results of a quantitative evaluation of the structure after initial soaking and deformation until failure in a temperature range of 900-1150°C and a strain rate of 0.1 and 1.0 s^{-1} are presented in Figs. 14-17. In the structure of the alloy after initial soaking 1100°C/2h and deformation in the investigated temperature range at a strain rate 0.1 s^{-1}, monotonous growth of the grain average area $\bar{A}$ is observed from a value 16 μm² at 900°C to 198 μm² at 1150°C (Fig. 14). Up to the deformation temperature of 1100°C, the dynamically recrystallized grains are approximately equiaxial (δ = 0.99÷1.12), whereas at the highest torsion temperature, 1150°C, they are elongated (δ = 1.31) (Fig. 15). An increase of the strain rate to 1.0 s^{-1} induces a certain reduction of the recrystallized grain size.

Also, in this case, in the investigated range of deformation temperatures, a monotonous growth of the grain average area $\bar{A}$ was observed, from 12 μm² at 900°C to 87 μm² at 1100°C (Fig. 14). In the analyzed range of torsion temperatures of 900÷1100°C, the dynamically recrystallized grains are approximately equiaxial (δ= 1.00÷1.09) (Fig. 15).

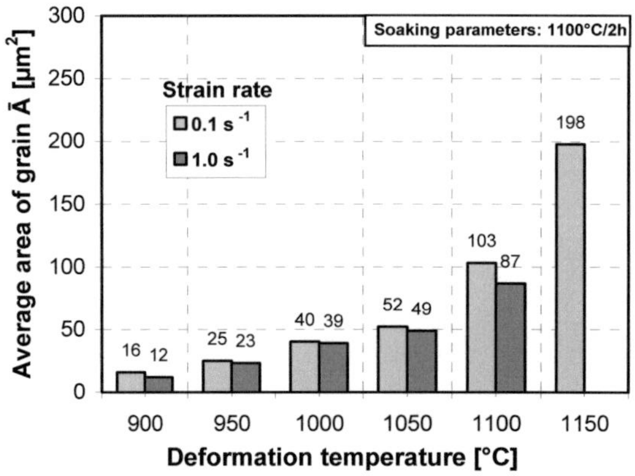

Fig. 14. The effect of deformation temperature on the average area of recrystallized grain after torsion at a rate of 0.1 and 1.0 s^{-1}. Initial soaking: 1100°C/2h

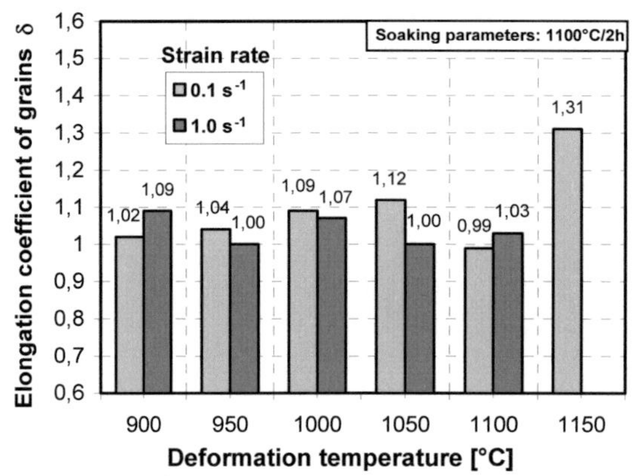

Fig. 15. The effect of deformation temperature on the elongation coefficient of recrystallized grain after torsion at a rate of 0.1 and 1.0 s^{-1}. Initial soaking: 1100°C/2h

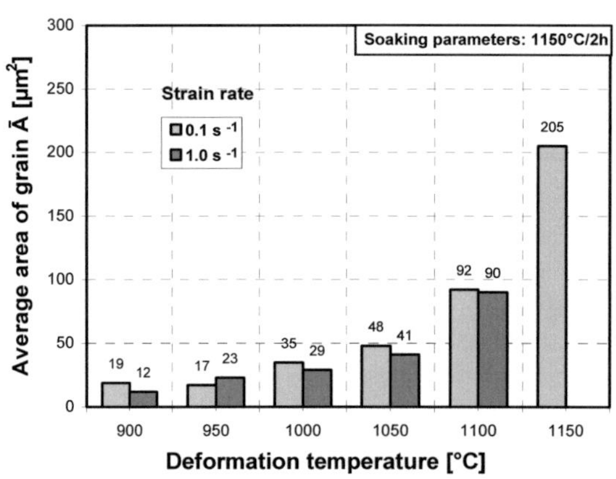

Fig. 16. The effect of deformation temperature on the average area of recrystallized grain after torsion at a rate of 0.1 and 1.0 s^{-1}. Initial soaking: 1150°C/2h

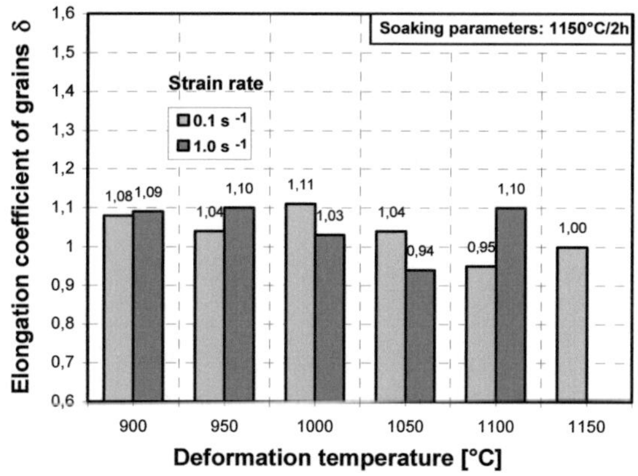

Fig. 17. The effect of deformation temperature on the elongation coefficient of recrystallized grain after torsion at a rate of 0.1 and 1.0 s^{-1}. Initial soaking: 1150°C/2h

An increase of the initial soaking temperature to 1150°C/2h and alloy deformation within the range of 900÷1150°C at a rate of 0.1 s^{-1} induces a similar reduction in size of the grain plane section area from 19 μm^2 at 900°C to 205 μm^2 at 1150°C (Fig. 16). An increase of the strain rate of samples to 1.0 s^{-1} within the deformation temperature range of 900÷1100°C causes a further reduction in the grain average area within the range from 12 to 90 μm^2. For both strain rates, the grains after dynamic recrystallization are approximately equiaxial (δ = 0.94÷1.11) (Fig. 17).

A comparison of the size of recrystallized grain in the Fe-Ni alloy after deformation at a strain rate of 0.1 and 1.0 s^{-1} for two variants of initial soaking at 1100°C/2 h and 1150°C/2 h is presented in Fig. 18 and 19. After deformation at a strain rate of 0.1 s^{-1} within the temperature range of 900÷1150°C for both variants of initial soaking, a similar grain size $\bar{A}$ was obtained in the range from 16 to 205 μm^2 (Fig. 18). A higher strain rate of 1.0 s^{-1} within the temperature range of 900÷1100°C allows obtaining a slightly higher degree of grain refining in the range from 12 to 90 μm^2 (Fig. 19). Thus, it can be concluded that the initial microstructure of an alloy after initial soaking has no significant influence on the size of recrystallized grain after plastic deformation.

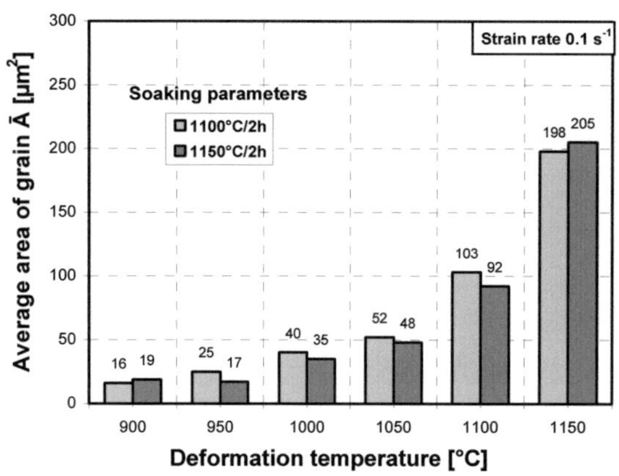

Fig. 18. The effect of deformation temperature on the average area of recrystallized grain after initial soaking of the alloy at 1100°C/2 h and 1150°C/2 h

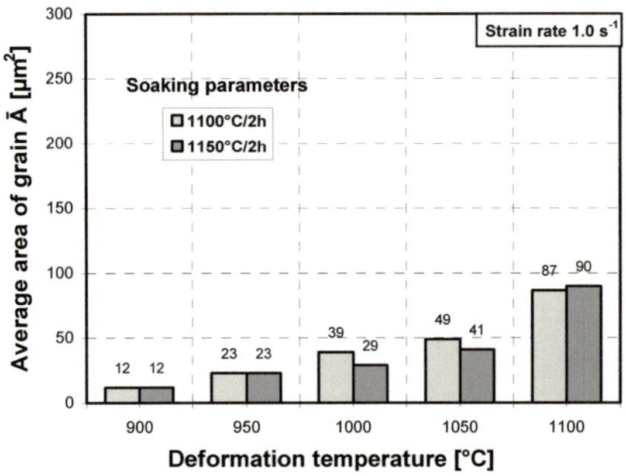

Fig. 19. The effect of deformation temperature on the average area of recrystallized grain after initial soaking of the alloy at 1100°C/2 h and 1150°C/2 h

The average grain plane section area of samples deformed under the same conditions but with a different initial grain size after initial soaking is similar. The average size of recrystallized austenite grain depends mainly on the deformation temperature and, to a lesser degree, on the strain rate applied for the alloy (Fig. 20 and 21).

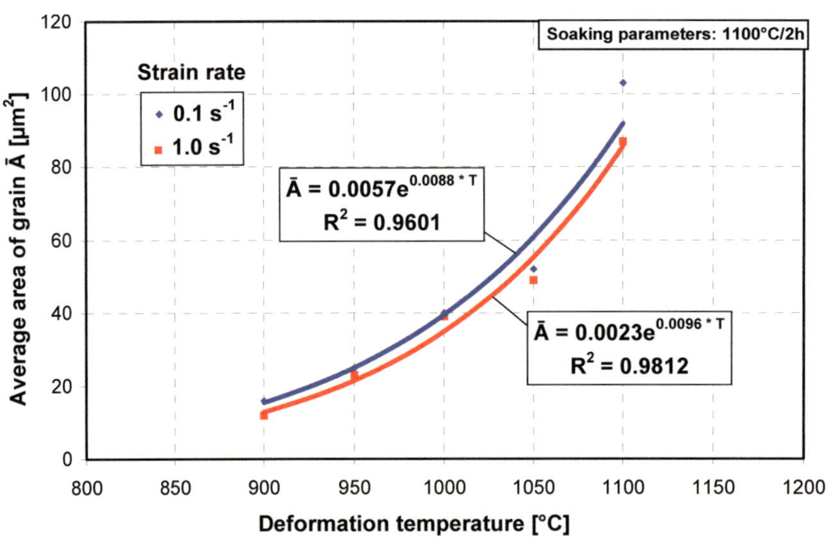

Fig. 20. Relationship between the average grain area after recrystallization versus deformation temperature and strain rate. Initial alloy soaking: 1100°C/2 h

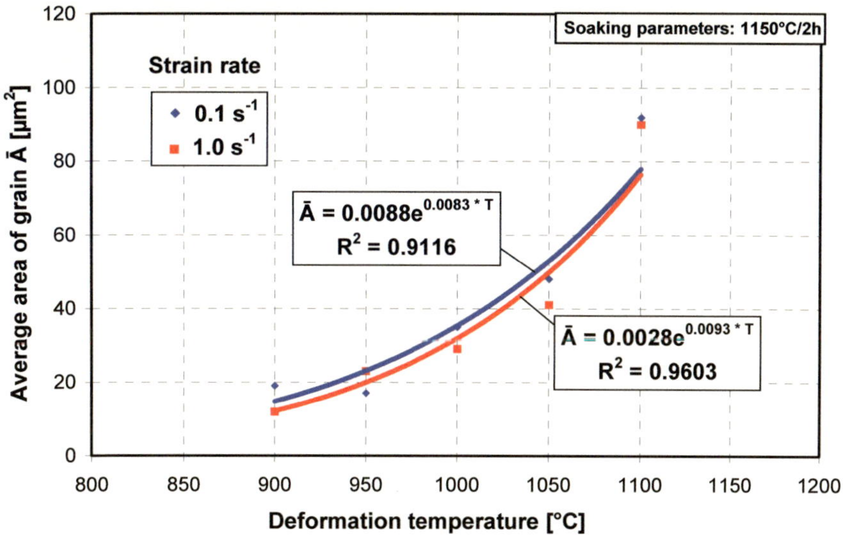

Fig. 21. Relationship between the average grain area after recrystallization versus deformation temperature and strain rate. Initial alloy soaking: 1150°C/2 h

An assessment of the degree of influence of the deformation temperature and strain rate on the average grain plane section area of the recrystallized Fe-Ni alloy was obtained after introducing the Zener-Hollomon Z parameter. As appears from the dependencies developed, the fine-grained microstructure of the alloy after deformation was obtained more easily after initial soaking at a temperature of 1100°C/2h when compared to soaking at 1150°C/2 h (Fig. 22). This is evidenced by lower Z parameter values for the deformation after initial soaking at 1100°C/2 h: a low strain rate and a moderate deformation temperature.

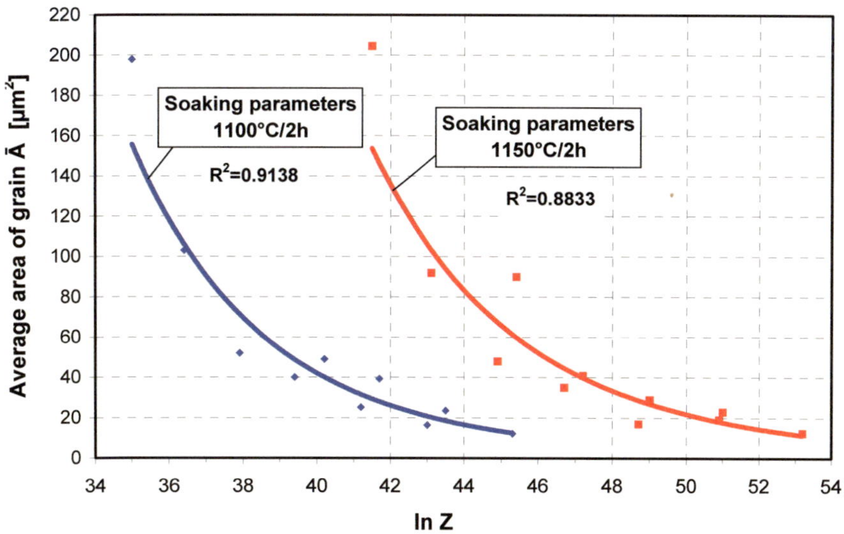

Fig. 22. Relationship between the Zener-Hollomon parameter and the average grain area after recrystallization. Initial alloy soaking: 1100°C/2 h and 1150°C/2 h

For both variants of initial soaking of the alloy, relationships were determined between the average plane section area of the recrystallized grain $\bar{A}$ and the Z parameter (eq. 16 and 17):

- for the alloy after initial soaking 1100°C/2h:

$$\bar{A} = 9.3 \times 10^{5} \times Z^{-0.250} \ [\mu m^2] \quad (16)$$

- for the alloy after initial soaking 1150°C/2h:

$$\bar{A} = 2.0 \times 10^{6} \times Z^{-0.224} \ [\mu m^2] \quad (17)$$

The determined relationships are essential where the critical value of a significance test for the direction factor of regression line "p" is less than 0.05. In the analyzed case, the "p" factor value for both variants of initial soaking of the alloy, i.e. at 1100°C/2h and 1150°C/2h, equaled 3.08×10^{-5} and 5.03×10^{-5}, respectively, which indicates the significance of the determined relationships.

3.4 Substructure of hot-deformed alloy

The recovery and dynamic recrystallization processes during hot plastic deformation of the Fe-Ni alloy cause changes in the dislocation and subgrain substructures. An analysis of the microscopic examination results allows affirming that the nature and extent of changes in the alloy substructure was dependent on the deformation temperature and strain rate, and the conditions of initial soaking. After initial soaking (1100°C/2 h) and deformation at a temperature of 900°C and a rate of 0.1 and 1.0 s^{-1}, effects of dynamic recovery and dynamic recrystallization (Fig. 23 and 24) were observed in the alloy microstructure. In grain areas with major defects of the austenite, a cellular dislocation substructure and subgrains with different dislocation densities were formed.

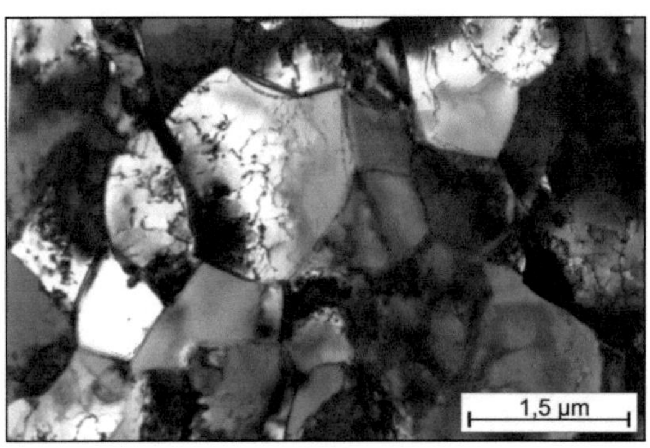

Fig. 23. Alloy microstructure after soaking at 1100°C/2 h and deformation at 900°C/0.1 s^{-1}. Subgrains and recrystallized grains

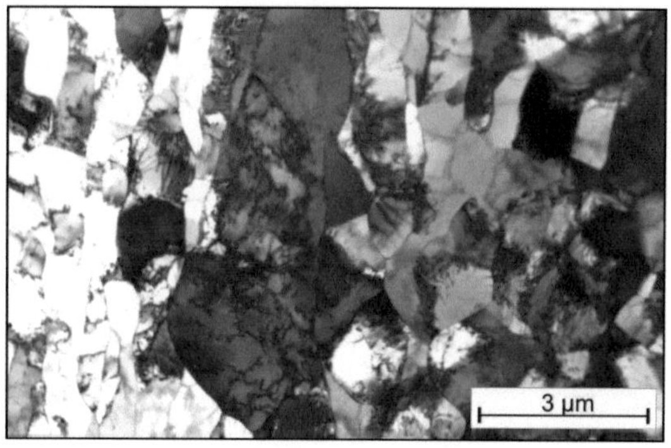

Fig. 24. Alloy microstructure after soaking at 1100°C/2 h and deformation at 900°C/1.0 s^{-1}. Subgrain structure formation

At a higher deformation temperature, e.g. 950°C, the proceeding recovery processes were accompanied by intensive dynamic recrystallization (Fig. 25 and 26). In the alloy substructure, dynamically recrystallized grains were observed next to the subgrains (Fig. 25). In the samples deformed at a higher rate, 1.0 s^{-1}, at this temperature the fraction and size of dynamically recrystallized microregions increase (Fig. 26).

The alloy deformed within the temperature range of 1000÷1050°C is characterized by a microstructure typical of a dynamically recrystallized material (Fig. 27 and 28). The austenite microstructure is composed predominantly of recrystallized grains free of dislocations (Fig. 27). Further perfecting of the substructure is observed in the neighbouring subgrains, as evidenced by equiaxiality of the subgrains and the decreasing density of dislocations inside them (Fig. 28). It was found that a higher strain rate (1.0 s^{-1}) leads to a growth of the subgrain and a reduction of the dislocation density.

Fig. 25. Alloy microstructure after soaking at 1100°C/2 h and deformation at 950°C/0.1 s^{-1}. Subgrains and recrystallized grains

Fig. 26. Alloy microstructure after soaking at 1100°C/2 h and deformation at 950°C/1.0 s^{-1}. Regions of recrystallized austenite and subgrains

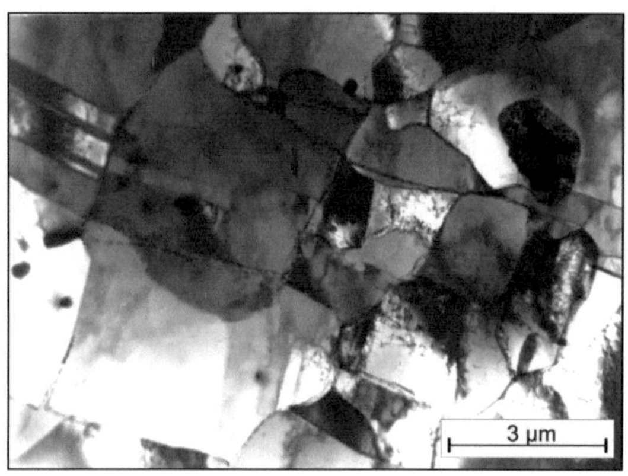

Fig. 27. Alloy microstructure after soaking at 1100°C/2 h and deformation at 1000°C/0.1 s^{-1}. Recrystallized grains with twins and subgrains

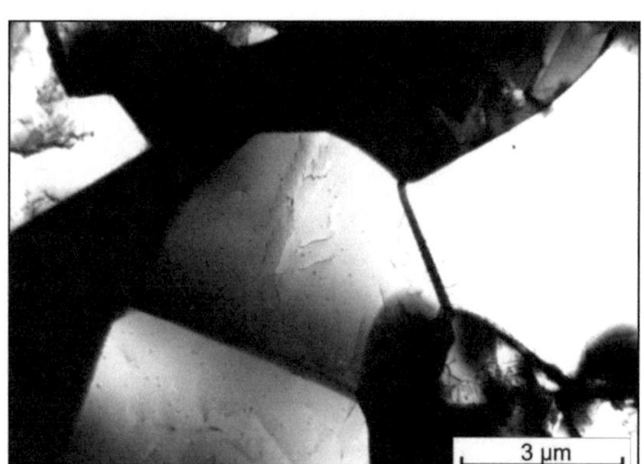

Fig. 28. Alloy microstructure after soaking at 1100°C/2 h and deformation at 1050°C/1.0 s^{-1}. The process of further perfecting of the subgrain structure

In the alloy substructure, after deformation at the highest temperature of 1100÷1150°C, a reincrease of the dislocation density and repolygonization were observed (Fig. 29). At a higher strain rate of 1.0 s^{-1}, areas with recrystallized grain free of dislocations are dominant in the alloy substructure (Fig. 30).

An increase of the initial soaking temperature to 1150°C/2h inhibits the dynamic recovery and recrystallization processes (Fig. 31 and 32). In the regions with austenite subgrain after deformation at a temperature of 900°C at a rate of 0.1 s^{-1}, deformation microtwins appear (Fig. 31). The subgrains being formed, especially at a high strain rate of 1.0 s^{-1}, have an elongated shape and different dislocation densities (Fig. 32).

At a higher deformation temperature, e.g. 1000°C, and a strain rate of 0.1 and 1.0 s^{-1}, the alloy substructure reconstruction is inhomogeneous. It is characterized by the presence of areas where structural changes are inhibited and accelerated (Fig. 33 and 34).

Effects of dynamic recovery (Fig. 33) and dynamic recrystallization (Fig. 34) were found there. The growth of new grains in the dynamic recrystallization process proceeded through the coalescence of subgrains and their subsequent growth (Fig. 35). The bending of the grain boundary towards areas with a higher dislocation density indicates the direction of the boundary movement. The principal mechanism of the coalescence includes reactions between dislocations which lead to disappearance of the dislocation boundary and formation of grain from the combination of several neighboring subgrains.

The dislocation density in the subgrain area does not decrease significantly when increasing the deformation temperature to 1100°C compared to a lower deformation temperature (Fig. 36 and 37). The dynamic deformation phenomena are accompanied by a continuous process of structural reconstruction of the material, i.e. repolygonization. It consists in re-saturation of subgrains with dislocations and their rearrangement together with the creation of new subboundaries and walls of a polygonal type (Fig. 36).

Deformation of the alloy at a higher rate of 1.0 s^{-1} was accompanied in the substructure by dislocation rearrangement with the formation of polygonal walls and a cellular substructure (Fig. 37). The phenomenon of repolygonization in the austenite grains was observed at the highest deformation temperature (1100÷1150°C).

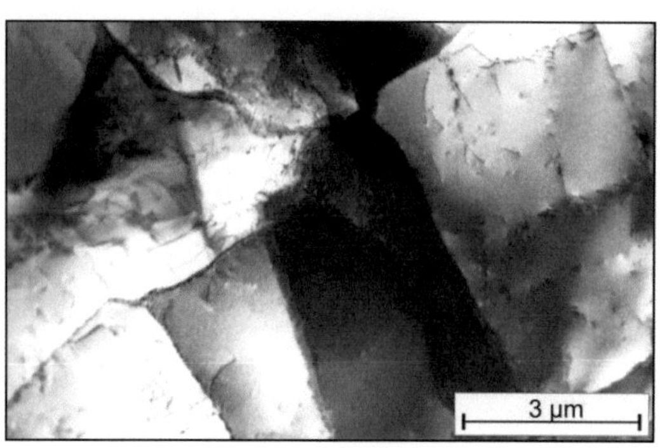

Fig. 29. Alloy microstructure after soaking at 1100°C/2 h and deformation at 1100°C/0.1 s^{-1}. The effects of repolygonization in austenite subgrains

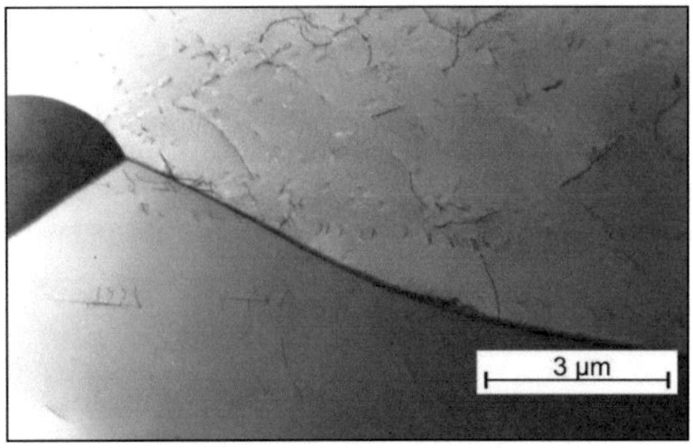

Fig. 30. Alloy microstructure after soaking at 1100°C/2 h and deformation at 1100°C/1.0 s^{-1}. Large austenite subgrains with a low dislocation density

Fig. 31. Alloy microstructure after soaking at 1150°C/2 h and deformation at 900°C/0.1 s^{-1}. Subgrain structure and grains with microtwins

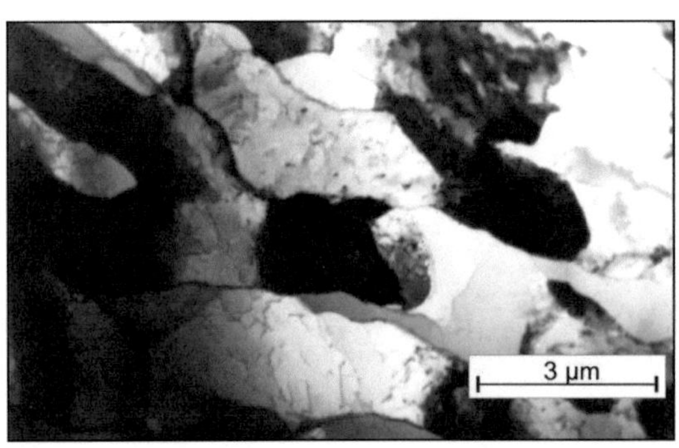

Fig. 32. Alloy microstructure after soaking at 1150°C/2 h and deformation at 900°C/1.0 s^{-1}. Non-equiaxed, austenite subgrains

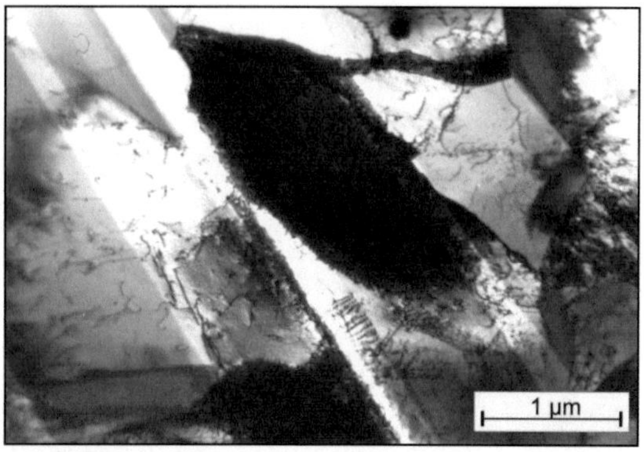

Fig. 33. Alloy microstructure after soaking at 1150°C/2 h and deformation at 1000°C/0.1 s^{-1}. The effects of polygonization and deformation microtwins

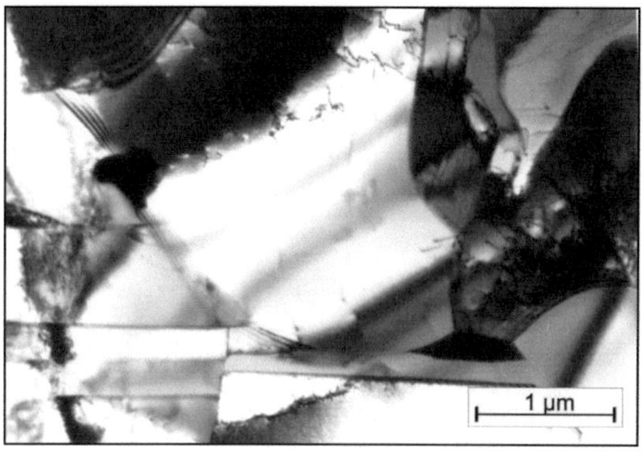

Fig. 34. Alloy microstructure after soaking at 1150°C/2 h and deformation at 1000°C/1.0 s^{-1}. The grains after dynamic recrystallization

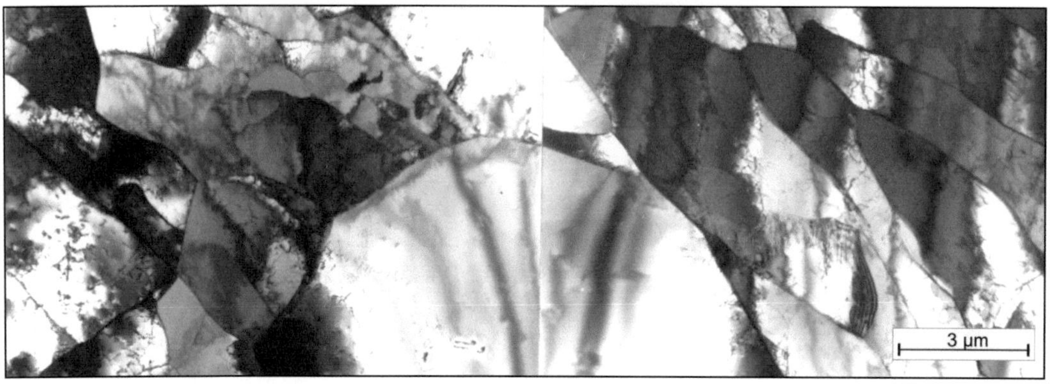

Fig. 35. Alloy microstructure after soaking at 1150°C/2 h and deformation at 1000°C/0.1 s^{-1}. The recrystallized grain formation as a result of coalescence of subgrains

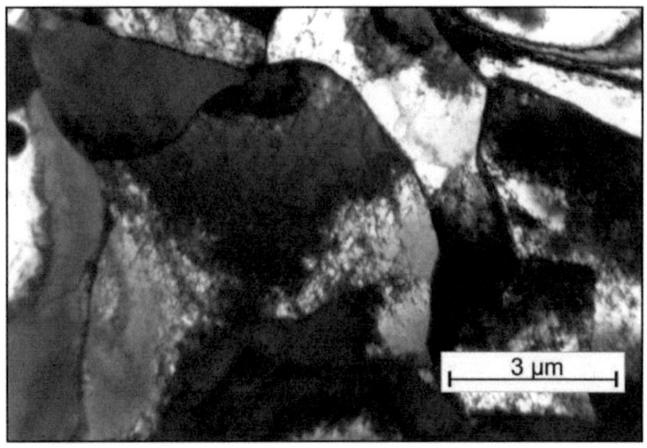

Fig. 36. Alloy microstructure after soaking at 1150°C/2 h and deformation at 1100°C/0.1 s^{-1}. The formed subgrain structure with dislocations

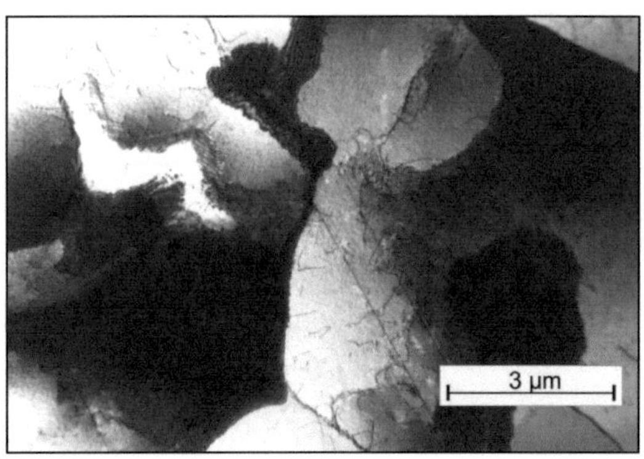

Fig. 37. Alloy microstructure after soaking at 1150°C/2 h and deformation at 1100°C/1.0 s^{-1}. Austenite repolygonization and a cellular dislocation structure

The course of changes in the subgrain sizes depending on the temperature deformation and strain rate for the two variants of initial soaking of the alloy is shown in Fig. 38 and 39. It was found that an increase in the alloy deformation temperature from 900 to 1150°C results in a growth of the subgrain.

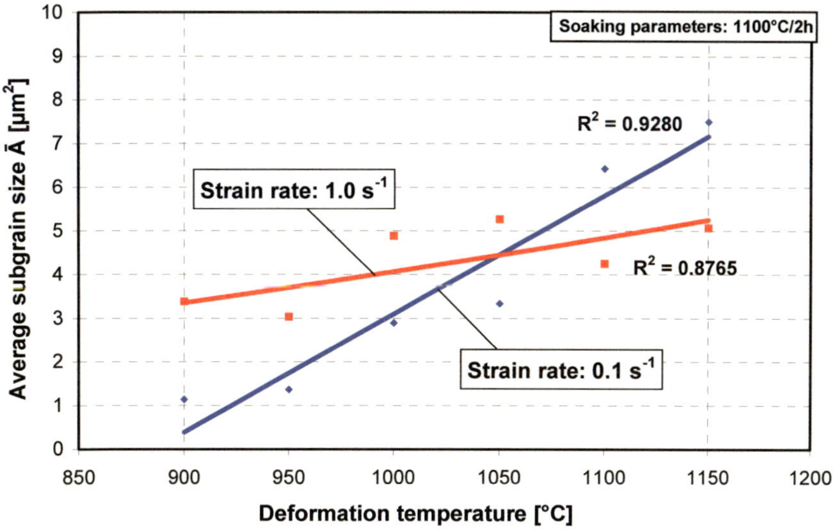

Fig. 38. The effect of temperature deformation and strain rate on the average subgrain size. Initial alloy soaking: 1100°C/2 h

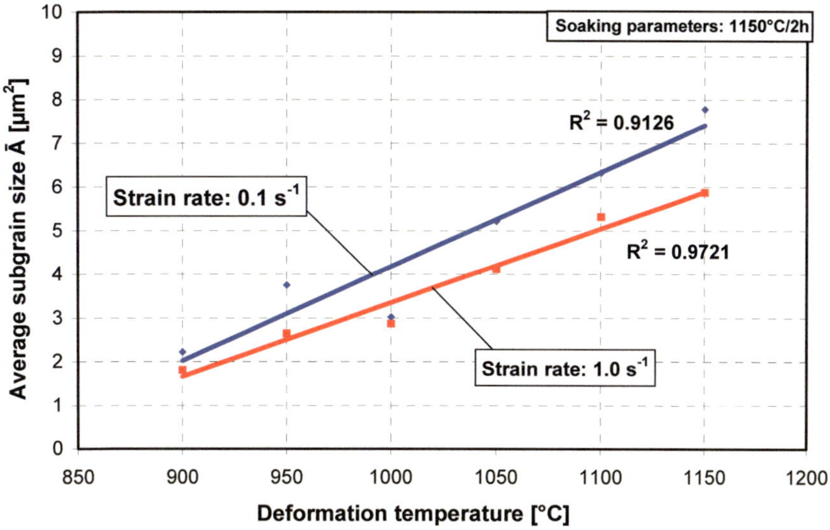

Fig. 39. The effect of temperature deformation and strain rate on the average subgrain size. Initial alloy soaking: 1150°C/2 h

However, no influence was observed of the conditions of initial soaking on the subgrain size. The average area of the subgrain plane section $\bar{A}$ varied within the range from 1.0 μm² to 7.8 μm² for both variants of initial soaking. The influence of the strain rate on the subgrain size was more significant, in particular for the initial soaking at 1100°C/2 h (Fig. 38). More intensive changes in the subgrain size were observed at a low strain rate of 0.1 s^{-1}, which can be explained by a higher cumulative deformation in the samples.

The dislocation density depending on the deformation temperature and strain rate for the two variants of initial soaking of the alloy is shown in Fig. 40 and 41. An increase of the deformation temperature was accompanied by a decreasing dislocation density. No significant influence was found of the initial soaking parameters on the dislocation density.

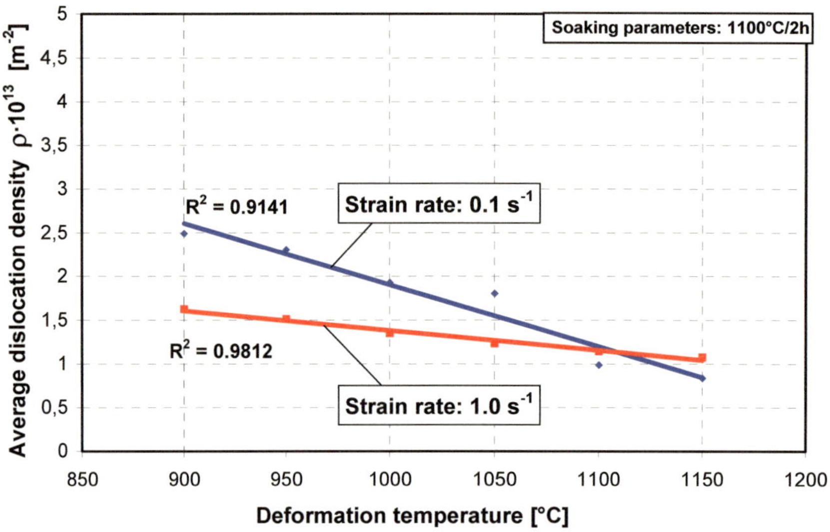

Fig. 40. The effect of temperature deformation and strain rate on the average dislocation density. Initial alloy soaking: 1100°C/2 h

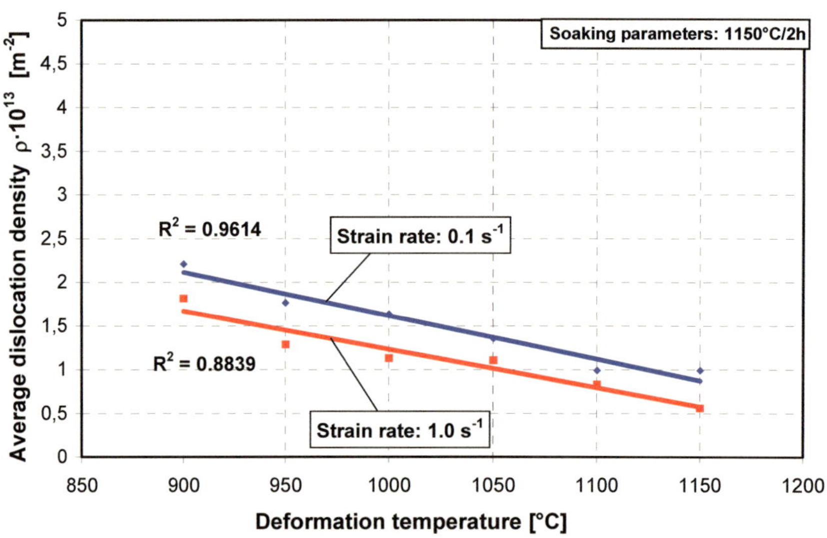

Fig. 41. The effect of temperature deformation and strain rate on the average dislocation density. Initial alloy soaking: 1150°C/2 h

For both variants of initial soaking, the dislocation density varied within a narrow range from 0.8×10^{13} m^{-2} to 2.5×10^{13} m^{-2}. The gradual reduction in the dislocation density observed in the samples as the deformation temperature increased from 900 to 1150°C shows a continuous process of substructure reconstruction and redeformation. For both variants of initial soaking of the alloy, higher dislocation densities were obtained for the lower strain rate (0.1 s^{-1}), which can be explained by a higher cumulative deformation in the material.

4. Summary

The work analyzes the relationships between the conditions of hot plastic deformation and deformability and microstructure of an austenitic Fe-Ni superalloy precipitation-strengthened by phase γ' type. The hot torsion tests carried out in the range of temperature of 900÷1150°C, at a strain rate of 0.1 s^{-1} and 1.0 s^{-1} allowed determining the influence of the initial soaking conditions and deformation parameters on technological plasticity of the Fe-Ni superalloy, as well as on the strengthening and structure reconstruction processes. It was found that optimal values of the maximum yield stress σ_{pp} and threshold deformation ε_f, as well as the required fine-grain microstructure, were obtained for the alloy after initial soaking at 1100°C/2 h and deformation at a rate of 0.1 s^{-1} in the temperature range of 1050÷950°C. The increase of yield stress, σ_{pp}, and the decrease of deformability of the alloy, ε_f, as the initial soaking temperature was rising up to 1150°C/2 h, with the deformation rate increasing to 1.0 s^{-1}, was associated with a growth of the initial grain size and the degree of austenite saturation with alloying elements. As a result of these processes, the stacking fault energy (SFE) of the austenite increased and so did the ability of the material to strengthen. This, in turn, led to an increase of activation energy of the hot plastic deformation process from the value Q = 441.8 kJ/mol (after initial soaking at 1100°C/2 h) to Q = 518.7 kJ/mol (after initial soaking at 1150°C/2 h).

An analysis of the flow curves and the examination results of the Fe-Ni alloy microstructure and substructure revealed dynamic recovery, recrystallization and repolygonization, occurring consecutively in the course of hot deformation. None of the detected stages of changes in the alloy structure constituted an independent process. Their course depended on both, the deformation parameters (T, $\dot{\varepsilon}$) and the initial soaking conditions. The growth of new grains in the dynamic recrystallization process took place through coalescence of subgrains and their subsequent growth. For both variants of initial soaking of the alloy, the analyzed quantitative indicators of the substructure depended fundamentally on the deformation temperature and, to a lesser degree, on the strain rate. The average size of subgrains $\bar{A}$ increased from 1.0 μm² to 7.8 μm² as the deformation temperature rose from 900°C to 1150°C. The average dislocation density ρ decreased gradually in the range from 2.5×10^{13} m^{-2} to 0.8×10^{13} m^{-2} as the deformation temperature rose in the range of 900÷1150°C.

The dynamic recrystallization proceeding in the Fe-Ni alloy during hot plastic deformation caused high refinement of the material structure. The average area $\bar{A}$ of recrystallized grains increased as the deformation temperature rose and it changed in the analyzed range of deformation parameters from 16 μm² to 205 μm², which, with reference to the initial grain size, meant refinement of the alloy structure of the order of 10÷370×. The average size of recrystallized austenite grain depended mainly on the deformation temperature and, to a lesser degree, on the strain rate. No significant influence was found of the initial grain size on the size of the dynamically recrystallized grain after plastic deformation. The existence of an exponential dependence between the average area of recrystallized austenite grain and the deformation temperature, as well as of an involutive dependence on the Zener-Hollomon parameter Z was shown.

5. Acknowledgment

The present work was supported by the Polish Ministry of Science and Higher Education under the research project No 7 T08A 038 18.

6. References

Bywater K.A. & Gladman T. (1976). Influence of composition and microstructure on hot workability of austenitic stainless steels, *Metals Technology*, Vol. 3 (1976), pp. 358-368

Cwajna J.; Maliński M. & Szala J. (1993). The grain size as the structural criterion of the polycrystal quality evaluation, *Materials Engineering*, Vol. XIV (1993), pp. 79-88

Ducki K.J.; Hetmańczyk M. & Kuc D. (2006). Quantitative description of the structure and substructure of hot-deformed Fe-Ni austenitic alloy, *Materials Science Forum*, Vol. 513 (2006), pp. 51-60

Ducki K.J. (2010). Microstructural aspects of deformation, precipitation and strengthening processes in austenitic Fe-Ni superalloy. *Monograph. Copyright by Silesian University of Technology* (2010), pp. 1-136, ISBN 978-83-7335-721-1

Hadasik E. (2005). Methodology for determination of the technological plasticity characteristics by hot torsion test. *Archives of Metallurgy and Materials*, Vol. 50 (2005), pp. 729-746

Hansen N. (1998). Microstructure and properties of deformed metals, *Materials Engineering*, Vol. XIX (1998), pp. 108-115

Härkegård G. & Guédou J.Y. (1998). Disc Materials for Advanced Gas Turbines, Proceedings of the 6th Liége Conference: *Materials for Advanced Power Engineering*, 1998, pp. 913-931

Head A.K.; Humble P.; Clarebrough L.M.; Morton A.L. & Forwood C.T. (1973). Computed Electron Micrographs and Defects Identification, In: *Defects in Crystalline Solids*, Amelinckx S., Gevers R., Nihoul G. (Ed.), 1973, pp. 39-47

Klaar H.J.; Schwaab P. & Österle W. (1992). Round Robin Investigations into the Quantitative Measurement of Dislocation Density in the Electron Microscope, *Praktische Metallographie*, Vol. 29 (1992) (1), pp. 3-26

Kohno M.; Yamada T.; Suzuki A. & Ohta S. (1981). Heavy disk of heat resistant alloy for gas turbine, *Internationale Schmiedetagung 1981, Verein Deutscher Eisenhüttenleute*, Düsseldorf, Vol. 12 (1981), pp. 4.1.1-4.1.22

Koul A.K.; Immarigeon J.P. & Wallace W. (1994). Microstructural control in Ni-base superalloys, In: *Advances in high temperature structural materials and protective coatings*, National Research Council of Canada, Ottawa, 1994, pp. 95-125

McQueen H.J. & Ryan N.D. (2002). Constitutive analysis in hot working, *Materials Science and Engineering*, Vol. A322 (2002), pp. 43-63

Schindler I. & Bořuta J. (1998). Utilization Potentialities of the Torsion Plastometer. *Published by Department of Mechanics and Metal Forming, Silesian University of Technology* (1998) pp. 1-106, ISBN 83-910722-0-7

Sellars C.M. (1998). Role of computer modelling in thermomechanical processing, *Materials Engineering*, Vol. XIX (1998), pp. 100-107

Szala J. (1997). Computer program Quantitative Metallography, *Edited by Department of Materials Science, Silesian University of Technology* (1997)

Zener C. & Hollomon J.H. (1944). Plastic flow and rupture of metals, *Transactions of the ASM*, Vol. 33 (1944), pp. 163-235

Zhou L.X. & Baker T.N. (1994). Effects of strain rate and temperature on deformation behaviour of IN 718 during high temperature deformation, *Materials Science and Engineering*, Vol. A177 (1994), pp. 1-9

7

Physical Metallurgy and Drawability of Extra Deep Drawing and Interstitial Free Steels

Kumkum Banerjee
Research and Development Department, Tata Steel Ltd., Jamshedpur, India

1. Introduction

The aim of this review is to present the underlying physical metallurgy for the development of aluminium killed (Extra deep drawing--EDD) and interstitial free (IF) steels, their recrystallization texture and its subsequent impact on the formability of these steels. A plethora of literature is available and a number of review articles have appeared previously **(Hutchinson, 1984; Ray et al, 1994)**. These contributions dealt broadly with the development of cold rolled and annealed textures till 1994 and since then further advances in research had been made on the subject and the present article is intended to provide the progresses made on the subject till date, while also giving a critical review of the subject as a whole.

The automotive industry aims to reduce the weight of outer-body car panels while maintaining strength, formability and dent resistance. However, conventional high strength sheet steels have insufficient formability to meet the drawing requirements of today's more complex outer-body car panels. In the recent years low and ultra low carbon steels like extra deep drawing aluminum killed, interstitial free, interstitial free high strength and bake hardening steels are known for their formability and are extensively used for the auto bodies.

Texture is an important parameter of steel sheets as it induces plastic anisotropy that can be beneficial to drawability of steels (**Hosford & Backholen, 1966; Lankford et al., 1950; Yoshicla, 1974**). The anisotropy is conveniently measured in terms of r_m-value that is the ratio of true width strain to true thickness strain determined through standard tensile tests. r_m-value varies essentially with respect to rolling direction of the sample. Thus, an average of the r-values is taken as r_m, which is expressed through the expression--$(r_0 + 2r_{45} + r_{90})/4)$ – termed as 'normal anisotropy'--where the subscripts, 0, 45 and 90 refer to the tensile specimens with parallel to, 45° and 90° to the rolling direction of the steel sheet. Isotropic steels have r_m-value around 1 while steels suitable for deep drawing applications should have r_m-value 1.8 (Holie, 2000).

High r_m-values correlate well with good deep drawability (**Lankford et al., 1950).** Good drawability also diminishes the edge splitting tendency during hole- expansion tests (**Klein & Hitchler, 1973**). The favourable texture for good deep drawability is a large fraction of the grains oriented with {111} planes parallel to the plane of a sheet **(Whiteley & Wise, 1962**). To ensure satisfactory drawability in these steels, i. e. to increase the depth of drawing and avoid the crack during deep drawing process and at the same time to make the edge on the top of a drawn cup smooth without the phenomenon of earing, the deep drawing sheet is required to possess high plastic anisotropy, r_m and low normal anisotropy, Δr. In other words, to maximize r_m-value and minimize Δr-value, {111}<112> and {111}<110> components of γ-fiber **(Figure 1) (Kestens et al., 1996)** are the ideal crystallographic textures for deep drawing steel, because the correct texture gives the proper orientation of slip system so that the strength in the thickness direction is greater than that in the plane of the sheet. If {100} plane parallels rolling plane, the strength is lowest in the thickness direction of sheet. This, in turn, adversely influences the formability of the sheet. The {111}/{100} intensity ratio is reported to be linearly related to r_m **(Held, 1965)** and can easily be determined using X-ray diffractometer measurements of the (222) and (200) lines.

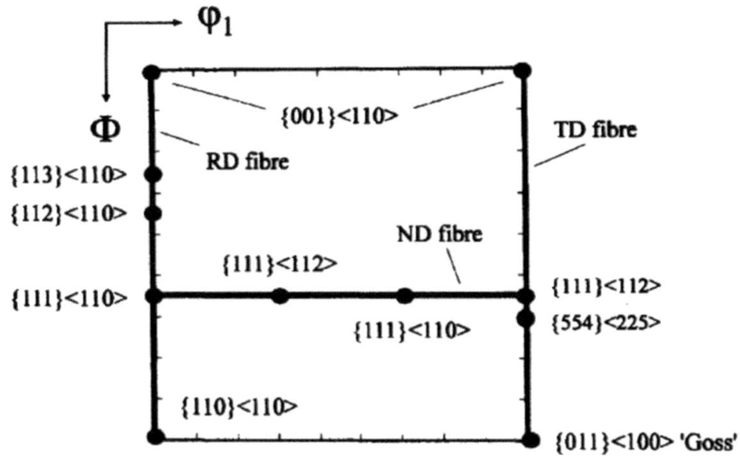

Fig. 1. φ2=45° section of Euler space showing the ideal bcc rolling and recrystallization texture components **(Kestens et. al., 1996)**

2. Recrystallization texture

In several cases, after only small cold deformation, nearly random textures are produced. However, on annealing after very heavy cold reductions, a strong recrystallization texture is usually obtained that may involve the partial retention of the deformation texture but quite often a very different but very strong new texture forms. Thus, the changes in texture that occur during the recrystallization process can be massive while compared to previous texture of the deformed state and in some cases, replaced by an entirely different texture **(Doherty et al., 1997)**.

Two major theories exist for the formation of recrystallization texture-- described as 'oriented nucleation' and 'oriented growth' (**Doherty et al., 1988 & Samajdar, 1994**). Oriented nucleation is the hypothesis that explains, grains with an orientation that dominates the fully recrystallized texture, nucleate more frequently than do grains of all other orientations. In other words, the oriented nucleation theory assumes orientation selection in the nucleation process based on the orientation dependence of the deformation energy stored in the grains during cold rolling (Tsunoyama, 1998). The high stored energy in (111) grains is considered to be responsible for the (111) orientation nucleation and growth. However, the energy stored in (110) grains is higher than that in (111) grains and another mechanism is required for the preferential development of (111) orientation. To describe the orientation nucleation theory quantitatively, for example for the most discussed case of the formation of 'cube' texture after the recrystallization of heavily rolled FCC metals such as Cu or Al, the fraction of grains, by number, within a selected misorientation, of 10 or 15° from exact cube, α_c, must be normalized by the fraction expected in a random grain structure, α_r (Doherty, 1985). The condition for a strong 'oriented nucleation' effect is that: $\alpha=\alpha_c / \alpha_r$. That is, the frequency of the formation of the new cube grains is much higher than the expected random frequency, so many of the grains will have the special orientation.

While, the oriented growth theory is based on the orientation dependence of the grain boundary mobility **(Tsunoyama, 1998).** In this theory, orientation relations between recrystallization nuclei and the deformed matrix is responsible for the texture development. However, no significant experimental evidence of oriented growth has been obtained for IF steels until now, even with modern techniques like EBSD (Electron Back-Scattered Diffraction). The oriented growth factor, β, is determined by the relative sizes $\overline{d_c} / \overline{d_r}$ of the cube to the average grains (**Doherty, 1985**). That is, there is a strong oriented growth effect if: $\beta= \overline{d_c} / \overline{d_r} >> 1$ (**Martin, et al., 1997**).In the opinion of Doherty et al (**Doherty et al., 1997**), the two theories of oriented nucleation and oriented growth should be renamed as: (i) the grain frequency effect; and (ii) the grain size effect, respectively. The reason behind the change was i) the nucleation involves only the growth of a particular subgrain and the terms, oriented nucleation and oriented growth are often taken to indicate specific mechanisms for the frequency or size advantage. Thus, the usage of "frequency" and "size effect" helps avoid such confusion.

The steel recrystallization texture is of major industrial importance. It is found that recrystallization in a cold-worked low carbon steel is mainly controlled by the oriented nucleation theory/grain frequency effect that is governed by the orientational dependence of the stored deformation energy (**Hölscher et al, 1991.**) The two key recrystallization texture components in steel are {110}<1$\bar{1}$0> and {554}<22$\bar{5}$>. The latter component is just a few degrees away from another recrystallization texture component {111}<112> (**Hatherly & Hutchinson, 1979**). During recrystallization two major changes take place. The orientation {001}<110> and the orientation spread surrounding the partial α-fiber texture gets eliminated after annealing and some redistribution of intensity in the fiber texture with {111} planes parallel to sheet.

The strength of {111} texture determines the drawability of low carbon and extra low carbon steels. The strength of {111} texture in turn is influenced by chemistry of the steel (**Perera et al., 1991; Wilshynsky-Dresler et al., 1995**) and the prior technological processing steps, such as hot rolling (Wilshynsky-Dresler et al., 1995 & Perera et al., 1991), cold rolling (**Perera et al., 1991**) and annealing (**Perera et al., 1991, Wilshynsky-Dresler et al., 1995**). Many studies have been made on the effect of process conditions and the following principles are obtained for the development of (111) recrystallization texture (**Tsunoyama, 1998**):

1. increasing coarseness of precipitates in hot bands;
2. decreasing grain sizes of hot band;
3. increasing cold reduction rate;
4. increasing annealing temperature.

However, these studies are not sufficient to make clear the mechanism of (111) texture development.

Thus, it is important to have a thorough understanding of the underlying physical metallurgy involved so that the desired recrystallization texture is obtained by suitably controlling the prior processing steps to result in formable grade of steels.

3. Processing of aluminium killed EDD steels

In annealed EDD, three types of microstructures are possible depending upon the stage while AlN precipitate forms from Al & N in solid solution (Auburn & Rocquet, 1973):

i. equiaxed grain structure is obtained while AlN either forms during coiling or after recrystallization during annealing.
ii. The transition zone recrystallized microstructure forms while recrystallization and AlN precipitation occur simultaneously.
iii. elongated or pancake grain structure is obtained while AlN forms prior to recrystallization during annealing. The AlN precipitates form on the defects of cold rolled structure and thus, recrystallization is retarded. A remarkable enhancement of {111}<uvw> crystallographic orientation occurs as nucleation occurs more rapidly in grains of this type (**Beranger et al., 1996**)

EDD steel sheets are produced by either batch or continuous annealing of cold-rolled steel sheets containing carbon up to about 0.05% and Mn up to about 0.2% (**Sarkar et al., 2004**). However, the physical processes involved in these processes are different. Thus, batch and continuous annealing processes will be detailed separately in the following sections.

3.1 Hot band texture

The hot band texture of such steels is reported to be nearly random with rotated cube component {001}<110> being approximately 2 times random (2XR) (**Heckler & Granzow, 1970**).The recrystallization of austenite during hot rolling is reasonably fast and gets completed prior to the transformation to ferrite. Further, in EDD steels no other texture component remains present after hot rolling indicating the fact that the austenite did not have any deformation texture component prior to transformation to ferrite (**Ray et al., 1990).**

3.2 Cold rolled texture

The cold reduction has an important role in dictating the grain morphology after annealing, texture and mechanical properties. **Figure 2a -2c** depict the effect of cold reduction on grain size, r_m-value (drawability) Δr (planer anisotropy—earing) for an EDD grade steel (C: 0.034%, Mn: 0.21%, Al: 0.06% and N: 0.005%) (**Hebert et al., 1992**). Thus, an optimized cold reduction must be taken into account while high deep drawability as well as minimum earing is desired.

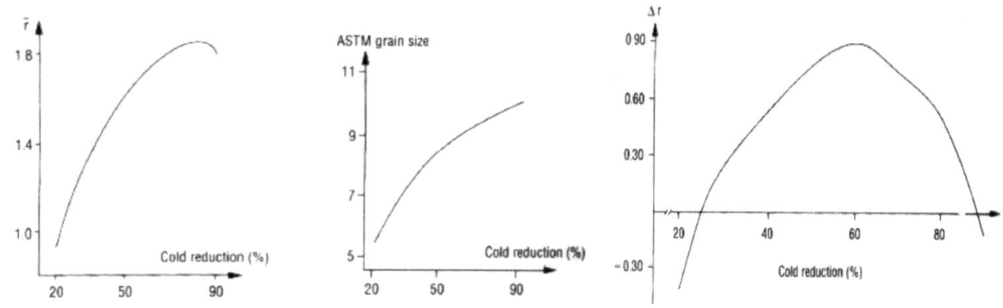

Fig. 2. (a) Variation of $\bar{r}$ (r_m-value) with cold work, (b) variation of grain size with cold work and (c) variation of Δr with cold work for the EDD grain steel (Hebert et al., 1992).

With increasing cold reduction the steel develops both partial α-fiber <110>||RD and γ-fiber {111} ||ND. The transformed {001}<110> component also strengthens noticeably. With the increase in cold reduction from 60% to 80%, the strongest texture component shifts from {111}<10> to {112}<110> **(Heckler & Granzow, 1970)**

3.3 Batch annealing

Since the pancake grain structure provides favourable texture for high deep drawability, the processing parameters are set to obtain pancake structure, while transition zone structure is avoided since the latter results coarse grains (ASTM<8) leading to reduced ductility and the risk of orange peel formation during drawing. In order to ascertain that the nitrogen remains in solution, the AlN that forms in the cast material requires to be dissolved during slab reheating **(Auburn & Rocquet, 1973; Meyzaud et al., 1974).** Usually, soaking temperatures of the order of 1200-1250°C are necessary. In addition, the recombination of Al with N also needs to be prevented during cooling and coiling after hot rolling. To attain this, the finish rolling temperature must be high enough and above Ar3 **(Figure 3) (Beranger et al., 1996)** followed by fast cooling in the AlN precipitation range in association with low coiling temperature (<600°C), **Figure 4 (Beranger et al., 1996)** to avoid poor ductility and drawability in the annealed steel.

3.4 Continuous annealing

Continuous annealing lines combine several processes including cleaning, annealing, over aging or galvannealing, and sometimes temper rolling, in one continuous operation. In continuous annealing due to high heating rate, recrystallization during annealing occurs at higher temperature than batch annealing and the precipitation of AlN occurs after recrystallization with nitrogen previously in solution. Thus, the nucleation of preferred oriented grains is hindered and due to nitride precipitation, subsequent growth of the recrystallized grains is also restricted. This causes the development of unfavourable texture. Further, the presence of carbides and carbon in solution during recrystallization also assist in the formation of unfavourable texture for drawing.

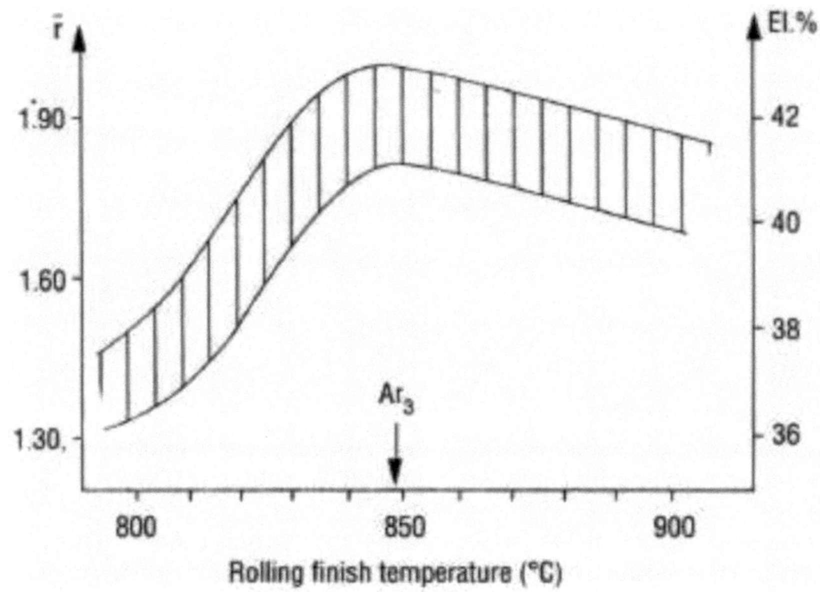

Fig. 3. Variation of $\bar{r}$ (r_m-value) and ductility with finish rolling temperature for a batch annealed EDD steel (**Beranger et al., 1996**).

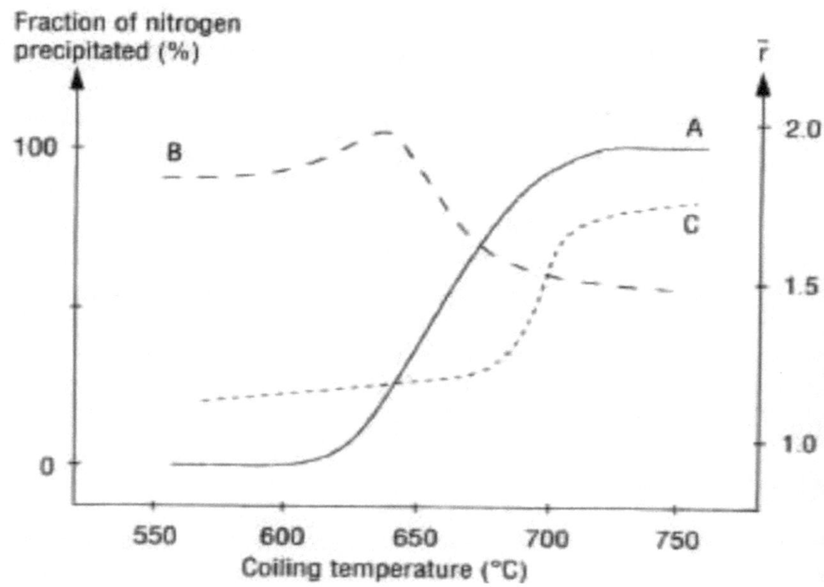

Fig. 4. Variation with coiling temperature of (A) AlN precipitation,(B) the r_m-value in batch annealing and (C) the r_m-value in continuous annealing for a low carbon low manganese EDD steel.(Beranger et al., 1996).

Therefore, in continuous annealing it is endeavoured so that AlN is precipitated prior to annealing, by high temperature coiling after hot rolling. Otherwise, by reheating the as-cast slabs at a temperature too low to take the nitrides back into solution.

However, high temperature coiling has two disadvantages (**Beranger et al., 1996**):

- It causes variation in mechanical properties in the product due to difference in cooling rate between middle and end regions of the coil. A selective coiling technique in which coiling the ends at higher temperature is performed can reduce the property heterogeneities.
- May cause abnormal grain growth for a certain combination of finish rolling and coiling temperatures which causes defect in both hot and cold rolled sheets.

Thus, to obtain favourable texture and improved deep drawability and therefore, to increase annealed grain size, it is required that the rate of nucleation of recrystallized grains is reduced, which can be done by lowering of recrystallization temperature. This can be achieved by (i) reducing the carbon content, alloying elements and impurity elements and (ii) increasing the stored energy of deformation by higher cold deformation, adjusting the composition or hot rolling parameters to obtain desired distribution of hot band precipitates **(Lebrun et. al, 1981)** .Control of the dissolved carbon and carbide contents is achieved by lowering the carbon content of the steel, overageing after continuous annealing, coarsening of the cementite particles and reducing the rate of redissolution of the carbides during annealing. High temperature coiling also promotes coarsening of the carbides present.

Thus, contrary to batch annealing high temperature coiling improves texture and drawability for EDD steel in continuous annealing (Figure 4). In batch and continuous both, a temper rolling is recommended after annealing to remove yield point elongation and thus to avoid stretcher strains in the final product.

4. Recrystallization texture and formability for EDD steel

4.1 Heating rate effect

While the Al and N are kept in solution prior to annealing, the microstructure varies with heating rate and mechanical properties and r_m-value are strongly dependent on heating rate during annealing as represented by **Figure 5 (Beranger et al., 1996)**. Batch annealing involves placing sheet steel coils in a gas fired furnace with a controlled atmosphere. Batch annealing cycles normally involve slow heating up to about 700 °C. Slow heating after cold rolling is normally necessary to allow adequate time for the Al to diffuse, forming clusters or precipitates before recrystallization commences. Thus, low heating rate leads to the precipitation of AlN during recovery that helps generate strong {111} texture after recrystallization. The precipitation of AlN takes place at a lower temperature and this is followed by recrystallization of the steel at a higher temperature (Takahashi & Okamoto, 1974). The optimum heating rate up to the precipitation stage to obtain highest r_m-value was calculated by Takahashi and Okamoto (**Takahashi & Okamoto, 1974**): $\text{Log (PHR)} = 18.3 + 2.7 \log((\text{Al}) (\text{N}) (\text{Mn})/R_{CR})$, where PHR is the peak heating rate in Kh^{-1} corresponding to the peak in r_m-value, (Al), (N) and (Mn) are solute concentration in weight percent and and R_{CR} is the percentage reduction via cold working. The holding temperature is always below Ac1 that varies in the range of about 650-720°C (**Beranger et al., 1996**). The coils are then slowly cooled at 10°C/hour (Takahashi & Okamoto, 1974) and the process takes several days in batch annealing.

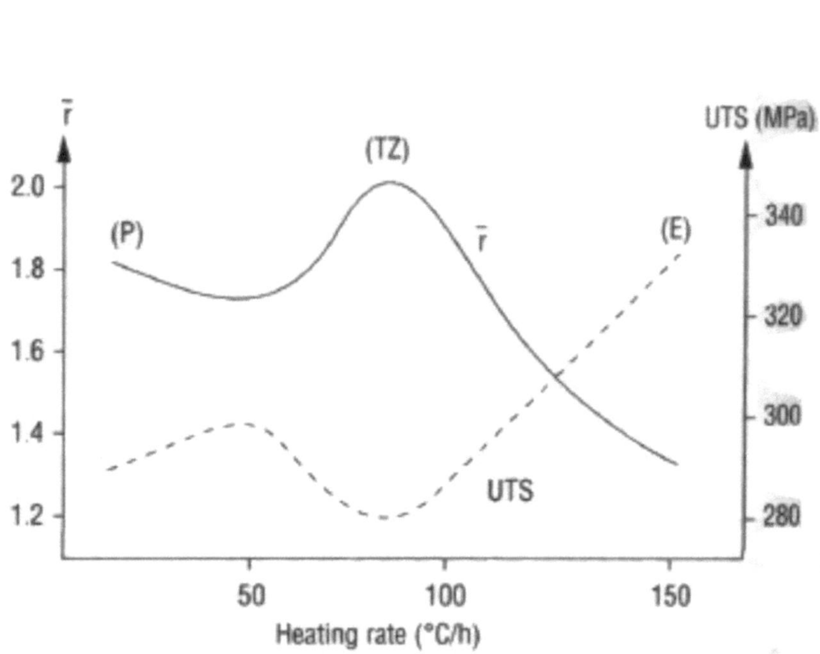

Fig. 5. Variation in r-bar (r_m) UTS values and microstructure with heating rate for batch annealed aluminium killed steel (Beranger et al., 1996).

Kozeschnik et al (**Kozeschnik et al., 1999**) calculated the logarithmic intensity ratio I_{111}/I_{100} at various annealing heating rates for the EDD steel. **Figures 6** (heating rate 50K/h) and **7** (heating rate -120K/h) show the effect of heating rate variation on the logarithmic intensity. Due to diminishing effective particle/recrystallization interaction time, the absolute value of the maximum intensity ratio is significantly decreased at higher heating rates. This effect is due to the shorter time available for aluminum nitride precipitation prior to the start of the recrystallization process. A higher supersaturation is needed in order to keep the two competing mechanisms balanced. Figure 8 shows the calculated variation of the I_{111}/I_{100} ratio with different heating rates. The qualitative comparison of the calculated logarithmic X-ray intensity ratios given in **Figure 8** and the experimental data for the r_m-value of Al-killed steel processed at low coiling temperature in Figure 9 (Hutchinson, 1984) shows good agreement.

Sarkar et al (**Sarkar et al., 2004**) also studied the effect of annealing heating rate on 70% cold rolled 0.05-C-0.19Mn-0.008 S-0.051Al EDD steel. In their work they selected heating rates of 50 and 70°C/h to the intermediate annealing temperatures of 550 and 600°C, and 15 and 30°C.h from intermediate annealing temperature to the final annealing temperature of 700°C. The soaking time at the intermediate and final annealing temperatures was half an hour. Tensile test and formability test inferred that the Annealing cycle that consisted of heating rate, 50°C/h to 600°C intermediate annealing temperature and 30°C/h up to 700°C resulted in the best combination of mechanical and formability properties with enhanced r_m of 1.92 and plane strain forming limit of 35%. The attractive combination of properties with high formability was attributed to the huge number of small spherical carbide precipitates that resulted in pure ferrite devoid of solute carbon.

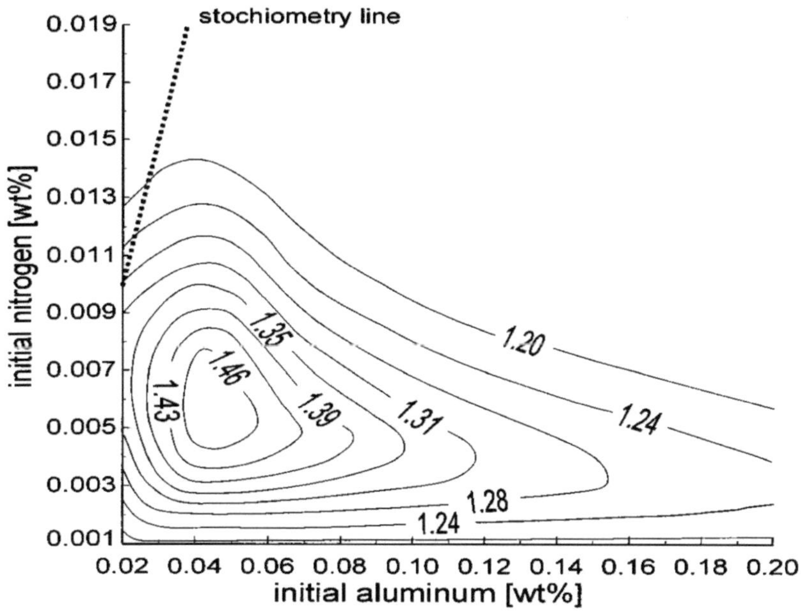

Fig. 6. Calculated logarithm of the intensity ratio (111)/(100) for a coiling temperature of 550 °C and a heating rate of 50 K/h as a function of initial amount of aluminum and nitrogen. **[Kozeschnik et al., 1999]**

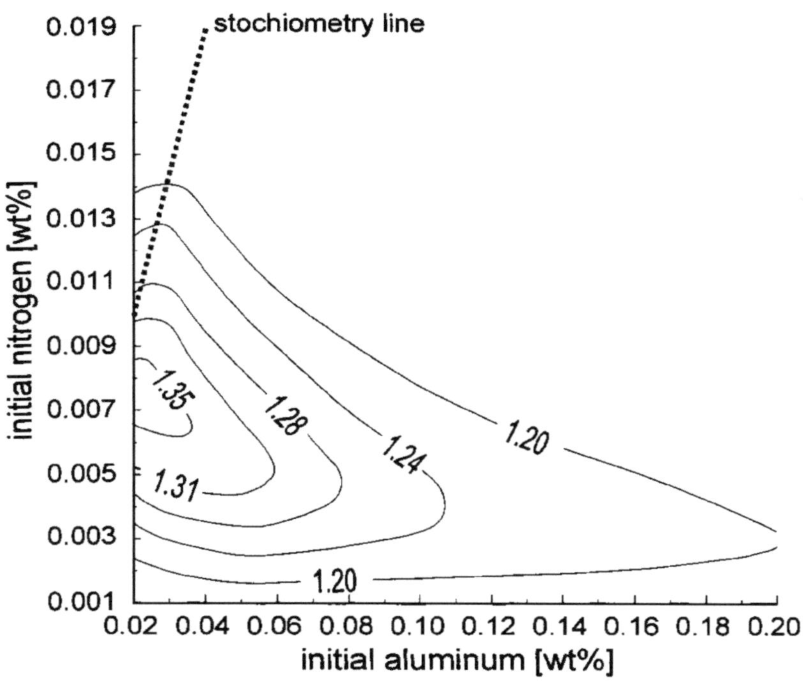

Fig. 7. Calculated logarithm of the intensity ratio (111)/(100) for a coiling temperature of 550 °C and a heating rate of 120 K/h as a function of initial amount of aluminum and nitrogen. **[Kozeschnik et al., 1999]**

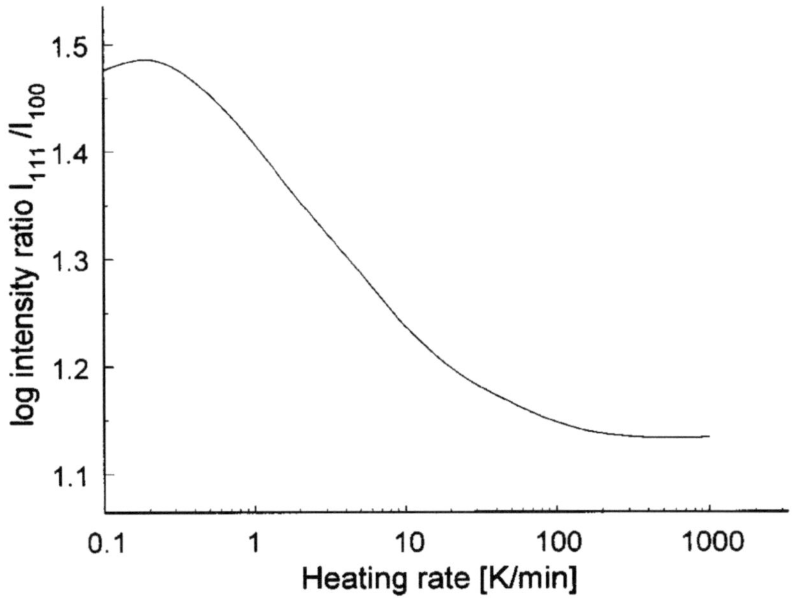

Fig. 8. Calculated logarithm of the intensity ratio (111)/(100) as a function of the heating rate during annealing for Al_{solute} = 0.03 wt pct and N_{solute} =0.005 wt pct. **(Kozeschnik et al., 1999).**

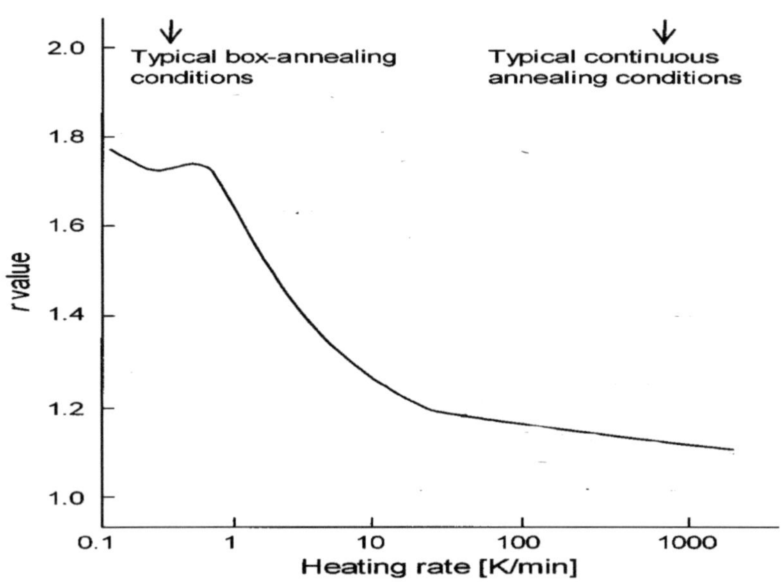

Fig. 9. Variation of mean plastic strain ratio, r_m with heating rate for EDD steel with low coiling Temperature (**Hutchinson,1984**).

4.2 Manganese and sulfur effect

Figure 10 (a, b) **(Ray & Jonas, 1990)** shows the comparison between the ODFs obtained from high Mn-high S (0.31%Mn, 0.018%S) and low Mn-low S (020%Mn, 0008%S) cold rolled batch annealed EDD steels, respectively. With the advances of steel making technology with desulphurization, the required Mn content to tie up with sulpher also reduced. The above technology in association with somewhat higher cold reduction, helped develop stronger γ-fiber (Fig 10b) that led to r_m, 1.8. While the r_m value obtained for the high Mn-high S steel (Fig 10a) was 1.5.

In the case of Al-killed steel, the optimized sheet properties are attained as a result of interaction between two processes: aluminium nitride precipitation and recrystallization (**Schulz, 1949**). Apart from pancaking of grains, the number distribution and morphology of carbides also play a role in formability. Smaller, spherical, and larger in number of carbides give good formability, because in this way the ferrite contains less C (ferrite is purer), which helps formability. Therefore, in order to improve the property of deep drawing sheet it is important to control every processing step effectively.

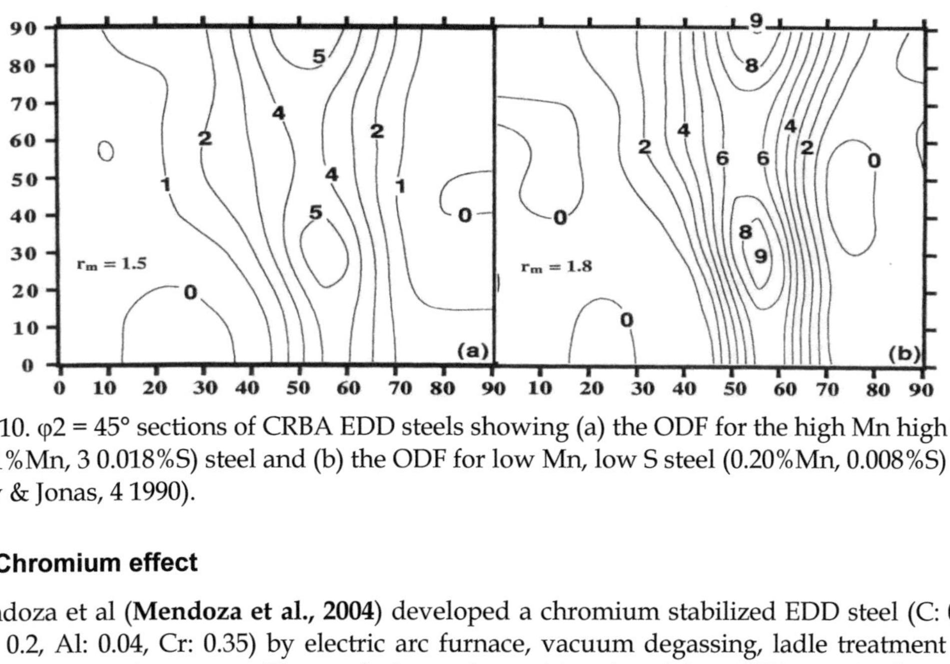

Fig. 10. φ2 = 45° sections of CRBA EDD steels showing (a) the ODF for the high Mn high S (0.31%Mn, 3 0.018%S) steel and (b) the ODF for low Mn, low S steel (0.20%Mn, 0.008%S) (Ray & Jonas, 4 1990).

4.3 Chromium effect

Mendoza et al (**Mendoza et al., 2004**) developed a chromium stabilized EDD steel (C: 0.02, Mn: 0.2, Al: 0.04, Cr: 0.35) by electric arc furnace, vacuum degassing, ladle treatment and continuous casting route. The steel sheet after cold reduced by ~85% was isothermally annealed under a protective argon atmosphere and was annealed at 700 ◦C from 1 to 300 s. The heating and cooling rate of annealed sheet specimens were ∽10 and ∽80 ◦C/s, respectively.

Figure 11 shows the SEM-micrograph of the as-cold rolled specimen. As may be observed, the ferrite grains are flattened, and inside the grains some shear bands can be observed. These in-grain shear bands corresponded to the narrow regions of intense shear that carry large strains during deformation and appear to become the major deformation mode (**Park et al.,2000**). In some interstitial free-steels also, it was observed that in-grain shear bands were inclined at angles of 30°–35° to the rolling plane (Barrett & Jonas, 1997). For instance, Barnett and Kestens (**Barnett et al., 1999**) reported that the increasing density and severity of these in-grain shear bands lead to a bulk recrystallization texture dominated by {1 1 1}<1 1 2> near the normal direction–rolling direction (ND–RD) for low carbon, ultra low carbon and interstitial free steels. In the present chromium stabilized EDD steel, the shear bands were inclined ∽32° with respect to the rolling plane and were almost parallel to each other within the individual grains.

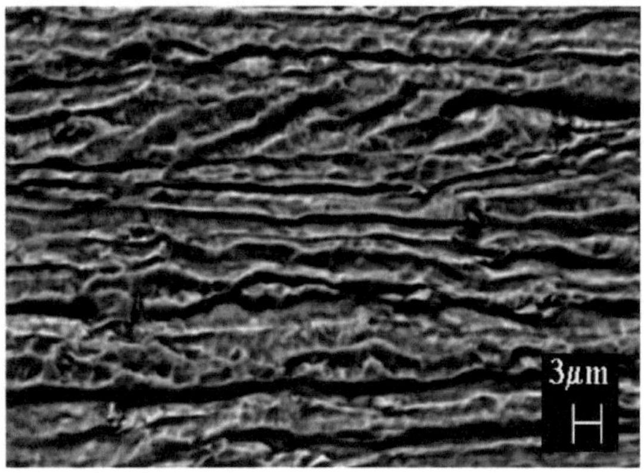

Fig. 11. Hot band microstructure of chromium stabilized EDD in hot rolled plates with flattened ferrite grains closely parallel to the sheet plane and inside the grains, some shear bands are observed (**Mendoza et al., 2004**).

Figure 12 shows {1 0 0} pole figures for the steel in the hot rolled, cold rolled and annealed conditions. It may be observed that in the cold rolled sheet, a {5 5 4}<2 2 5> component was noticed near {1 1 1}<1 1 2> texture. While, in the annealed sheet an appreciably strong texture than in the as cold rolled condition was observed from {5 5 4}<2 2 5> to {2 1 1}<0 1 1>. The microstructure obtained in the annealed condition was partially recrystallized, which was attributed to the presence of chromium-carbides precipitates that helped retard the recrystallization rate of the low carbon Al-killed/Cr-stabilized steel and which eventually was instrumental in the formation of {1 1 1}<1 1 2> textures.

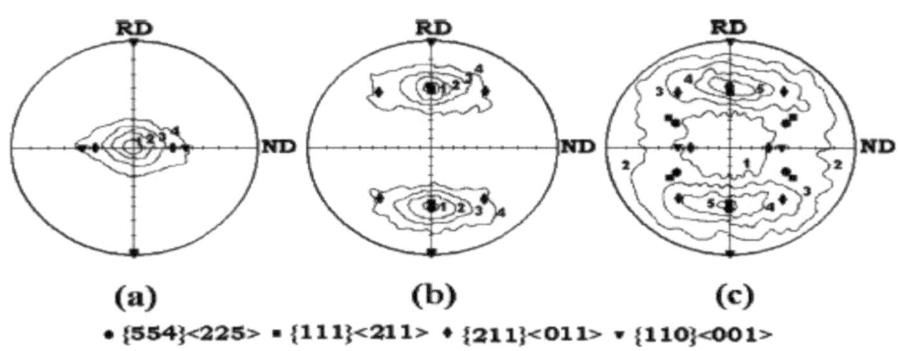

Fig. 12. {1 0 0} pole figures for the chromium stabilized EDD steel in (a) hot rolled, (b) cold rolled and (c) annealed conditions (**Mendoza et al., 2004**).

The mechanical properties of the annealed chromium stabilized EDD sheet are depicted in **Figure 13.** The use of chromium instead of niobium or titanium to stabilize the low carbon steel was effective in slowing down the recrystallization rate, thus, enhancing the formation of {1 1 1}<1 1 2> textures and achieving $\bar{r}$ values (r_m- values) >2 (Mendoza et al., 2004).

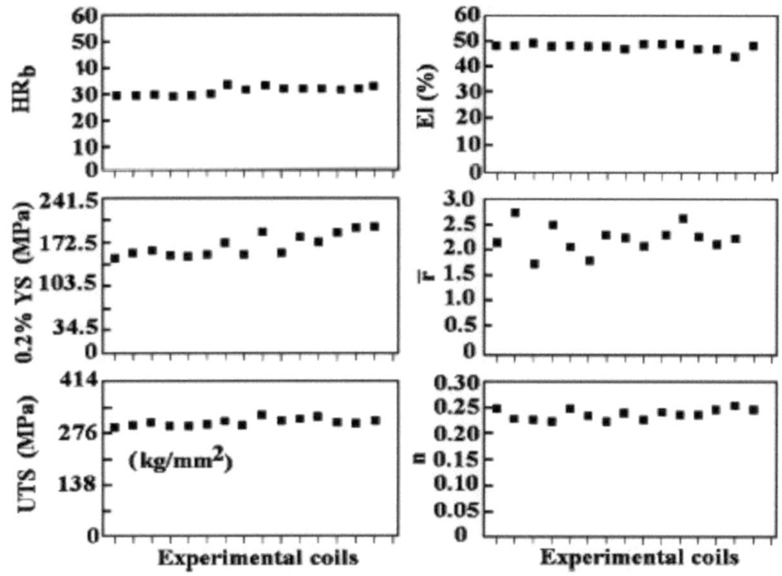

Fig. 13. Mechanical properties of annealed coils. (a) Hardness Rockwell b, (HRb), (b) 0.2% of yield strength, (YS), (c) tensile strength, (TS), (d) percent of elongation, (El), (e) the average plastic anisotropy value, ($\bar{r}$) and (f) the strengthening hardening exponent, n (**Mendoza et al., 2004**).

4.4 Phosphorous effect

It was observed by researchers that rimmed steel with P addition and by annealing the cold rolled sheet in decarburization furnace improved formability and thus studies were made on EDD steels to examine the interactive effect of P and AlN precipitation on formability of the steel sheets (Beranger., et al).

4.4.1 Heating rate effect on phosphorous added EDD steel

Rephosphorized Al-killed (EDD-P) steels show the similar dependence on heating rate of grains to conventional Al-killed steels (**Ono et al., 1982**). However, the smaller grain size, lower grain elongation ratio, lower r_m-value, faster recrystallizatio rate observed in the EDD-P steel were all attributed to the effect of phosphorous on the precipitation site of AlN. If phosphorus weakens the retardation effect of AlN on recrystallization, the restriction on nucleation of less favourable orientations, other than the {111} nucleus by AlN, which is usually observed in EDD steels will be relieved, resulting in the reduction of (111} intensity and an increase in the {100} intensity. In EDD-P steel {111}<112> orientation is observed unlike {111}<110> in the conventional EDD, which was attributed to the modification of cold rolled microstructure by phosphorous.

Figure 14 (Ono et al., 1982) shows the change in X-ray integrated intensity during heating. From the early stage to the half way of recrystallization, a rapid increase in the {222} intensity in association with a rapid decrease in the {200} and {110} intensities were observed in the EDD-P steel (Steel—1: C: 0.05, Mn-0.24, P:0.069) as well as in the conventional one (Steel-4: C:0.05, Mn:0.26, P:0.016). However, at the end of recrystallization, the EDD-P steel showed a lower {222} intensity and a higher {200} intensity than the conventional one. There was a marginal difference in the {110} intensity between them.

4.4.2 Manganese effect on phosphorous added EDD steel

An optimum combination of phosphorus and manganese content in steel can render a strong {111} texture through a simulated batch annealing cycle. As per Hu and Goodman **(Hu & Goodman, 1970)** and Hughes and Page **(Hughes & 1971)** the interaction between manganese and carbon would affect the recrystallization kinetics and thus, recrystallization textures. Further, Matsudo et al. (**Matsudo et al, 1984**) reported the detrimental effect of the Mn-C interaction on the drawability. On the other hand, it is widely known that steels containing phosphorus are likely to show a banded structure with segregated P-bearing ferrite, suggesting phosphorus and carbon in steel can act as repulsive elements to each other.

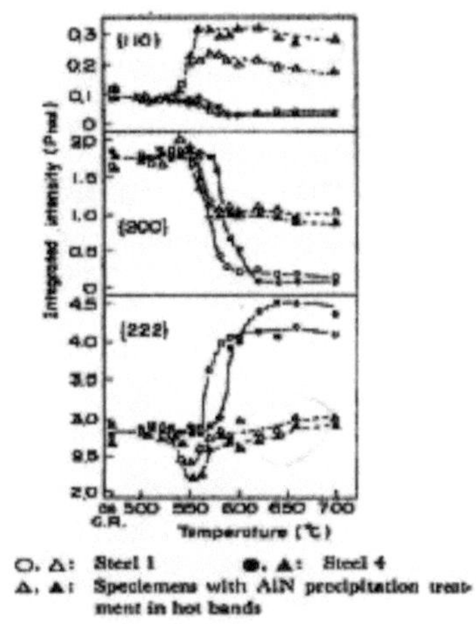

Fig. 14. Change in integrated intensity of Steels 1 and 4 during continuous heating at a peak heating rate 50°C/hr. **(Ono et al., 1982).**

Hu (**Hu, 1977**) reported that the drawability of low-carbon steel containing 0.067 pct P diminished more gradually than that of P-free steel with increasing manganese content up to 0.3 pct.

However, the work conducted by Chung et al (**Chung et al., 1987**) to examine the effect of manganese on the development of {111} recrystallization textures of P-containing low-carbon Al-killed steel sheet demonstrated that the phosphorous addition could modify the Mn-C interaction favourably for achieving improved drawability in EDD steels. **Figure 15** shows the (200) pole figures for the 0.1 pct P and P-free steels annealed at 973 K for 3 hours. The P-containing steels manifested {554} (225)-type texture and the texture was extremely strong in the 0.78 pct Mn-0.1 pct P steel. On the other hand, in the P-free steels the texture was not as strong at 0.5 pct Mn as it was at 0.06 pct Mn.

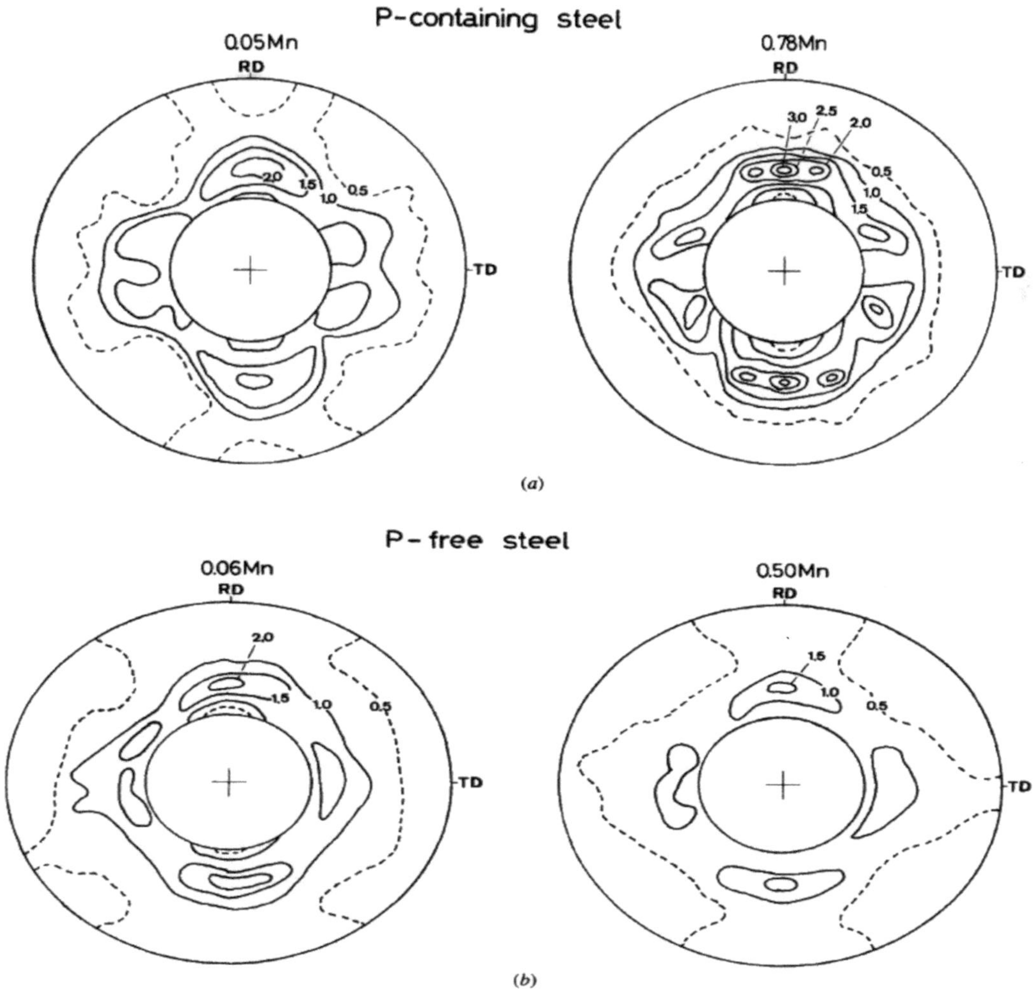

Fig. 15. (200) pole figures for (a) 0.1 pct P steels normalized, cold rolled, and annealed at 973 K for 3 h and (b) P-free steels normalized, cold rolled, and annealed at 973 K for 3 h (**Chung et al., 1987**).

To confirm the effectiveness of phosphorus addition to a high manganese steel in order to get a high r_m-value, the 0.2 pct P-1.2 pct Mn steel was employed. The recrystallization texture, as shown in **Figure 16**, had strong {554}(225) components which were favorable for a high r_m-value. The r_m-value and other mechanical properties are shown in **Table 1 (Chung et al., 1987).** The high manganese P-containing steel had a tensile strength of 430 MPa and an r_m-value of 2.0.

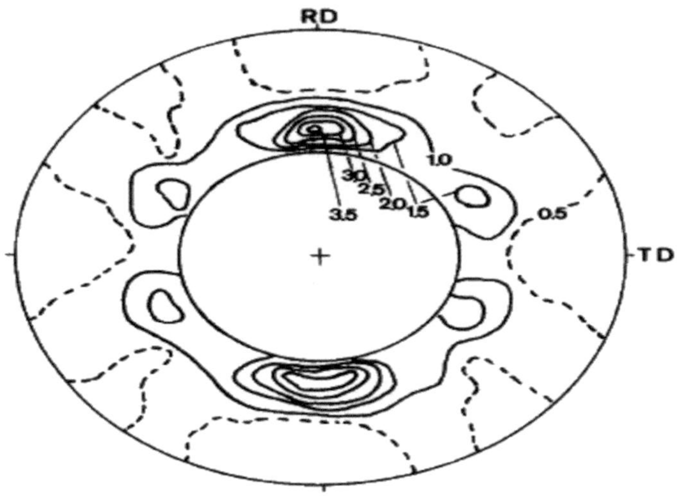

Fig. 16. (200) pole figure of 0.2 pct P-1.2 pct Mn steel preaged at 843 K for 24 h, cold rolled 75 pct, and annealed at 973 K for 4 h (**Chung et al., 1987**).

The micrographs of the P-free steel and the P-containing steel hot rolled and aged are shown in **Figures 17** and **18**, respectively. In the P-free steels, the precipitates that were identified as AlN dispersed within the grain at 0.06 pct Mn, whiles the precipitates were barely noticed at 0.5 and 1.0 pct Mn. On the other hand, in the P-containing steels, precipitates which might be $(FeMn)_3C$ and/or A1N were dispersed uniformly in the ferrite matrix only at the 1.5 pct Mn level. From the results, it must be noted that the fine precipitates of nitrogen and/or carbon disperse uniformly with decreasing manganese in the P-free steels, and vice versa with increasing manganese in the P-containing steels.

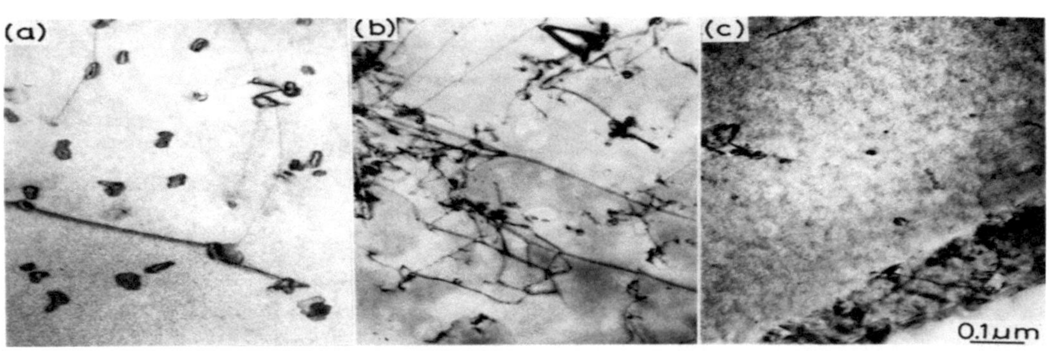

Fig. 17. Transmission electron micrographs of P-free Al-killed steels aged at 833 K for 1000 h after hot rolling. (a) 0.06 pct Mn, (b) 0.5 pct Mn, and (c) 1.0 pct Mn **(Chung et al., 1987).**

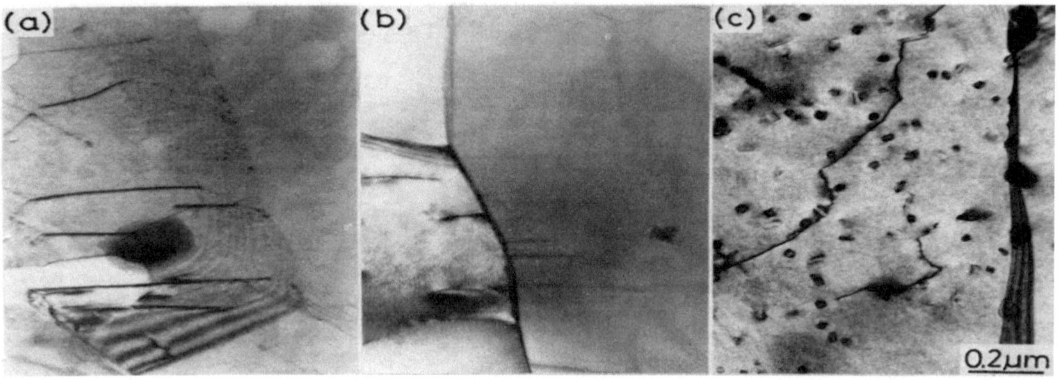

Fig. 18. Transmission electron micrographs of 0.2 pct P steels aged at 843 K for 24 h after hot rolling. (a) 0.1 pct Mn, (b) 0.5 pct Mn, and (c) 1.5 pct Mn **(Chung et al., 1987).**

Y.S. (MPa)	T.S. (MPa)	Uniform El. (Pct)	Total El. (Pct)	r_0	r_{45}	r_{90}	$\bar{r}$
262	434	24	33	1.60	2.09	2.25	2.01

Table 1. Mechanical properties of 0.2 pct P-1.2 pct Mn Al-Killed steel preaged at 8543 K, cold rolled 75 pct, and annealed at 973 K for 4 hrs.

Thus, a strong recrystallization texture with {554} (225) texture component and an uniform dispersion of $(FeMn)_3C$ in ferrite matrix can be developed through batch annealing of a P-containing EDD- steel if the steels contain high P and high-Mn .

5. On metallurgy of ultra-low carbon interstitial-free (IF) steels

Interstitial free steels are highly formable due to their low carbon and nitrogen content of ((wt%) < 0.003% and < 0.004%, respectively). The C and N are tied to Ti, Nb etc as precipitates. A limited excess of titanium or niobium relative to carbon, nitrogen and sulfur contents has a favourable influence on mechanical properties. During hot strip rolling, the level of interstitial elements, such as C and N, remnant from the steel making process, can be reduced by combining them with the stabilizing elements. The application of these steels is in the rear floor pan, front, rear door inners, etc.

5.1 Titanium stabilized IF steels

Titanium is very effective in scavenging nitrogen, sulphur, and carbon, readily forming TiN during casting and TiS during slab reheating . Subsequently, while the nitrogen and sulphur are scavenged, remaining Ti ties up with TiC during coiling. The minimum amount of titanium required for full stabilization, based on a stoichiometric approach, is **(Tither & Stuart, 1995)**:

$$Ti_{stab} = 4C+3.42N+1.5S$$

It was proposed that excess Ti addition than that required to combine with all C, N, and S was beneficial to achieve high r values (**Gupta & Bhattacharya, 1990**). Excess Ti (Ti*) is given by :

$$Ti^{*}=Ti_{total}-(4C+3.42N+1.5S)$$

However, it is also commented that an excess Ti content can be linked to the incidence of surface streaking. The frequency of appearance of this type of defect can be minimized using Nb in combination with Ti. Titanium only intierstitial free steels are the least susceptible to compositional changes and process variation. This is attributed to the precipitation of Ti compounds at higher temperatures (TiN, TiS) that play less of a role in subsequent lower temperature processing (Krauss et al., 1991). Tsunoyama et al (**Tsunoyama et al, 1988**) considered three precipitation mechanisms suggested by various authors: a) TiS may provide a preferential site for TiC nucleation. Thus, lowering the S content can retard the precipitation of TiC b) with decreasing Ti content, precipitation of $Ti_4C_2S_2$ occurs in place of TiS and controls carbon stabilization. The stabilization of C by $Ti_4C_2S_2$ is preferred compared to TiC precipitation as $Ti_4C_2S_2$ precipitates are larger, remove solute carbon from the matrix earlier, and are more stable.

The precipitation mechanism of titanium in these steels is summarized in **Figure 19** (**Tither, 1994**). It is postulated that titanium nitride particles formed during slab casting acted as nucleation sites for the precipitation of TiS and $Ti_4C_2S_2$. The small amount of carbon remaining is precipitated as TiC. During reheating of the slab, solution of carbosulphide occurs, leaving only TiS and TiN. Cooling of the strip to the inctercritical (austenite -ferrite transformation) temperature region during hot rolling transforms TiS to $Ti_4C_2S_2$ by the absorption of titanium and carbon.

5.2 Niobium stabilized IF steels

In Nb IF steels, Nb forms carbonitride precipitates. The aluminium addition during the killing process also reacts with N to form AlN. However, the favourable solubility product of AlN compared to NbCN, leads to preferential precipitation of AlN at a higher critical temperature. This reduces the N available for the precipitation as NbCN. The solubility products are dependent on bulk chemistry, temperature, and precipitate composition. The precipitates sequence in such steels can be considered as (**Holie, 2000**)

$$Al+N=AlN$$

$$Mn+S=MnS$$

$$Nb+C+N=NbC$$

The solute Nb on grain boundaries introduces site competition for the elements like phosphorous, which enables lower ductile to brittle transition temperatures, poorer elongation and r_m- values than Ti IF steels. The niobium steels have higher recrystallisation temperature (750 - 800°C) than titanium stabilized IF steels (**Baker et al., 2002**) and it was reported that the temperature of recrystallisation of Nb IF could be up to 50 K higher than that of the Ti IF steels. This in turn requires higher finish rolling and annealing temperatures. The higher recrystallization temperature is attributed to the low temperature formation of niobium carbides, producing fine particles that readily retard grain boundary movement during annealing.(**Tsunoyama, 1990**). The temperature can be reduced by lowering C content. In addition, recrystallisation temperature can be reduced to enhance drawability, using higher coiling temperature, so that precipitates are coarsened and subsequent pinning effect is reduced.

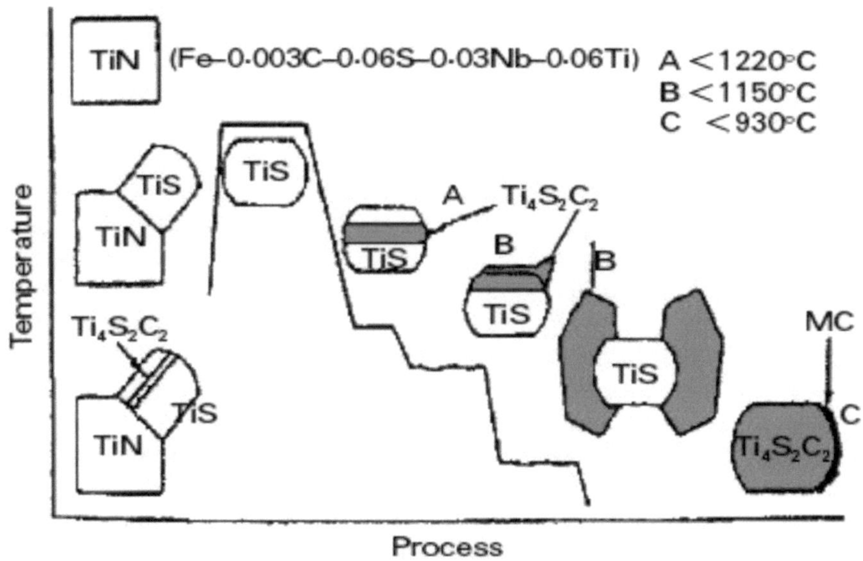

Fig. 19. Schematic mechanism of carbosulphide and carbide formation in titanium stabilized IF steels (**Tither et al., 1994**).

5.3 Titanium and niobium stabilized IF steels

Titanium in association with niobium added IF steels have the advantages of both Ti only and Nb only IF steels with reduced disadvantages. This combination provides the best combination of properties (**Tokunaga 1990**). The levels of Ti and Nb additions that are required to stabilize Ti +Nb IF steels are given by (**Holie, 2000; Tokunaga & Kato, 1990)**

$$(\%Nb)-(7.75(\%Ti)-3.42(\%N)-1.5(\%S))/4=0$$

If insufficient Ti is added such that the level of addition falls below that required for stabilization, TiN forms first and any S that is left behind after the formation of TiS is consumed as MnS. If the steel is understabilised by the insufficient addition of Ti and/or Nb then excess C exists in solid solution and a bake hardenable steel may be produced.

For IF steels, the increase in r_m-value with degree of cold work is continuous over the whole range employed in industrial production (**Beranger et al**).An increase in annealing temperature and/or holding time improves the deep drawability due to grain growth and recrystallization texture enhancement. Further, these grades have very high Ac1 temperatures, enabling continuous annealing at temperatures of ~850°C, or even higher. For such steels, the metal has no yield point elongation and no requirement of overaging treatments in continuous annealing, due to the absence of interstitials C and N.

5.4 Phosphorous effect on IF steels

One of the ways of increasing the strength of the IF steels is addition of solid solution strengtheners viz. P, Si and Mn to the steel (**Rege et al., 2000)**. It has been found that P is the most potent and cost effective solid solution strengthener that increases strength without appreciably affecting the drawability of the steel **(Katoh et al., 1985; Tokunaga & Kato, 1990)**. Mn, on the other hand, significantly deteriorates drawability and ductility (**Irie et al. 1981; Katoh et al., 1985; Tokunaga & Kato, 1990**), and Si adversely affects coating adhesion (**Nishimoto et al., 1982)**. Thus, P is the preferred addition to increase the strength of IF-steel. However, it is found that phosphorus tends to segregate at the grain boundaries or precipitate out of the matrix during the recrystallization annealing. The phosphorus segregations reduce the cohesive strength of the grain boundaries, which leads to the secondary cold work embrittlement (CWE) and reduces the resistance of the steel to brittle fracture or more precisely, intergranular fracture **(El-Kashie et al. 2003; Rege & DeArdo, 1997).** The CWE is defined as the susceptibility of the sheet material to intergranular fracture during the secondary work of deeply drawn part or while in service. Cao et al (**Cao, 2005**) had reported that the segregation of phosphorus occurred when the P>0.07 wt pct in the IF steel. It is also found that the batch annealing leads to the higher phosphorus than the continuous annealing. The segregation behavior of phosphorus in the Ti and Ti+Nb IF steels was studied by Rege et al (**Rege et al., 2000**). It was found that the segregation of phosphorus to the ferrite grain boundaries occurred not only during the coiling stage of the thermo-mechanical processing, but also during the cold rolling and annealing process. The steels with higher phosphorus content showed higher ductile to brittle transition temperature, i.e, lower resistance to cold work embrittlement (CWE). The problem of CWE is more often encountered while annealing time is long, which enables P to segregate in the grain boundary (**Yasuhara, 1996**).

The CWE can be avoided by :

- controlling the chemical composition—i. e. by partial stabilization of carbon in IF steels or by addition of B/Nb - It is believed that P and C both segregate to the grain boundary and compete for the available sites, and C reduces the grain boundary segregation and embrittlement by P. Furthermore, C enhances the grain boundary cohesion and counteracts the embrittlement this way. B also plays the same role as that of C. Hence P-C or P-B site competitive process was expected to minimize the CWE phenomena. In Nb added steel it is believed that CWE is lower than that in Ti-stabilized steel that has been attributed to the partial stabilization of C in Nb-containing steel.
- grain boundary engineering—i. e. by grain boundary character distribution—by suitable annealing cycles low angle and low -CSL boundaries can be produced by avoiding the development of continuous random boundary network and which help reduce CWE.

6. Recrystallization texture and drawability of IF steels

Cold rolling plays an important role in the formation of favourable textures for deep drawing during annealing, however, has little effect on other properties. The variation of r-value with cold reduction in the three common IF steel types, Ti-stabilized, Nb-stabilized and Ti-Nb stabilized is shown in **Figure 20** (**Tokunaga & Yamada, 1985**). Ti -

Nb steels show the highest r-value for equivalent cold reduction. This was attributed to the fact that the precipitates formed during hot rolling and coiling were not large enough to compromise the r-value during annealing. A reduction of 90% produced the highest r-value in all the steels, however, these reductions were rarely achieved in practice, 80% being more common.

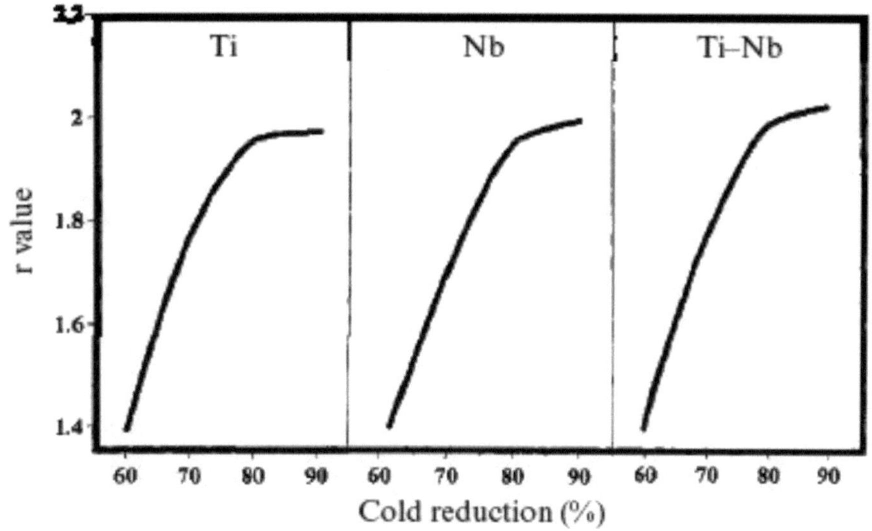

Fig. 20. Effect of cold reduction on r-value for titanium, titanium-niobium, and niobium IF steels (**Tokunaga & Yamada, 1985**).

Mendoza et al (**Mendoza, 2000**) studied the hot rolled precipitation behaviour of Ti-stabilized IF steels and mechanical properties and reported about improved r_m-value of 2-2.14. While Banerjee et al (**Banerjee et al., 2008**) vividly studied the precipitation of hot-band and annealed Ti stabilized IF steels to optimize hot and cold spot temperatures during annealing for improving drawability of the steels to >1.9. The hot bands of the as-received steels showed two types of texture development: (a) shear texture, manifested by {225}<554>, {110}<100>, and {113}<332> and (b) cold deformation type texture with a strong and incomplete α- fiber consisting of the components {001}<110>, {112}<110>, and {111}<110>, and the γ-fiber components, {111}<110> {111}<112>. The presence of a substantial amount of coarse equiaxed ferrite grains at the surface and finer grains at the central region was attributed to the outcome of finish rolling in the two-phase α+ γ region. On the other hand, the deformed grains at the surface and finer grains at the central region another hot band was the result of rolling in the single- phase α region. Figure 21 (Banerjee 21 et al., 2008) shows the φ2= 45° orientation distribution functions for various combinations of cold reduction and batch annealing temperatures to obtain improved r_m-value in an optimized processing condition. The poor texture and r_m-values at the annealing temperature of 660 °C were associated with the precipitation of fine precipitation of FeTiP-type within the grains and at the grain boundaries. This study had shown that for the given chemical composition of the Ti-IF steel, the optimized condition for cold spot temperature in batch-annealing cycle was 680 °C, preceded by 80 pct cold reduction, which resulted in an r_m-value of as high as 2.29.

The influence of the texture development in Ti-added (0.03, 0.05 and 0.07 wt%) IF on r_m-value was investigated by Kim et al (**Kim et al., 2005**). It was intended to determine the optimized Ti content for the promotion of deep drawability in the IF steels. For the IF steel with the composition of 0.0025C, 0.070Mn, 0.002N and 0.007S, the optimum Ti content was found to be 0.05wt%.

Juntunen et al (**Juntunen et al., 2001**) investigated the continuous annealing parameters in laboratory scale on drawability of Ti+Nb stabilized IF and IF-HS steels and it was reported that r_m-value could be enhanced by about 13% simply by adjusting the annealing conditions. The annealing cycle with maximum studied temperature produced the sharpest γ-fiber and highest r_m-value. While, Ruiz-Aparicio et al (**Ruiz-Aparicio, et al., 2001**) studied the evolution of the transformation texture in two $Ti_4C_2S_2$-stabilized interstitial-free (IF) steels (Ti and Ti/Nb) as a function of different thermomechanical processing parameters. Analysis showed that the $Ti_4C_2S_2$-stabilized steels stabilized were not very sensitive to the processing conditions employed in the study. The study also revealed that, under conditions of large deformations and coarse austenite grain sizes, the main components of the transformation textures are the beneficial {111}||ND orientations.

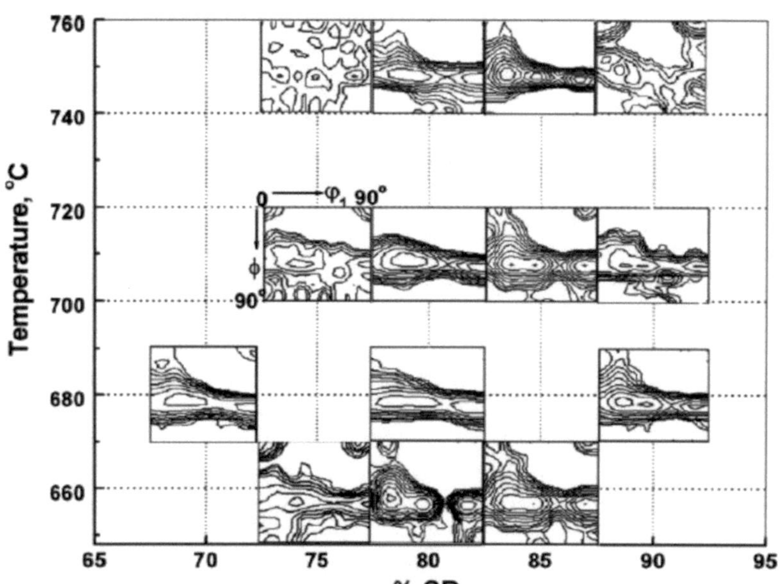

Fig. 21. ODFs of φ2 = 45° section at various processing conditions for a particular batch annealing temperature at various cold reductions. Maximum intensity for **660 °C**: 75 pct-3.53, 80 pct -8.43, and 85 pct-6.43; for **680°C** (hot spot): 70 pct-9.68, 80 pct-12.13, and 90 pct-10.38; for **710 °C**: 75 pct-2.95, 80 pct-8.09, 85 pct-6.89, and 90 pct-4.13; and for 750 °C: 75 pct-2.0, 80 pct-10.77, 85 pct-9.46, 90 pct-3.32 (**Banerjee et al., 2008**).

The effects of electric field annealing on the development of recrystallization texture and microstructure in a Ti+Nb stabilized cold rolled IF steel sheet were studied by He et al (**He et al., 2003**) means of ODF analysis and optical microscopy to assess the drawability response of the steel. Specimens of size 50 mm X 20 mm were cut from the sheet with the longitudinal direction parallel to the rolling direction. They were then subjected to isothermal annealing at different temperatures ranging from 650 to 800 °C for 15 min, respectively with or without a DC electric field of 200 V/mm. The annealing treatments were done in a nitrogen atmosphere and at a heating rate of 5 °C/min to the chosen peak temperatures. The external electric field was applied by placing the specimens (positive electrode) in the middle of two parallel stainless steel sheets (negative electrode) that were 2 cm apart. The experimental arrangement is shown in **Figure 22**.

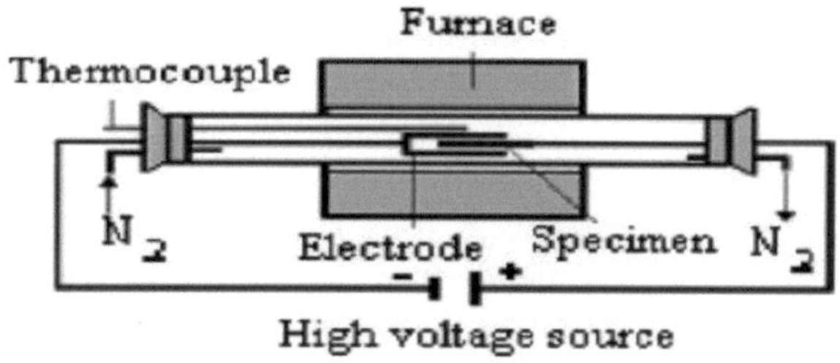

Fig. 22. Experimental arrangement for annealing with electric field (**He et al., 2003**).

Heating rate annealing experiments were carried out by Muljono et al (**Muljono et al., 2001**) to study the effect of heating rate on the recrystallization kinetics, grain size and texture of steels with a range of carbon levels (0.003–0.05% C). The steels were cold-rolled to 70% reduction and subsequently annealed at heating rates from 50 to 1000°C/sec to temperatures in the range 600 to 900°C. **Figure 23** shows that, for the 0.02 and 0.003% C steels, the {111}||ND texture increases in strength with increased heating rates up to 200°C/sec and maintains a plateau thereafter. Both steels exhibit similar trends and only the relative strength of the γ-fibre differs. The grain size and texture results are in general agreement with work by Hutchinson and Ushioda (**Hutchinson & Ushioda, 1984**). The {111}||ND components of the recrystallization texture increased at rates up to 200°C/sec due to enhanced nucleation at grain boundary sites.

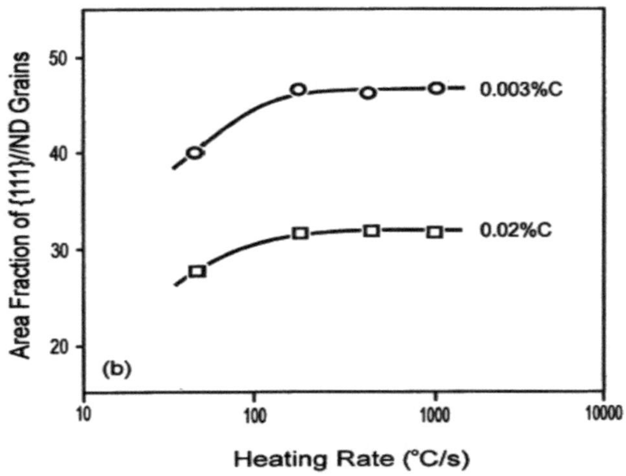

Fig. 23. Effect of heating rate on strength of recrystallization texture, given as the area fraction of grains within 15° of{111}||ND (**Muljono et al., 2001).**

During electric field annealing, specimens work as anode and the applied electric field reduces the lattice defect energy by lowering the shield effect that would decrease the driving force for recrystallization **(Cao et al.,1990; Conrad, 1989; Wang, 2000)**. Thus, although the application of electric field generally reduces the driving force for nucleation and grain growth, the nuclei with random orientations are much more restricted than that of the γ-nuclei. Consequently, the electric field annealing yields a high nucleation rate for the γ-nuclei that lead to a relatively strong γ-texture after complete recrystallization. From Figures 24 and 26 **(He et al, 2003)** it can be noted that both kinds of specimens annealed with and without application of electric field, exhibited a similar tendency in the development of recrystallization textures, i.e. the α-fiber was weakened and the γ-fiber was strengthened with increasing annealing temperature. In addition the Figures 4 & 6 also depict that the application of electric field (200 V/mm) during annealing may promote the development of the γ-fiber (ND||<111>) recrystallization texture of the cold-rolled IF steel sheet, which is beneficial to the deep-drawability. While, Figures 25 and 27 **(He et al, 2003)** indicate that the recrystallization was noticeably retarded intensively by electric field annealing under the investigated conditions.

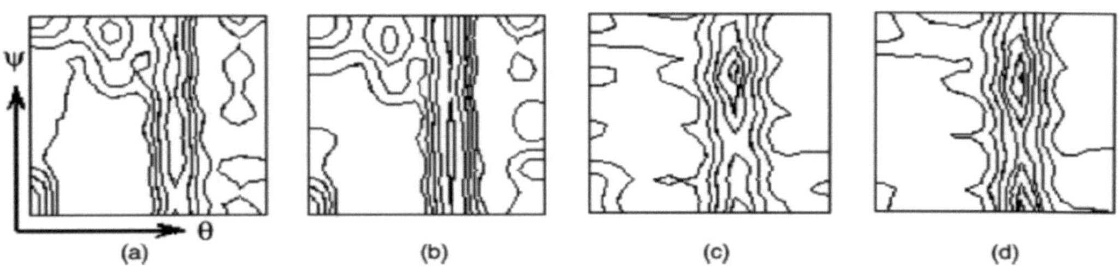

Fig. 24. ϕ= 45° sections of the ODFs (levels: 1; 2; 3; . . .) for the specimens annealed without an electric field at (a) 650 °C, (b) 700 °C, (c) 750 °C and (d) 800 °C **(He et al, 2003)**.

In an innovative finding, Jeong communicated **(Jeong, 2000)** that r value was markedly improved by reducing the carbon content from 0.0035 (Steel B) to 0.0009 pct. (Steel A) in Ti stabilized IF steels that were treated with 0.25 pct Si, 1.25 pct Mn, and 0.09 pct P to attain tensile strength of 400 MPa **(Figure 28a)**. The difference in r_m-value between two steels is 0.2 to 0.3 at all annealing temperatures **(Figure 28b)**. Steel A containing 0.0009 pct carbon showed a high r_m-value of about 1.6 while annealed at 800 °C to 860 °C, corresponding to deep drawing quality (DDQ). The r_m-value increased to 1.85 for extra deep drawing quality (EDDQ) grade with increasing annealing temperature to 890 °C. It was thus a remarkable finding as the steel was high strength steel with tensile strengths of 400 MPa or higher. This result indicated that while the carbon content decreases below 0.001 pct, superior formability of EDDQ grade could be achieved in high strength steel with tensile strength of 400 MPa or higher.

The highest $\bar{r}$ (r_m-values) in steels A and B were obtained in the specimens annealed at 890°C while the coarsening of the ferrite grain was remarkable **(Figure 28c,d)**. In order to find out the reason for the effect of carbon on the r_m-value, (200) pole figures for the annealed sheets were measured. The comparison of (200) pole figures of steels A and B showed that with the decrease in the carbon content, {554}(225} near the ND//(111} texture became stronger that was responsible for the improvement of r_m- value (Jeong, 2000).

Song et al (**Song et al., 2010**) worked on phosphorous segregation and phosphide precipitation on grain boundaries in association with drawability of rephosphorised IF steel .The cold rolled steel was annealed at 810°C for 90 to 600 sec in a protected environment. The **Table 2 (Song et al., 2010)** illustrates yield strength, tensile strength and r_m-value reduce while n (work hardening index) value increases with the annealing time from 180 to 600 s. The phosphorus concentration at grain boundary is shown in **Figure 29 (Song et al., 2010).** It was observed that as annealing time increased, the phosphorus concentration increased from 0.22 to 0.8 wt pct. The phosphorus concentration at the grain boundary is 20 times higher than that in the matrix for the sample annealed for 600 s, which greatly reduces the steel strength.

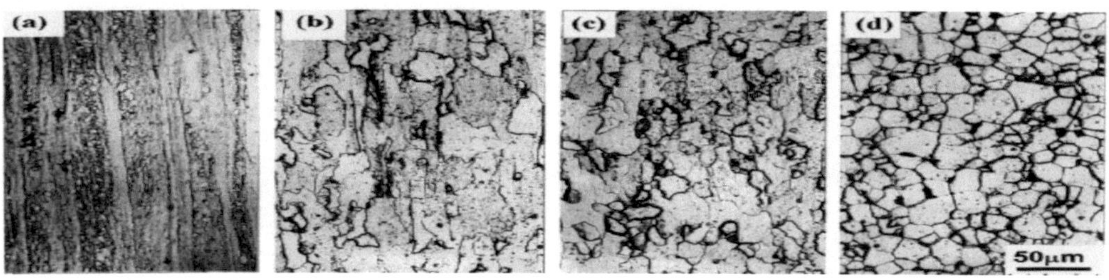

Fig. 25. Microstructures of the specimens annealed without an electric field at (a) 650 °C, (b) 700 °C, (c) 750 °C and (d) 800 °C (**He et al, 2003)**

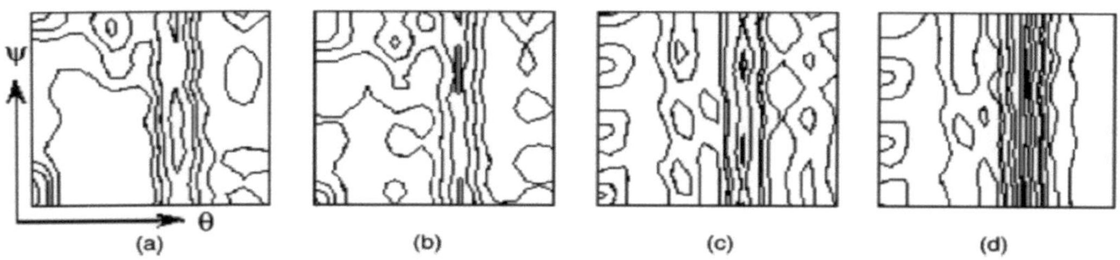

Fig. 26. ϕ= 45° sections of the ODFs (levels: 1; 2; 3; . . .) for the specimens annealed with an electric field at (a) 650 °C, (b) 700 °C, (c) 750 °C and (d) 800 °C (**He et al, 2003**).

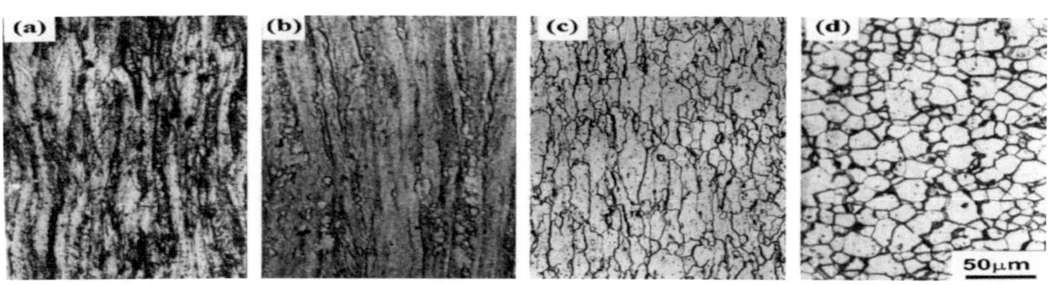

Fig. 27. Microstructures of the specimens annealed with an electric field at (a) 650 °C, (b) 700 °C, (c) 750 °C and (d) 800 °C (**He et al, 2003).**

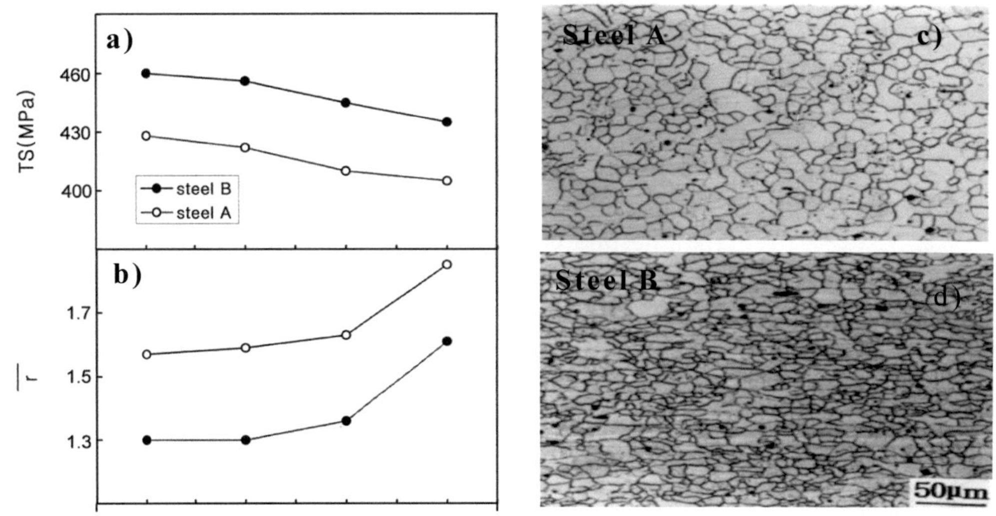

Fig. 28. Effects of carbon and annealing temperature on (a) tensile strength (b) $\bar{r}$ (r_m-value) and (c) & (d) ferrite grain sizes after annealing at 890ºC for Steels A and B (**Jeong, 2000**).

Annealing time/s	Yield strength/MPa	Tensile strength/MPa	r	n
180	158.1	354.9	2.098	0.2376
360	155.4	346.8	1.939	0.2936
600	141.1	322.7	1.688	0.3046

Table 2. Mechanical properties of the annealed samples for different times

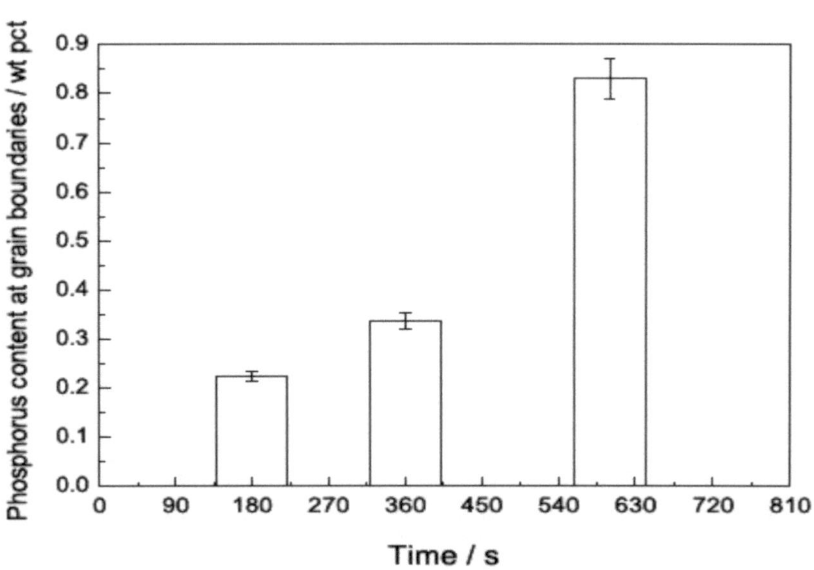

Fig. 29. Phosphorus content at grain boundaries in the rephosphorised IF steels annealed for different time 5 (**Song, et al., 2010**)

Kang et al **(Kang et al., 2007)** studied on the effects of aluminium on interstitial free high strength steel for the improvement of drawability (**Figure 30**). 78%cold rolled samples were annealed in an infrared-ray heating furnace. The annealing cycle consisted of heating the specimens to 830°C at a constant heating rate of 7°C/s and held at temperature for 30 s and then cooled to RT. Aluminum content more than 0.10 wt% improved the formability of the IF-HS. Texture analyses showed that the {111}||ND fiber (γ-fiber) was intensified, and <110>||RD (α-fiber) was weakened, with the increase of aluminum content. Recrystallization was completed earlier in the steel with the high aluminum content and the grain size of the annealed sample was larger than the steel containing lower aluminium. It was confirmed thorough the SANS analysis that the size of the precipitates in the sample with higher aluminum content was larger and their number was much fewer than in the sample with lower aluminum content. It appears that the high aluminum content in IF-HS containing Mn, P, Ti and Nb improved the scavenging effect of Ti or Nb and thus purified the iron matrix.

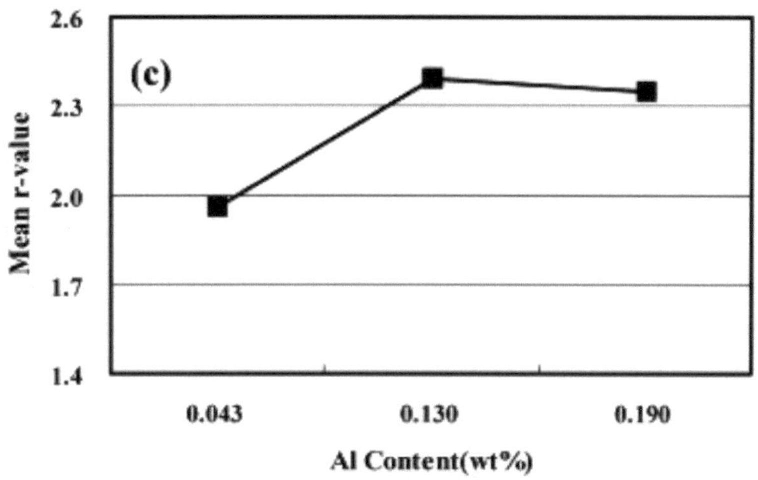

Fig. 30. Effect of aluminum content on mean r-value of the interstitial free high strength steel containing Mn, P, Ti and Nb (Kang et al., 2007) **(Kang et al., 2007)**

7. Metallurgy of bake hardening steels

Bake hardening is a diffusion controlled process involving the migration of solute carbon atoms within the iron lattice. The diffusion of these atoms is affected by heat treatment time and temperature and the amount of solute present in the steel. Factors such as grain size and dislocation density may also have an influence.

7.1 Mechanism of bake hardening

The yield strength increase from bake hardening is accompanied by the return of the yield point and yield point elongation; there may also be a slight increase in tensile strength and decrease in elongation. To attain such increased strength, the following criteria must be met: **(Matlock et al., 1998)**

i. mobile dislocations must be present in the steel.
ii. there must be sufficient concentration of solute in the steel to pin these dislocations.
iii. the solute must be mobile at the aging temperature.
iv. dislocation recovery must be sufficiently slow to prevent softening.

The driving force for pinning is a reduction in lattice energy. Both impurity atoms and dislocations induce lattice strains in the iron matrix and these strains can be relaxed if the interstitial atoms diffuse to the vicinity of dislocations (**Mizui et al., 1990**).Bake hardening steel is amply used by the automotive industry for the outer panel of cars. This steel grade is characterized by low yield strength prior to final manufacturing process and a remarkably enhanced yield strength of the finished product in association with excellent deep drawability. The increase in yield strength during bake hardening of steel occurs due to the blocking of otherwise mobile dislocations by forming Cottrell atmosphere of solute carbon or iron-carbide precipitates during the baking operation after painting at 170°C for about 20 minutes (**Kozeschnik et al., 1999**).

There are several types of bake hardening steel in production today. Individual grades are determined by the processing technology available and the properties required in the final product. These steels can be loosely grouped into two main categories, bake hardening EDD (aluminium killed) steels and bake hardening IF steels.

7.2 Bake hardening EDD steels

The EDD steel grades contain carbon content of the level~ 0.01%. If the carbon is allowed to remain in solution, room temperature aging occurs. Thus, carbon levels must be controlled by suitable annealing practices. These steels can exhibit problems during galvannealing, however, their main disadvantage is poor formability due to the presence of relatively high carbon contents (**Baker et al., 2002**). Careful chemistry control during steelmaking is therefore crucial to ensure suitable amounts of carbon remain in solution in the final product.

During slow cooling of batch annealed EDD grade steels almost complete precipitation of the carbon occurs and thus the remaining solute carbon is insufficient to cause bake hardening. To obtain bake hardening effect in batch annealed EDD, very low carbon grades of C~ 0.01% with some elements, like phosphorous, that increase carbon concentration in solution are employed. (**Beranger et al.**).

While in the case of continuous annealing, the carbon level in solid solution at the end of annealing is quite high and the solute carbon can produce strengthening (70-80 MPa) by bake hardening (steel book). Care must be taken, however, with overaging practice during annealing, to ensure an appropriate amount of carbon (15-25 ppm) is left in solution in the final product.

7.3 Bake hardening IF steels

IF steels contain very low amounts of total carbon (~0.004%) (**Baker et al., 2002**).. In these steels all the interstitial elements are removed from solution by addition of carbide and nitride formers such as aluminium, titanium, niobium. These steels do not exhibit bake hardening as they have no interstitial elements in solution. However, the chemistry and processing of these steels, can be adjusted to leave 15 -25 ppm carbon in solution, to render bake hardening effect for increasing strength by 30 - 60 MPa . There are several types of IF steels and in the following sections Ti-only, Nb-only and Ti+Nb steels will be discussed for bake hardening.

7.3.1 Titanium stabilized interstitial free bake hardening steels

This mechanism of precipitation applies only when some titanium remains in solution: in just stabilized or understabilized chemistries, formation of TiC and $Ti_4C_2S_2$ is inhibited because of the low titanium content (Baker et al., 2002). This successive precipitation of carbides and carbosulphides makes it difficult to control the amount of titanium available for carbon stabilization and thus the solute carbon content. However, a bake hardening product can be produced from a titanium chemistry by inhibiting or avoiding TiC formation, so that the total carbon content remains in solution and is available for bake hardening.

Work by Tanioku et al. (**Tanioku et al., 1991**) and Kojima et al. (**Kojima et al., 1993)** showed that this can be achieved by controlling total carbon at 15 -25 ppm and titanium at ~ 0.01%. The manganese content must be kept low (~ 0.3%) to prevent the formation of MnS in preference to TiS and the slab reheat temperature must be high (~ 1200°C) to prevent the precipitation of $Ti_4C_2S_2$. Another method for producing titanium IF bake hardening steels, discussed by Tsunoyama et al. (**Tsunoyama et al, 1998**) relied on reducing the sulphur level to minimize formation of TiS, as the role of TiS as a heterogeneous nucleation site for the precipitation of TiC can lead to the reduction in solute carbon and hence bake hardening response. Kawasaki et al. (**Kawasaki et al., 1991**) used a philosophy, involving a reduced sulphur level (0.005%) in association with an increase in manganese to 1.0% and thus, the formation of $Ti_4C_2S_2$ was suppressed, leaving carbon in solution and producing a bake hardening steel.

7.3.2 Niobium stabilized interstitial free bake hardening steels

Niobium is a strong carbide former that can stabilize carbon as NbC when added according to the stoichiometric ratio:(%Nb)=7.75(%C) (**Baker et al., 2002**).).

Nitrogen is stabilized by the addition of aluminium to form AlN. This is more stable and thus forms at higher temperatures than Nb(C,N), so it can be assumed that all niobium is available for carbide formation. The relative simplicity of carbon stabilization in these steels makes them ideal for the production of bake hardening grades. Control of solute carbon in the niobium bearing bake hardening steels can be achieved in two ways. First, insufficient niobium can be added fully to stabilize the carbon, leaving 15 - 25 ppm in solution after steelmaking. This methodology requires tight chemistry control during steelmaking and, because of the presence of solute carbon throughout subsequent processing, the rm-value can suffer.

The second method requires the full stabilization of carbon during steelmaking. Solute carbon is then liberated by solution of NbC during annealing. By annealing at high temperatures (800-850°C) and cooling at 420 Ks^{-1},15 -25 ppm carbon can be retained in solution (**Irie et al., 1982**). Since the carbon is fully stabilized until the end of annealing, r_m-

values in steels of this type are comparable with those of traditional IF grades. High temperature annealing of this kind can, however, lead to shape problems in the strip such as heat buckling. Some continuous annealing lines can operate at these high temperatures, but they are beyond the limits of conventional hot dip galvanizing lines. Steelmakers must therefore assess their own production capabilities before deciding on a suitable processing route for the niobium IF based bake hardening steels.

By reducing the sulphur level and increasing manganese, the formation of TiS can be suppressed, leaving all titanium available for nitrogen stabilization. The precipitation of $Ti_4C_2S_2$ can also be suppressed in this way, so carbon is controlled by niobium alone. As with the Nb only compositions, carbon content can be controlled either by understoichiometric addition of niobium, or by high temperature annealing and controlled cooling, depending on the capabilities of individual steelmakers. These steels have been widely researched and are the choice of many manufacturers.

7.3.3 Titanium and niobium stabilized interstitial free bake hardening steels

By reducing the sulphur level and increasing manganese, the formation of TiS can be suppressed, leaving all titanium available for nitrogen stabilization (**Baker et al., 2002**).). The precipitation of $Ti_4C_2S_2$ can also be suppressed in this way, so carbon is controlled by niobium alone. As with the Nb only compositions, carbon content can be controlled either by understoichiometric addition of niobium, or by high temperature annealing and controlled cooling, depending on the capabilities of individual steelmakers. These steels have been widely researched and are the choice of many manufacturers.

8. Recrystallization texture and drawability of bake hardening steels

Kitamura et al. (**Kitamura et al., 1994**) had presented a completely different methodology for the production of titanium based interstitial free bake hardening steels. The theory suggests that as the steel sheet absorbs carbon by annealing in a carburizing atmosphere, the Ti/C ratio decreases, eventually resulting in some solute carbon in the matrix. A bake hardening response of 20 -50 MPa was achieved in this way without compromising r value (**Figure 31**) (**Kitamura et al., 1994**). Through this process, the reduction in r-value due to solute carbon in solution for interstitial free bake hardening steels can be eliminated as the steel remains fully stabilized by titanium and the excess carbon is introduced by carburizing atmosphere during annealing.

Xiaojun and Xianjin (**Xiaojun & Xianjin, 1995**) developed a new technology (the details were not mentioned) to improve the drawability of Ti+ Nb stabilized interstitial free high strength bake hardened steel. The r_m-value of the experimental sheet treated by the new technology is as high as 2.67, and this is the highest r_m-value published so far for phosphorus-added high strength and deep drawing sheet steels with increased strength due to bake hardening. Compared to conventional technology **(Figure 32a),** the new technology annealing rolling texture **(Figure 32b)** exhibited strong {111} components and weak {100}. The crystal orientations corresponding to the peak values of orientation concentrations of the texture were found to be changed from conventional (111)(112) orientations to (111)(011) orientations for the new technology (Figure 33).

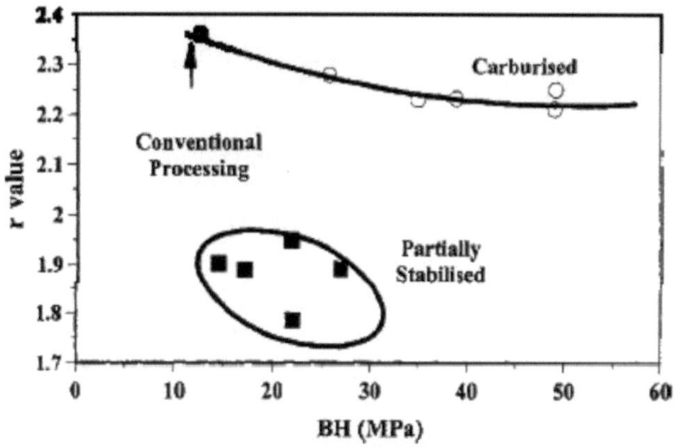

Fig. 31. Effect of processing on relationship between bake hardenability and r value for Ti-stabilized IF steel (**Kitamura et al., 1994**).

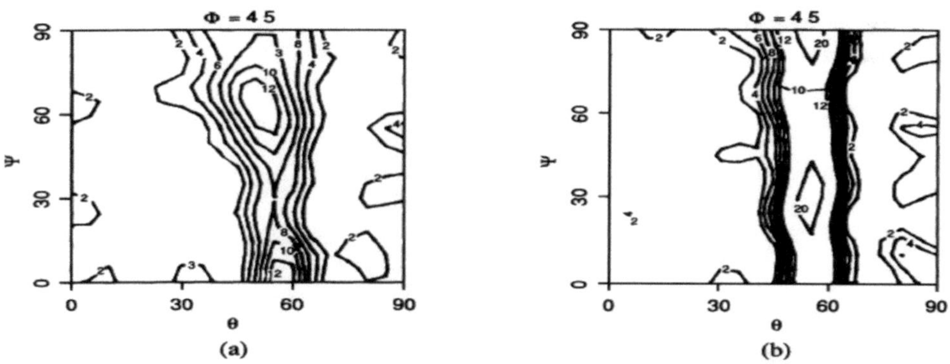

Fig. 32. ϕ=45° sections of ODFs of annealing textures obtained by two technologies respectively: (a) conventional processing and (b) new processing (**Xiaojun & Xianjin, 1995**).

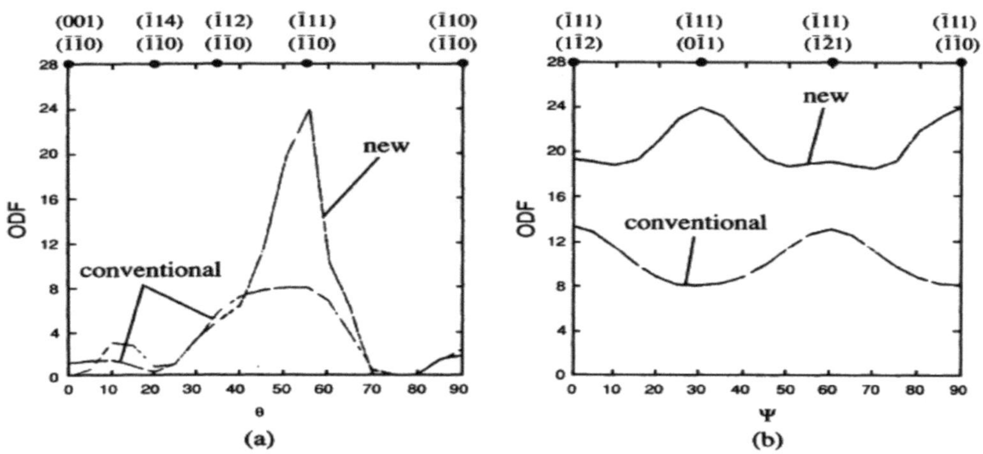

Fig. 33. Comparison of annealing textures obtained by two technologies respectively. (a) α-fiber axis textures (b) γ-fiber axis textures (**Xiaojun & Xianjin, 1995**).

Bake hardenability and drawability of IF steels with under- to over-stoichiometric atomic ratios of Ti/N and Nb/C were studied by Storojeva et al (http://www.cbmm.com.br/portug/sources/techlib/report/novos/pdfs/stabiliza.pdf). They observed that the r_m-value increased with a larger ferrite grain **(http://www.cbmm.com.br/portug/sources/techlib/report/novos/pdfs/stabiliza.pdf)** thus a high annealing temperature is favorable for good cold formability. The authors also reported that the r_m-value of the steels were higher that had lower solute carbon in the hot bands (**Storojeva et al., 2000**). This is confirmed (**Figure 34**) by just comparing the Ti+Nb containing steels, where the r_m-values of the steels with the high solute carbon in hot strip (12-14 ppm) were remarkably lower than those of the steels without any solute carbon in hot strip. However, the r_m-value of Ti-free, just Nb-containing steel with the high solute carbon content (14 ppm) in the hot strip was almost as high as in the Ti+Nb steels without any solute carbon. Thus, titanium free IF steel exhibits a lower recrystallization start temperature and by this means enhances r_m-value in the final product, allowing compensation of the negative effect of solute carbon in the hot band. **Figure 35 (http://www.cbmm.com.br/portug/sources/techlib/report/novos/pdfs/stabiliza.pdf)** summarizes test results of the r_m-value and BH-effect for annealing temperatures up to 840°C. It indicates, that a BH-effect >30 MPa together with r >1.7 can be obtained with the Ti-free, just Nb containing steel.

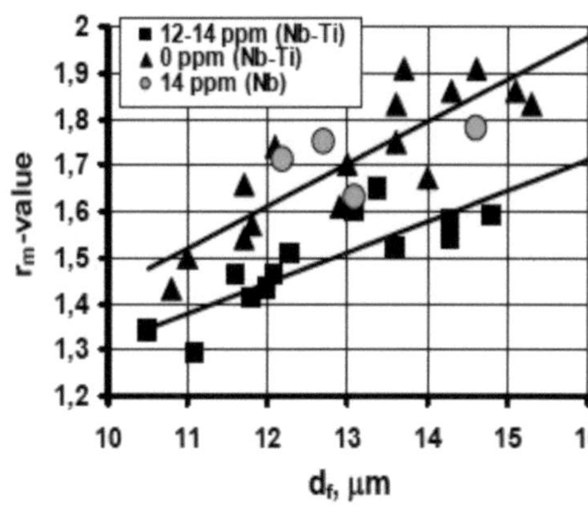

Fig. 34. r_m- value of steels with various solute carbon in hot band of IF steels.

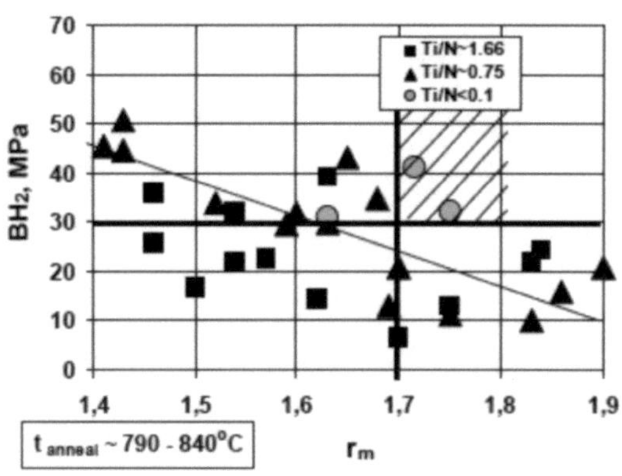

Fig. 35. r_m-value – BH-effect for Ti containing IF steels.

9. Recrystallization texture and drawability of warm rolled steels

The demand for thin and ultra thin rolled products led to an intensive search for alternatives to conventional hot and cold rolling processes. A number of studies were undertaken to investigate the advantages of rolling at temperatures between 850 and 500°C that is known as warm rolling (**Harlet et al., 1993**). Because of the lower reheat temperatures, warm rolling leads to lower production costs than hot rolling and requires significantly lower loads than cold rolling (**Harlet et al., 1993**). The latter factor means that higher reductions per pass can be produced by warm rolling than by cold rolling due to the lower plasticity of steel at room temperature. In conventional hot rolling, the finish rolling temperature is above the Ar3 temperature and in warm rolling finish rolling is made in the ferrite phase.

In the past years many researchers have worked on low, ultra carbon steels (**Sakata et al., 1997**) using warm rolling. The products obtained by ferritic hot rolling can be divided into two kinds according to the coiling temperature (**Mao, 2004**). One is a thin gauge soft and ductile hot rolled strip obtained by high temperature coiling for direct application that could be considered as a substitute for the conventional cold rolled and annealed sheet, and the other is a strained thin gauge hot strip gained by low temperature coiling for cold rolling and annealing, during which recrystallization texture strengthens by accumulating strains from hot rolling and cold rolling reductions.

Sa´nchez-Araiza et al (**Sa´nchez-Araiza, 2006**) reported on the texture changes in low carbon steel during recrystallization and established the nucleation and growth mechanisms applicable to warm-rolled quantitatively.

The requirement necessary for the development of (111) texture is the accumulation of strain in the matrix and to achieve uniform accumulation of strain through the thickness of sheet steel, lubricant is required in addition to the optimization of Ti and Nb concentration **(Figure 36) (Sakata et al., 1997)**. In 1996, Kawasaki Steel Corporation constructed a new hot strip mill in the Chiba works, where sheet bars are welded between the coil box and the finish mill to accomplish fully continuous rolling. This 'endless hot strip mill' makes lubricant rolling practical. The r_m-value achieved using warm rolling technique is 2.9. This value is noteworthy since the best value obtained in the conventional process does not exceed 2.6.

Wang et al (**Wang et al., 2007**) studied drawability of Ti-stabilized IF steel by finish rolling the steel at 760°C in the ferritic region with lubrication and coiling at 740 and 400°C. The evolution of texture for both the high and low temperature coiling is shown in in **Figures 37-40.** The optical micrographs of the test steels in hot rolled and high temperature coiled status, cold rolled as well as annealed one are shown in **Figure 37**. It can be seen from **Figure 37(a)** that after high temperature coiling, the deformed microstructure vanished and the recrystallization microstructure is characterized by uniform and equiaxed grains. **Figure 37(b)** shows that after cold rolling, the grains can not be discerned and the obvious characteristics of the cold rolled microstructure is the formation of the in-grain bands denoted by the arrow. As shown in **Figure 37(c),** deformed microstructure disappears and there are small and elongated grains after annealing. In order to obtain more equiaxed grains, the annealing temperature or the annealing time should be increased.

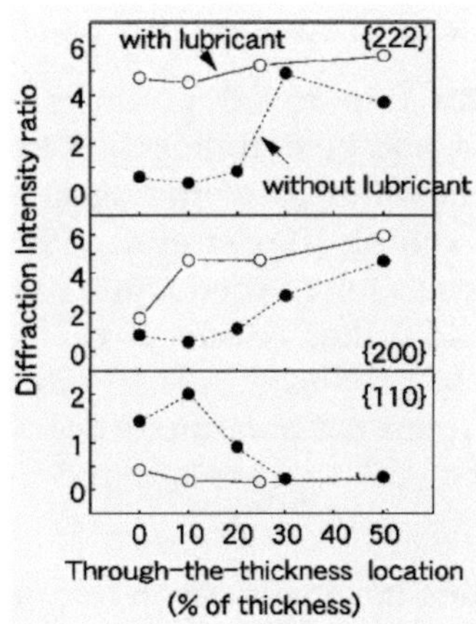

Fig. 36. Effect of lubricant on texture distribution through the thickness in ferrite rolled sheet steel; with finish rolling temperature 700 °C, cold reduction 50%, and annealing time 20s at 850° C (**Sakata 16 et al., 1997**)

Fig. 37. Optical microstructures of samples in the condition of high temperature coiling: (a) hot rolled (b) cold rolled and (c) annealed samples (**Wang et al., 2007**).

Figure 38 shows the ϕ=45° ODF sections for high temperature coiled hot band, cold rolled and annealed sample. It is clear that after ferritic rolling and high temperature coiling, the most prominent texture intensity is along the γ-fiber and the maximum is at {111}<112>. Moreover, its characteristics are the same as that of the annealed texture in the condition of low temperature coiling, indicating that the hot band after high temperature coiling can be considered as a substitute for the conventional cold rolled and annealed sheet **(Wang et al., 2007).**

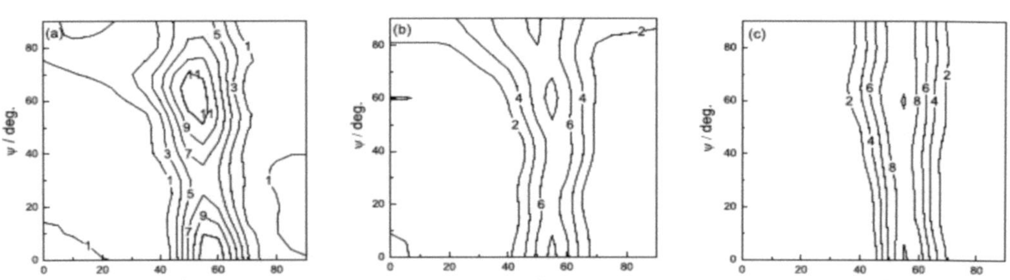

Fig. 38. ϕ=45° ODF sections in the condition of high temperature coiling: (a) hot rolled, (b) cold rolled and (c) annealed samples (**Wang et al., 2007**).

The optical micrographs of the low temperature coiled hot band, cold rolled as well as annealed one are shown in **Figure 39**. It is evident that a completely deformed microstructure is produced after hot rolling and low temperature coiling. Straighter grain boundaries and thinner deformation bands form after cold rolling. After annealing, the ferrite grains recrystallize completely, and small and uniform grains develop. **Figure 40** shows $\phi 2=45^{o}$ODFs of low temperature coiled hot band, cold rolled and annealed samples. The texture of hot band includes a strong α-fiber whose peak is at {001}<110> as well as a weak γ-fiber whose main component is {111} <110>. The components in the α-fiber intensify and the intensity of {111}<112> in the α fiber changes little after cold rolling. A complete γ-fiber with the peak at {111}<112> develops and the components in γ-fiber weaken evidently after annealing.

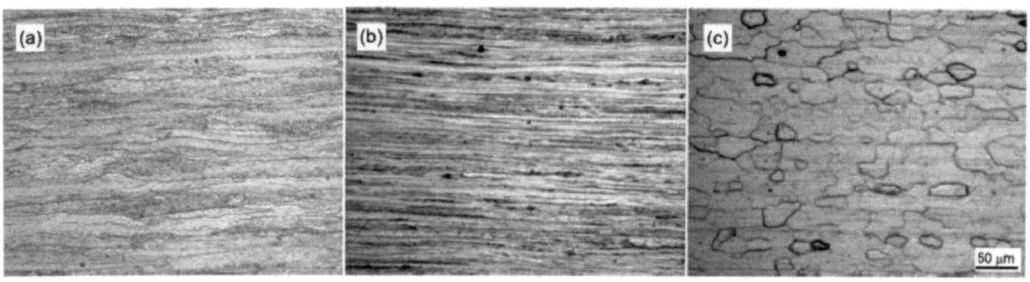

Fig. 39. Optical microstructures of samples in the condition of low temperature coiling: (a) hot rolled (b) cold rolled and (c) annealed samples (**Wang et al., 2007**).

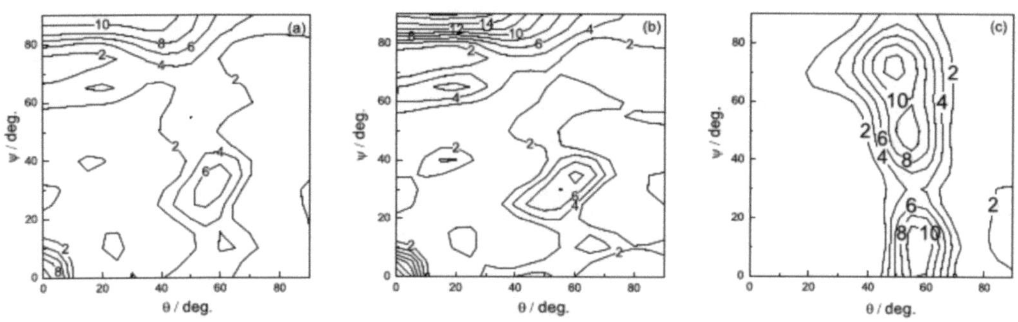

Fig. 40. $\phi=45^{o}$ ODF sections in the condition of low temperature coiling: (a) hot rolled, (b) cold rolled and (c) annealed samples (**Wang et al., 2007**).

In a study on the effect of chemical composition and ferritic hot rolling on the formation of texture in the hot rolled and coiled, cold rolled, and cold rolled and annealed Ti and Ti+Nb added IF steels, it was reported that high r_m-values could be obtained at the same intensity levels of {111}<110> texture component if the grain growth was inhibited by a suitable addition of microalloying elements. (**Tiitto et al., 2004**). While as per the conventional wisdom it is known that grain growth during annealing is beneficial for drawability as it increases the volume fraction of grains with the {111} texture, leading to higher r_m-value. It was claimed that due to the grain refining effect of Nb, a small addition of the element (100 ppm) to the Ti alloyed IF steel increased the r_m-values in the annealed condition by about 15% at the same {111}<110> intensity levels. Thus, ferritic hot rolling seemed to be beneficial only if it contributed to the development of a strong intensity of the γ -fiber texture in association with a uniform and small grain size during annealing. Further they added, hot deformation of Nb and Ti alloyed steels at a high temperature (870°C) in the ferritic region led to higher {111}<110> intensities and higher r- values in the annealed condition than hot deformation at a lower temperature (800°C).

The development of the γ-fibre in annealed steels was linked (**Barnett & Jonas, 1997a; Duggan et al., 2000**) to the presence of a high volume fraction of grains containing shear bands after warm rolling. These bands appear to be the nucleation sites for recrystallised grains of the desirable orientation. After warm rolling, many of the grains in interstitial-free (IF) steels contain such shear bands and the steels then exhibit good forming characteristics after annealing (**Barnett, 1998**). By contrast, in low carbon steels, the presence of carbon in solid solution leads to dynamic strain aging during warm rolling and to high positive rate sensitivities. The latter prevent the formation of high densities of in-grain shear bands, leading to a lack of nucleation sites for the γ-fibre during annealing.

Timokhina et al **(Timokhina et al., 2004)** studied the effect of in grain shear bands on the volume fraction of favourable γ-fibre in IF, low carbon (LC) and LC with Cr, P and B added steels. Shear bands are usually contained within single deformed grains and are tilted by 20–35° with respect to the rolling plane. There are four types of in-grain shear bands: long (5-40 μm long--±15-40° to rolling direction), short (0.5-15 μm long), intense short (0.4-7 μm, ±5-45°) and intense long (continuous wavy lines--±15-40° to rolling direction) (**Barnett & Jonas, 1997b**) as shown in **Figure 41.** All the grains containing shear bands were characterized by zones of grain boundary displacement or stepping that provide evidence for local flow along the bands (**Barnett, 1996**). The presence of moderate amounts of long shear bands in IF steels was attributed to the formation of the γ-fibre after annealing (**Barnett & Jonas, 1997a**).

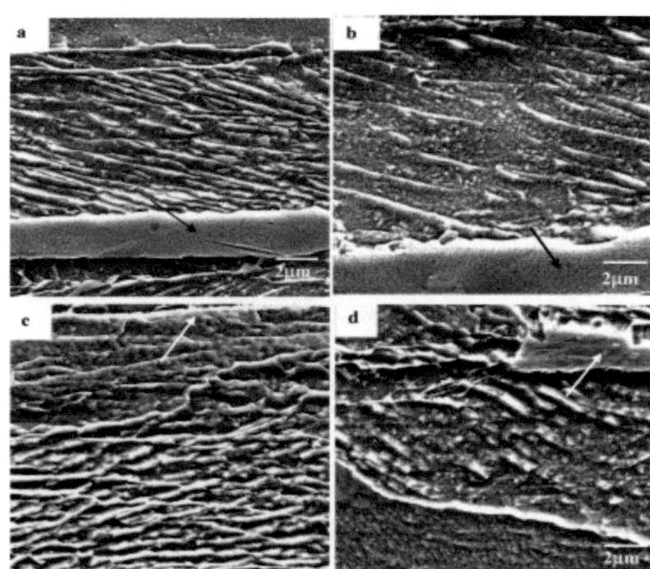

Fig. 41. SEM micrographs of (a) long shear bands (b) short shear bands, (c) intense long shear bands and (d) intense short shear bands. Arrows indicate the displacement zones (**Barnett & Jonas, 1997b**).

In their study, Timokhina et al (**Timokhina et al., 2004**) reheated the sample to 1050°C for 30 mins and subsequently warm rolled by 65% in a single pass pilot mill. Temperatures of 640°C and 710°C and average strain rates of $30s^{-1}$ were employed followed by quenching. To establish grain orientation with different sheer bands, the grains were marked with hardness tester prior to electron back scattered diffraction analysis. The addition of alloying elements was observed to affect different types of shear band and the formation of γ -fibre. The IF steel was characterized by the largest number of grains with long shear bands with γ -fibre, 65 % and 67 % in the steels rolled at 640°C and 710°C respectively. Increasing the amount of carbon and decreasing the grain size, as in the LC steel, increased the proportion of grains with the short shear bands. Further, the proportion of grains containing intense long shear bands was higher in the LC steel than in the IF grade. The addition of carbon perceptibly decreased the volume fraction of grains with γ-fibre. The addition of chromium led to the formation of similar volume fractions of grains with short, long and intense long shear bands at both rolling temperatures. This led to an increase in the number of grains with beneficial γ -fibre to 46–47% compared to 25–26 % in the LC steel. However, in a subsequent study Pereloma et al (**Pereloma et al., 2004**) reported that although the formation of chromium carbides in the microstructures of the LC-Cr and LC-Cr+P steels removed carbon from solid solution and in this way slightly increased the fraction of γ-fiber nuclei formed at shear bands, the effect led to a reinforcement of the ND component (γ-fiber) only in the early stages of recrystallisation. The strength of γ-fiber gradually deteriorated with the progress of recrystallisation. The reasons put forward for this undesirable development were i) the retarding effect of the carbides on the mobility of grain boundaries during growth, ii) the influence of the particles on nucleus rotation during annealing and iii) the absorption of γ-fiber nuclei by other components during grain coalescence and growth. Thus, further study of the texture behaviour during grain coalescence and growth was required.

The addition of boron suppressed the formation of long, short and intense long shear bands and assisted in the formation of intense short shear bands that, in turn, decreased the volume fraction of grains with γ -fibre. The addition of phosphorus, on the other hand, increased the long and short shear band frequency and the proportion of grains with γ -fibre. In the Cr–B–Ti modified steel, the short, long and intense long shear bands were absent and they were replaced by intense short shear bands.

Jing et al (**Jing et al., 2011**) analyzed the grain boundary and microtexture characters of a rephosphorised high strength IF-steels under 700 and 800°C warm-rolled temperatures to observe the effect on deep drawability. It was found that while the samples were rolled at 700°C, more γ-fiber texture components, {111}<112>, {111}<110>, {554}<225>, low angle and CSL grain boundaries were formed that were beneficial for deep-drawability. However, the samples that were rolled at 800°C, manifested more α-fiber texture components and high angle grain boundaries that led to inferior deep drawing property. The average r-value was 1.32 for the samples rolled at 700°C and 1.05 for those that were rolled at 800°C, respectively.

Recrystallization texture investigation for IF-Ti and Ti stabilized IF-HS was conducted by Wang et al (**Wang et al., 2006**) under ferritic hot rolling and high-temperature coiling. Comparing with the completely recrystallized textures of the ordinary IF steel, the textures of the high-strength IF steel were of deformation type. This was attributed to the high phosphorous content in the high-strength IF steel that prevented recrystallization during the coiling process. For the ordinary IF steel, the texture components were mainly very weak rotated cube component {001}<110>at the surface, and partial α-fiber with key orientation {223}<110>orientation and <111>| |ND texture at the mid-section and 1/4-section. While, for the high-strength IF steel, the texture components were orientation (Goss) at the surface and a sharp α-fiber extending from {001}<110> to {223}<110〉 in association with a Weak <111>| |ND texture at the mid-section and 1/4-section.

In another work by Ferry et al. (**Ferry et al., 2001**) on ultra low carbon steel (0.0036C, 0.03Ni, 0.019Mn, 0.03Al, 0.004Ti and 0.003N) it was found that the hot deformation microstructure had a strong influence both on the kinetics of recrystallization and texture development during cold rolling and annealing. In particular, a warm-deformed ferrite microstructure (lower finish deformation temperature (FDT)) recrystallized most rapidly to produce a strong <111>| |ND recrystallization texture (γ -fibre) as it produces as-strained α in combination with the additional strain by cold rolling, which result in rapid recrystallization. while an initial coarse-grained ferrite microstructure recrystallized most sluggishly to produce a strong {001}<110> texture due to copious nucleation of grains at shear bands, which is consistent with previous studies (**Hutchinson, 1984; Muljono, 2001**). **Figure 42** (**Ferry et al., 2001**) shows φ_2=45° ODF sections in the fully recrystallized cold rolled and annealed samples following hot deformation at three significant finish rolling temperatures: (a) 920°C (fine, equiaxed ferrite), (b) 850°C (coarse ferrite) and (c) 600°C (warm deformed ferrite). The maximum intensity of the two most dominant recrystallization texture components, {001}<110> and {111}//ND, as a function of FDT are given in **Figure 43** (**Ferry et al., 2001**). It can be seen that the development of the strongest {111}//ND CRA texture is favoured when FDT is: (i) greater than 870°C (which produces fine-grained ferrite by transformation), and (ii) below, 800°C (which also produces fine-grained ferrite but with an additional true strain of 0.8 prior to cold rolling). Thus, it is indicated that warm rolling has a significant influence on final texture after cold rolling and annealing and warm rolling is capable of strengthening the {111}//ND (γ-fiber) recrystallization texture that is the favourable texture for drawability in the production of formable ultra low carbon steel sheets.

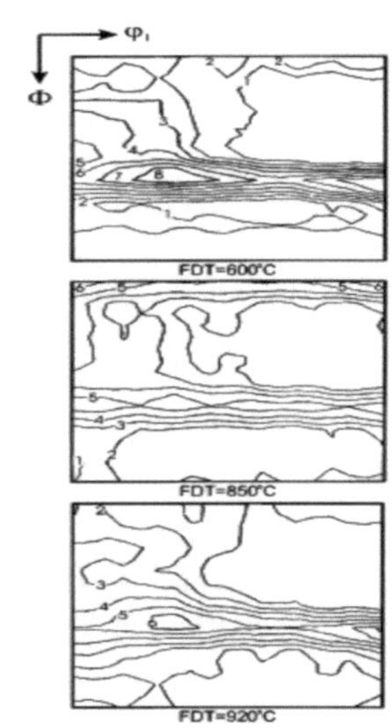

Fig. 42. φ_2=45° sections in Euler space showing recrystallization textures of the ultra low carbon steel with FDTs 600°C, 850°C and 920°C (contours: 1, 2, 3...random) (**Ferry et al., 2001**).

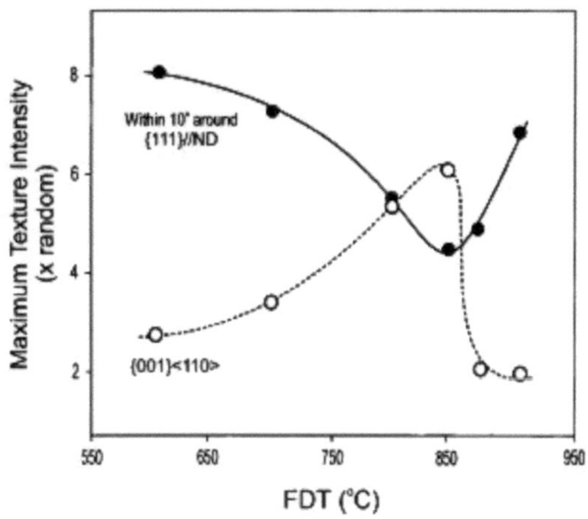

Fig. 43. Effect of FDT on maximum texture intensity of {001}<110> and within 10° around <11l> | | ND (γ -fibre) of the annealed ultra low carbon steel (the maximum intensity is taken from the each calculated ODF) (**Ferry et al., 2001**).

10. References

Auburn, Ph. & Rocquet P. (1973). Mem. Sci. Rev. Met., Vol. LXX, No. 4, p. 261.

Baker L. J., Daniel S. R & ParkerJ. D. (2002). Mater. Sci. Technol. Vol. 18, No. April, pp. 355-367.

Banerjee K., Verma A. K. & Venugopalan T.(2008). Metall. and Mater. Trans. A, Vol. 39, No. June, pp. 1410-1425.

Barnett M. R. & Jonas J. J. (1997). ISIJ Int., Vol. 37 pp. 697-705.

Barnett M. R. & Jonas J. J. (1997). ISIJ Int., Vol. 37, pp. 706-714.

Barnett M. R. & Jonas J. J. (1999).ISIJ Int., Vol. 39 pp. 856-873.

Barnett M. R. (1996).Ph.D. Thesis, McGill University, p. 52.

Barnett M. R. (1998). ISIJ Int., Vol. 38 pp. 78-85.

Barnett M.R. & Kestens L.(1999). ISIJ Int., Vol. 39, pp. 923-929.

Barrett M.R. & Jonas J.J.(1997) , ISIJ Int., Vol. 37, pp. 706-714.

Beranger G., Henry G & Sanz G. 1996.The book of Steel, Springer-Verlag, USA, 2-85206-981-18 (1994), USA, pp. 935-951.

Cao W. D. , Lu X. P. , Sprecher A. F. & Conrad H. (1990). Mater. Lett., Vol. 9, p. 193.

Cao S. Q. (2005). Grain Boundaries and the Evolution of Texture in Interstitial-free (IF) Steels, Ph.D. Thesis, Shanghai Jiao Tong University.

Chung J. , Era H. & Shimizu M. (1987). Metall. Trans. A, Vol. 18, No. June , pp. 957-968.

Conrad H. , Guo Z. , Sprecher A. F. (1989). Scripta Metall .Vol. 23, pp. 821-823.

Doherty R.D. (1985). Scr. Metall., Vol. 19, pp. 927-930.

Doherty R.D. , Hughes D.A. , Humphreys F.J. , Jonas J.J., Juul Jensen D., Kassner M.E., King W.E., McNelley T.R. McQueen H.J. & Rollett A. D. (1997). Mater. Sci. Eng. A, Vol. 238, pp. 219–274.

Duggan B. J. , Liu G. I. , Ning H. , Tse Y. Y. (2000). Proc. Int. Conf. Thermomechanical Processing of Steels, IOM Communications Ltd. London, UK, pp. 365 - 371.

El-Kashif E., Asakura K. & Shibata K. (2003). ISIJ Int., Vol. 43, No. 12, pp. 2007-2014.

Ferry M., Yu D. & Chandra T. (2001). ISIJ Int., Vol. 41, No. 8, pp. 876–882.
Gupta I. and Bhattacharya D. (1990). Metallurgy Of Vacuum Degassed Steel Products, ed. R. Pradhan, TMS, Warrendale, PA, TMS, pp 43– 72.
Harlet P. , Beco F. , Cantinieaux P. , Bouquegneau D. , Messien P. & Herman J. C. (1993). Int. Symp. on Low C Steels for the 90's, eds. R. Asfahani & G. Tither, TMS-AIME, Warrendale, PA, p. 389.
Hatherly M. & Hutchinson W. B. (1979). An introduction to textures in metals, The Institution of Metallurgists, Monograph 5.
He C.S. , Zhang Y.D., Wang Y.N. , Zhao X. , Zuo L. & Esling C. (2003). Scripta Mater., Vol. 48, pp. 737–74.
Hebert V., Louis P. , Zimmer P. & Delaneau P. (1992), CESSID document 92154.
Heckler A. J. & Granzow W. G. (1970). Metall. Trans A, Vol. 1, pp. 2089-94.
Held J. F. (1965). Mechanical Working and Steel Processing IV, ed.D. A. Edgecombe, American Institute of Mining, Metallurgical and Petroleum Engineers, New York, p. 3.
Holie S. (2000). Mater. Sci .Technol., Vol 16, pp. 1079-1093.
Hölscher M. , Raabe D. & Lu¨cke K. (1991). Steel Res., Vol. 62, No. 12, pp. 567-75.
Hook R. E. (1990). Metallurgy of vacuum-degassed steel products (ed. R. Pradhan), 1990, Warrendale, PA, Metallurgical Society AIME, p.263.
Hosford W. F. & Backholen W.A. (1964). Fundamentals of deformation processing, Syracuse, Press, New York. P. 259.
Hu H. & Goodman S.R. (1970). Metall. Trans., Vol. 1, pp. 3057-64.
Hu H. (1977). Metall. Trans. A, Vol. 8, pp. 1567-75.
Hughes I.F. & Page E.W. (1971). Metall. Trans., Vol. 2, pp. 2067-75.
Hutchinson W. B. (1984). Intl. Mater. Reviews, Vol. 29, No. 1, pp. 25-42.
Hutchinson W.B. & Ushioda K. (1984). Scand. J. Met., Vol. 13, p. 269-284.
Irie T. , Hashiguchi H. , Satoh S. , Konoshi M. , Takahashi K. & Hashimoto M (1981).Trans. Iron Steel Inst. Jpn., Vol. 21, No. 11, pp. 793-801.
Irie T. , Satoh S. , Yasuda A. & Hashimoto O. (1982). Metallurgy of Continuously Annealed Sheet Steel, TMS, Warrendale, PA, TMS, pp. 155-171.
Jeong W. C. (2000). Metall. and Mater. Trans. A, Vol. 31, No. April, pp. 1305-1307.
Jing C. , Wang M. , Liu X. , Tan Q. , Wang Z. & Han F. (2011). Mater. Sci. Forum, Vol. 682, pp. 71-74.
Juntunen P. , Raabe D., Karjalainen P. , Kopio T. & Bolle G. (2001). Metall. & Mater. Trans. A, Vol. 32, No. Aug, pp. 1989-95.
Kang H. , Garcia C. I. , Chin K. & DeArdo A. J. (2007). ISIJ Int., Vol. 47 No. 3, pp. 486–492.
Katoh H. , Takechi H. , Takahashi N. & Abe M. (1985). Int. Conf. on Technology of Continuously Annealed Cold Rolled Sheet Steels, ed. Pradhan. R, Proc. . TMS-AIME, Warrendale, PA, USA, pp. 37-60.
Kawasaki K., Senuma T. & Sanagi S. (1991). Processing, Microstructure and Properties of Microalloyed and other Modern HSLA Steels, ISS, Warrendale, PA, pp. 137 - 144.
Kestens L.., Jonas J. J., Van Houtte P. & Aernoudt E. (1996). Textures and Microstructures, Vol. 26-27, pp. 321-335.
Kim S., Choi I., Park I. & K. Cho (2005). Mater. Sci. Forum, Vol. 475-479, pp. 475-479.
Kitamura M. , Tsukatani I. & Inoue T.(1994) : ISIJ Int., Vol. 34, No. 1, pp. 115 - 122.
Klein A. J. & Hitchler, E. W. (1973). Met. Eng. Q., Vol. 13, pp. 25—27.

Kojima N. , Mizui N.& Tanioku T. (1993). Sumitomo Search, Vol. 45, No. 5, pp.12 - 19.

Kozeschnik E. , Pletenev V. , Zolotorevsky N. & Buchmayr B. (1999) , Metall. & Mater Trans. A, Vol. 30, No. June, pp. 1663-1673.

Krauss G., Wilshynsky D. O. & Matlock D. K. (1991). Interstitial Free Steel Sheet: Processing, Fabrication and Properties, eds. L. E. Collins and D. L. Baragar, CIM/ICM1, Ottawa, pp. 1- 14.

Lankford W. T. , Snyder S. C. & Bauscher J. A. (1950). Trans.AS1/I, Vol. 42, pp.1197—1232.

Lebrun J. L., Maeder G. & Parniere P.(1981) : Proc. 6th Intl. Conf on Texture of Materials, Vol. 2, Tokyo, The Iron and Steel Institute of Japan, p. 787.

Mao X. (2004). Iron Steel, Vol. 39, No. 5, p. 71.

Martin, J.W., Doherty R. D. , Cantor B. (1997). Stability of Microstructure in Metallic Systems (2nd edition)., Cambridge University Press, Cambridge.

Matlock D. K., Allan B. J. & Speer J. G. (1998). Proc. Conf. Modern LC and ULC Sheet Steels for Cold Forming Processing and Properties, ed. W. Bleck, Aachen, Verlag Mainz, pp. 265 - 276.

Matsudo K. , Osawa K. & Kutihara K. (1984). Technology of Continuously Annealed Cold-Rolled Sheet Steel, ed. R. Pradhan, TMS-AIME, Michigan, pp. 3-36.

McQueen H.J. , Rollett A.D. (1997). Materials Science and Engineering, Vol.A238, pp. 219–274.

Mendoza R. , Huante J. , Alanis M., Gonzalez-Rivera C. & Juarez-Islas J. A. (2000). Mater. Sci. Eng A, Vol. 276, pp. 203-209.

Mendoza R. ,Alanis M. , Aramburo G. , Serrania F. & Juárez-Islas J.A. (2004). Mater. Sci. Eng.A, Vol. 368, pp. 249–254.

Meyzaud Y.& Parniere P.(1974). Mem. Sci. Rev. Met., Vol. LXXI, No. 7-8, pp. 423.

Mizui N. , Okamoto A. & Tanioku T.(1990). Proc. LTV/SMI Technology Exchange Meeting, Ltv Steel/Sumitomo Metal Industries.

Muljono D. , Ferry M. & Dunne D.P.(2001). Mater. Sci. Eng. A, Vol. 303, pp. 90–99.

Nishimoto A. , Inagaki J. & Nakaoka K.(1982).Tetsu-to-Hagne, Vol. 68, pp.1404-1410.

Ono S. , Shimomura T., Osawa K. & Matsudo K.(1982). Transaction ISIJ, Vol. 22, pp. 732-738.

Park Y.B. , Kestens L.& Jonas J.J. (2000). ISIJ Int., Vol. 40, pp. 393-401.

Pereloma E. V. , Timokhina I. B. , Nosenkov A. I. & Jonas J. J. (2004). Metallurgija, Vol. 43, No. 3, pp. 149-154.

Perera M., Saimoto S. & Boyd D. (1991). Interstitial Free Steel Sheet: Processing, Fabrication and Properties, eds.. L. E. Collins and D. L. Baragar, ;Ottawa, CIM/ICM, pp. 55 - 64.

Ray R. K. & Jonas J. J. (1990). Int., Mater. Rev., Vol. 35, No. 1, pp. 1-36.

Ray R. K., Jonas J. J. & R. E. Hook (1994). Intl. Mater. Reviews, Vol. 39, No. 4, pp. 129-172.

Rege J. S. , . Garcia C. I & DeArdo A. J.(1997). Proc. 39th Mechanical Working and Steel Processing, Vol. 35, ISS, Warrendale, PA, USA, pp. 149-158.

Rege J. S., Hua C., Garcia I. & DeArdo A. J. (2000). ISIJ Intl., Vol. 40, No. 2, pp.191-199.

Ruiz-Aparicio L.J. , Garcia C.I. & Deardo A.J. (2001) , Metall.and Mater. Trans. A, Vol. 32, No. September, pp. 2325-2334.

Sa´nchez-Araiza M. , Godet S. , Jacques P.J. & Jonas J.J. (2006). Acta Mater. , Vol. 54, pp. 3085–3093.

Sakata K. , Matsuoka S. , Obara T. , Tsunoyama K. & Shiraishi M.(1997). Materia. Japan, Vol. 36 No. 4 p. 376.
Samajdar I. (1994).Ph.D. Thesis, Drexel University.
Sarkar B., Jha B. K. & Deva A. (2004). J. Mater. Eng. and Perform., Vol. 13, No. 3, pp. 361-36
Schulz L. G. (1949). J. Appl. Phys. Vol. 20, No.11, pp.1030-33.
Song X., Yuan Z. Jia J. , Wang D. , Li P. & Deng Z (2010). J. Mater. Sci .Technol. Vol. 26, No. 9, pp.793-797.
Storojeva L. , Escher C. , Bode R. , K. Hulka & Yakubovsky O. (2000). IF Steels 2000, ISS, Warrendale, PA, p. 289.
http://www.cbmm.com.br/portug/sources/techlib/report/novos/pdfs/stabiliza.pdf
Takahashi M., Okamoto, A. (1974). Sumimoto Met., Vol. 27,pp. 40-49.
Tanioku T. , Hobah Y. , Okamoto A. & N. Mizui (1991). SAE Technical Paper 910293, Society of Automotive Engineers, Warrendale, PA, USA.
Tiitto K. M. , Jung C. , Wray P. , Garcia C. I. & DeArdo A. J. (2004). ISIJ Int., Vol. 44, No. 2, pp. 404-413.
Timokhina I. B. , Nosenkov A. I. , Humphreys A. O. , J. J. Jonas & Pereloma E. V.(2004). ISIJ Int., Vol. 44, No. 4, pp. 717-724.
Tither G. & Stuart H. (1995). HSLA Steels '95', ed. L. Guoxun et al.., Chinese Society for Metals, Beijing, pp. 22-31.
Tither G., Garcia C. I., Hua M. & Deardo A. J. (1994). Int. Forum for. Physical Metallurgy in IF Steels, Iron and Steel Institute of Japan, Tokyo, pp. 293-322.
Tokunaga Y. & Yamada M. (1985). Method for the Production of Cold Rolled Steel Sheet Having Super Deep Drawability, US Patent 4,504,326.
Tokunaga Y. & Kato H. (1990). Metallurgy of Vacuum Degassed Products , TMS, Warrendale, PA, pp. 91 -108.
Tsunoyama K. (1998). Phys. Stat. Sol. (A), Vol. 167, No. 427, pp. 427-433.
Tsunoyama K. , Sakata K., Obara T. , Satoh S. , Hashiguchi K. & Irie T.(1988). Hot and Cold Rolled Sheet Steels, eds. R. Pradhan and G. Ludkovsky, TMS, Warrendale, PA, pp. 155 - 165.
Tsunoyama K., Satoh S., Yamasaki Y. & Abe H. (1990). Metallurgy of Vacuum Degassed Products, 1990, TMS, Warrendale, PA, pp. 127 -141.
Wang Y. N. , He C. S. , Zhao X. , Zuo L., Zhi Q. Z. & Liang Z. D (2000). Acta Metall Sinica, Vol. 36, No. 2, p. A126.
Wang Z. , Guo Y. , Xue W., Liu X. & Wang G.(2007). J. Mater. Sci. Technol., Vol. 23 No.3, pp. 337-341.
Wang Z.D. , Guo Y.H. , Sun D.Q. , Liu X. H. & Wang G.D. (2006). Mater. Charact., Vol. 57, No. 4-5, pp. 402-407.
Whiteley R. L. & Wise D. E (1962). Flat rolled products III, Interscience, New York, pp. 47—63.
Wilshynsky-Dresler D. O., Matlock D. K. & Krauss G. (1995): ISS Mech. Work. Steel Process. Conf., 1995, 33, pp. 927 - 940.
Wilson D. V. (1966). J. Inst. Met., Vol. 94, pp. 84—93.
Xiaojun G. & Xianjin W. (1995). Textures and Microstructures, Vol. 23, pp. 21-27.
Yasuhara E., Sakata K. , Furukimi O. & Mega T. (1996): Proc. 38th Mechanical Working and Steel Processing, Vol,. 34, Cleveland, Ohio, USA, pp. 409-415.
Yoshicla K. at. (1974): Deep Drawing Research Group, Proc. 8th Biennial IDDIG Congr. Gothenburg, 1974, pp. 258—268.

8

The Failure Mechanism of Recrystallization – Assisted Cracking of Solder Interconnections

Toni T. Mattila and Jorma K. Kivilahti
Aalto University
Finland

1. Introduction

The typical user environment load spectrum varies significantly between different electronic applications but changes in temperature are involved in nearly all of them. Owing to the increasing number of integrated high-performance functions, smartphones and handheld computers, for example, can experience significant changes in temperature during normal operation. The changes in temperature are typically caused either by internally generated heat dissipation or changes in the temperature of the environment. Today the maximum temperatures inside modern portable electronic products are in the range of 60-70 ^{0}C but occasionally they can even rise above 90 ^{0}C [1,2].

The thermomechanical reliability of electronic component boards has been one of the most studied aspects in the field for several decades. Sustained interest in this topic has endured primarily because: (a) the power densities and heat dissipation of novel electronic devices have increased; (b) electronic devices are being designed for use in ever harsher environments, such as the engine compartments of automobiles, and (c) newly developed materials whose long-term behavior in electronic applications is still unknown are continuously being introduced into electronic assemblies.

Thermomechanical strains and stresses in electronic assemblies are produced when the thermal expansion and contraction of materials is restricted. The standard thermal cycling tests extend the temperature range of electronic devices under normal operating conditions in order to accelerate the accumulation of failures. The maximum extreme temperatures in some of the standards are set to -65 ^{0}C and +150 ^{0}C, but the conditions of -45 ^{0}C and +125 ^{0}C with 15- to 30-minute dwell times are most commonly used [3-5]. The coefficients of thermal expansion (CTE) of most printed wiring boards (PWB) are much higher than those of most packages. For instance, the CTE of the most commonly used PWB base material, FR–4, is about 16-17 x 10^{-6}/˚C [6], whereas that of silicon is only 2.5 x 10^{-6}/˚C [7]. Furthermore, because of the large volume fraction of silicon in electronic packages (see Fig. 1a), packages have much higher rigidity than PWBs and, consequently, as the component boards are exposed to changes in temperature, strains and stresses are concentrated in the solder interconnections between the packages and the PWB, as the natural expansion/contraction of the PWB is restricted by the packages (see Fig. 1b). Therefore, the reliability of most electronic products under changing thermal conditions is determined by the ability of the solder interconnections to withstand thermomechanical loads. Thermomechanical strains

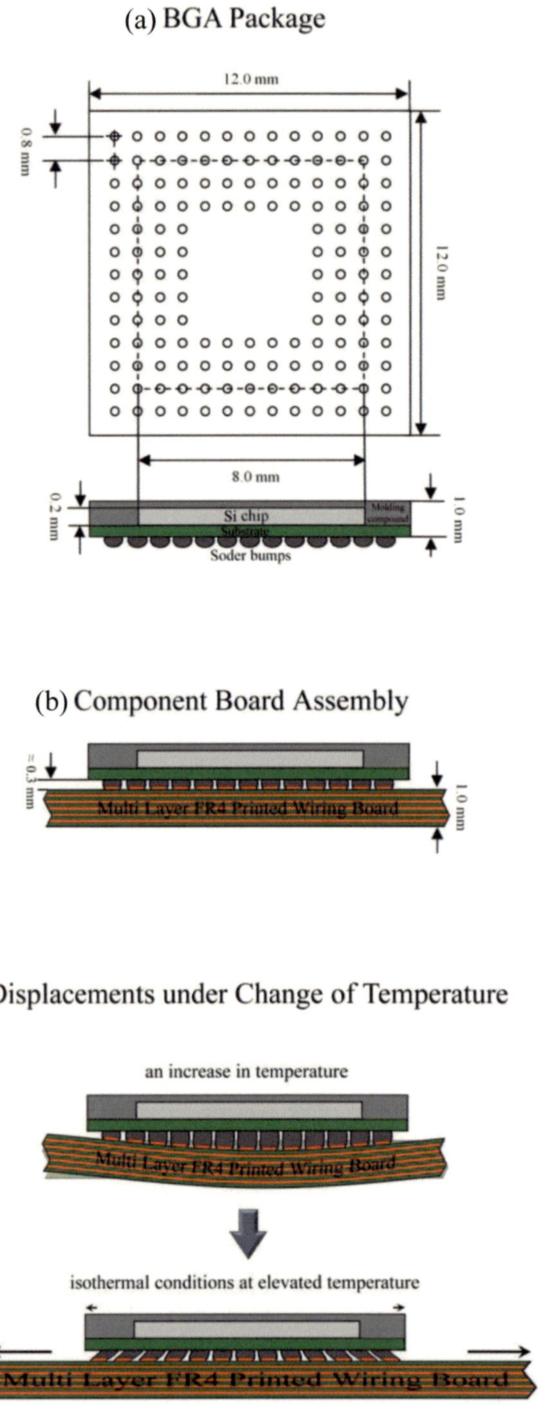

Fig. 1. a) A Ball Grid Array (BGA) package commonly used in high-density electronic devices; b) structure of the component board assembly; c) formation of strains and stresses under an increase in temperature.

and stresses in solder interconnections are built up to different extents, of course, at all length scales, ranging from the submicron intermetallic particles submerged in the tin matrix of the tin-rich solder interconnections to the structural dimensions of a product. However, as the CTEs of most printed wiring boards is many times higher than that of most packages, the influence of strains and stresses at higher length scales (i.e. the board or the product level) are more influential.

The mechanical properties of solder interconnections are, of course, dependent on the microstructures formed during soldering but the fact that they are not stable and tend to change distinctly during the operation of products makes the evaluation of reliability a challenging task. Thus, detailed understanding of the evolution of microstructures under thermal cycling conditions and their influences on the failure of solder interconnections is essential because it can provide us with the means not only to improve reliability but also to develop more efficient and meaningful methods for the reliability evaluation of and lifetime prediction.

Despite it being a popular topic of academic research for decades, we are only beginning to understand the complexities related to the failure mechanisms of solder interconnections under cyclic thermomechanical loading. Justification for this perhaps surprising statement lies in the fact that it was only recently that the widely used tin-lead solder alloys were replaced with new lead-free materials and this change has made a comprehensive re-assessment of reliability necessary [8,9]. It is well known that the reliability of solder interconnections is strongly influenced by the microstructures formed during soldering and their evolution during use, but it is particularly interesting to observe that the microstructural changes in tin-rich lead-free solders are markedly different from those observed in tin-lead solders, where failure takes place as a result of the heterogeneous coarsening of tin and lead phases and eventual cracking of the bulk solder (see, e.g., [10,11]). The microstructural observations of failed tin-silver-copper solder interconnections have indicated that the microstructures of solder interconnections change distinctly before cracking but by a different mechanism, namely cracking that is assisted by recrystallization (see, e.g. [12,15]).

In this chapter we discuss the failure mechanisms of tin-rich lead-free solder interconnections of electronic component boards from the perspective of the evolution of microstructures. The focus is placed on the identification of the factors driving the microstructural evolution in lead-free interconnections that creates the preconditions for the energetically feasible intergranular propagation of cracks through the interconnections. The microstrucutural approach to the reliability of electronics is useful, particularly because many failure mechanisms are related to the inevitable evolution of microstructures that takes place during the normal operation of products. Owing to the extensive research carried out over the last decade we know the microstructures and mechanical properties of many lead-free solder compositions well, but the evolution of microstructures during the operation of products has still gained little attention. However, before going into details of the changes in microstructures one should have a generalized understanding of the microstructures to be discussed when they are in an as-solidified state. Therefore we will begin this chapter with a brief overview of the solidification and microstructures of tin-rich lead-free solder interconnections formed during solidification. After that the restoration of solder interconnections and the conditions under which the recrystallization is initiated are discussed. The onset and progress of recrystallization in solder interconnections will be discussed in detail. The failure of the solder interconnections is discussed in a separate section to emphasize the fact that the propagation of cracks takes place after the change in microstructures as a result of recrystallization and that recrystallization significantly enhances their propagation in the solder interconnections. Finally, a method to predict the changes in microstructure and, ultimately, to predict the lifetime of solder interconnections is presented and discussed.

2. As-solidified microstructures of tin-rich solder interconnections

The majority of lead-free solders are based on tin (Sn), with a few alloying elements. Silver (Ag) and copper (Cu) are the most common major alloying elements but alloys with minor additions of elements such as nickel, antimony, indium, germanium, manganese, bismuth, zinc, or rare earth elements are also commercially available. However, for the sake of simplicity let us consider the three-component SnAgCu alloy with near-eutectic composition, which is the most commonly used solder alloy in the electronics industry.

The eutectic composition of the tin-silver-copper alloy is about Sn3.4Ag0.8Cu [13,14]. Nearly all of the solder compositions used in the electronics industry, such as the composition of Sg3.0Ag0.5Cu that is most widely used today, have silver and copper concentrations below the eutectic concentration. For such compositions, the tin-rich phase (i.e., high-tin solid solution) is formed first at the beginning of solidification and its morphology strongly affects the solidified microstructure. Owing to the high tin content of the near-eutectic SnAgCu alloys (more than 95 wt-% Sn), the solidification and generated microstructures of the interconnections are most significantly influenced by the solidification of the tin-rich phase, even though in practice the tin-rich phase may or may not always be the first phase to form during solidification. This is because the dissolution of the contact metallizations changes the nominal composition of the solder and can thereby influence solidification, as will be pointed out shortly.

Figures 2a-b show an example of a cross-section of a near-eutectic SnAgCu solder interconnection in the as-solidified condition as imaged by employing the optical (cross-polarized) microscopy and the scanning electron microscopy. The boundaries between the contrasting areas in Figure 2b are high-angle boundaries between the matrices of solidification colonies[1] (the orientation difference between adjacent regions is quite large, larger than about 15°). A uniformly oriented cellular solidification structure of tin is enclosed within the colony boundaries. There is a low-angle orientation difference between the cells enclosed by the high-angle colony boundaries. The cellular structure of tin is clearly distinguishable as cells are surrounded by eutectic regions (see Fig. 2c-d). It is also noticeable that the solder interconnections, such as that shown in Figure 2, are commonly composed of a few solidification colonies of relatively uniformly oriented tin cells. [12,15] Similar observation have been made by other groups also [16,17-21].

[1] We have chosen to use the term "solidification colony" in order to emphasize the fact that under the reflow conditions employed, a cellular structure is generated in which the difference in crystal orientations between individual cells is small (few degrees or less). The use of this term also helps us to make a verbal distinction between the as-solidified microstructures and the recrystallized grains.

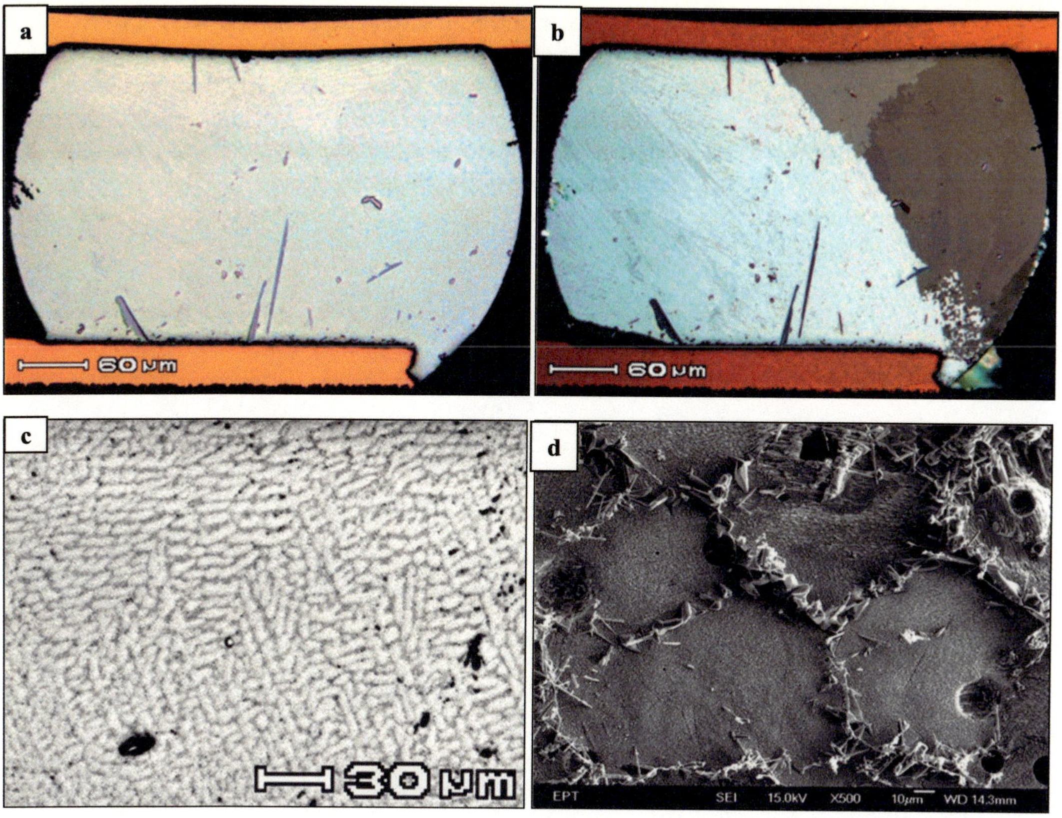

Fig. 2. The as-reflowed microstructures of a near-eutectic SnAgCu interconnection: (a) optical bright field image of a cross-section; (b) a cross-polarized light image of the same cross-section highlights the colony boundaries (high angle boundaries); (c) an SEM micrograph showing the cellular structure within a colony, where the cells are separated by low-angle boundaries; (d) the cell structure of a solidification colony is emphasized by the small intermetallic particles that surround the tin cells (the sample has been selectively etched).

Dendritic morphologies of the tin-rich phase are also reported in the literature; see e.g. [22-25]. It is evident that this phase can solidify in different morphologies depending on the solidification conditions (e.g., cooling rate and metallizations in contact with the solidifying solder) and the nominal composition of the solder. The volume of the solder interconnections obviously influences the microstructures formed during solidification, as a dendritic structure of the tin-rich phase is more often observed in the case of large solder volumes, such as cast dog-bone or lap-joint pull test specimens (prepared for the mechanical characterization of solders) or packages with a relatively large (≈ 1 mm) bump diameter (e.g. [16,26-28]). However, the as-solidified microstructure in the near-eutectic solder interconnections in our studies has consistently been cellular, regardless of the compositions of the near-eutectic SnAgCu paste or bump alloy, contact pad metallization, package type or dimensions, or the setup parameters of our full-scale forced convection reflow soldering oven.

In order to rationalize the formation of the observed microstructures, it is useful to examine the solidification of solder interconnections with the help of equilibrium phase diagrams. It should be kept in mind, however, that the equilibrium diagrams do not contain information about either the effect of the cooling rate or the morphology of the phases and, thus, the solidification structures are examined as equilibrium solidification[2]. Figure 3 presents the tin-rich corner of the SnAgCu phase diagram with the isothermal lines representing the liquidus temperatures, and the primary phase regions of the tin-rich phase, Cu_6Sn_5 and Ag_3Sn.

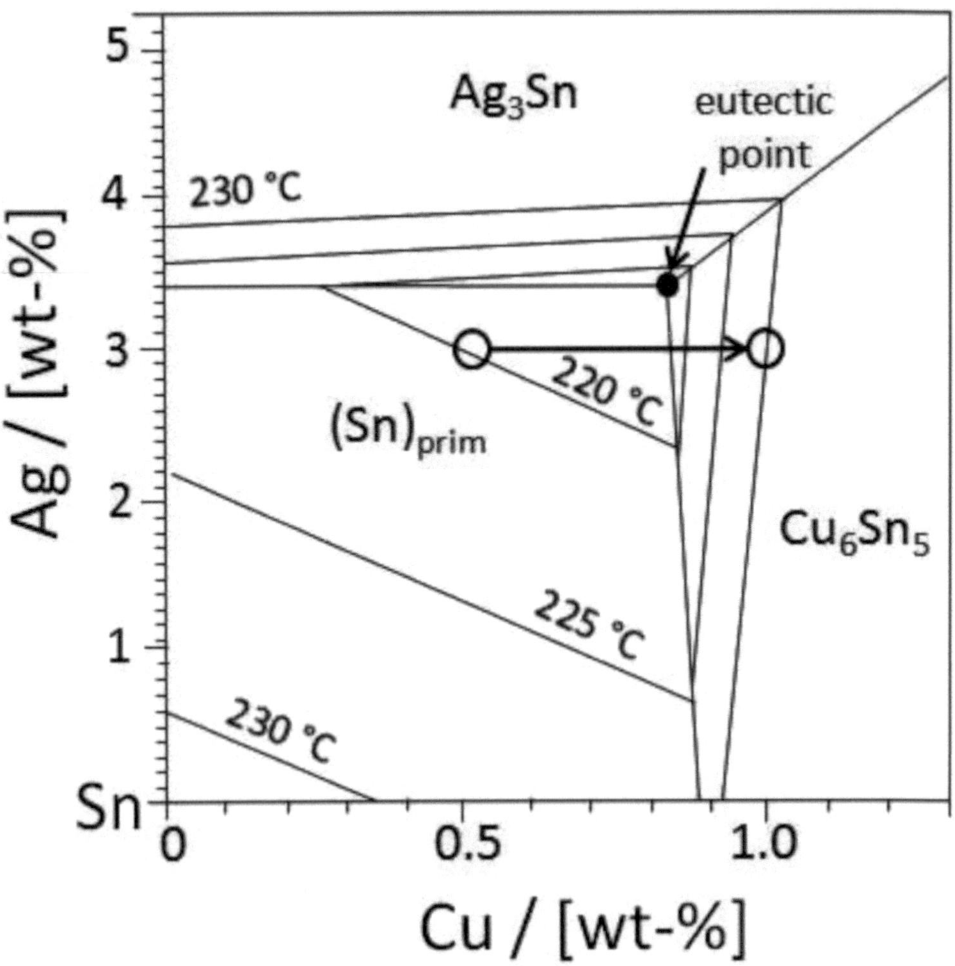

Fig. 3. Tin-rich corner of the SnAgCu phase diagram with isothermal lines and primary phase regions [29]. The arrow between the two circles represents the change in nominal composition (circle) owing to copper dissolution.

[2]For example, in order to take the undercooling (which will be discussed shortly) into consideration, one should extrapolate the liquidus surfaces of the Cu_6Sn_5 and Ag_3Sn phase to lower temperatures and lower (or remove) the liquidus surface of the $(Sn)_{prim.}$ phase.

Let us consider, for example, the composition of Sn3.0Ag0.5Cu, which is a very commonly used alloy in the reflow soldering of component boards. The solidification of this composition starts with the formation of the tin-rich solution phase when the interconnections are cooled down from the peak reflow temperature to below the liquidus temperature of about 220 °C. The secondary phase, namely Cu_6Sn_5 or Ag_3Sn, is formed only after the nominal composition of the remaining liquid meets the curve of two-fold saturation, after which the solidification of the interconnections proceeds by the binary eutectic reaction (liquid transforms to $(Sn)_{eut}$ + the secondary phase; see [15] for more details).

It should also be noticed that in practice the microstructures formed on other pad metallizations can differ notably, even though the same solder compositions are used (see, e.g., [15,30,31]). The dissolution of the contact pads or pad metallizations of packages and printed wiring boards during reflow changes the composition of the molten solder. The influence of contact metallizations depends primarily on the dissolution rate and reactivity of the metallizations. For example, the dissolution rate of copper in near-eutectic SnAgCu solder is about 0.07 µm/s, but that of Ni is more than an order of magnitude lower and can be considered negligible [15,32,33]. Thus, practically all nickel that is dissolved into liquid solder is used in the reaction to form intermetallic layers. However, in the cases where the solder is in direct contact with copper pads (i.e., no protective coating or organic soldering preservative is used on the copper soldering pads) the dissolution rate of copper from the pads into typical BGA solder interconnections (with a bump diameter of about 0.5 mm) is high enough to lift the copper concentration above 1 wt-% during soldering (see Fig. 3). This change in the nominal composition of the liquid can change the primary phase formed during solidification from the tin-rich phase to Cu_6Sn_5. Therefore, the as-solidified microstructures on the copper pads often show large amounts of primary Cu_6Sn_5 (hexagonal) tubes or rods in the microstructure that are absent from the microstructures of interconnections that are soldered on slow dissolution rate metallizations, such as nickel. Furthermore, interconnections soldered on copper pads, as opposed to those soldered on nickel, typically show more numerous and larger Cu_6Sn_5 particles embedded at the boundaries between the tin cells that are formed in the binary solidification, as the composition of the liquid moves a greater distance along the eutectic valley (along the curve of two-fold saturation).

There is an important consequence related to the increased copper content: the relatively large primary phase needles or particles dispersed in the solder interconnections can influence the evolution of solder interconnection microstructures. The non-coherent high-angle boundaries between the Cu_6Sn_5 crystals and tin matrix provide good nucleation sites for the recrystallization. It has been previously demonstrated that the second phase particles can accelerate the nucleation of recrystallization in common structural alloys [34,35]. Readers interested in particle-stimulated nucleation of recrystallization can refer e.g. to [36,37].

However, as pointed out earlier, in practice the solidification process departs somewhat from that of equilibrium solidification. Taking the undercooling into account would result in the nucleation of the Cu_6Sn_5 as a primary phase at even lower copper concentrations. Figure 4 illustrates an as-solidified microstructure of the Sn3.0Ag0.5Cu solder interconnection. It is interesting to observe that the primary Cu_6Sn_5 needles in the micrographs have the tendency to nucleate at the free surfaces of the molten interconnection, most probably on oxide

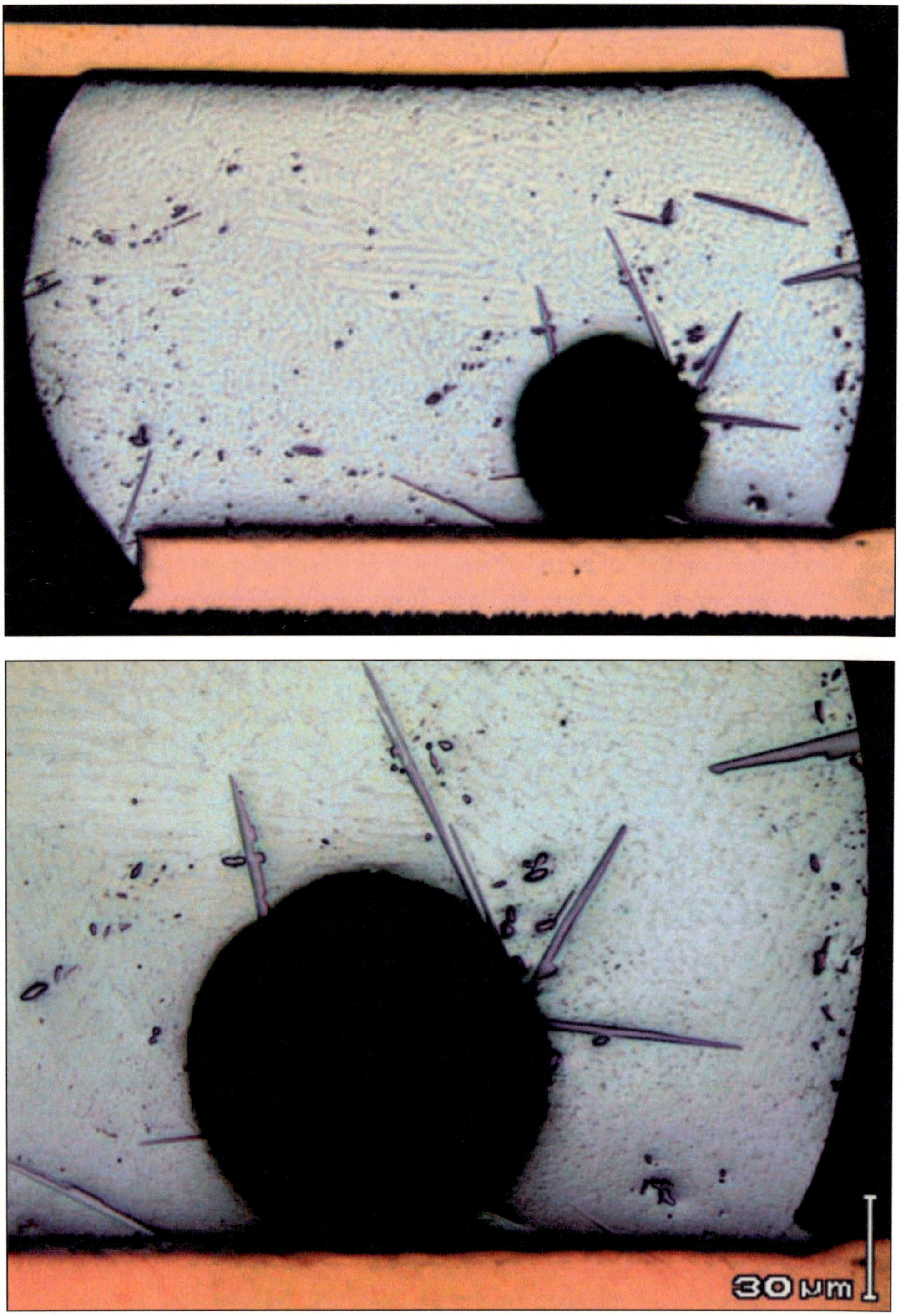

Fig. 4. As-solidified microstructure of a Sn3.0Ag0.5Cu solder interconnection illustrate how the primary Cu_6Sn_5 particles have nucleated on oxide particles on the surface of the liquid.

particles on the liquid surfaces, instead of the package or the PWB side interfaces. It should also be mentioned, without going into detail, that it seems as if the solidification of the tin-rich phase is controlled by the kinetics of heterogeneous nucleation at the surface of (Cu_6Sn_5) intermetallic layers [38]. Darveaux et al. observed experimentally that SnAgCu solder alloys with a higher copper concentration exhibit a higher amount of undercooling than those with a lower concentration [39].

Before we move on to the evolution of microstructures under operating conditions, we would like to point out a few aspects to consider in more detailed investigations of the failure mechanisms of recrystallization-assisted cracking of solder interconnections. It is particularly noteworthy that the Cu_6Sn_5 or the Ag_3Sn phase can nucleate with minimum undercooling in the liquid SnAgCu interconnections [40-42] but the nucleation of the tin-rich phase results in a significantly wide range of undercooling that can extend up to 60 ^{0}C [43-48]. The large amounts of undercooling indicate apparent difficulties in the nucleation of tin crystals in the liquid, which can be one of the reasons why there are very often only few orientations of the Sn-rich phase observed on a cross-section of solder interconnections. The tendency to form only a few large crystals, which can be several hundred micrometers in diameter, has been observed in interconnections of various length scales, ranging from about 100 μm (the diameter of a Flip Chip interconnection) to millimetre scale (lap-joint specimens used in material characterization) [15-20,28,49]. Furthermore, the fact that sometimes neighboring regions of a cross-section share a twinning relationship (indicated by a misorientation angle of about 60° between adjoining regions) suggests that these crystals originate from a common nucleus and, thus, the number of different crystals can be even smaller that the amount determined by the commonly employed qualitative method of cross-polarized light microscopy [16,20,21].

What has been stated above indicates clearly that mechanical behavior of solder interconnections is most probably quite different from that of a "normal" polycrystalline material. As has been pointed out and is currently being studied by many authors, the fact that the physical and mechanical properties of the tin-rich phase exhibit significant anisotropic behaviour[3] can cause severely non-homogeneous deformation and the formation of internal stresses in the solder interconnections [52-54]. As the internally produced strains and stresses are combined with the higher-level strains and stresses, as caused by the differences in the coefficients of thermal expansion of printed wiring boards and packages, it is clear that the thermomechanical response of the solder interconnections becomes very complex and unique to each solder interconnection. Furthermore, the grain boundary cracking should not occur in the as-solidified structure due to the absent of high angle boundaries (other than those between colonies). This can lead to unpredictable failure sites, as reported in [54,55]. Therefore, when stress is applied to interconnections having this kind of microstructure, they undergo microstructural evolution before fractures can propagate. Investigations of the microstructures of failed solder interconnections have indicated that the microstructures formed during solidification are not stable and will change notably during the operation of products [15,17,18,56-63].

---

[3] The coefficient of thermal expansion along the c-axis of the tetragonal unit cell {c/a ratio of about 0.5} is twice that along the other two axes {a and b-axis}; the elastic modulus along the c-axis is only about 0.6 times that along the other two axes [[50],[51]]

3. The role of recovery and recrystallization in the failure mechanisms of solder interconnections

The reliability of electronic devices is commonly assessed by employing standard thermal cycling tests that place the extreme temperatures in the range of about - 45 °C to +125 °C. The thermomechanical stresses formed in the solder interconnections under these conditions are high enough to cause instantaneous plastic deformation of the commonly used near-eutectic SnAgCu solders [63]. Furthermore, this whole temperature range remains above the 0.3-0.4 homologous[4] temperature range of the solders, which is the temperature range above which the time-dependent deformation of metals becomes significant. Thus, time spent at either elevated or lowered temperatures allows diffusion creep processes to transform the elastic strain part of the total strain into inelastic strain[5]. The energy stored during deformation acts as the driving force for the evolution of the microstructures.

The initiation of microstructural changes in solder interconnections is localized because of the highly non-uniform distribution of strains inside the solder interconnections. Figure 5 shows the calculated strain energy density distribution in the cornermost solder interconnection of the component board assembly shown in Figure 1. It should be pointed out that there is a difference in the rate of strain energy accumulation during thermal cycling between the solder regions on the opposite sides of the interconnections and, therefore, changes in the microstructure are observed first on the package side regions of the interconnections, where the inelastic deformation is more extensive.

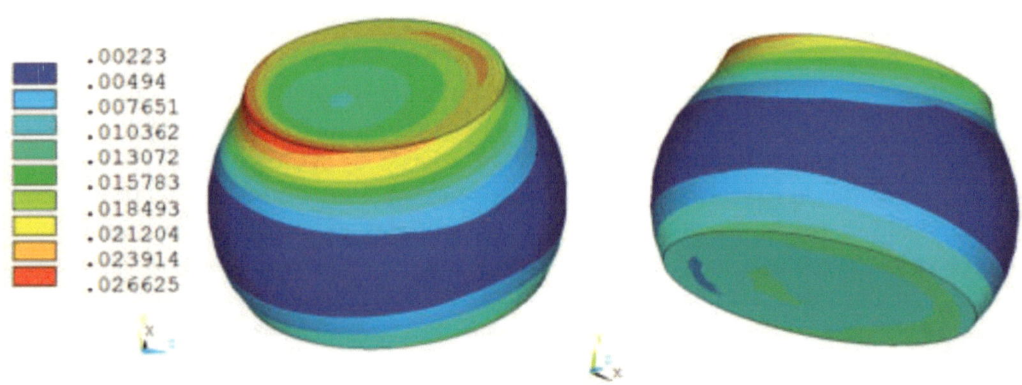

Fig. 5. Calculated strain energy density of the cornermost solder interconnections of the component board assembly sketched in Figure 1 [63,64].

---

[4]Defined as the ratio of the prevailing temperature to the melting point of a solvent metal; both expressed in Kelvin.

[5]Here we consider 'total strain' = elastic + inelastic strain = elastic + plastic + creep strain.

In studies reported in more detail elsewhere, the evolution of microstructures was investigated as a function of thermal cycles by taking out samples at fixed intervals during the course of the test and inspecting them for the development of microstructures and failures [64,65]. The results showed that the evolution of microstructures in the strain concentration regions commenced with gradual evolution of the cellular solidification structure, but after some time, i.e., the incubation period, the microstructures changed discontinuously by recrystallization. A similar observation has been reported in [66].

3.1 Restoration of tin-rich solder alloys

When solder interconnections are deformed plastically, a part of the work is stored in solders as lattice defects, mainly in the strain fields of dislocations. The increased internal energy of deformed solder acts as the driving force for the competing restoration processes, recovery and recrystallization. It is well known that the degree of restoration by recovery depends on the stacking fault energy of the solder alloy. At the time of writing there is little information in the literature about the recrystallization behavior of tin-based solder alloys but since the near-eutectic SnAgCu alloy contains more than 95 wt-% of tin, recrystallization studies on pure tin can be considered indicative, bearing in mind that the alloying elements in solid solutions, as well as small precipitates, do affect the restoration processes. Creep studies carried out with high-purity tin have suggested that the stacking fault energy of tin is high [67,68]. The recovery is very effective in high-stacking fault energy metals, such as aluminum and iron, as a result of the efficient annihilation of dislocations by cross slip and climb. Therefore one can expect the restoration of high-tin solder alloys to take place to a large extent by recovery. Gay et al. and Guy have observed that pure tin (99.995% purity) recrystallizes at room temperature even after a modest deformation (reductions of a few percent) [69,70]. However, in a more recent study Miettinen concluded that that even highly deformed (up to a 50% reduction) near-eutectic SnAgCu solders do not recrystallize statically when annealed at 100 °C after deformation at room temperature [71]. Korhonen et al. also failed to observe recrystallization in dynamic fatigue tests performed at room temperature [72]. All these results indicate that recovery is effective also in near-eutectic SnAgCu solder alloys. Because recovery and recrystallization are competing processes, the progress of recovery can reduce the driving force of recrystallization significantly and recrystallization may not always initiate. On the other hand, it is well documented that near-eutectic SnAgCu interconnections do recrystallize under dynamic loading caused by changes in temperature (between -45 °C and +125 °C), as well as under power cycling conditions (between room temperature and +125 °C) [15,17,18,56-63,73]. Thus, it seems that near-eutectic SnAgCu solder interconnections recrystallize only under restricted loading conditions: dynamic loading conditions where the strain hardening is more effective than the recovery.

Figure 6a shows a micrograph of the recrystallized microstructure on the package side neck region of a thermally cycled interconnection taken by employing optical microscopy and cross-polarized light. Figure 6b shows an electron backscatter diffraction orientation map of the same surface. The black lines in the figure represent the boundaries where the crystal orientation of the adjacent grains exceeds 30° and correspond well to the grain boundaries visible in the optical micrograph in Figure 6a.

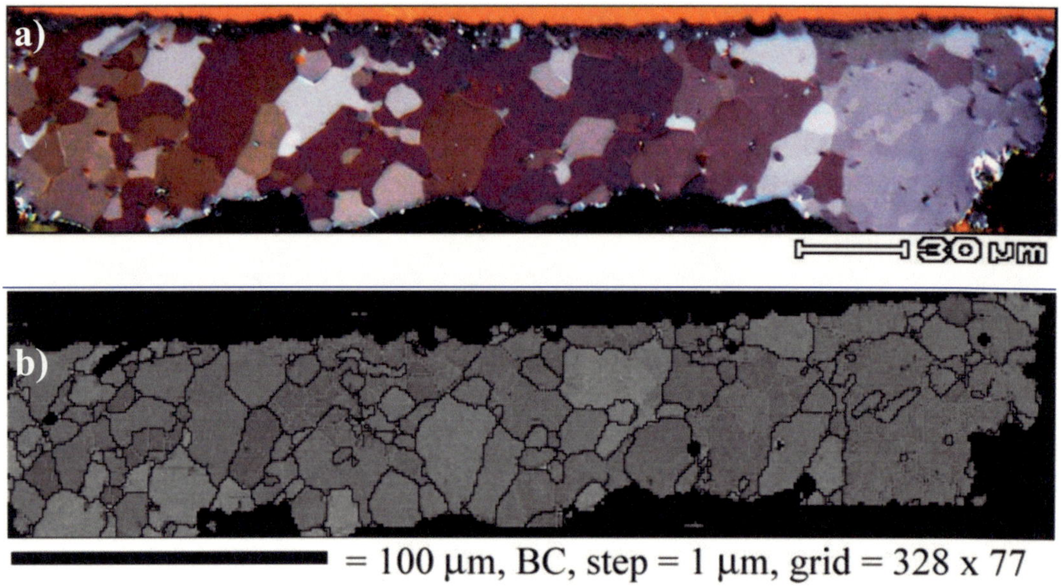

Fig. 6. a) Optical micrograph showing the recrystallized structure on the package side interfacial region of the SnAgCu solder interconnection taken with polarized light;
b) EBSD graph of the same location as in a) showing boundaries with large misorientation (larger than 30°C) between the adjacent grains with black lines.

3.2 Early phase of evolution: Effects of recovery and coarsening of the microstructures

Figure 7 shows a collage of micrographs that are all taken from the same solder interconnection. It should be noticed that we are using the same interconnection to exemplify the features of microstructural evolution that take place consequently in the stress concentration regions of the solder interconnections. The evolution of microstructures on the PWB side interfacial region is much slower than the evolution on the package side of the interconnection as a result of the less extensive plastic deformation per cycle (see Fig. 5). Even though this interconnection has experienced more than 3000 cycles and failed from the package side interfacial region, the microstructures visible on the PWB side of the interconnections are very similar to those on the package side interfacial regions about 1000 to 1500 cycles earlier.

As already discussed, the as-solidified microstructures of tin-rich solder interconnections are typically composed of relatively few large tin colonies distinguished by high-angle boundaries. The cellular structure of the tin-rich colonies is clearly visible within the high-angle boundaries as the individual tin cells are surrounded by eutectic regions composed of fine Cu_6Sn_5 and Ag_3Sn particles. As shown in Figure 7b the cellular structure of the primary tin-rich phase is still visible in the less deformed regions on the solder interconnections, such as in the middle, even after the component board has been thermally cycled until failure. However, in the strain concentration regions the tin cells begin to rearrange by the gradual coalescence of the tin cells and coarsening of the intermetallic particles (see Fig. 7c-d), during which the eutectic structures around the tin cells gradually disappear and a network

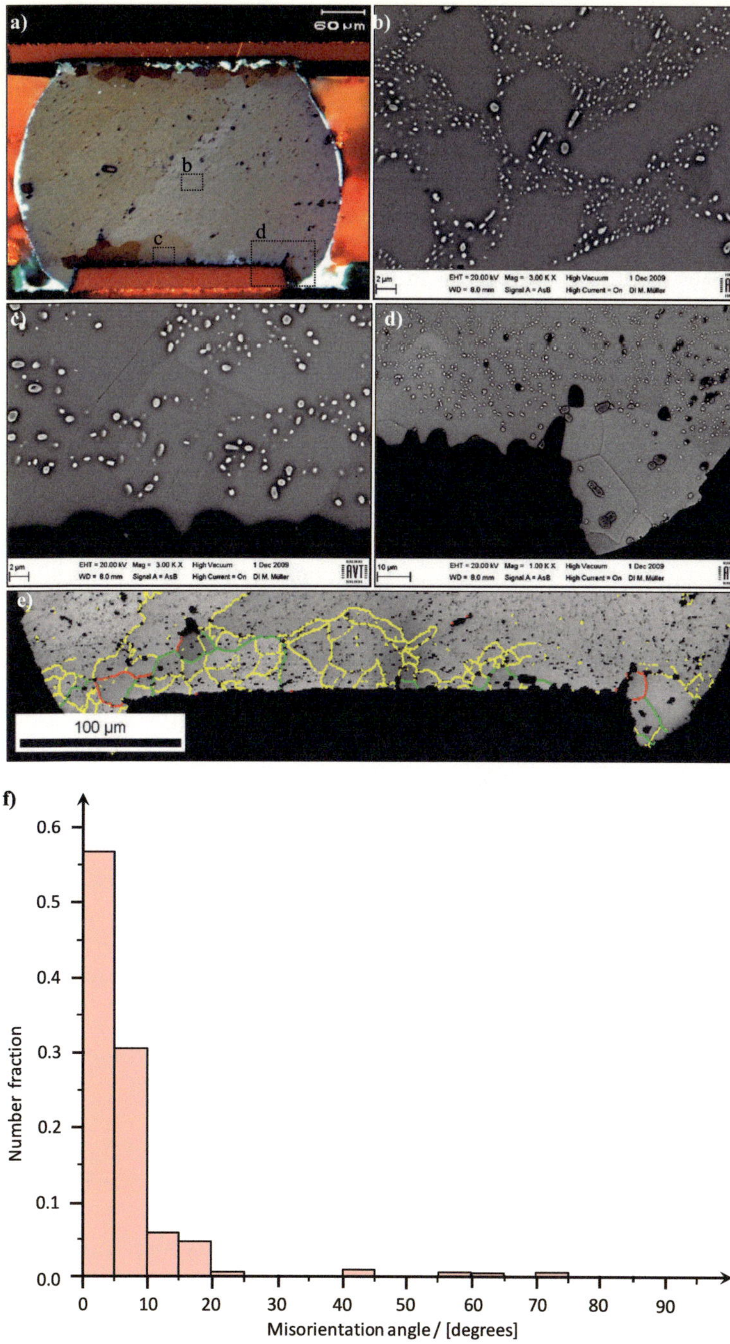

Fig. 7. a) Cross-polarized light micrograph of a solder interconnection that has failed under thermal cycling; b-d) magnifications from the regions indicated in (a); e) EBSD map of the cross-section that shows boundaries with a misorientation < 15° by yellow lines, 15°-45° by green lines, and > 45° by red lines; f) histogram distribution of grain boundaries over the region shown in (e). [64,65]

of low-angle boundaries produced by recovery emerges. Figure 7e shows an EBSD map of the PWB side interfacial region of the same cross-section as shown in Figure 7a. The yellow lines represent the low-angle boundaries, i.e., boundaries where the crystal orientation between the adjacent grains is *well* below 15° (see also Fig. 7f). The green and red lines represent high-angle boundaries where the misorientation between adjacent grains is higher than 15°, the green line those between 15° and 45°, and the red line those above 45°. The figure shows that in addition to the coarsening of the intermetallic particles, the formation of additional low-angle boundaries (additional to those between the cells of the Sn-rich phase formed during solidification) takes place in the regions of high strain energy density. These changes were observed to initiate early in the course of the thermal cycling tests, within about 50 to few hundred thermal cycles. It is also noteworthy that the coarsening of the intermetallic particles is strong in the regions near the high-angle grain boundaries, while the regions near the low-angle boundaries still include finer particles, comparable to the bulk of the solder ball (see Fig. 7d-e). This can be expected as the diffusion is much faster along the high-angle boundaries than it is along the low-angle boundaries.

3.3 Later phase of evolution: Transformation of the microstructure by recrystallization

In the course of further thermal cycling, the microstructures keep evolving gradually in the manner already described until these regions change discontinuous into more or less equiaxed grain structures by recrystallization. This change in the microstructures is first observed in the edge regions (i.e. corner regions of a cross-section) of the solder interconnections on the package side of the interconnections (the regions with the highest strain; see Fig.6) and, after the initiation of recrystallization, the recrystallized volume gradually expands from the edges toward the center, across the interconnections near the package side interfacial region of the interconnections. The incubation time of recrystallization varies significantly from one interconnection to another and even from one package to another (the same location of the interconnection) under the same loading conditions. The first indications of recrystallization in the BGA packaged board assemblies shown in Figure 1 were observed after about 500 thermal cycles but it can take up to about 2000 cycles until recrystallization is consistently observed in every interconnection in the corner regions of the packages.

Figure 8 shows a typical example of a failed interconnection, where the cracking of the solder interconnection is accompanied by a distinct change in the microstructure by recrystallization. The micrograph in Figure 8a is an optical micrograph that shows the crack path distinctly and the micrograph in Figure 8b is a cross-polarized light image of the same location as Figure 8a. The comparison of the micrographs shows that the propagation path of the crack is enclosed entirely within the recrystallized region of the interconnection. Figure 8c shows an EBSD map of the same cross-section as shown in Figures 8a-b. The colour lines represent the misorientation between the adjoining regions: yellow below 15°, green between 15° and 45°, and red above 45°. The image illustrates well the fact that the cracked region on the package side interface of the solder interconnections shows primarily high misorientations and that the high-angle grain boundaries are located very close to the crack path while, in general, the misorientations become smaller with increasing distance from the crack region. Figure 8d shows a histogram distribution of grain boundaries with different orientation over the region shown in Figure 8c (see also Fig 7f). As can be seen, the region still contains a high number of low-angle boundaries (caused by recovery) but a large

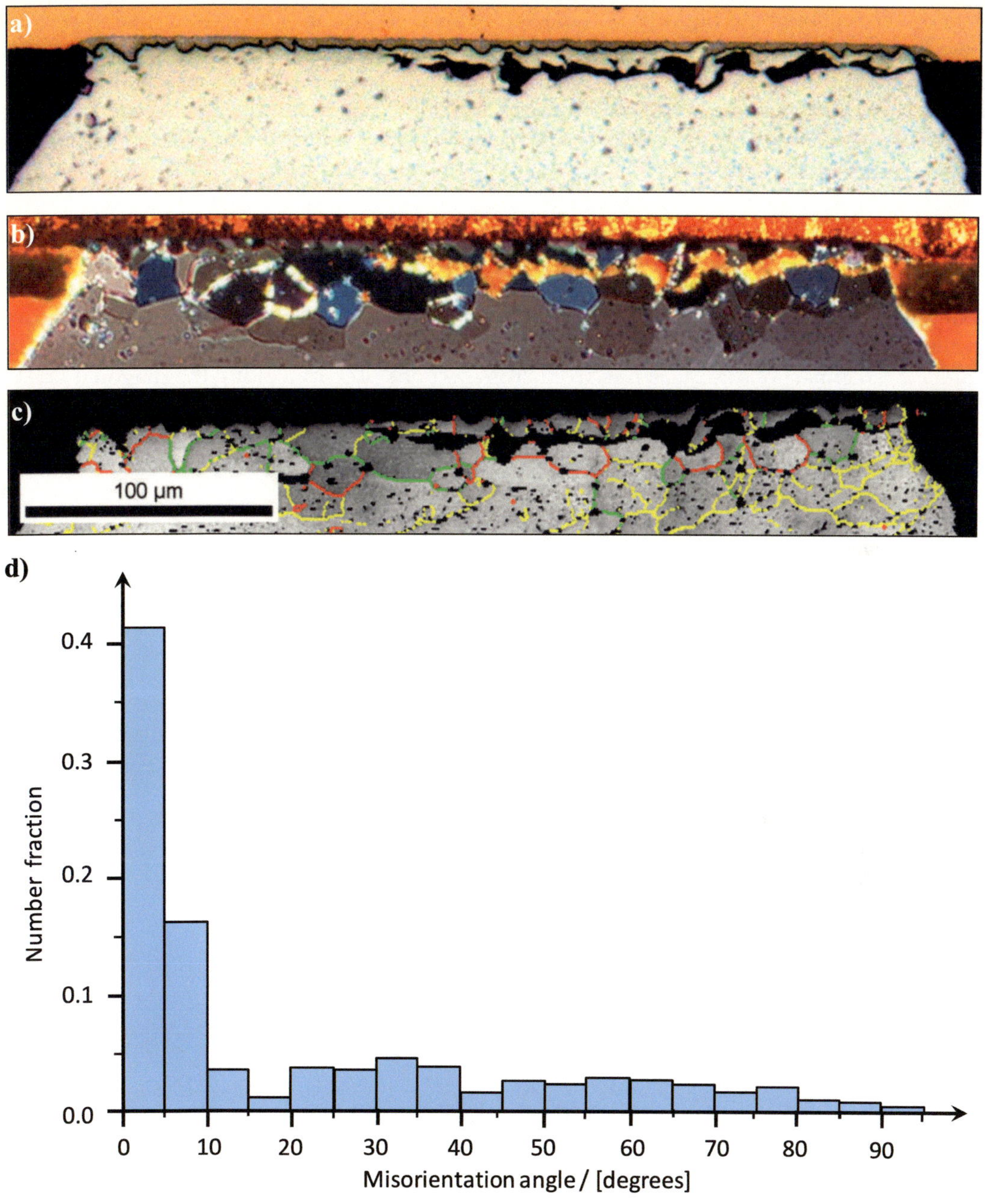

Fig. 8. a) an optical micrograph of a failed solder interconnection shows the crack path clearly (the thin black line between the gray intermetallic layer and the white solder is a contrast effect caused by specimen preparation); b) a cross-polarized light image of the same location as (a) highlights the recrystallized grains as caused by the cyclic deformation; c) EBSD map of the cross-section that shows boundaries with misorientation < 15° by yellow lines, 15°-45° by green lines, and > 45° by red lines; d) histogram distribution of grain boundaries over the region shown in (c). [64,65]

number of higher-angle boundaries (caused by recrystallization) have emerged. After the initiation of recrystallization in the strain concentration regions at the edges of the solder interconnections, the recrystallized volume gradually expands over the diameter of the interconnections, and cracks follow the expansion of the microstructurally changed volume. It was also observed that cracks rarely propagate outside the recrystallized volume of the solder interconnections.

4. Cracking of recrystallized solder interconnections

Work presented in more detail elsewhere focused on the evaluation of the nucleation time and propagation rate of cracks in BGA component board assemblies under different thermal cycling conditions [74,75]. Figure 9 shows the average crack lengths of the most critical solder interconnections as measured from cross-sections prepared along the diagonal line of the package as a function of thermal cycles. The different lines represent different cycling conditions (TS = thermal shock, TC = thermal cycling; the accompanying value is the dwell time of the profile in minutes).

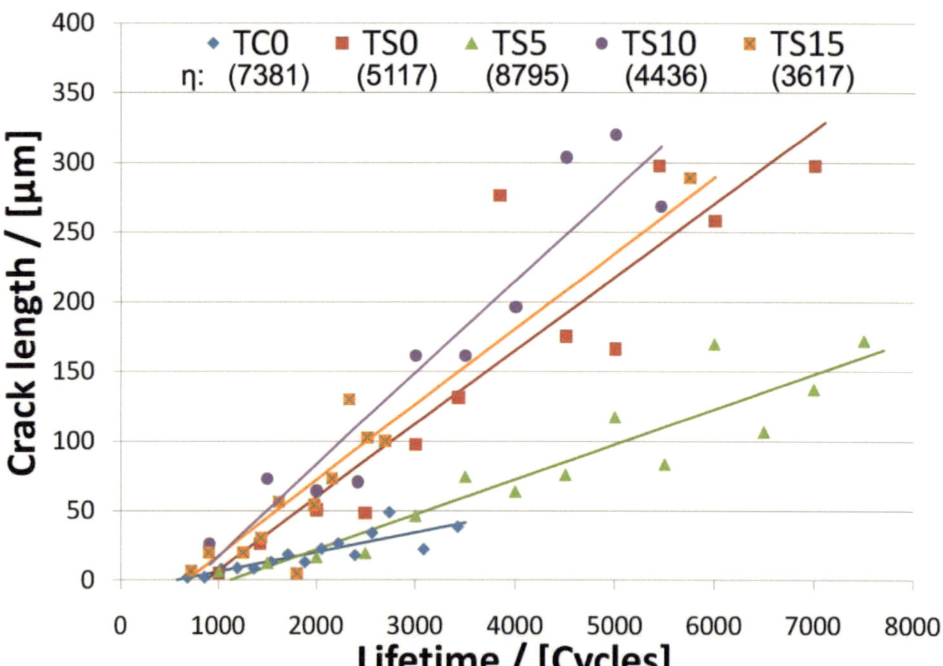

Fig. 9. Measured average crack lengths as a function of the number of thermal cycles (the numbers in parentheses are estimates of the Weibull characteristic lifetimes [η]). [74,75]

More detailed examinations showed that the nucleation of cracks in the interconnections of the BGA board assemblies took place within a relatively narrow range, between about 1000 and 1500 cycles, regardless of the dwell time or ramp rate used in the thermal cycling tests. However, the propagation rate of cracks without the influence of recrystallization was very slow. This conclusion was made based on the comparison of the measured crack lengths in interconnections that were removed from the thermal cycling oven at the same time and that showed or did not show recrystallization. Thus, the primary failure mechanism under thermomechanical fatigue involves the formation of a continuous network of grain boundaries by recrystallization that enables cracks to nucleate and propagate intergranularly through the solder interconnections (see Fig. 10).

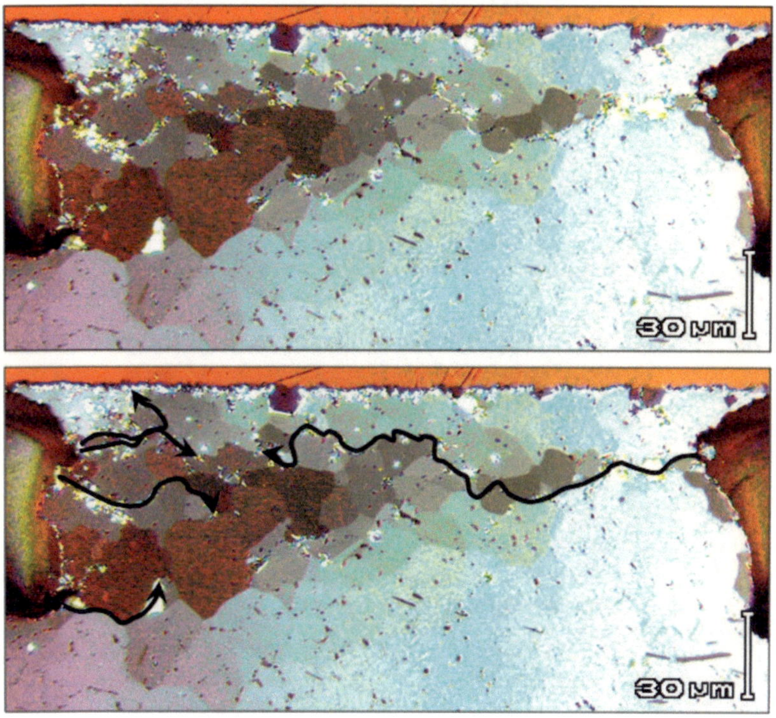

Fig. 10. Cracks propagate intergranularly between the recrystallized grains: a) cross-section of a recrystallized and cracked solder interconnection; b) the same image as in (a) but with superimposed crack paths.

The fractographic examinations illustrate the influence of recrystallization on crack propagation. Figure 11 shows the fractrographs of failed solder interconnections. The fracture surfaces exhibit a globular appearance as a result of the propagation of cracks between the recrystallized grains. Fatigue striations were occasionally observed on the inspected fracture surfaces. Although they were quite uncommon, they indicate that cracks can also propagate transgranularly under conditions when the intergranular propagation is not benignant, namely when the recrystallized grain size is larger or the stress state, orientation, and geometry of the grains are not in favor of cracking along the grain boundaries. The networks of the grain boundaries formed by recrystallization evidently provide favorable paths for cracks to propagate intergranularly with less energy consumption in comparison with transgranular propagation. It can also be expected that the cohesion between the recrystallized tin grains is lowered by grain segregation of impurities as well as locally by intermetallic particles. Furthermore, the mechanical anisotropy of the (recrystallized) tin grains can also enhance the nucleation and propagation of microcracks along their boundaries as the value of the coefficient of thermal expansion of tin single-crystal in the [100] = [010] directions is about two times that in the [001] direction [50,51].

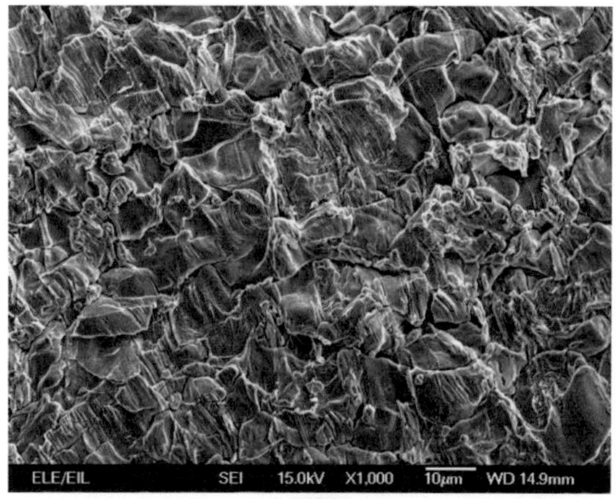

a) TS10

b) TS0

c) TC0

Fig. 11. Fractographs of the interconnections viewed from the component side [75]. Note the several secondary cracks visible in all micrographs (striations are not visible here).

As discussed in the previous chapter, the range of the incubation periods of recrystallization in tin-rich solder interconnections is relatively large, in the range of 500 to 2000 cycles in the case of the BGA interconnections in our study. Thus, the nucleation of cracks can take place before or after the change in microstructures by recrystallization. However, the failure analyses showed that the formation of the networks of grain boundaries by recrystallization had influenced the propagation of cracks in all electrically failed interconnections. The range of 1000-1500 thermal cycles required to produce distinguishable small cracks (nuclei) equals about 25-30% of the average lifetime of the component board assemblies and, thus, cracks are in the propagation stage for about three quarters of the lifetime of component boards. Therefore, it is evident that the cracking of tin-rich solder interconnections is controlled by the rate of recrystallization.

On the basis of the results presented above we can draw two conclusions: 1) the nucleation of cracks in solder interconnections is primarily dependent on the number of load reversals. In other words, nucleation is relatively insensitive of microstructural features and their evolution, as well as the parameters of the loading conditions; 2) the rate of crack propagation is dependent on the expansion rate of the recrystallized volume. In other words, the lifetimes of solder interconnections are primarily controlled by the onset and expansion of recrystallization.

5. An approach to lifetime prediction based on the evolution of microstructures

A thorough understanding of the restoration process in solders can allow the development of methods for improved lifetime estimation that are based on the evolution of microstructures. Work presented in [76] describes an approach to lifetime prediction based on the competing nature of the restoration processes: under conditions in which the strain hardening is more effective than recovery, the cyclic deformation accumulates the stored energy above a critical value, after which the recrystallization can initiate. The total stored energy of the system consists of the grain boundary energy and the volume defect energy (mainly line and point defects). The stored energy is released through the nucleation and growth of new strain-free grains (grains with low defect density), which gradually consume the strain-hardened matrix of high defect density. Li et al. have developed a multiscale model based on this principle for predicting the microstructural changes of recovery, recrystallization, and grain growth in solder interconnections subjected to dynamic loading conditions [77,78]. The approach developed in this work is based on the principle that the stored energy of the solder is gradually increased during each thermal cycle. When a critical value of the energy is reached, recrystallization is initiated. It is assumed that, even though recovery consumes a certain amount of the energy, the net change in the energy per cycle is always positive as experimental investigations have shown that newly recrystallized grains consistently appear sooner or later under various thermal cycling conditions. The stored energy is released through the nucleation and growth of new grains, which gradually consume the strain-hardened matrix of high defect density.

The approach is realized by combining Monte Carlo simulations with finite element calculations. The Monte Carlo method is employed to model the mesoscale microstructure and the finite element method to model the macroscale non-homogeneous deformation (see

Fig. 12). The non-homogeneous volume stored energy distribution in solder interconnections is scaled from the finite element model results and mapped onto the lattice of the Monte Carlo model. The quantitative prediction of the onset of recrystallization is carried out with the help of the Monte Carlo simulation.

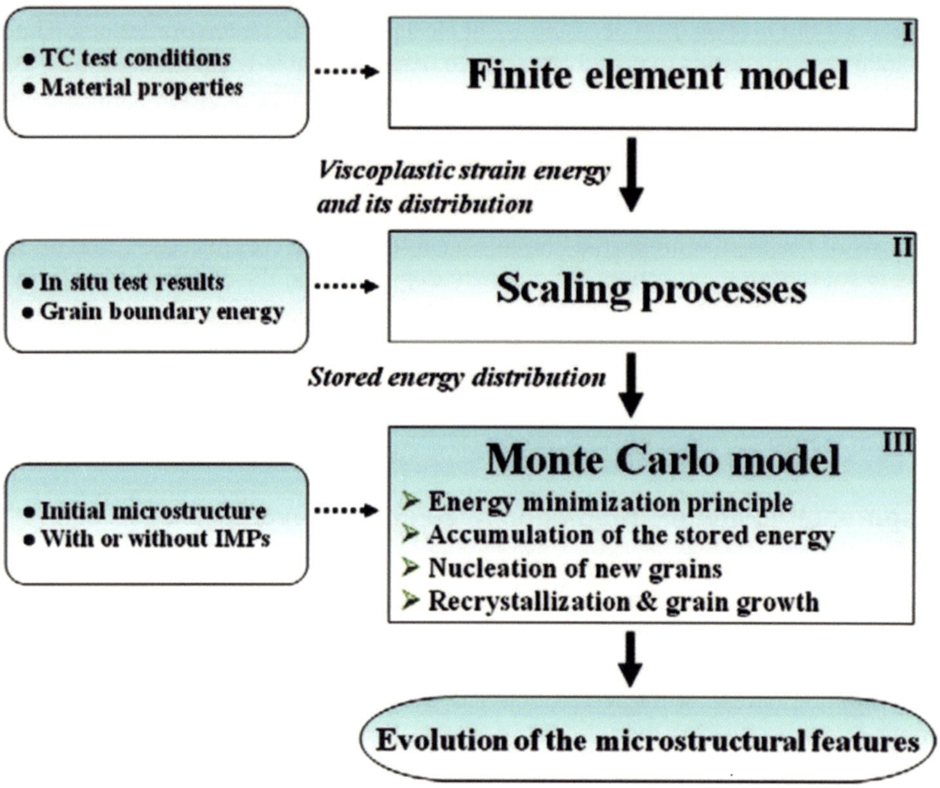

Fig. 12. Flow chart for the simulation of microstructural changes in solder interconnections [77,78]. Acronym TC stands for 'thermal cycling'.

In the Monte Carlo lattice, two adjacent sites with different grain orientation numbers are regarded as being separated by a grain boundary, while a group of sites with the same orientation number are considered a single grain. The total stored energy of the system under consideration consists of the grain boundary energy and the volume defect energy. Each site contributes an amount of stored energy to the system, and each pair of dissimilarly oriented neighboring sites contributes a unit of grain boundary energy to the system. The recrystallization process is modeled by randomly introducing nuclei (small embryos with zero stored energy) into the Monte Carlo lattice at a constant rate. An non-recrystallized site will become recrystallized if (a) the volume stored energy of the chosen sites is larger than the critical stored energy, and (b) the total energy of the system is reduced. If the selected site is recrystallized, it is considered as a contribution to the grain growth process.

The computational results were compared with the experimentally observed microstructural changes in solder interconnections subjected to thermal cycling tests. The results of the microstructural simulations carried out in this work can be summarized as follows: the incubation period of the recrystallization is about 1000 thermal cycles under the particular cycling conditions. The recrystallization is initiated in the corner regions on the package side of the solder interconnections and the expansion of the recrystallized region is controlled by the volume stored energy distribution. The expansion takes place first along the package side interfacial region toward the center of the interconnections. After that the recrystallized region expands toward the rest of the interconnections.

The incubation period of the recrystallization, the expansion of the recrystallized region, and the rate of increase of the recrystallized fraction are in good agreement with the experimental observations of thermally cycled component boards. This method predicts reasonably well the incubation period and the growth rate of the recrystallization, as well as the expansion of the recrystallized region. Figure 13 shows a comparison of predicted microstructural evolution with experimental evolution. The onset of recrystallization is a useful criterion to determine when the material models for the as-solidified microstructures are not valid anymore and, therefore, crack nucleation and propagation should be taken into account.

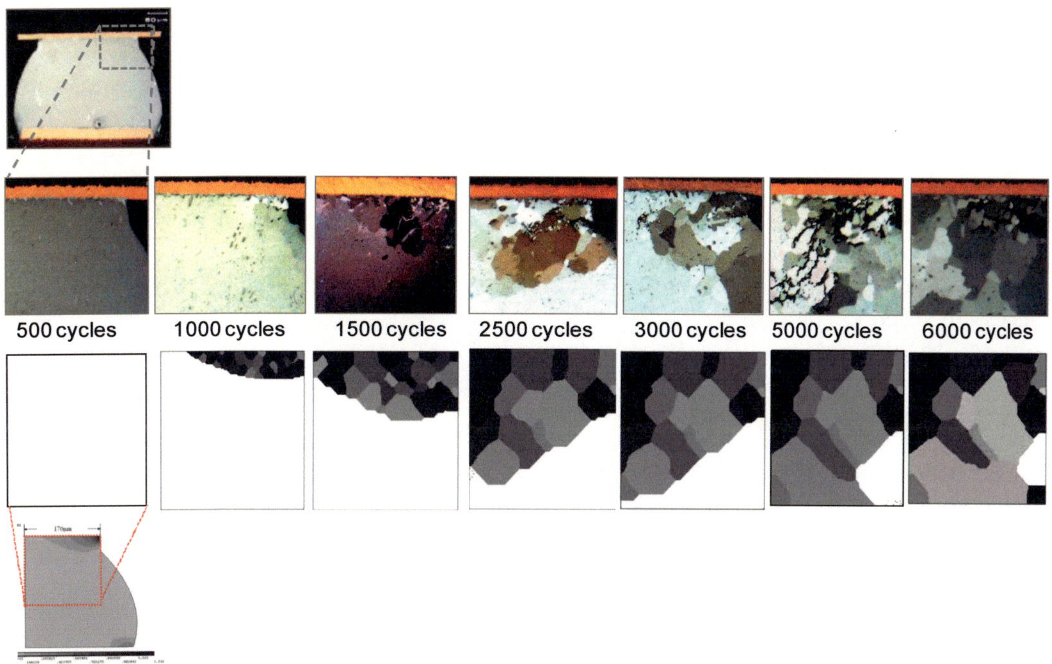

Fig. 13. Observed and simulated microstructural changes of solder interconnections with an increasing number of thermal cycles [77].

In a more recent work Li et al. [78] have developed the method to take into consideration the fact that the nucleation of new grains of low defect density is more likely at the boundaries of high misorientation, such as the boundary between the intermetallic particles and the tin matrix. In particular, the coarse primary intermetallic particles can generate localized stress concentrations under an applied load because of their dissimilar mechanical properties with respect to the tin matrix. The intermetallic particles are introduced in the Monte Carlo model as inert particles that also do not move or grow. In practice the fine and uniformly distributed intermetallic particles would suppress the recrystallization to some extent by influencing the motion of the grain boundaries of recrystallizing grains but this is not (yet) considered in the model.

6. Conclusions

The cracking of the near-eutectic SnAgCu interconnections under thermal cycling conditions occurs through the bulk of the solder interconnections after a change in the microstructure by recrystallization. The cumulative increase in the stored energy during each deformation cycle provides the driving force for the recrystallization. After the solidification structure has changed into a more or less equiaxed grain structure, there is a continuous network of new high-angle boundaries providing favorable sites for cracks to nucleate and propagate with less energy consumption in comparison with the cracking of the as-soldered microstructure. The decrease in the stored energy in near-eutectic SnAgCu solders is assumed to take place very effectively by the recovery resulting from the high stacking fault energy of tin. Therefore, the recrystallization is initiated under well-defined loading conditions: dynamic loading where strain hardening is more effective than recovery.

The recrystallization of solder interconnections under thermomechanical (or in cyclic power) loading is an important phenomenon for the following reasons: being an experimentally observable indicator of microstructural evolution, the recrystallization enables one to establish a correlation between the field use loading conditions and those produced in accelerated reliability tests. Furthermore, the theoretically well-known phenomena of recovery and recrystallization can provide the means to incorporate the effects of microstructural evolution into lifetime prediction models, which are being increasingly employed to reduce the amount of reliability testing. Finally, by controlling the recrystallization of solder interconnections, for example by alloying, one may discover new solutions to improve the reliability of soldered electronic devices.

7. Acknowledgements

The authors wish to thank the following people for their valuable contribution to the work presented in this chapter: Mr. Jussi Hokka, Dr. Jue Li, Dr. Maik Mueller (TU Dresden, Germany), Mr. Otso Ratia, and Dr. Hongbo Xu. The authors would also like to thank the following people for their much appreciated help and the numerous discussions over the years: Ms. Pirjo Kontio, Ms. Sini Niiranen, Ms. Johanna Koivisto, Ms. Riitta Viitala, Dr. Hongtao Chen, Dr. Erkki Heikinheimo, Mr. Simo Miettinen, Dr. Tomi Laurila, and Dr. Vesa Vuorinen. Special thanks go to Prof. Mervi Paulasto-Kröckel and Prof. Klaus-Juergen Wolter (TU Dresden, Germany) for their favorable support for this work. The financial support from the Academy of Finland (decision number 123922), the Finnish Funding Agency for Technology and Innovation (decision numbers 40135/07, 662/06, and the ELMO program), and the Finnish electronics industry is gratefully acknowledged.

8. References

[1] J. Karppinen, T. T. Mattila, and J. K. Kivilahti, "Formation of thermomechanical interconnection stresses in a high-end portable product," *The Proceedings of the 2nd Electronics System Integration Technology Conference*, London, UK, September 1-4, 2008, IEEE/EIA CPMT, (2008), pp. 1327-1332.

[2] J. S. Karppinen, J. Li, T. T. Mattila, and M. Paulasto-Kröckel, "Thermomechanical reliability characterization of a handheld product in accelerated tests and use environment," *Microelectronics Reliability*, (in print).

[3] IEC 60068–2–14 Ed. 5.0 b: 1984, "Environmental testing - part 2: tests. Test N: change of temperature," International Electrotechnical Commission, (1984), 34 p.

[4] JESD22-A104C, "Temperature Cycling," Jedec Solid State Technology Association, (2005), 16 p.

[5] IPC-TM-650 rev. A, "Thermal Shock and Continuity, Printed Board," The Institute for Interconnecting and Packaging Electronic Circuits, (1997), 2 p.

[6] C. F. Coombs Jr., Printed Circuits Handbook, 5th ed., New York, (2001), McGraw-Hill, 1200 p.

[7] Y. S. Touloukian and C. Y. Ho, Thermal Expansion: Metallic Elements and Alloys, New York, (1975), IFI/Plenum, 316 p.

[8] Directive 2002/95/EC of the European Parliament and of the Council on the Restriction of the Use of Hazardous Substances in Electrical and Electronic Equipment (RoHS), Jan. 27th, 2003.

[9] Directive 2002/96/EC of the European Parliament and of the Council on Waste of Electrical and Electronic Equipment (WEEE), Jan. 27th, 2003.

[10] D. R. Frear, "Microstructural evolution during the thermomechanical fatigue of solder joints," in M.J. Cieslak, M.E. Glicksman, S. Kang, and M.E Glicksman, *The Metal Science of Joining*, The Minerals, Metals & Materials Society, pp. 191-2000.

[11] J. W. Morris Jr., D. Tribula, T. S. E. Summers, and D. Grivas, "The role of microstructure in thermal fatigue of Pb-Sn solder joints," in John H. Lau, *Solder Join Reliability*, New York, 1991, Van Nostrand Reinold, pp. 225-265.

[12] T. T. Mattila, T. Laurila, and J. K. Kivilahti, "Metallurgical factors behind the reliability of high density lead-free interconnections," in E. Suhir, C. P. Wong, and Y. C. Lee, *Micro-and Opto-Electronic Materials and Structures: Physics, Mechanics, Design, Reliability, Packaging*, Springer Publishing Company, New York, 2007, (1), pp. 313-350.

[13] W. Q. Peng, "Lead-free electronic assembly based on Sn–Ag–Cu solders," Espoo, licentiate thesis, Helsinki University of Technology, (2001), p. 124.

[14] K.-W. Moon, W. J. Boettinger, U. R. Kattner, F. S. Biancaniello, and C. A. Handwerker, "Experimental and thermodynamic assessment of Sn–Ag–Cu solder alloys," *Journal of Electronic Materials*, 29, 10, (2000), pp. 1122-1136.

[15] T. T. Mattila, V. Vuorinen, and J. K. Kivilahti, "Impact of printed wiring board coatings on the reliability of lead-free chip-scale package interconnections," *Journal of Materials Research*, 19,11, (2004), pp. 3214-3223.

[16] A. LaLonde, D. Emelander, J. Jeannette, C. Larson, W. Rietz, D. Swenson, and D. W. Henderson, "Quantitative metallography of β-Sn dendrites in Sn3.8Ag0.7Cu ball grid array solder balls via electron backscatter diffraction and polarized light microscopy," *Journal of Electronic Materials*, 33, 12, (2004), pp. 1545-1549.

[17] D. Henderson, J. J. Woods, T. A. Gosseling, J. Bartelo, D. E. King, T. M. Korhonen, M. A. Korhonen, L. P. Lehman, E. J. Cotts, S. K. Kang, P. Lauro, D.-Y. Shih, C. Goldsmith, and K. J. Puttliz, "The microstructure of Sn in near eutectic Sn-Ag-Cu alloy solder joints and its role in thermomechanical fatigue," *Journal of Materials Research*, 19, 6, (2004), pp. 1608-1612.

[18] S. Terashima and M. Tanaka, "Thermal fatigue properties of Sn-1.2Ag-0.5Cu-xNi Flip Chip interconnects," *Materials Transactions*, 45, 3, (2004), pp. 681-688.

[19] S. K. Kang, P. A. Lauro, D.-Y. Shih, D. W. Henderson, and K. J. Puttlitz, "Microstructure and mechanical properties of lead-free solders and solder joints used in microelectronic applications," *IBM Journal of Research and Development*, 49, 4/5, (2005), pp. 607-620.

[20] A. U. Telang, T. R. Bieler, J. P. Lucas, K. N. Subramanian, L. P. Lehman, Y. Xing, and E. J. Cotts, "Grain-boundary character and grain growth in bulk tin and bulk lead-free solder alloys," *Journal of Electronic Materials*, 33, 12, (2004), pp. 1412-1423.

[21] L. P. Lehman, S. N. Athavale, T. Z. Fullem, A. C. Giamis, R. K. Kinyanjui, M. Lowenstein, K. Mather, R. Patel, D. Rae, J. Wang, Y. Xing, L. Zavalij, P. Borgesen, and E. J. Cotts, "Growth of Sn and intermetallic compounds in Sn-Ag-Cu solder," *Journal of Electronic Materials*, 33, 12, (2004), pp. 1429-1439.

[22] Z. G. Chen, Y. W. Shi, Z. D. Xia, and Y. F. Yan, "Study on the microstructure of a novel lead-free solder alloy SnAgCu-RE and its soldered joints," *Journal of Electronic Materials*, 31, 10, (2002), pp. 1122-1128.

[23] Maik Müller, Steffen Wiese, and Klaus-Jürgen Wolter, "Influence of cooling rate and composition on the solidification of SnAgCu solders," *The Proceedings of the 1st Electronics Systemintegration Technology Conference*, Dresden, Germany, September 5 - 7, 2006, IEEE/EIA CPMT, (2006), pp. 1303-1311

[24] O. Fouassier, J.-M. Heintz, J. Chazelas, P.-M. Geffroy, and J.-F. Silvain, "Microstructural evolution and mechanical properties of SnAgCu alloys," Journal of Applied Physics, 100, 043519 (2006), pp. 043519-1 - 043519-8.

[25] J. Gong, C. Liu, P. P. Conway, and V. V. Silberschmidt, "Crystallographic structure and mechanical behaviour of SnAgCu solder interconnects under a constant loading rate," *The Proceedings of the 57th Electronic Component and Technology Conference*, Reno, NV, May 29- June 1, 2007, IEEE/EIA CPMT, (2007), pp. 677-683

[26] P. Lauro, S. K. Kang, W. K. Choi, and D.-Y. Shih, "Effects of mechanical deformation and annealing on the microstructure and hardness of Pd-free solders," *Journal of Electronic Materials*, 32, 12, (2003), pp. 1432-1440.

[27] M. Krause, M. Mueller, M. Petzold, S. Wiese, and K. J. Wolter, "Scaling effects on grain size and texture of lead free interconnects - Investigations by electron backscatter diffraction and nanointendation," *The Proceedings of the 58th Electronic Component and Technology Conference*, Orlando, FL, May 27-30, 2008, IEEE/EIA CPMT, (2008), pp. 75-81.

[28] T.-M. K. Korhonen, P. Turpeinen, L. P. Lehman, B. Bowman, G. H. Thiel, R. C. Parkes, M. A. Korhonen, D. W. Henderson, and K. J. Puttlitz, "Mechanical properties of near-eutectic Sn-Ag-Cu alloy over a wide range of temperatures and strain rates," *Journal of Electronic Materials*, 33, 12, (2004), pp. 1581-1588.

[29] W. Peng, K. Zeng, and J. Kivilahti, "A literature review on potential lead-free solder systems," Espoo, Helsinki University of Technology, Report Series HUT-EPT-1, (2000), 53 p.

[30] S. Chada, R. A. Fournelle, W. Laub, and D. Shangguan, "Copper substrate dissolution in eutectic Sn-Ag solder and its effect on microstructure," *Journal of Electronic Materials*, 29, 10, (2000), pp. 1214-1221.

[31] M. O. Alam, Y. C. Chan, and K. N. Tu, "Effect of 0.5 wt% Cu addition in Sn-3.5%Ag solder on the dissolution rate of Cu metallization," *Journal of Applied Physics*, 94, 12, (2003), pp. 7904-7909.

[32] W. G. Bader, "Dissolution of Au, Ag, Pd, Pt, Cu and Ni in a Molten Sin-Lead Solder," *Welding Journal*, 48, 12, (1969), pp. 551s-557s.

[33] W. G. Bader, "Dissolution and formation on intermetallics in the soldering process," *The proceedings of the Conference on Physical Metallurgy and Metal Joining*, St. Louis, MO, Oct. 16-17. 1980, Warrendale, USA.

[34] W. C. Leslie, T. J. Michalak, and F. W. Aul, "The annealing of cold-worked iron," in C. W. Spencer and F. E. Werner, *Iron and Its Dilute Solid Solutions*. New York, 1963, Interscience Puhlishers, pp. 103-119.

[35] R. W. Cahn, "Recovery and recrystallization," in R. W. Cahn, *Physical Metallurgy*, Amsterdam, 1965, North-Holland Publishing Company, pp. 925–987.

[36] F. J. Humphreys and M. Hatherly, *Recrystallization and Related Annealing Phenomena*, 2nd ed., Oxford, 2004, Elsevier Ltd., 574 p.

[37] R.D. Doherty, D. A. Hughes, F. J. Humphreys, J. J. Jonas, D. Juul Jensen, M. E. Kassner, W. E. King, T. R. McNelley, H. J. McQueen, and A. D. Rollett, "Current issues in recrystallization," *Materials Science and Engineering A*, 238, (1997), pp. 219 – 274.

[38] H. Yu and J. K. Kivilahti, "Nucleation kinetics and solidification temperatures of SnAgCu interconnections during reflow process," IEEE Transactions on Components and Packaging Technologies, 29, 4, (2006), pp. 778 - 786.

[39] R. Darveaux, C. Reichman, and P. Agrawal, "Solidification behavior of lead free and tin lead solder bumps," *The Proceedings of the 60th Electronic Component and Technology Conference*, Las Vegas, NV, June 1-4, 2010, IEEE/EIA CPMT, (2010), pp. 1442-1447.

[40] J. S. Kang, R. A. Gagliano, G. Ghosh, and M. E. Fine, "Isothermal solidification of Cu/Sn diffusion couples to form thin-solder joints," *Journal of Electronic Materials*, 31, 11, (2002), pp. 1238-1243.

[41] J.-M. Song, J.-J. Lin, C.-F. Huang, and H.-Y. Chuang, "Crystallization, morphology and distribution of Ag_3Sn in Sn–Ag–Cu alloys and their influence on the vibration fracture properties," *Materials Science and Engineering A*, 466, (2007), pp. 9-17

[42] D. W. Henderson, T. Gosselin, A. Sarkhel, S. K. Kang, W.-K. Choi, D.-Y. Shih, C. Goldsmith, and K. J. Puttlitz, "Ag_3Sn plate formation in the solidification of near ternary eutectic Sn–Ag–Cu alloys," *Journal of Materials Research*, 17, (2002), pp. 2775-2778.

[43] J. H. Perepezko, D. H. Rasmussen, I. E. Anderson, and C. R. Loper, Jr., "Undercooling of low-melting point metals and alloys," The *Proceedings of the International Conference on Solidification and Casting of Metals*, Sheffield, England, July 1977, Sheffield Metallurgical and Engineering Association / University of Sheffield / the Metals Society, (1979), pp. 169-174.

[44] S. Wiese, E. Meusel, and K. J. Wolter, "Microstructural dependence of constitutive properties of eutectic SnAg and SnAgCu solders," *The Proceedings of the 53rd Electronic Components and Technology Conference*, 27 May-30 May, 2003, New Orleans, LA, IEEE EIA/CPMT, (2003), pp. 197-206.

[45] S. K. Kang, W. K. Clioi, D.-Y. Shih, D. W. Henderson, T. Gossefin, A. Sarkliel, C. Goldsmith, and K. J. Puttlitz, "Formation of Ag_3Sn plates in Sn-Ag-Cu alloys and optimization of their alloy composition," *The Proceedings of the 53rd Electronic Components and Technology Conference*, 27 May-30 May, 2003, New Orleans, LA, IEEE EIA/CPMT, (2003), pp. 64-70.

[46] D. Swenson, "The effects of suppressed beta tin nucleation on the microstructural evolution of lead-free solder joints," *Journal of Material Science: Materials in Electronics*, 18, 1-3, (2007), pp. 39-54.

[47] S. K. Kang, G. C. Moon; P. Lauro, and D.-Y. Shih, "Critical Factors Affecting the Undercooling of Pb-free, Flip-Chip Solder Bumps and In-situ Observation of

Solidification Process," *The Proceedings of the 57th Electronic Component and Technology Conference*, Reno, NV, May 29-June 1, 2007, IEEE/EIA CPMT, (2007), pp. 1597-1603.

[48] J. W. Elmer, E. D. Specht, and M. Kumar, "Microstructure and in situ observations of undercooling for nucleation of β-Sn relevant to lead-free solder alloys," *Journal of Electronic Materials*, 39, 3, (2010), pp. 273-282.

[49] M. A. Matin, W. P. Vellinga, and M. G. D. Geers, "Thermomechanical fatigue damage evolution in SAC solder joints," *Materials Science and Engineering A*, 445-446, (2007), pp. 73-85.

[50] N. S. Brar and W. R. Tyson, "Elastic and plastic anisotropy of white tin," *Canadian Journal of Physics*, 50, 19, (1972), pp. 2257-2264.

[51] *Metals Handbook*, Volume 1 - Properties and Selection of Metals, 8th ed., New York, 1961, American Society for Metals, 1300 p.

[52] K. N. Subramanian, "Role of anisotropic behavior of Sn on thermomechanical fatique and fracture of Sn-based solder joints under thermal excursions," *Fatigue and Fracture of Engineering Materials and Structures*, 30, 5, (2007), pp. 420-431.

[53] M. A. Matin, E. W. C. Coenen, W. P. Vellinga, and M. G. D. Geers, "Correlation between thermal fatigue and thermal anisotropy in a Pb-free solder alloy," *Scripta Materialia*, 53, (2005), pp. 927-932.

[54] T. R. Bieler, H. Jiang, L. P. Lehman, T. Kirkpatrick, E. J. Cotts, and B. Nandagopal, "Influence of Sn grain size and orientation on the thermomechanical response and reliability of Pb-free solder joints," *IEEE Transactions on Components and Packaging Technologies*, 31, 2, (2008), pp. 370-380.

[55] T. R. Bieler, B. Zhou, L. Blair, A. Zamiri, P. Darbandi, F. Pourboghrat, T.-K. Lee, and K.-C. Liu, "The role of elastic and plastic anisotropy of Sn on microstructure and damage evolution in lead-free solder joints," *The Proceedings of the 2011 IEEE International Reliability Physics Symposium*, 10-14 April, 2011, Monterey, CA, IEEE, (2011), pp. 5F.1.1-5F.1.9.

[56] S. Terashima, K. Takahama, M. Nozaki, and M. Tanaka, "Recrystallization of Sn grains due to thermal strain in Sn-1.2Ag-0.5Cu-0.05N solder," *Materials Transactions, Japan Institute of Metals*, 45, 4, (2004), pp. 1383-1390.

[57] S. Dunford, S. Canumalla, and P. Viswanadham, "Intermetallic morphology and damage evolution under thermomechanical fatigue of lead (Pb)-free solder interconnections," The Proceedings of the 54th Electronic Components and Technology Conference, June 1-4, 2004, Las Vegas, NV, USA, IEEE/EIA/CPMT, (2004), pp. 726-736.

[58] L. Lehman, S. Athavale, T. Fullem, A. Giamis, R. Kinyanjui, M. Lowenstein, K. Mather, R. Patel, D. Rae, J. Wang, Y. Xing, L. Zavalij, P. Borgesen, and E. Cotts, "Growth of Sn and intermetallic compounds in Sn-Ag-Cu solder," *Journal of Electronic Materials*, 3, 12, (2004), pp. 1429-1439.

[59] P. Limaye, B. Vandevelde, D. Vandepitte, and B. Verlinden, "Crack growth rate measurement and analysis for WLCSP Sn-Ag-Cu solder joints," The proceedings of the SMTA international annual conference, Chicago, IL, September 25-29, (2005), pp. 371-377.

[60] L. Xu and J. H.L. Pang, "Intermetallic growth studies on SAC/ENIG and SAC/CU-OSP lead-free solder joints," *Thermal and Thermomechanical Phenomena in Electronics Systems*, (2006), pp.1131-1136.

[61] A. U. Telang, T. R. Bieler, A. Zamirini, and F. Pourboghrat, "Incremental recrystallization/crain growth driven by elastic strain energy release in a thermomechanically fatigued lead-free solder joint," *Acta Materialia*, 55, (2007), pp. 2265-2277.

[62] J. J. Sundelin, S. T. Nurmi, and T. K. Lepistö, "Recrystallization behavior of SnAgCu solder joints," *Materials Science and Engineering A*, 474, (2008), pp. 201-207.

[63] J. Li, J. Karppinen, T. Laurila, and J. K. Kivilahti, "Reliability of Lead-Free solder interconnections in Thermal and Power cycling tests," *IEEE Transactions on Components and Packaging Technologies*, 32, (2009), pp. 302-308.

[64] H. T. Chen, M. Mueller, T. T. Mattila, J. Li, X.W. Liu, K.-J. Wolter, and M. Paulasto-Kröckel, "Localized Recrystallization and Cracking of Lead-Free Solder Interconnections under Thermal Cycling," *Journal of Materials Research*, 25, 16, (2011), pp. 2103-2116.

[65] T. T. Mattila, M. Mueller, M. Paulasto-Kröckel, and K. J. Wolter "Failure mechanism of solder interconnections under thermal cycling conditions," *The Proceedings of the 3rd Electronic System-Integration and Technology Conference*, Berlin, Germany, September 13-16, 2010, IEEE/EIA CPMT, (2010), pp. 1-8.

[66] B. Zhou, T. T. Bieler, T. K. Lee, and K.C. Liu, "Crack development in a low-stress PBGA package due to continuous recrystallization leading to formation of orientations with [001] parallel to the interface," *Journal of Electronic Materials*, 39, (2010), pp. 2669 -2679.

[67] D. Hardwick, C. M. Sellars, and W. J. McG. Tegart, "The occurrence of recrystallization during high-temperature creep," *Journal of the Institute of Metals*, 90, (1961), pp. 21-22.

[68] D. McLean and M. H. Farmer, "The relation during creep between grain–boundary sliding, sub-crystal size, and extension," *Journal of Institute of Metals*, 85, (1956), pp. 41-50.

[69] P. Gay and A. Kelly, "X-ray studies of polycrystalline metals deformed by rolling. II. Examination of the softer metals, tin, zinc, lead and cadmium," *Acta Crystallographica*, 6, (1953), pp. 172-177.

[70] A. G. Guy, *Elements of Physical Metallurgy*, 2nd ed., London, 1960, Addison-Wesley Publishing Company Inc., 528 p.

[71] S. Miettinen, *Recrystallization of Lead-free Solder Joints under Mechanical Load*, Master's Thesis (in Finnish), Espoo, (2005), 84 p.

[72] T. M. Korhonen, L. Lehman, M. Korhonen, and D. Henderson, "Isothermal fatigue behavior of the near-eutectic Sn-Ag-Cu alloy between -25°C and 125°C," *Journal of Electronic Materials*, 36, 2, (2007), pp. 173-178.

[73] T. Laurila, T. T. Mattila, V. Vuorinen, J. Karppinen, J. Li, M. Sippola, and J. K. Kivilahti, "Evolution of microstructure and failure mechanism of lead-free solder interconnections in power cycling and thermal shock tests," *Microelectronics Reliability*, 47, 7, (2007), pp. 1135-44.

[74] T. T. Mattila, H. Xu, O. Ratia, and M. Paulasto-Kröckel, "Effects of thermal cycling parameters on lifetimes and failure mechanism of solder interconnections," *The Proceedings of the 60th Electronic Component and Technology Conference*, Las Vegas, NV, June 1-4, 2010, IEEE/EIA CPMT, (2010), pp. 581-590.

[75] H. Xu, T. T. Mattila, O. Ratia, and M. Paulasto-Kröckel, "Effects of thermal cycling parameters on lifetimes and failure mechanism of solder interconnections," *IEEE Transactions on Manufacturing and Packaging Technologies – a Special Issue*, (in print). Invited paper.

[76] T. T. Mattila and J. K. Kivilahti, "The role of recrystallization in the failure mechanism of SnAgCu solder interconnections under thermomechanical loading," *IEEE Transactions on Components and Packaging Technologies*, 33, 3, (2010), pp. 629-635.

[77] J. Li, T. T. Mattila, and J. K. Kivilahti, "Multiscale simulation of recrystallization and grain growth of Sn in lead-free solder interconnections," *Journal of Electronic Materials*, 39, 1, (2010), pp. 77-84.

[78] J. Li, H. Xu, T. T. Mattila, J. K. Kivilahti, T. Laurila, and M. Paulasto-Kröckel, "Simulation of dynamic recrystallization in solder interconnections during thermal cycling," *Computational Materials Science*, 50, (2010), pp. 690-697.

9

Mathematical Modeling of Single Peak Dynamic Recrystallization Flow Stress Curves in Metallic Alloys

R. Ebrahimi and E. Shafiei
Department of Materials Science and Engineering, School of Engineering, Shiraz University, Shiraz, Iran

1. Introduction

Deformation of metals and alloys at temperatures above 0.5 T_m is a complex process in which mechanical working interacts with various metallurgical processes such as dynamic restoration including recovery and recrystallization, and phase transformation for polymorphous materials. The understanding of these processes, however, enables the behavior of the metals and alloys. Recent developments have been described in several review papers. These reviews were in agreement that metals and alloys having relatively low values of stacking fault energy (SFE) could recrystallize dynamically, whereas those of high SFE including bcc metals and alloys which behave in a manner similar to fcc metals of high SFE recovered dynamically only during high temperature deformation. So that, according to microstructural evolutions, material response can principally be divided into two categories in hot deformation: dynamic recovery (DRV) type and dynamic recrystallization (DRX) type.

The process of recrystallization of plastically deformed metals and alloys is of central importance in the processing of metallic alloys for two main reasons. The first is to soften and restore the ductility of material hardened by low temperature deformation that occurring below about 50% of the absolute melting temperature which leads to lower forces. The second is to control the microstructure and mechanical properties of the final product.

The analysis of metal forming process such as hot rolling, forging and extrusion has been dependent on various parameters including constitutive relation which contributes to stress-strain curves at high temperatures, shape of workpiece and product, shapes of tools, friction, temperature, forming speed, etc. In such parameters, the constitutive equation is one of the most important factors which have an effect on solution accuracy.

A number of research groups have attempted to develop constitutive equations of materials and suggested their own formulations by putting the experimentally measured data into one single equation. Misaka and Yoshimoto proposed a model which gives the flow stress of carbon steels as a function of the strain, strain rate, temperature and carbon content. Shida's model takes account of the flow stress behavior of the steels in austenite, ferrite and in the two-phase regions. Voce suggested an approximate equation of stress–stain curve considering the dynamic recrystallization. Finally, Ebrahimi et al. obtained a mathematical model according to the phenomenological representation of the shape of the stress-strain curves that consists an additional constant. Due to the importance of flow stress estimation of metals and alloys at high temperatures, the required forces for the deformation processes, dimensional accuracy of final products and simulation of processes; it is necessary to provide a model in order to eliminate the limitations of previous models to some extent. The contexts of this section could generally be classified into two main parts, at the beginning, some previous mathematical models and basic concepts in relation with dominate processes during hot deformation of metals and alloys, such as dynamic recovery (DRV) and dynamic recrystallization (DRX) will be reviewed and finally, by concluding from existing mathematical models related to prediction of single peak flow stress curves, a new model will be introducd. With regards to the importance of macroscopic data from mechanical tests as compared to microscopic ones from metallurgical investigation, due to its less time and cost consuming nature, mathematical and macroscopic aspects of DRX process are considered in this model.

2. Basic concepts

2.1 Dynamic recovery (DRV)

The basic mechanisms of dynamic recovery are dislocation climb, cross-slip and glide, which result in the formation of low angle boundaries as also occurs during static recovery. However, the applied stress provides an additional driving pressure for the movement of low angle boundaries and those of opposite sign will be driven in opposite directions, and this stress-assisted migration of dislocation boundaries may contribute significantly to the overall strain.

Such migration results in some annihilation of dislocations in opposing boundaries and Y-junction boundary interactions and these enable the subgrains to remain approximately equiaxed during the deformation. In-situ SEM deformation experiments have shown that some reorientation of subgrains may also occur during hot deformation. The subgrains can therefore be considered to be transient microstructural features.

The processes of work hardening and recovery lead to the continual formation and dissolution of low angle boundaries and to a constant density of unbound or 'free' dislocations within the subgrains. After a strain of typically 0.5 to 1, the subgrain structure often appears to achieve a steady state. The microstructural changes occurring during dynamic recovery are summarized schematically in figure 1.

2.2 Dynamic recrystallization (DRX)

In metals which recovery processes are slow, such as those with a low or medium stacking fault energy (copper, nickel and austenitic iron), dynamic recrystallization may take place when a critical deformation condition is reached. A simplified description of the phenomenon of dynamic recrystallization is as follows. As shown in figures 2, new grains originate at the old grain boundaries, however, as the material continues to deform, the dislocation density of the new grains increases, thus reducing the driving force for further growth, and the recrystallizing grains eventually cease to grow. An additional factor which may limit the growth of the new grains is the nucleation of further grains at the migrating boundaries.

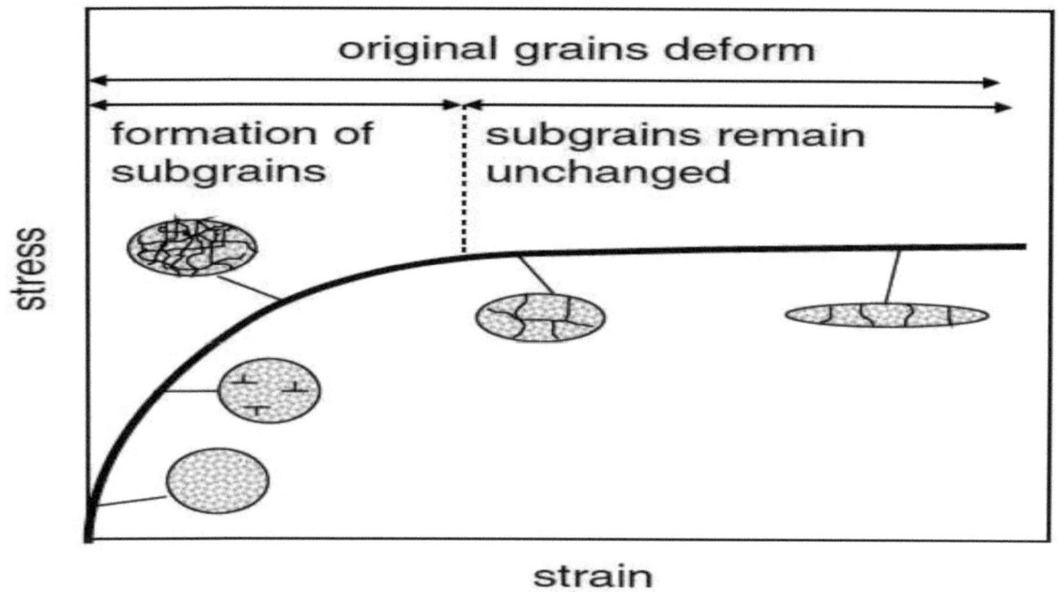

Fig. 1. Summary of the microstructural changes which occur during dynamic recovery.

This type of dynamic recrystallization, which has clear nucleation and growth stages, can be classified as a discontinuous process. There are other mechanisms which produce high angle grain boundaries during high temperature deformation and which may be considered to be types of dynamic recrystallization such as continuous dynamic recrystallixation (CDRX).

The general characteristics of dynamic recrystallization are as follows (Humphreys et al., Elsevier):

- As shown in figure 3, the stress-strain curve for a material which undergoes dynamic recrystallization generally exhibits a broad peak that is different to the plateau, characteristic of a material which undergoes only dynamic recovery (fig.1). Under conditions of low Zener–Hollomon parameter, multiple peaks may be exhibited at low strains, as seen in figure 3.
- A critical deformation is necessary in order to initiate dynamic recrystallization.
- The critical deformation decreases steadily with decreasing stress or Zener–Hollomon parameter (description in section 3.2), although at very low strain rates (creep) the critical strain may increase again.
- The size of dynamically recrystallized grains increases monotonically with decreasing stress. Grain growth does not occur and the grain size remains constant during the deformation.
- The flow stress and grain size are almost independent of the initial grain size, although the kinetics of dynamic recrystallization are accelerated in specimens with smaller initial grain sizes.
- Dynamic recrystallization is usually initiated at pre-existing grain boundaries although for very low strain rates and large initial grain sizes, intragranular nucleation becomes more important.

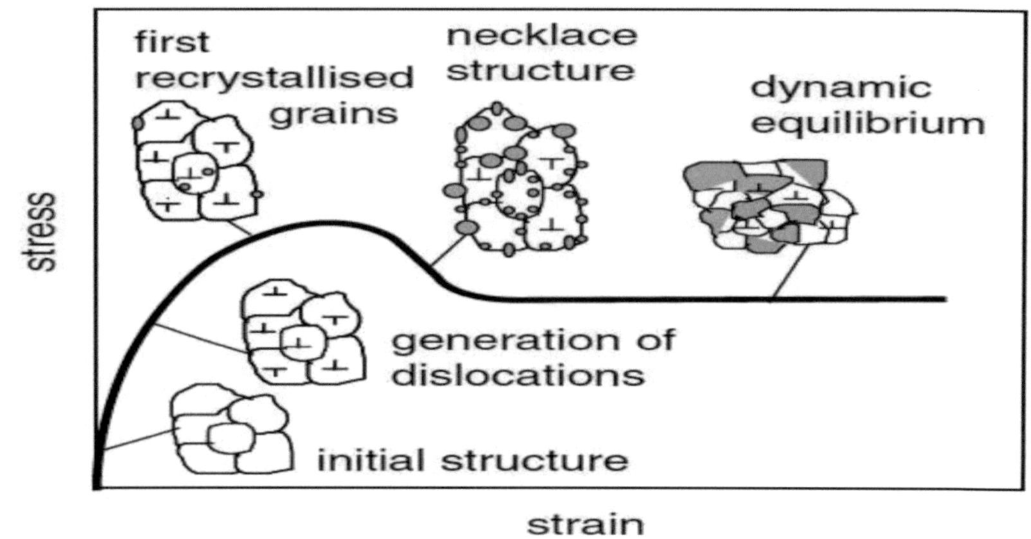

Fig. 2. Evolution of microstructure during hot deformation of a material showing DRX.

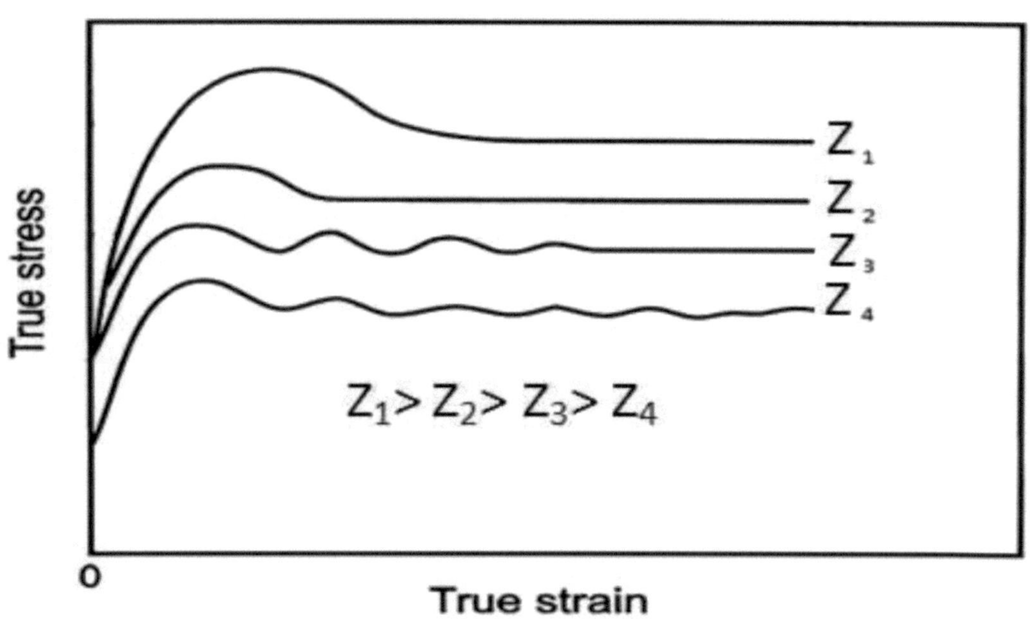

Fig. 3. The effect of Zener Hollomon parameter on the stress- strain curve.

2.3 Test methods

There are many different types of test methods, such as hot torsion, compression, tension, cam plastometer and drop hammer for developing the constitutive equation of the materials behavior. Among them, hot torsion and compression tests have widely been used. Hot torsion test has been used by many researchers to formulate constitutive equation of materials subject to a large strain because it has a forte in simulating the multi-pass deformation, in comparison with axisymmetric compression test. In the case of compression test, high friction at the interface of material and stroke head results in barreling during test. Thus, the compression test has a limitation in generating flow stress curve when the material undergoes large strain. Also Kim et al. have shown that the flow stress obtained from compression test is higher than that from torsion test, although the general shapes of the measured curves are similar. The differences between the stresses were approximately in the range of 10–20%.

These differences might be attributed to the following reasons: compression test has the glowing frictional forces at the ram-specimen interface as the test progress, while, there is no frictional effect in torsion test. In compression test, it is difficult to achieve constant strain rate and isothermal condition during test, whereas to control them in torsion test is accurate.

2.4 Flow curves

For metals with high stacking fault energy which experience DRV, the flow stress curves increase with strain in the initial deformation and reach constant in consequence of attaining the balance between work hardening and DRV (saturation stress, σ_s). For metals with DRX, initially the flow stress increases with strain which is being dominated by work hardening, and as DRX takes place upon critical strain (ε_c), the flow stress begins to decrease after reaching certain peak value. When the equilibrium is reached between softening due to DRX and work hardening, the curves drop to a steady state region (σ_{ss}). As shown in figure 3, the stress strain curves of a dynamically recrystallizing material may be characterized by a single peak or by several oscillations. Luton and Sellars have explained this in terms of the kinetics of dynamic recrystallization. At low stresses, the material recrystallizes completely before a second cycle of recrystallization begins, and this process is then repeated. The flow stress, which depends on the dislocation density, therefore oscillates with strain. At high stresses, subsequent cycles of recrystallization begin before the previous ones are finished, the material is therefore always in a partly recrystallized state after the first peak, and the stress strain curve is smoothed out, resulting in a single broad peak. Fig. 4 shows a schematic representing of dynamic recovery and a single peak dynamic recrystallization.

2.5 Strain hardening rate versus stress

The change in the slope ($\Theta=d\sigma/d\varepsilon$) of the stress-strain curve with stress can be a good indication of the microstructural changes taking place in material. All of the Θ-σ curves for a particular alloy originate from a common intercept Θ_0 at $\sigma=0$. The Θ-σ curves have three segments, two of them are linear. The first linear segment decreases with stress for initial strain to the point where subgrains begin to form with a lower rate of increase in DRV. The curve gradually changes to the lower slope of the second linear segment up to the point where the critical stress σ_c is attained for initiating the dynamic recrystallization. The curve then drops off rather sharply to $\Theta=0$ at peak stress. Extrapolation of the second linear segment of the Θ-σ curve intercepts the σ axis at the saturation stress. This would be the shape of the Θ-σ curve if dynamic recrystallization were absent with only dynamic recovery as the restorative mechanism operation, Fig.5.

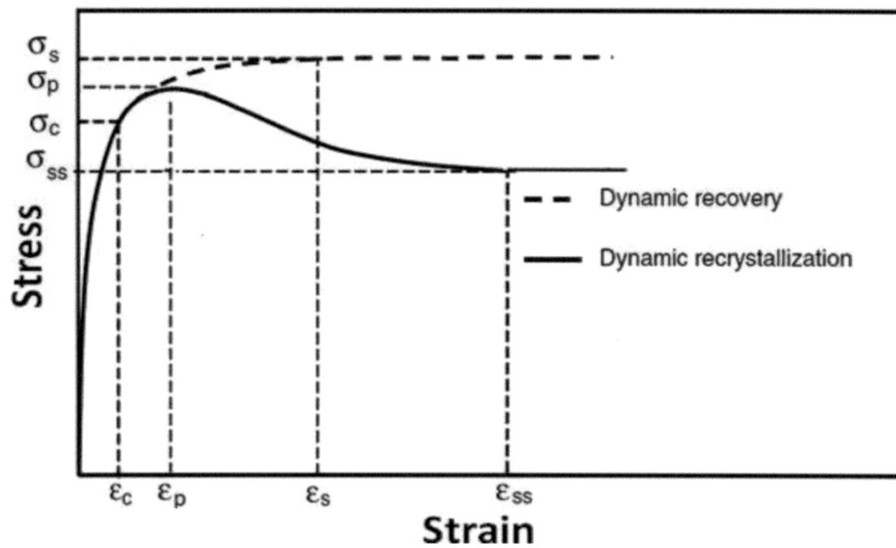

Fig. 4. Schematic representing of DRV and DRX.

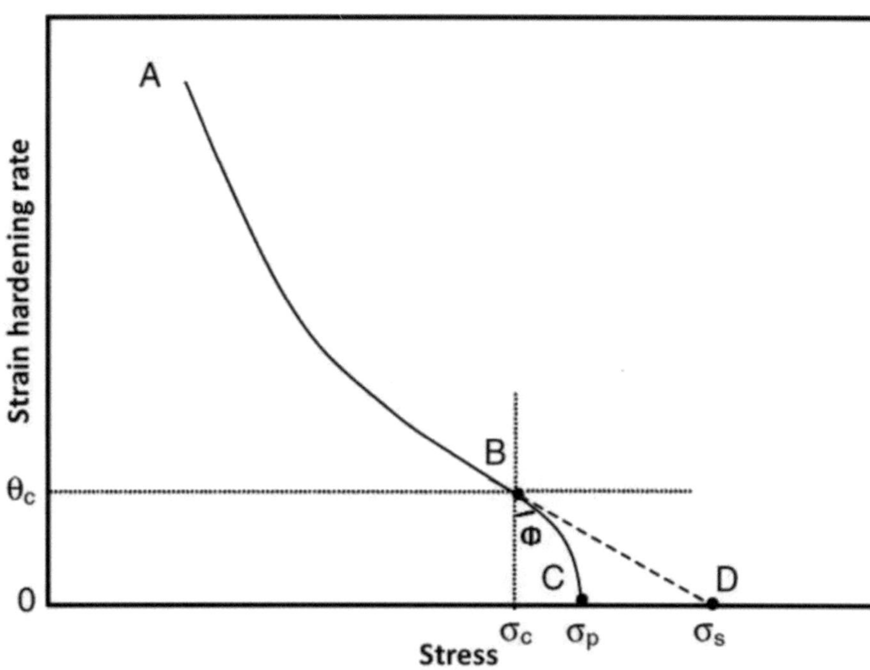

Fig. 5. Changes in the slope of the stress-strain curve with stress.

3. A review on presented models

Considerable researches have been carried out to model flow stress curves at high temperatures based on both empirical and mathematical models. Empirical models which are based on trial and error mainly obtained by repeating the mechanical tests at different conditions, and among these models, Shida and Misaka models could be mentioned. The most important limitation of these types of models is that they are valid for specific conditions and cannot be generalized, whereas in mathematical models, in order to estimate the flow curve, the material response are modeled first and then the empirical data is used to verify the prediction. Among these models, Cingara, Kim et.al and Ebrahimi et.al could be mentioned. According to the presented models, some limitations could be considered in this category, too. For instance, the presented model by McQueen and Cingara for DRX is only valid up to the peak stress (σ_P) and does not consider the softening behavior due to DRX, although a complement model has been developed to predict the softening behavior due to DRX.

Mathematical modeling of flow stress curves can be divided into three main topics as follows:

- Mathematical models which characterize the initiation of DRX.
- Mathematical models which predict characteristic points of flow stress curves.
- Mathematical models which predict flow stress curves.

It is important to mention that, due to less time and cost consuming nature of mathematical and macroscopic models, only this aspect of DRX process is considered in this chapter.

3.1 Initiation of dynamic recrystallization

The critical strain for initiation of DRX could be determined by metallography. However, this technique requires extensive sampling before and after the critical strain. Furthermore, phase changes during cooling from hot working temperature alter the deformed structure, which in turn render difficulties for metallographic analysis. Also this technique requires a large number of specimens deformed to different strains. On the other hand, the critical strain thus obtained is not precise.

Several attempts have been made to predict the initiation of DRX. For example, Ryan and McQueen observed that the presence of a stress peak at a constant strain rate flow curve leads to an inflection in plots of strain hardening rate, θ, versus stress, σ. Moreover, the points of inflection in θ-σ plots where the experimental curves separate from the extrapolated lower linear segments give critical conditions for initiation of DRX. Later, Poliak and Jonas have shown that this inflection point corresponds to the appearance of an additional thermodynamic degree of freedom in the system due to the initiation of DRX.

3.1.1 Determination of critical stress

The simple method of Najafizadeh and Jonas was used for determination of the critical stress for initiation of DRX. The inflection point is detected by fitting a third order polynomial to the θ-σ curves up to the peak point as follows:

$$\theta = A\sigma^3 + B\sigma^2 + C\sigma + D \tag{1}$$

where *A*, *B*, *C*, and *D* are constants for a given set of deformation conditions. The second derivative of this equation with respect to σ can be expressed as:

$$\frac{d^2\theta}{d\sigma^2} = 6A\sigma + 2B \tag{2}$$

At critical stress for initiation of DRX, the second derivative becomes zero. Therefore,

$$6A\sigma_c + 2B = 0 \Rightarrow \sigma_c = \frac{-B}{3A} \tag{3}$$

An example of θ-σ curves and its corresponding third order polynomial are shown in Fig. 6. Therefore, this method is used to determine the value of critical stress at different deformation conditions. Using the flow curves, the values of critical strain are determined (Najafizadeh et al. 2006).

According to Fig. 7, the normalized critical stress and strain for 17-4 PH stainless steel could be presented as:

$$\frac{\sigma_c}{\sigma_p} = 0.89 \tag{4}$$

$$\frac{\varepsilon_c}{\varepsilon_p} = 0.467 \tag{5}$$

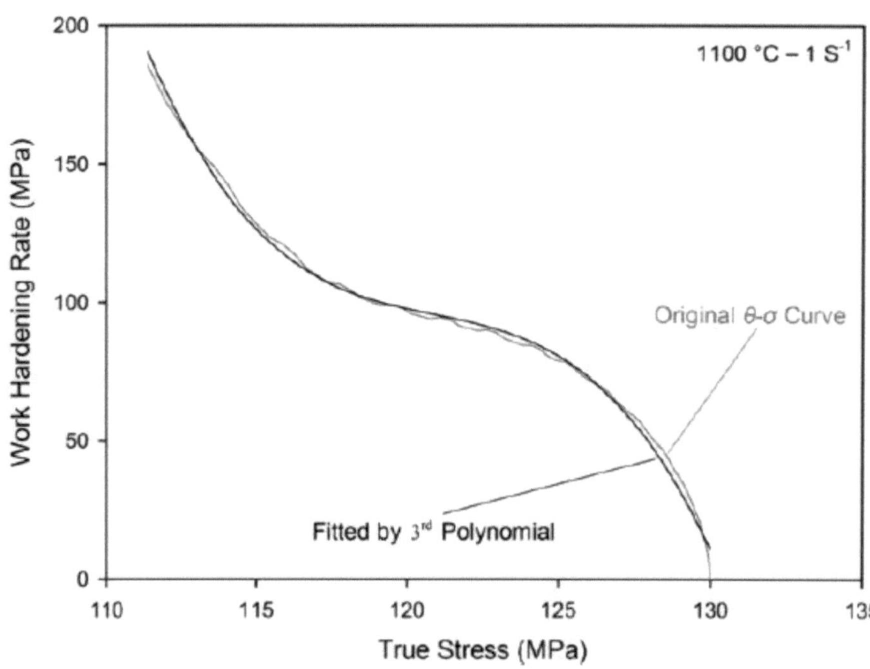

Fig. 6. The θ-σ curve of 17-4 PH Stainless steel and its corresponding third order polynomial.

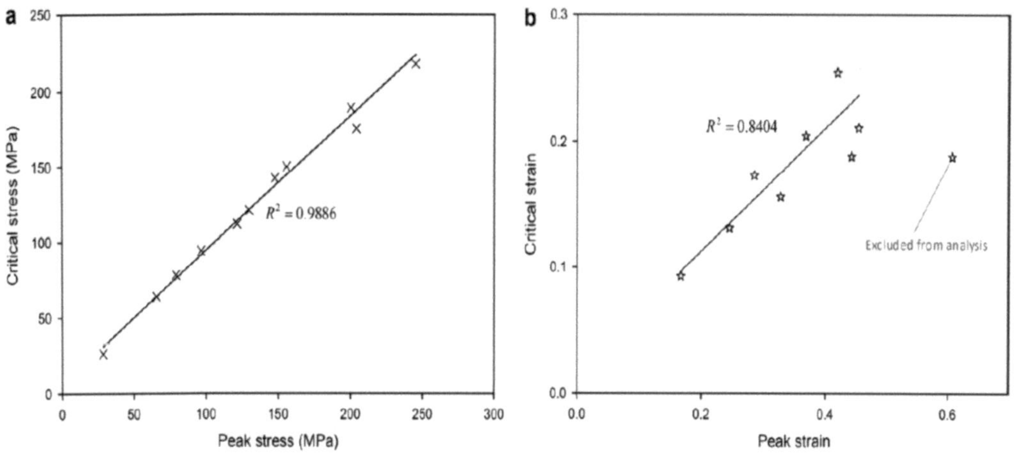

Fig. 7. Critical stress and strain versus (a) peak stress and (b) peak strain.

3.1.2 Determination of critical strain, (Poliak and Jonas)

Poliak and Jonas have shown that the inflection in plots of lnΘ-ε and lnΘ-σ curves can also be used for determination of initiation of DRX. The procedure of Section 3.1.1 is used to determine the values of critical strain at different deformation conditions (Mirzadeh et al., 2010).

According to figure 8, the normalized critical strain could be presented for 17-4 PH stainless steel:

$$\frac{\varepsilon_c}{\varepsilon_p} = 0.47 \tag{6}$$

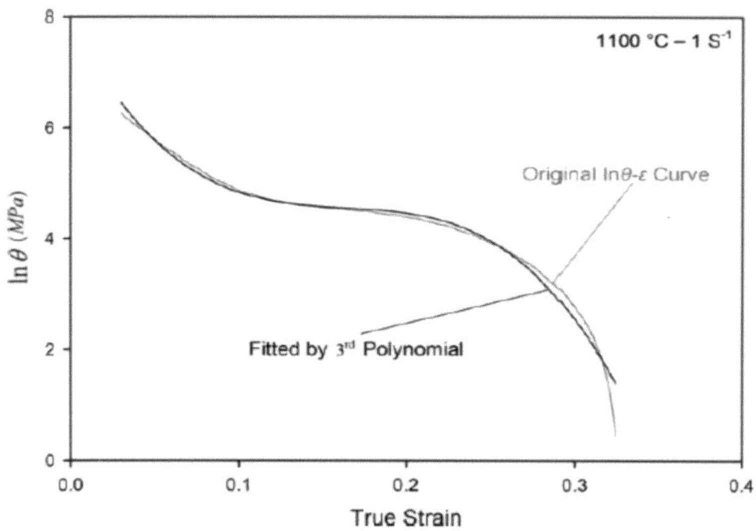

Fig. 8. The lnΘ-ε curve for 17-4 PH stainless steel and its corresponding third order polynomial.

3.1.3 Determination of critical strain (Ebrahimi and Solhjoo)

The flow stress up to the peak stress was modeled using the Cingara and McQueen equation as shown below:

$$\sigma = \sigma_p[(\varepsilon / \varepsilon_p)\exp(1-\varepsilon / \varepsilon_p)]^C \tag{7}$$

Where constant C is an additional parameter to make the results acceptable and is obtained from logarithmic form of Eq.7. The derivative of the true stress with respect to true strain yields the work hardening rate, Θ. Therefore, the Θ formula using Eq.7 may be expressed as:

$$\theta = C\sigma(1/\varepsilon - 1/\varepsilon_p) \tag{8}$$

In order to determine the critical strain, the second derivative of Θ with respect to σ must be zero. By solving this equation, the critical strain as a function of peak strain will obtain as follows:

$$\frac{\varepsilon_c}{\varepsilon_p} = \frac{\sqrt{1-C}-(1-C)}{C} \tag{9}$$

The results showed a good agreement with the experimentally measured ones for Nb-microallyed steel (Ebrahimi et al., 2007).

3.2 Kinetic equations (prediction of single points)

The influence of temperature and strain rate on peak stress was analyzed by the following equations which were originally developed for creep but have found applicability in the high strain rates in hot working:

$$A'\sigma_p^{n''} = \dot{\varepsilon}\exp\left(\frac{Q_{HW}}{RT}\right) = Z \tag{10}$$

$$A''\exp(\beta\sigma_p) = \dot{\varepsilon}\exp\left(\frac{Q_{HW}}{RT}\right) = Z \tag{11}$$

$$A'''\left[\sinh(\alpha\sigma_p\right]^n = \dot{\varepsilon}\exp\left(\frac{Q_{HW}}{RT}\right) = Z \tag{12}$$

where $A', A'', A''', n, n'', Q_{HW}, a, \beta$ and R are constants and Z is the Zener Hollomon. Q_{HW} is the activation energy related to hot working and R is the universal gas constant. A power law plot of log σ versus log Z gives linear segments only at low stress, indicating the limited applicability of Eq. (10). However in a plot of σ against log Z linearity is lost at low stress, showing that Eq. (11) is an inadequate fit for the entire range of stresses. Eq. (12) is a more general form of Eq. (10) and Eq. (11) reducing to Eq. (10) at lower stresses ($\alpha\sigma<0.8$) and Eq. (11) at higher stresses ($\alpha\sigma>1.2$). The validation of these equations to predict characteristic points of flow curves were approved by many authors (Cingara et al., 1992).

3.3 Experimental models

3.3.1 Misaka's model

Misaka and Yoshimoto have utilized the following double-power constitutive equation to determine flow stress associated with the processing of steels:

$$\sigma_{Misaka} = 9.8\exp\left(0126 - 1.75[C] + 0.594[C]^2\right) + \left(\frac{2851 + 2968[C] - 1120[C]^2}{T + 273}\right)\varepsilon^n \dot{\varepsilon}^m \quad (13)$$

Application range of this formula is as follows; carbon content: ≈1.2%, temperature: 750–1200 °C, reduction (natural strain): 50% and strain rate: 30–200 s^{-1}. Misaka's equation was updated by including effects of solution-strengthening and dynamic recrystallization.

The updated Misaka's constitutive equation is:

$$\sigma_{Misaka,Updated} = \left(f\sigma_{Misaka}\right)\left(1 - X_{DRX}\right) + k\sigma_{ss}X_{DRX} \quad (14)$$

$$f = 0.835 + 0.51[Nb] + 0.098[Mn] + 0.128[Cr]^{0.8} + 0.144[Mo]^{0.3} + 0.175[V] + 0.01[Ni] \quad (15)$$

(σ_{Misaka})updated indicates the flow stress of steels containing multiple alloying additions. σ_{ss} is steady state stress, K =1.14 is a parameter that converts flow stress to mean flow stress, and X_{DRX} is volume fraction of dynamic recrystallization. It might be useful for practical purpose but its mathematical base is weak. The factors for Mn, Nb, V and Ni are linear although the terms for Cr and Mo are nonlinear. Devadas et al. compared the predicted flow stress data for a low alloy steel with measured data from a cam-plastometer. They showed that Misaka's model overestimated the flow stress (Kim et al., 2003).

3.3.2 Shida's equation

Shida's equation is based on experimental data obtained from compression type of high strain rate testing machines. Shida then expressed the flow stress of steels, σ, as a function of the equivalent carbon content (*C*), the strain (ε), the strain rate ($\dot{\varepsilon}$) and temperature (*T*) as followings:

$$\sigma = \sigma_d(C,T)f_w(\varepsilon)f_r\left(\dot{\varepsilon}\right) \quad (16)$$

$$\sigma_d = 0.28\exp\left(\frac{5}{T} - \frac{0.01}{C + 0.05}\right) \quad (17)$$

$$T[k] = \frac{T\left[{}^{o}C\right] + 273}{1000} \quad (18)$$

$$f_w(\varepsilon) = 1.3\left(\frac{\varepsilon}{0.2}\right)^n - 0.3\left(\frac{\varepsilon}{0.2}\right) \quad (19)$$

$$n = 0.41 - 0.07C \tag{20}$$

$$f_r\left(\dot{\varepsilon}\right) = \left(\frac{\dot{\varepsilon}}{10}\right)^m \tag{21}$$

$$m = (-0.019C + 0.126)T + (0.076C - 0.05) \tag{22}$$

where $f_w(strain)$ and f_r(strain rate) are functions dependent upon strain and strain rate, respectively. The formulation of Eq. (16) is based on assumption that flow stress increases with the strain rate and strain increased. The range of validity of the formula is quite broad. This formula is applicable in the range of carbon content: 0.07–1.2%, temperature: 700–1200°C, strain: ≈ 0.7 and strain rate: ≈100 s^{-1}.

3.3.3 Modified Voce's equation

In contrast to Misaka's and Shida's equations, Voce's equation can describe the flow stress over the wide range of strains and strain rates. The equation can express the dynamic softening portion of the flow stress curve by using Avrami equation:

$$\sigma_{WH+DRV} = \sigma_p\left[1 - \exp(-c\varepsilon)\right]^m \tag{23}$$

The coefficient, *C*, and work hardening exponent, *m*, are dependent on the deformation conditions. The parameters, *C* and *m* are normally taken as being a constant, however, it is a function of the deformation conditions (strain rate, temperature). Thus, the *C* and *m* are considered to be a function of dimensionless parameter, *Z/A*.

Also Kim et al. developed an equation by modifying Voce's constitutive equations accounting for the dynamic recrystallization as well as the dynamic softening. During thermomechnical processing, the important metallurgical phenomena are work hardening (WH), dynamic recovery (DRV) and dynamic recrystallization (DRX). Thus, the flow stress curve can be bisected with the WH + DRV region and the DRX region. For the evaluation of the WH and DRV region, Eq. 23 was used. Also For the region of DRX, the drop of flow stress was expressed as the following equation:

$$\sigma_{Drop} = \left(\sigma_p - \sigma_{ss}\right)\left[\frac{X_{DRX} - X_{\varepsilon_p}}{1 - X_{\varepsilon_p}}\right] \quad \text{for } \varepsilon > \varepsilon_p \tag{24}$$

Therefore, the flow stress can be expressed in subtraction form as follows:

$$\sigma = \sigma_{WH+DRV} - \sigma_{Drop} \tag{25}$$

where σ_{ss} is the steady state stress achieved at larger strains and $X_{\varepsilon p}$ the volume fraction of DRX at peak strain. X_{DRX} is the volume fraction of DRX at any strain.

Fig. 9 shows the measured and predicted flow stress curves in large strain range of AISI 4140 steel when the three different types of constitutive equations are used for prediction. The flow stress curves calculated by using Misaka's equation agree with the measured ones with some extent of error in comparison to the flow stress curve obtained from Shida's equation. Although, the stress-strain curves predicted by using the modified Voce's equation are in a good agreement with experimentally measured ones, it seems Misaka's equation does not reflect recrystallization behavior properly (Kim et al., 2003).

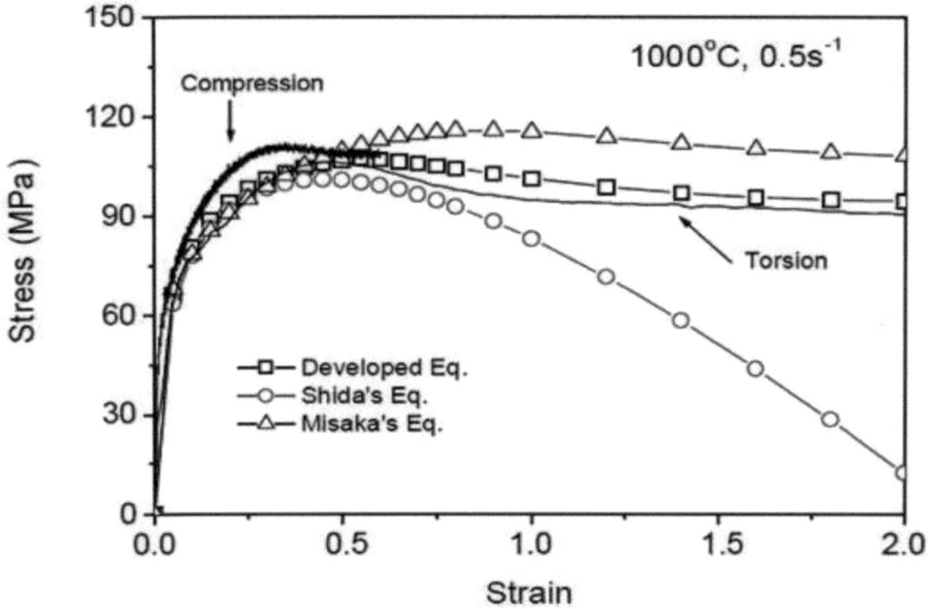

Fig. 9. Comparison of measured and predicted constitutive relations for AISI 4140 steel.

3.4 Mathematical models

3.4.1 Solhjoo's model

Solhjoo was considered linear estimation of θ-σ curve up to the peak stress as follows:

$$\frac{\partial \sigma}{\partial \varepsilon} = S_4 + A_1 \tag{26}$$

where S_4 is the slope of the line and A_1 is a constant. Due to the first assumption which considers a linear segment up to the peak, this model would not be able to show the critical strain. Using the maximum point of the θ-σ curve ($\sigma=\sigma_p, \theta=0$), value of constant A_1 is obtained to be $-S_4\sigma_p$. Solution of the differential Eq. (26) with respect to ε, using boundary condition $\sigma=\sigma_p$ at $\varepsilon=\varepsilon_p$is:

$$\sigma = \sigma_p \left[1 - \left(\frac{(\varepsilon_p - \varepsilon)}{\varepsilon_p} \right) \exp(S_4 \varepsilon) \right] \tag{27}$$

where σ_p is the peak stress and ε_p is the peak strain.

Determination of S_4 can be simply done using a linear plot of $\ln\left(\frac{\varepsilon_p}{\sigma_p}\frac{\sigma_p-\sigma}{\varepsilon_p-\varepsilon}\right)$ vs. ε; S_4 would be the slope of the line. As S_4 is very sensitive to strain rate, it is better to plot $\ln\left(\frac{\varepsilon_p}{\sigma_p}\frac{\sigma_p-\sigma}{\varepsilon_p-\varepsilon}\right)$ vs. ε for each sets of strain rates, and then the average value of the slopes determines S_4. Assuming S_4 as a constant shows a rough estimation of stress- strain curve. Since S_4 is more sensitive to strain rate than temperature, this parameter can be considered as a power law in form of:

$$S_4 = -C\dot{\varepsilon}^{E} \tag{28}$$

where E is strain rate and C and ε are constants. Using a plot of ln(-S4) vs. $\dot{\varepsilon}$, constants C and E can be determined. It should be considered that another limit of this model is its disability of prediction of flow stress at very low strains (less than 0.05) that the work hardening rate has very high values (Solhjoo, 2009).

3.4.2 Avrami's analysis

The DRX may be considered as a solid-state transformation and its kinetics can be modeled by the Johnson-Mehl-Avrami-Kolmogorov (JMAK) equation as follows:

$$X_{DRX} = 1 - \exp\left(-kt^n\right) \tag{29}$$

where X_{DRX} *and* t are the recrystallized volume fraction and DRX time, respectively. The effect of dynamic recovery (DRV) on flow softening was not considered. Therefore, this model is preliminary intended for materials with relatively low stacking fault energy. Moreover, for modeling the flow curves after the peak point of stress-strain curve, the initiation of DRX was intentionally considered at peak point. This assumption simplifies the Avrami analysis with acceptable level of accuracy. Therefore, Eq. (30) gives the magnitude of flow stress at each fractional softening:

$$\sigma = \sigma_p - \left(\sigma_p - \sigma_{ss}\right)X_{DRX} \tag{30}$$

3.4.3 Ebrahimi's model

This model is based on a phenomenological representation of the shape of the flow stress curves and the traditional theories for constitutive equations which incorporate the power law. Ebrahimi et al. considered the variations of the slope of flow stress curves as follows:

$$\frac{d\sigma}{d\varepsilon} = C_1\left(\sigma - \sigma_{ss}\right)\left(1 - \frac{\varepsilon}{\varepsilon_p}\right) \tag{31}$$

Where ε_p is the strain at the peak stress and σ_{ss} is the steady state stress. The term $(1-\varepsilon/\varepsilon_p)$ estimates variation of the stress- strain curve for $\sigma>\sigma_{ss}$. Solution of the differential Eq.(31) with respect to ε using boundary condition $\sigma=\sigma_p$ at $\varepsilon=\varepsilon_p$ results in:

$$\sigma = \sigma_{ss} + \left(\sigma_p - \sigma_{ss}\right)\exp\left[C_1\left(\varepsilon - \frac{\varepsilon_p}{2} - \frac{\varepsilon^2}{2\varepsilon_p}\right)\right] \quad (32)$$

If $\varepsilon=\varepsilon_k$ and $\varepsilon_k=k\varepsilon_p$ where k<1 and $\sigma_k>\sigma_{ss}$, then coefficient C_1 can be estimated from Eq. (33) as:

$$C_1 = \frac{2}{\left(k^2 - 2k + 1\right)\varepsilon_p}\ln\left(\frac{\sigma_p - \sigma_{ss}}{\sigma_k - \sigma_{ss}}\right) \quad (33)$$

Where σ_k is the stress calculated from Cingara equation. The equation represented by this model required the values of stress and strain at the peak and stress at the steady state zone. These parameters can be calculated by kinetic equations (Ebrahimi et al., 2006).

Fig. 10 shows the calculated flow curves by Cingara, Avrami and Ebrahimi et al. equations. At low Z, the predictions of Ebrahimi et al. equation are relatively accurate (Fig. 10a), but at high Z, this equation overestimates the flow softening of DRX (Fig. 10c). In other words, this equation seems to be suitable for ideal DRX behavior. The Ebrahimi et al. equation is based on a phenomenological representation of the shape of the stress-strain curves and the traditional theories for constitutive equations which incorporate the power law. Conversely, the Avrami equation is amenable to all deformation conditions as shown in Fig. 10. In fact, this equation is an adaptive one with two adjustable constants. As a result, it could be better fitted to hot flow curves. Therefore, the Avrami equation can be used for prediction of flow curves as shown in Fig. 10 (Mirzadeh et al., 2010).

3.4.4 Shafiei and Ebrahimi's constitutive equation

Using the extrapolation of DRV flow stress curves and kinetic equation for DRX, Shafiei and Ebrahimi proposed the following equation for modeling single peak DRX flow curves for $\varepsilon_c \le \varepsilon < \varepsilon_{ss}$

$$\sigma = \sigma_s - (\sigma_s - \sigma_c)\exp(C'(\varepsilon_c - \varepsilon)) - (\sigma_s - \sigma_{ss})X_{DRX} \quad (34)$$

Where C′ is a constant with metallurgical sense. According to the geometrical relations shown in Fig.5, the value of C′ can be formulized as Eq.s (35) and (36) . σ_c, σ_p, σ_{ss}, σ_s , ε_c , X_{DRX} are critical stress, peak stress, steady state stress, saturation stress , critical strain and volume fraction of DRX, respectively.

$$\tan\varphi = \frac{1}{C'} \quad (35)$$

$$\varphi = \frac{\pi}{2} + Arctg\left(\frac{\sigma_p\varepsilon_p}{\varepsilon_c(\varepsilon_c - \varepsilon_p)}\right) \quad (36)$$

As shown in Fig.11, the stress-strain curves predicted by using presented model are in a good agreement with experimentally measured ones for Ti-IF steel. In order to evaluate the accuracy of the model, the mean error was calculated. The mean error of flow stress is calculated at strains of 0.19-2 for every measurement under all deformation conditions. For all stress-strain curves, the mean errors are between -2.9% and 2.5%. The results indicate that the proposed model give a good estimate of the flow stress curves. Therefore, it can be deduced that the approach to obtain a constitutive equation applicable to large strain ranges was fruitful and this presented model might have a potential to be used where more precise calculation of stress decrement due to dynamic recrystallization is important. Moreover, this analysis has been done for the stress-strain curves under hot working condition for Ti-IF steel, but it is not dependent on the type of material and can be extended for any condition that a single peak dynamic recrystallization occurs.

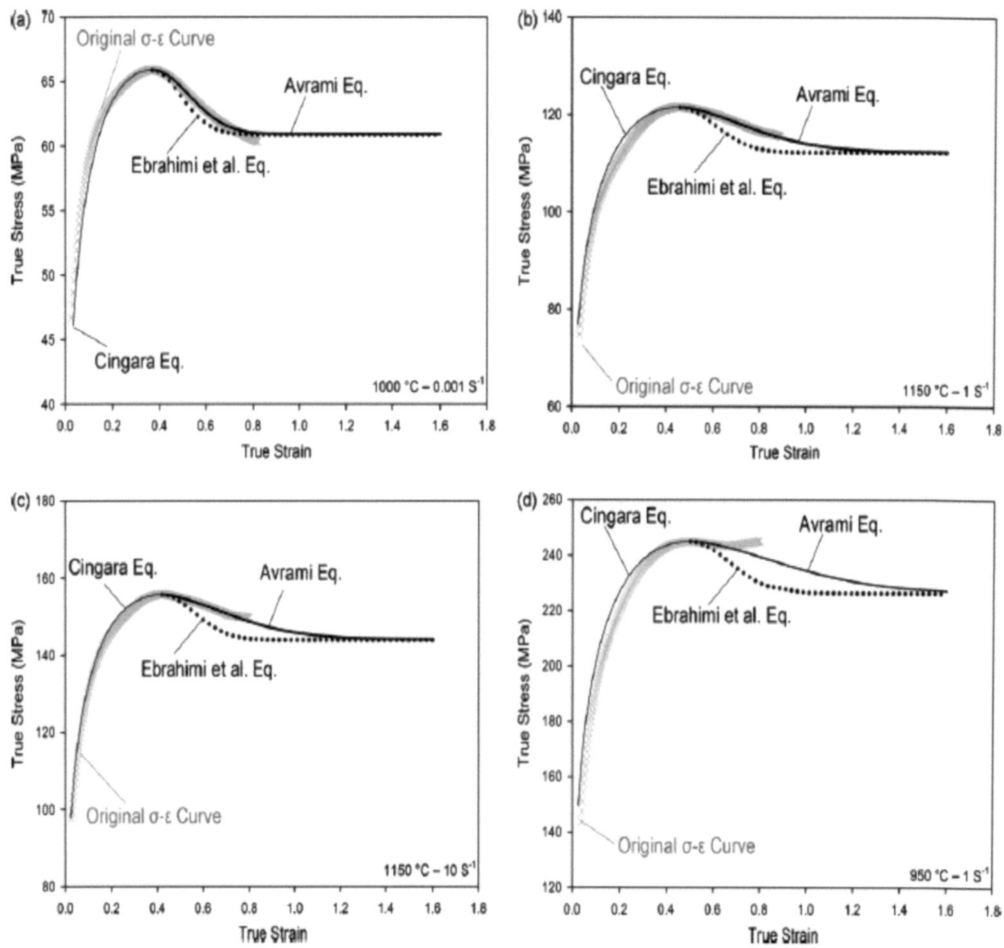

Fig. 10. Comparison between calculated and measured flow stress curves of 17-4 PH stainless steel.

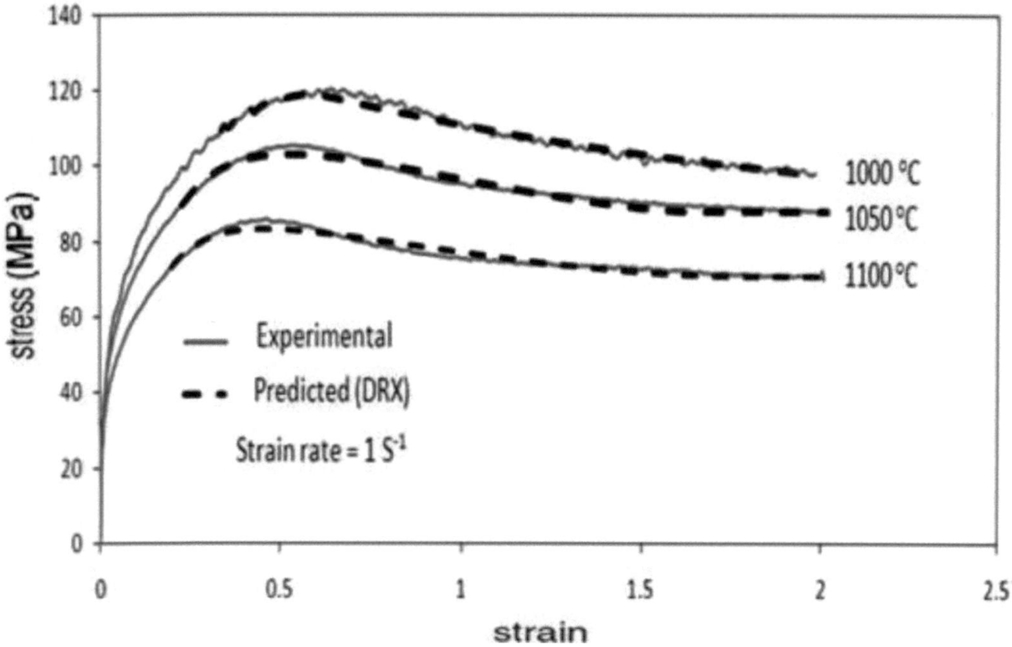

Fig. 11. Comparison of measured and predicted stress-strain curves of Ti-IF steel.

4. Summary

In this chapter, a review on recent model of single peak flow stress curves was presented. At first, the basic concepts on hot deformation and dynamic restoration, including affecting factors on dominated processes such as DRX and DRV and related microstructure evolutions were discussed. Then, an introduction on experimental models which are more capable in the field of industrials investigations such as Misaka, Shida and Voce constitutive equations followed by the details of mathematical models such as MC Queen, Ebrahimi, Solhjoo, Avrami and Shafiei- Ebrahimi constitutive equations were presented. In this case, the accuracy of these models as well as some limitations was evaluated in order to obtain the optimum working conditions.

5. References

Cingara, A. & McQueen, H. J. (1992). New method for determining sinh constitutive constants for high temperature deformation of 300 austenitic steel. Mater. Proc. And Tech, 36, 17-30.

Ebrahimi, R. & Solhjoo, S. (2007). Characteristic points of stress-strain curve at high temperature. ISSI, 2, 24-27.

Ebrahimi, R., Zahiri, S.H. & Najafizadeh, A. (2006). Mathematical modeling of the stress-strain curves of TI-IF steel at high temperature. Mater. Proc. And Tech., 171, 301-305.

Ebrahimi, R. (2003). Hot working of Ti-IF steel. Ph.D thesis. Isfahan University of Technology.

He, X. Yu, Zh. & Lai, X. (2008). A method to predict flow stress considering dynamic recrystallization during hot deformation. Comp. Mater. Sci., 44, 760-764.

Humphreys, F.J. & Hatherly, M. (2004). *Recrystallization and related annealing phenomena* (Second edition), (Elsevier, UK).

Imbert, C. A. C & McQueen, H. J. (2001). Peak strength, strain hardening and dynamic restoration of A2 and M2 tool steels in hot deformation. Mater. Sci. and Eng. A, 313, 88-103.

Imbert, C. A. C & McQueen, H. J. (2001). Dynamic recrystallization of A2 and M2 tool steels. Mater. Sci. and Eng. A, 313, 104-116.

Jonas, J. J., Quelennec, Jiang, L. & Martin, E. (2009). The avrami kinetics of dynamic recrystallization. Acta Mater. 137, 1748-2756.

Kim, S. I., Lee, Y. & Byon, S. M. (2003). Study on constitutive relation of AISI 4140 steel subject to large strain at elevated temperature. Mater. Proc. And Tech.,140, 84-89.

Liu, J., Cai, Zh. & Li, C. (2008). Modelling of flow stress characterizing dynamic recrystallization for magnesium alloys. Comp. Mater. Sci., 41,375-382.

Mirzadeh, H. & Najafizadeh, A. (2010). Extrapolation of flow curves at hot working conditions. Mater. Sci. and Eng. A, 527,1856-1860.

Mirzadeh, H. & Najafizadeh, A. (2010). Prediction of the critical conditions for initiation of dynamic recrystallization. Mater. And Des., 31, 1174-1179.

Najafizadeh, A. & Jonas, J. J. (2006). Predicting the critical stress for initiation of dynamic recrystallization. ISIJ Int., 46, 1679-1684.

Poliak, E. I. & Jonas, J. J. (2002). Initiation of dynamic recrystallization in constant strain rate hot deformation. ISIJ Int., 43, 684-691.

Shaban, M. & Eghbali, B. (2010). Determination of critical conditions for dynamic recrystallization of a microalloyed steel. Mater. Sci. and Eng. A, 527, 4320-4325.

Shafiei, E. & Ebrahimi, R. (2011). Mathematical modeling of single peak flow stress curves. Comp. Mater. Sci., submitted paper.

Solhjoo, S. (2009). Analysis of flow stress up to the peak at hot deformation. Mater. And Des.,30, 3036-3040.

Ueki, M., Horie, S. & Nakamura, T. (1987). Factors affecting dynamic recrystallization of metals and alloys. Mater. Sci. and Tech.,3, 329.

Verlinden, B., Driver, J., Samijdar, I. & Doherty, R. D. (2007). Thermomechanical processing of metallic materials. (Elsevier, Uk).

Zahiri, S. H., Davies, C. H. J. & Hodgson, P. D. (2005). A mechanical approach to quantify dynamic recrystallization in polycrystalline metals. Scripta Mater., 52, 299-304.

Zeng, Zh., Jonsson, S., Rovan, H. J. & Zhang, Y. (2009). Modelling the flow stress for single peak dynamic recrystallization. Mater. And Des.,30-1939-1943.

10

Phase Transformations and Recrystallization Processes During Synthesis, Processing and Service of TiAl Alloys

Fritz Appel

Institute for Materials Research, Helmholtz-Zentrum Geesthacht, Geesthacht, Germany

1. Introduction

Titanium aluminides alloys based on the intermetallic phases $\alpha_2(Ti_3Al)$ and $\gamma(TiAl)$ are one of the few classes of emerging materials that have the potential for innovative applications in advanced energy conversion systems whenever low density, good high-temperature strength and resistance against ignition and corrosion are of major concern [1]. The outstanding thermo-physical properties of the individual phases mainly result from the highly ordered nature and directional bonding of the compounds. However, two-phase $\alpha_2(Ti_3Al)+\gamma(TiAl)$ alloys exhibit a much better mechanical performance than their monolithic constituents $\gamma(TiAl)$ and $\alpha_2(Ti_3Al)$, provided that the phase distribution and grain size are suitably controlled. The synergistic effects of the two phases are undoubtedly associated with the many influences that the microstructure has on deformation and fracture processes. Constitution and microstructure are the result of phase transformations, ordering reactions and recrystallization processes, which occur during synthesis, processing and service. Many aspects of these mechanisms are intimately linked to defect configurations at the atomic level; thus standard techniques of metallography were often inadequate to provide the necessary information. This lack of information is addressed in the present article in that observations on recrystallization and phase transformations by high-resolution electron microscopy are presented. Particular emphasis will be paid on

i. the origin of microstructures
ii. heterogeneities in the deformed state and recovery behaviour
iii. atomic structure of crystalline and crystalline/amorphous interfaces
iv. misfit accommodation and coherency stresses.

2. Constitution and microstructure

2.1 Constitution

TiAl alloys of technical significance have the general composition (in at. %, as are all compositions in this paper)

$$Ti\text{-}(42\text{-}49)Al+X, \quad (1)$$

with X designating alloying elements, such as Cr, Nb, W, V, Ta, Si, B, and C [2]. When referred to the binary phase diagram (Fig. 1), the equilibrium phases for Al contents between 46 and 49 % are: the disordered solution phases hexagonal (h.c.p.) α(Ti), body centred cubic (b.c.c.) β(Ti), and the ordered intermetallic compounds γ(TiAl) with $L1_0$ structure, and $\alpha_2(Ti_3Al)$ with $D0_{19}$ structure [3]. Based on this constitution, different microstructures have been designed. The phase transformations involved in the microstructural evolution will be demonstrated for two examples, the classical lamellar structure of $\alpha_2(Ti_3Al)+\gamma(TiAl)$ alloys and a novel crystallographically modulated morphology occurring in multiphase alloys.

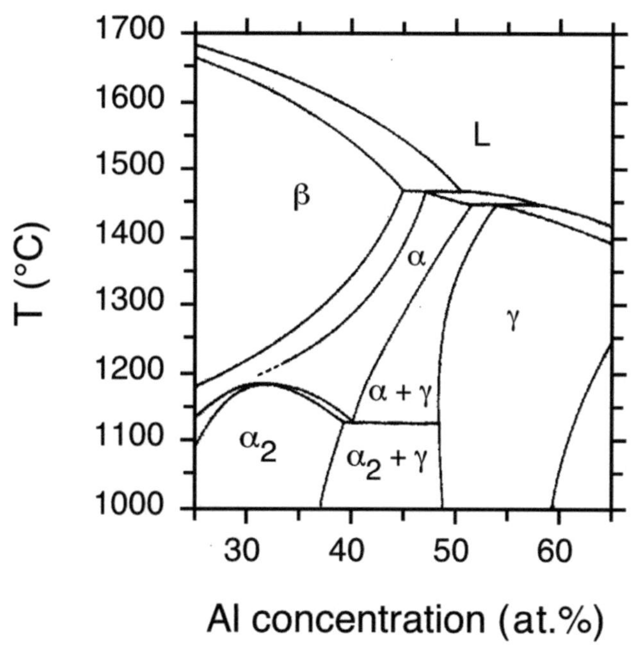

Fig. 1. Central portion of the Ti-Al-phase diagram in the region of technical interest [3].

2.2 Lamellar microstructure

The technologically most relevant $\alpha_2(Ti_3Al)+\gamma(TiAl)$ alloys contain a significant volume fraction of lamellar grains [1, 2]. The morphology of these grains represents a multilayer system made of two phases. It is well documented in the literature [4] that such a system could exhibit extraordinary mechanical properties when the layer thickness is small enough.

The so-called lamellar microstructure results from the precipitation of γ lamellae in either a disordered α or a congruently ordered α_2' matrix, following one of the transformation paths [3]:

$$\alpha \rightarrow \alpha_2' \rightarrow \alpha_2+\gamma \text{ or } \alpha \rightarrow \alpha_2+\gamma. \quad (2)$$

α_2' and α_2 have the same crystal structure but different composition. The exact pathway is still a matter of debate and could depend on alloy composition and thermal treatment. The orientation relationships between the α_2 and γ platelets are [5]

$$\{111\}\gamma \mid\mid (0001)\alpha 2 \text{ and } [1\bar{1}0]\gamma \mid\mid \langle 11\bar{2}0\rangle \alpha 2. \quad (3)$$

The length of the lamellae is determined by the size of the parent α/α_2 grain. The γ phase is formed as an ordered domain structure, as sketched in Fig. 2. This gives rise to six variants of the above orientation relationships. Thus, there are four types of lamellar interfaces: the α_2/γ interface and three distinct γ/γ interfaces that are typified by rotations of 60°, 120° and 180° between adjacent lamellae. A lamellar grain consists of a set of γ lamellae, which are subdivided into domains and interspersed by α_2 lamellae. The volume fraction of the two phases is controlled by the composition on the basis of the phase diagram and the processing conditions of the alloy. However, it should be noted that the decomposition of α phase into lamellar (α_2+γ) is sluggish and often cannot be established within the constraints of processing routes; thus the volume fraction of γ phase is less than equilibrium [6]. It might be expected that the Ti concentration in the γ phase increases when the alloy becomes richer in Ti until the maximum solubility of Ti in the γ phase is reached. This non-equilibrium phase composition may provide significant driving forces for structural changes, as will be outlined in the subsequent sections.

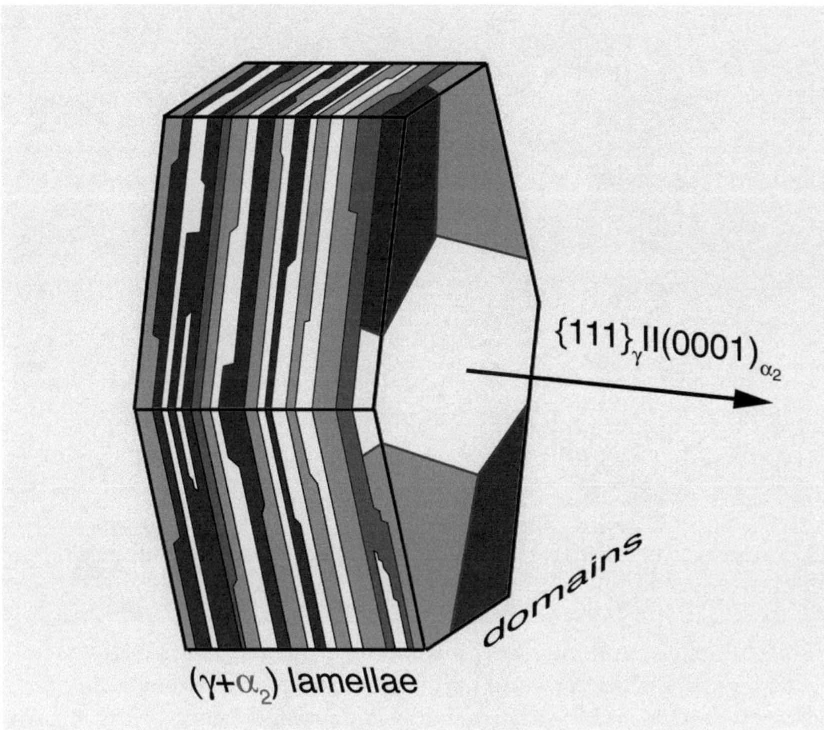

Fig. 2. Schematic drawing of a lamellar colony comprised of γ and α_2 platelets.

Among the various lamellar interfaces only the twin boundary is fully coherent, as the adjacent lattices are symmetrically oriented. At all the other interfaces the matching is imperfect, i.e., these interfaces are semicoherent. The mismatch arises from the differences in the crystal structure and lattice parameters and amounts to 1 to 2 %, depending on alloy composition and processing conditions. Different modes of mismatch accommodation have been discussed for lamellar TiAl alloys, which in broad terms correspond to the early models of misfitting interfaces [7]. Up to a certain point, the misfit strain could be solely taken up by elastic distortion, i.e., the lamellae are uniformly strained to bring the atomic spacings into registry. This homogeneous strain accommodation leads to coherent interfaces but introduces lattice distortions that are known as coherency strains. Hazzledine [4] has shown that the elastic misfit accommodation in lamellar (α_2+γ) alloys is only possible if the lamellae are very thin. The predicted critical thicknesses are $d_c \leq 8$ nm for the mismatched γ/γ interfaces and $d_c \leq 0.8$ to 3.9 nm for the α_2/γ interfaces, depending on the volume content of α_2 phase. Figure 3 shows a small Ti_3Al platelet embedded in γ phase with a thickness of 4.5 nm, which is just above the coherency limit predicted by Hazzledine [4]. While homogeneous strain accommodation is still recognizable, part of the misfit is already taken up by interfacial defects.

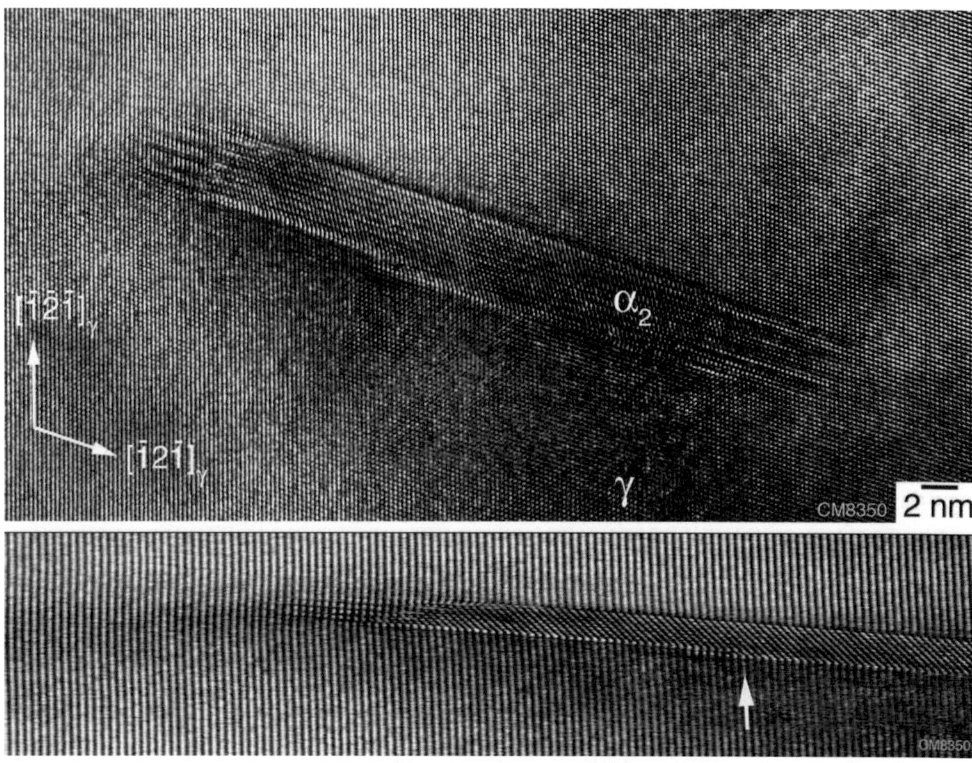

Fig. 3. High-resolution electron micrograph showing a small Ti_3Al platelet embedded in γ phase. Interfacial steps, dislocations and homogeneous elastic straining accommodate the misfit between the particle and the matrix. In the compressed image below, the elastic straining at the tip of the particle is readily visible by the distortion of the $(\bar{1}1\bar{1})$ planes. A dislocation compensating the misfit between the $(\bar{1}1\bar{1})_\gamma$ and $\{2\bar{2}01\}_{\alpha 2}$ planes is arrowed. Ti-46.5Al-4(Cr, Ta, Mo, B), sheet material.

The coherency strains raise the total energy of the system. Thus, for a sufficiently large misfit, or lamellar spacing, it becomes energetically more favourable to replace the coherent interface by a semicoherent interface, a situation that is referred to as a loss of coherency. As described in the early model of Frank and van der Merve [8] misfit dislocations partially take up the misfit, i.e., the atoms at the interface adjust their positions to give regions of good and bad registry. In other words, the misfit is concentrated at the dislocations.

Mismatch strains at interfaces can be also relieved by the formation of ledges or steps at the interface [9]. The introduction of mono-atomic steps significantly improve the atomic matching, thus preventing the disregistry from becoming large anywhere. Structural ledges may replace misfit dislocations as a way of retaining low-index terraces between the respective defects. The general view is that planar boundaries are favoured for large misfits and small Burgers vectors of the misfit dislocations, whereas stepped boundaries are favoured for small misfits and large values of the Burgers vector. Steps may range in scale from atomic to multi-atomic dimensions depending on energetic or kinetic factors. However, it is very often the case that an interfacial defect exhibits both dislocation- and step-like character, thus, comprising a more general defect that has been defined as a disconnection [10,11]. Because of its step character, disconnection motion along an interface transports material from one phase to the other, the extent of which is essentially determined by the step height. At the same time, the dislocation content of the disconnection leads to deformation. In this sense, disconnection motion couples interface migration with deformation. The different extents of symmetry breaking at the interfaces lead to a broad variety of disconnections [10, 11]. This is reflected in different step heights and dislocation contents, which eventually determine the function of disconnections in phase transformations. Disconnection models have been developed for a variety of diffusional and diffusionless phase transformations in crystalline solids; and an extensive body of literature has evolved. For more details the reader is referred to a review of Howe et al. [11]. Several authors [12-16] observed interfacial steps at α_2/γ interfaces with heights that were always a multiple of $\{111\}_\gamma$ planes. The most commonly observed two-plane steps were characterized as disconnections with the topological parameters $b=1/6[11\bar{2}]$, $t(\gamma)=1/2[\overline{112}]_\gamma$ and $t(\alpha_2)=[000\bar{1}]_{\alpha2}$ [17]. b is the Burgers vector of the dislocation component (parallel to the (111) interface) and $t(\gamma)$ and $t(\alpha_2)$ are vectors describing the ledge risers [11] of the disconnection. This type of disconnection has no Burgers vector component perpendicular to the interface.

The complexity of misfit accommodation at α_2/γ interfaces is illustrated in Fig. 4. The micrograph shows a γ(TiAl) lamella terminated within the α_2 phase of a two-phase alloy. The interface marked in the micrograph borders the crystal region in which the exact *ABC* stacking of the $L1_0$ structure is fulfilled. Outside of this exactly stacked region of γ phase, there is a two to three atomic plane thick layer in which neither the *ABC* stacking of the γ phase nor the *ABAB* stacking of the α_2 phase is correctly fulfilled. This becomes particularly evident at the tip of the γ lamellae and indicates a significant homogeneous straining of the lattice. This strain seems to locally relax by the formation of dislocations, as can be seen in the compressed form of the image. The other salient feature is the misfit accommodation by steps. A Burgers circuit constructed around the tip of the γ lamella results in a projected Burgers vector of $b_p=1/6[11\bar{2}]$. This indicates that the small misfit between the $(111)_\gamma$ and $(0002)_{\alpha2}$ planes is elastically taken up. The observed misfit accommodation is pertinent to the issue of how differential material flux during the $\alpha_2 \rightarrow \gamma$ transformation is accomplished.

There is ample evidence [18] of enhanced self-diffusion along dislocation cores. Likewise interfacial ledges are envisaged as regions where deviation from the ideal structure is localized and which may provide paths of easy diffusion. This gives rise to the speculation that it is mainly the tip of a newly formed γ lamella where the atomic composition between the two phases is adjusted during transformation.

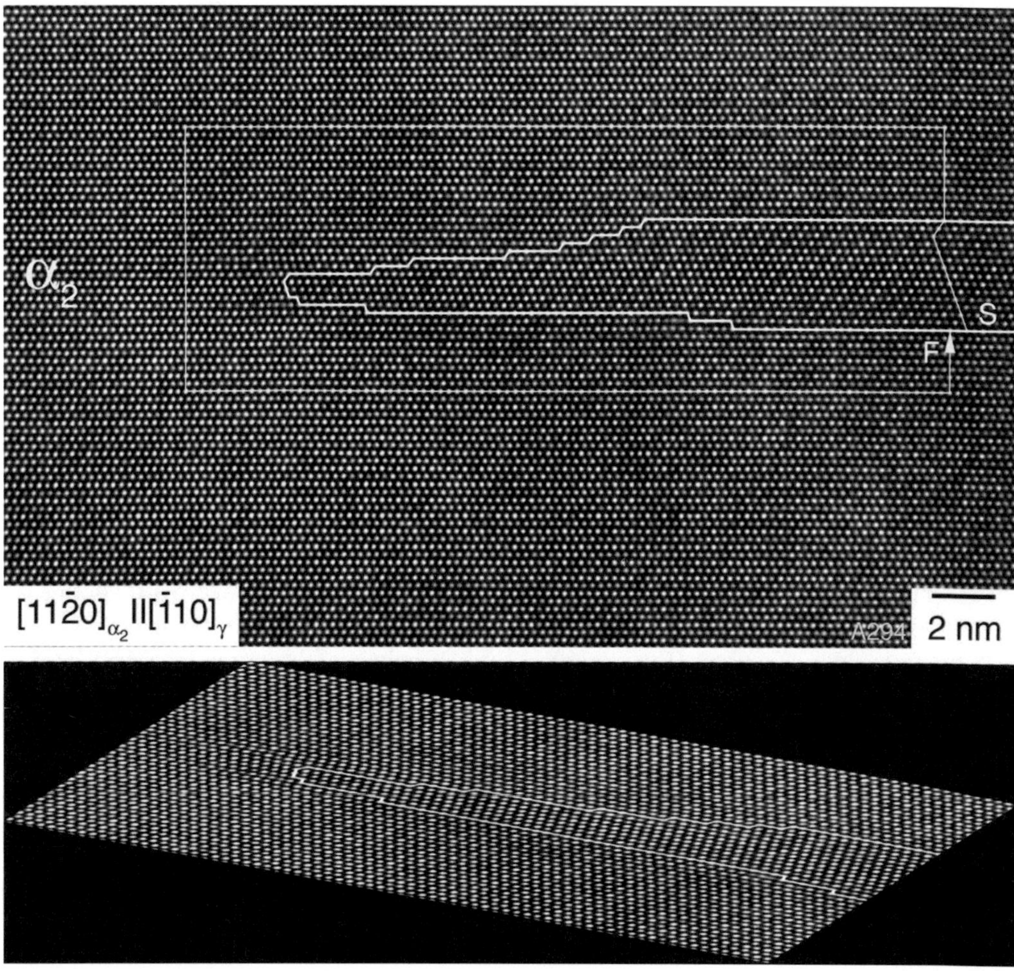

Fig. 4. A high-resolution micrograph of a γ(TiAl) lamella terminated within α_2 phase imaged down the $<10\bar{1}>_{\gamma}$ and $<11\bar{2}0>_{\alpha 2}$ directions. The thick white line marks the position of the interface. The salient feature is the misfit accommodation by steps. S and F denote the start and finish, respectively, of the Burgers circuit constructed around the tip of the γ lamella. After elimination of all the cancelling components in the initial circuit, and transforming the sequence of operations in the α_2 phase into the γ coordinate frame, the projected Burgers vector is $b_p=1/6[11\bar{2}]$. Note the dislocation in front of the γ tip. This can be recognized in the image below, which was compressed along the $(2\bar{2}01)_{\alpha 2}$ planes. Ti-Al-Nb, as cast.

In spite of the misfit accommodation by interfacial dislocations and ledges a significant elastic strain remains at the interfaces. The resulting coherency stresses were determined by convergent beam electron diffraction (CBED) [19] and by analyzing the configuration of dislocations emitted from the lamellar interfaces [20]. The investigations have shown that the residual coherency stresses are comparable with the yield stress of the material. From theory it is expected that the residual coherency stresses present in the individual lamellae are inversely proportional to the lamellar spacing λ_L [21]. However, when sampled over a sufficiently large volume, the average of the coherency strain was zero. Thus, the sign of the coherency strain alternates from lamella to lamella. In the design of lamellar alloys λ_L is often reduced in order to maximize the yield stress. At the same time the coherency stresses grow both in absolute magnitude and relative to the yield stress [4]. Thus, in high-strength alloys the coherency stresses can be very large and can affect deformation, phase transformations, recovery, and recrystallization in various ways, as will be described in the following sections.

2.3 Modulated microstructures

In an attempt to improve the balance of mechanical properties, a novel type of TiAl alloys, designated γ-Md, with a composite-like microstructure has been recently developed [22]. The design bases on the general composition

$$\text{Ti-(40-44)Al-8.5 Nb.} \tag{4}$$

The characteristic constituents of the alloy are laths with a modulated substructure that is comprised of stable and metastable phases. The modulation occurs at the nanometer scale and thus provides an additional structural feature that refines the material. As indicated by X-ray analysis the constitution of the alloys (4) involves the β/B2, α_2 and γ phases. Additional X-ray reflections could be attributed to the presence of two orthorhombic phases with B19 structure, (oP4, Pmma and oC16, Cmcm). However, a clear association with the various orthorhombic structures reported in the literature was not possible because of their structural similarity. It might be expected that the evolution of the constitution does not reach thermodynamic equilibrium; thus, the number of the transformation products may be larger than expected from the phase rule. The microstructure of the alloy is shown in Fig. 5. The characteristic features are laths with a periodic variation in the diffraction contrast, which intersperse the other constituents. As shown in Fig. 5a, the contrast fluctuations occur at a very fine length scale. The evidence of the high-resolution electron microscope observations is that a single lath is subdivided into several regions with different crystalline structures with no sharp interface in between. The high-resolution micrograph in Fig. 5b shows a lath adjacent to a γ lamella in $\langle 101]_\gamma$ projection, which can be used as a reference. The interface between the lath and the γ phase (designated as γ/T) consists mainly of flat terraces, which are parallel to the $(111)_\gamma$ plane. Steps of different heights often delineate the terraces. The modulated laths are comprised of an orthorhombic constituent, which is interspersed by slabs of β/B2 and a little α_2 phase (Fig. 6). Selected area diffraction of the orthorhombic constituent is consistent with the B19 phase, which can be described as the orthorhombic phase (oP4) or as a hexagonal superstructure of $D0_{19}$ (hP8). In the Ti-Al system, the B19 structure has already been observed by Abe et al. [23] and Ducher et al. [24]. The B19 structure is structurally closely related to the orthorhombic phase (oC16, Cmcm) with the ideal stoichiometry Ti_2AlNb, which among the intermetallic compounds is remarkable for its relatively good room temperature ductility [25].

In Figs. 5b and 6a the B19 structure is imaged in the $[010]_{B19}$ projection. As deduced from the high-resolution images, the orientation relationships between the constituents involved in a modulated lath with the adjacent γ are [22]

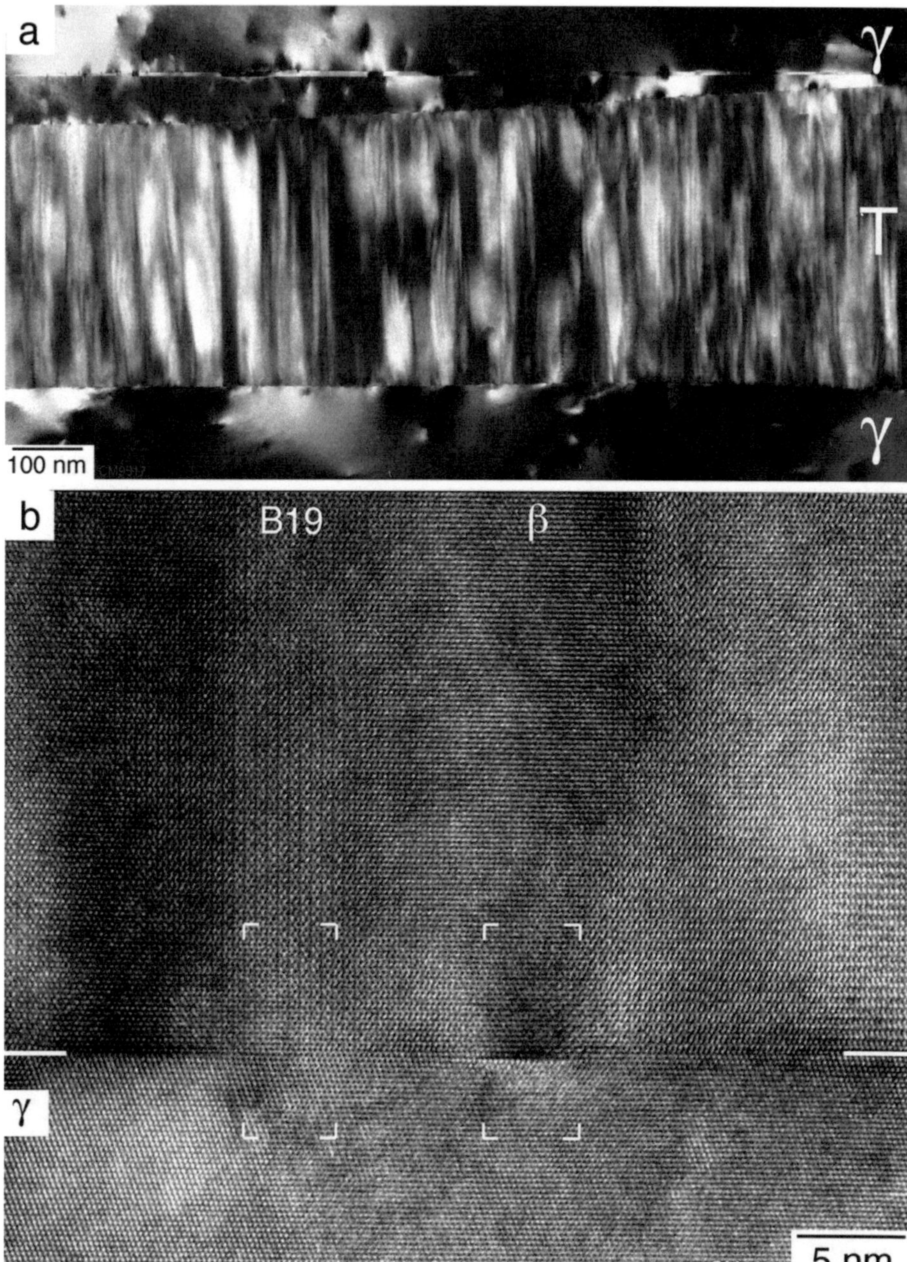

Fig. 5. TEM analysis of the modulated microstructure. (a) A modulated lath imaged by diffraction contrast. (b) High-resolution TEM micrograph of a modulated lath adjacent to a γ lamella.

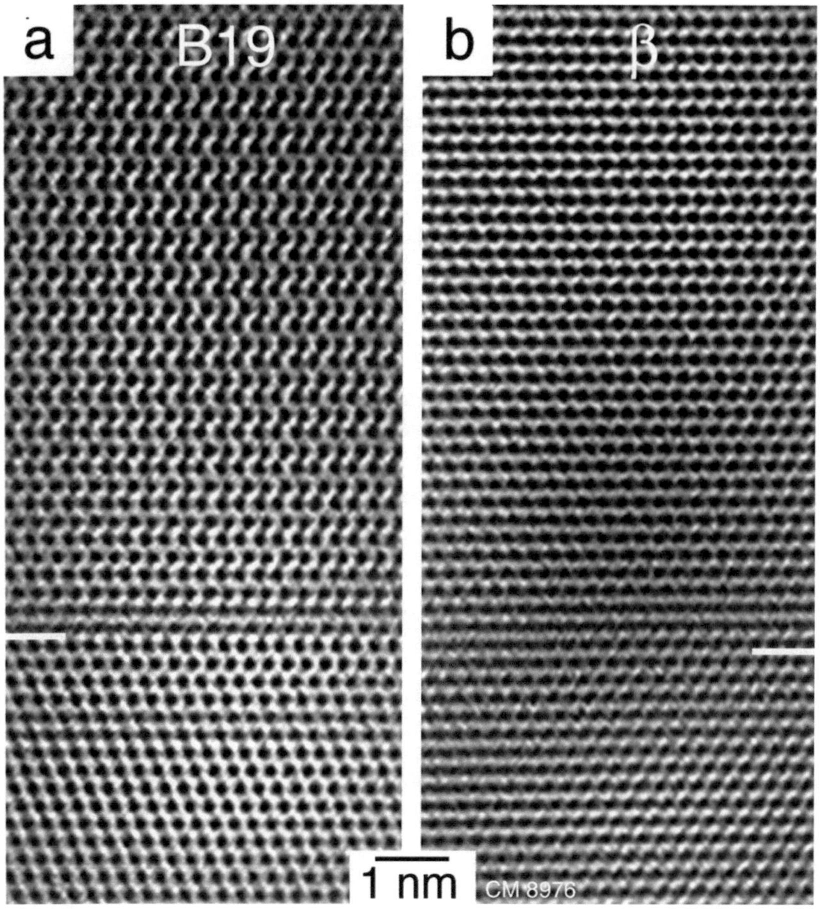

Fig. 6. High-resolution TEM evidence of the modulated microstructure. (a) and (b) Fourier-filtered images of the areas boxed in Fig. 5.

$$(100)_{B19} \;||\; \{110\}_{\beta/B2} \;||\; (0001)_{\alpha 2} \;||\; \{111\}_{\gamma}$$
$$[010]_{B19} \;||\; <111>_{\beta/B2} \;||\; <11\bar{2}0>_{\alpha 2} \;||\; <1\bar{1}0>_{\gamma};\ [001]_{B19} \;||\; <11\bar{2}>_{\gamma}. \quad (5)$$

In the diffraction pattern the presence of the periodic distortion is manifested by the existence of weak satellite reflections adjacent to the main reflections. The distance of the satellites from the main reflections is the reciprocal of the modulation wavelength, and the direction joining the satellites with their main reflections is parallel to the direction of the modulation vector. These observed features are reminiscent of a modulated structure, which in recent years have attracted considerable interest; for a review see [26]. A crystal structure is said to be modulated if it exhibits periodicities other than the Bravais lattice periodicities. These additional periodicities arise from one or more distortions, which increase to a maximum value and then decrease to the initial value. The modulation may involve atomic coordinates, occupancy factors or displacement parameters. The strain of the discontinuities is often relieved by a continuous and periodic variation of the physical properties of the product. First principle calculations of Nguyen-Manh and Pettifor [27] and Yoo and Fu [28] have shown that the β/B2 phase existing in TiAl alloys containing supersaturations of transition metals (Zr, V, Nb) is unstable under tetragonal distortion; a shear instability that was attributed to the anomalous (negative) tetragonal shear modulus. Specifically, B2 may transform by homogeneous shear to several low temperature orthorhombic phases, which can exist metastably. The energetically favourable transformations are [27]

$$B2(Pm3m) \rightarrow B19(Pmma) \rightarrow B33(Cmcm). \qquad (6)$$

At the atomic level, the B2 phase may transform to B19 by a shuffle displacement of neighboring $(011)_{B2}$ planes in opposite $[01\bar{1}]$ directions, as illustrated in Fig. 7. A subsequent displacement of neighboring $(011)_{B2}$ planes in the [100] direction generates the B33 structure [27]. In the system investigated here the predominant orthorhombic phase seems to be B19. It is tempting to speculate that the modulation of the laths is triggered by a periodic variation of the composition of the parent B2 phase, as occurs during spinodal decomposition. Clearly the mechanism requires further investigation.

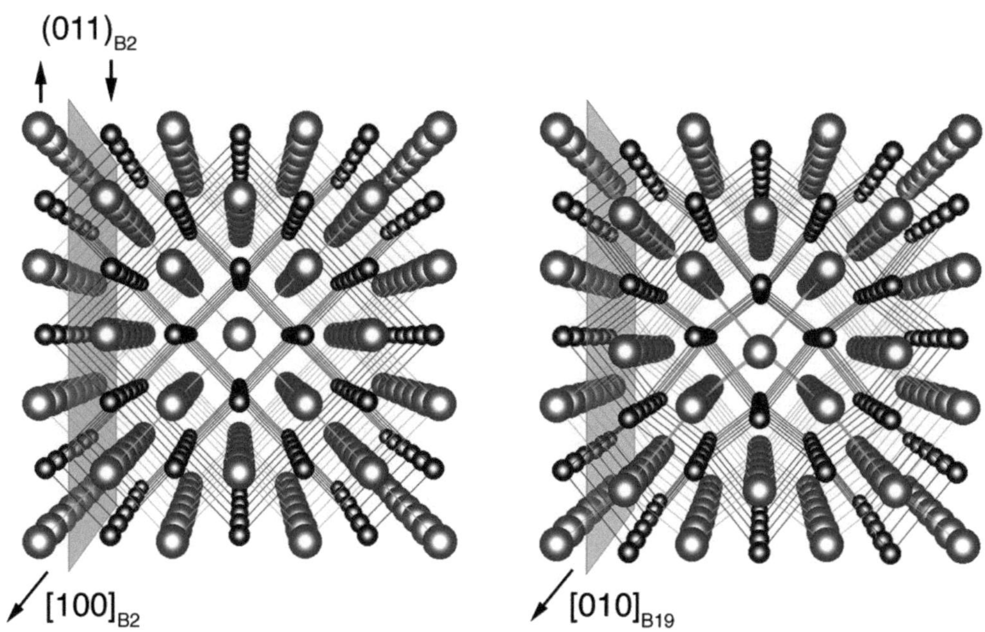

Fig. 7. Formation of the B19 structure from the parent B2 phase illustrated by perspective views of hard-sphere models. [100] projection of the B2 phase; arrow heads mark shuffle displacements of neighbouring $(011)_{B2}$ planes in opposite $[01\bar{1}]$ directions to form B19.

The modulated laths can apparently further transform into the γ phase, as demonstrated in Fig. 8. This process often starts at grain boundaries and proceeds through the formation of high ledges via distinct atomic shuffle displacements. As this transformation was frequently observed in deformed samples [22], it might be speculated that the process is stress induced and provides some kind of transformation toughening.

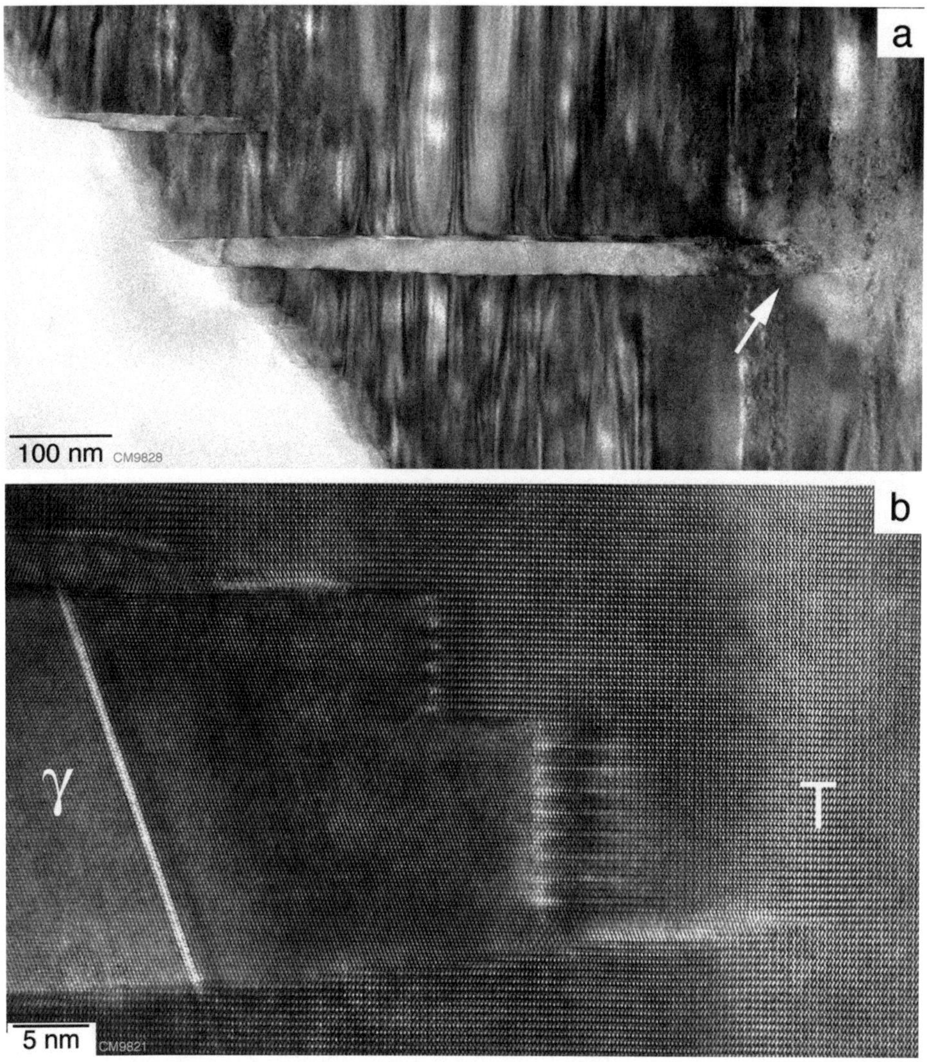

Fig. 8. Stress-induced transformation of a modulated lamella (T) into γ phase. (a) Generation of a γ lamella at a grain boundary. (b) Higher magnification of the area marked by the arrow in (a). Note the ledges at the interfaces.

3. Hot working

Titanium aluminide alloys are relatively brittle materials; attaining chemical homogeneity and refinement of the microstructure are therefore the most important prerequisites for engineering applications. To this end, large effort has been expended to establish wrought processing of TiAl alloys [29, 30]. Hot working of these alloys is generally impeded by a significant plastic anisotropy [31], low diffusivity [32] and the susceptibility to hot cracking [29]. While these aspects are well documented in the TiAl literature, there are many open questions about the elementary processes that determine dynamic recovery, recrystallization and phase transformations. These aspects are addressed in the present section.

3.1 The deformed state

A commonly held concept of hot working is that dynamic recovery and recrystallization are triggered by heterogeneities in the deformed state. Thus, the deformation mechanisms occurring in the majority phases γ(TiAl) and $\alpha_2(Ti_3Al)$ will briefly be discussed, for details see [1, 31]. Deformation of γ(TiAl) is mainly provided by ordinary dislocations with the Burgers vector b=1/2<110] and mechanical twinning along $1/6<11\bar{2}]\{111\}$. Since twinning shear is unidirectional, the operating twinning systems vary with the sense of the load and the loading direction. There are crystal orientations for which twinning is forbidden. To a lesser extent deformation is provided by the motion of superdislocations with the Burgers vectors b=<101] and $b=1/2<11\bar{2}]$. The superdislocations exhibit an asymmetric non-planar core spreading; this results in a high glide resistance, which is sensitive to the direction of motion, for a review see [33]. Thus, the superdislocations do not move as readily as ordinary dislocations and mechanical twins. Taken together, this glide geometry gives rise to a significant plastic anisotropy of the γ phase. Ti_3Al alloys have several potential slip systems, which in principle may provide sufficient shear components for the deformation of polycrystalline material. However, there is a strong predominance for prismatic slip; plastic shear with c components of the hexagonal cell is practically impossible [34]. This plastic anisotropy appears to be even more enhanced at intermediate temperatures because the pyramidal slip systems exhibit an anomalous increase of the critical resolved shear stress with temperature. Thus, the brittleness of polycrystalline Ti_3Al alloys can be attributed to the lack of independent slip systems that can operate at comparable stresses; hence the von Mises criterion is not satisfied. Due to this situation deformation of γ(TiAl)+$\alpha_2(Ti_3Al)$ alloys is mainly carried by the γ phase.

There are various processes that might lead to heterogeneities in the deformed state. As a specific example for TiAl alloys, deformation heterogeneities resulting from mechanical twinning will be outlined in more detail. At the beginning of deformation the slip path of the twins is essentially identical with the domain size or lamellar spacing. As soon as multiple twinning with non-parallel shear vectors is activated, extensive intersections among twin bands occur. The intersection of a moving deformation twin with a barrier twin is expected to be difficult because the incorporation of an incident twinning system into the barrier twin may no longer constitute a crystallographically allowed twinning system. Several authors [35-39] have analyzed the mechanism and have proposed crystallographic relations by which an incident twin could intersect a barrier twin. The high-resolution electron micrograph shown in Fig. 9 demonstrates a so-called type-I twin intersection, which is favoured if the $<\bar{1}10]$ intersection line is at 0° to 55° from the sample axis [35]. The structure is imaged along the common $<\bar{1}10]$ direction of the two twins; this can be recognized by the different contrast of the (002) planes, which are alternately occupied by Ti and Al atoms. On its upper side the vertical twin is thicker than on its lower side, thus this twin was considered as incident twin T_i; T_b is the barrier twin. The intersection leads to a significant deflection of the two twins, which, however, is more pronounced for the barrier twin. The difficulty in forming a twin intersection is manifested by the ledged interface between the incident twin and the matrix (arrows 1 and 3) and various dislocations that emerge at the intersection zone (arrows 2). Figure 10 shows the details of the intersection zone. The intersection zone remains in the $L1_0$ structure and seems to be relatively free of defects; however highly defective regions border it. The $(002)_{Tb}$ planes of the central zone

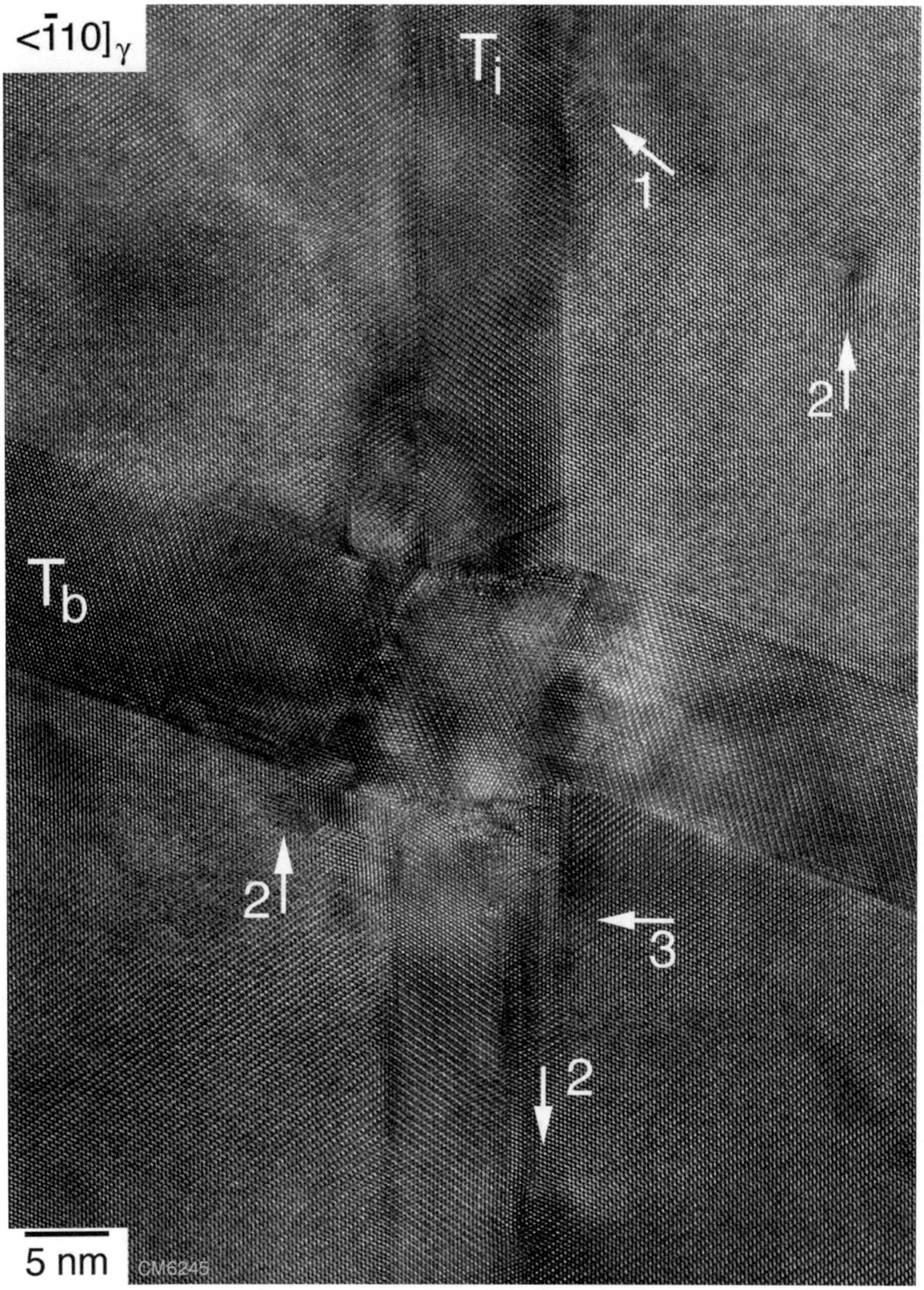

Fig. 9. Type-I intersection of two deformation twins observed after room temperature compression of a Ti-48.5Al-0.37C alloy. The structure is imaged along the common $<\bar{1}10]$ intersection line. The vertical twin T_i is considered to be the incident twin because its upper side is thicker than its lower side; T_b is the barrier twin.

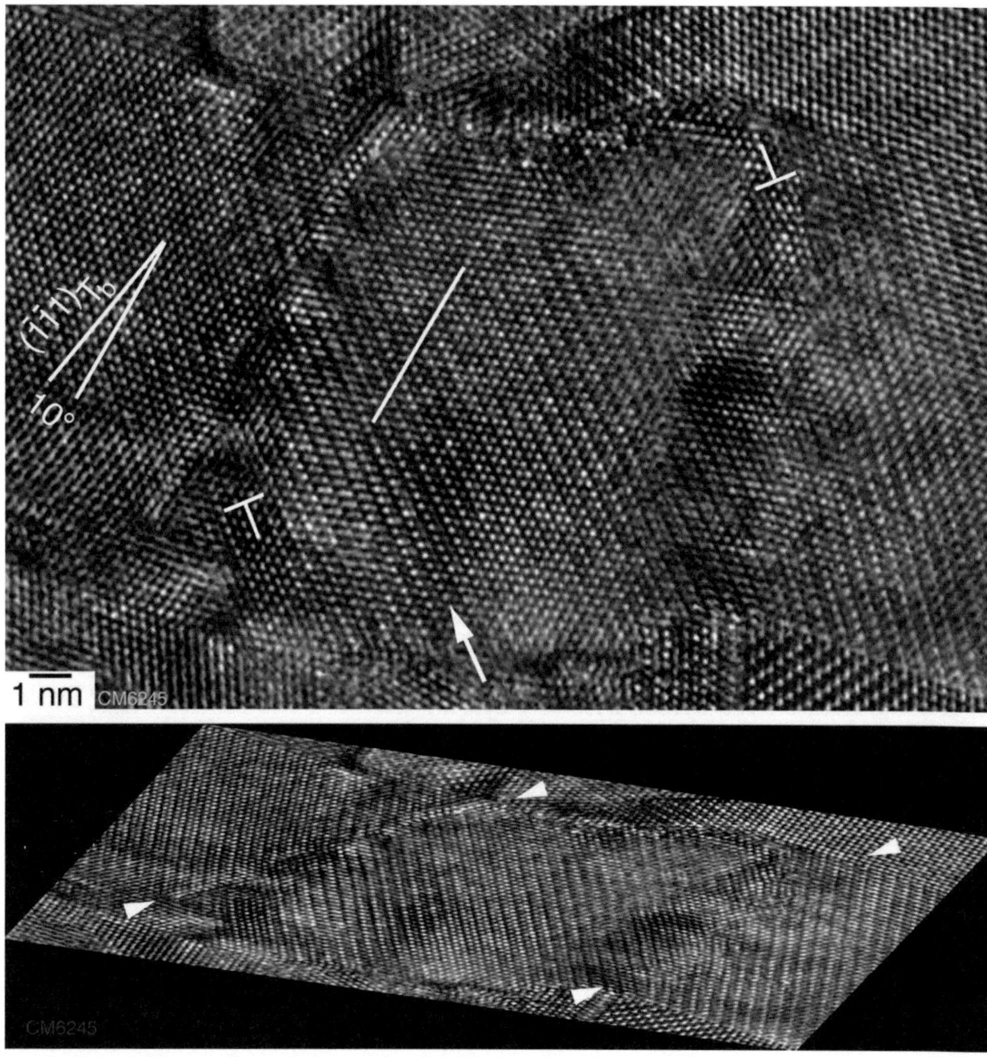

Fig. 10. Structural details of the intersection zone: translation of the twinning shear along the $(\bar{1}\bar{1}1)_{Tb}$ planes by $1/2<10\bar{1}]$ superpartials that become evident by extra $(002)_{Tb}$ planes. Two of the dislocations and their extra half planes are indicated by dislocation symbols. Note the anticlockwise rotation of the intersection zone with respect to the barrier twin by 10°, which is indicated by the traces of the respective $(\bar{1}\bar{1}1)_{Tb}$ planes. The compressed image below shows these features in more detail. Arrowheads mark the dislocation walls on either side of the intersection zone.

are not continuous with those of the barrier twin, but appear displaced along the $(\bar{1}\bar{1}1)_{Tb}$ planes of the barrier twin. This displacement is consistent with glide of $1/2<10\bar{1}]$ or $1/2<01\bar{1}]$ superpartials on $(\bar{1}\bar{1}1)_{Tb}$ planes. These dislocations become manifest by extra $(002)_{Tb}$ planes; a few of these dislocations and the orientation of their extra planes are indicated by dislocation symbols. It is tempting to speculate that these dislocations were generated under the high stress concentration acting at the corners of the intersection zone. It is worth adding that a shear accommodation by twinning along the $(\overline{11}1)_{Tb}$ planes is not possible because this would require anti-twinning operations. The close distance of these dislocations explains why the lattice of the intersection zone is not congruent with that of the barrier twin, but rotated by about 10° against the barrier twin. The image below is the micrograph compressed along the (002) planes and shows these details more clearly. The observation largely reflects the strong rotation field that was generated by the incident twin. The various dislocation reactions that could be involved in the intersection process probably give rise to dislocation emission; a few of these dislocations are marked with arrows 1 to 3. Twin intersections undoubtedly leave significant internal stresses and dense defect arrangements bordering the intersection zone. Under hot-working conditions these heterogeneities in the deformed state can be the prevalent sites for recrystallization. At elevated temperatures rearrangement of the dislocation walls surrounding the misoriented zone may occur by climb so that the misorientation with respect to the surrounding matrix increases. The intersection zone is transformed into a new grain of low internal energy, growing into deformed material from which it is finally separated by a high-angle boundary. This process is certainly driven by the release of stored energy. Thus, it might be expected that the structural heterogeneities produced by twin intersections act as precursor for recrystallization. Such processes are certainly beneficial for the conversion of the microstructure under hot-working conditions.

3.2 Dynamic recovery

There is good consensus that recovery and recrystallization are competing processes as both are driven by the stored energy of the deformed state. The extent of recovery is generally expected to depend on the stacking fault energy, which, in turn, determines dislocation dissociation. The ordinary dislocations, which mainly carry the deformation in γ(TiAl), have a compact core because dissociation would involve a high-energy complex stacking fault (CSF), which destroys the chemical environment of first neighbours in the fault plane. This compact core structure makes cross slip and climb of the ordinary dislocations relatively easy and is a good precondition for static and dynamic recovery. The most convincing evidence for climb of ordinary dislocations has also been obtained from in situ heating experiments performed inside the TEM [31]. The samples used were pre-deformed at room temperature to a strain of ε=3%; this introduced sufficient dislocations for observation and certainly a small supersaturation of intrinsic point defects. Fig. 11 demonstrates the change of the dislocation fine structure occurring upon heating by a sequence of micrographs. The vacancies produced during room-temperature deformation condense onto screw dislocations, causing them to climb into helices. TEM observations performed after high-temperature deformation have also revealed a remarkable instability of twin structures [31]. Due to these factors, the release of strain energy by recovery is probably relatively easy and can account for the large reduction of the flow stress that has been observed after annealing at moderately high temperatures. Recovery lowers the driving force for recrystallization; thus, a significant amount of prior recovery may retard the recrystallization kinetics. More details about deformation and recovery are provided in [40].

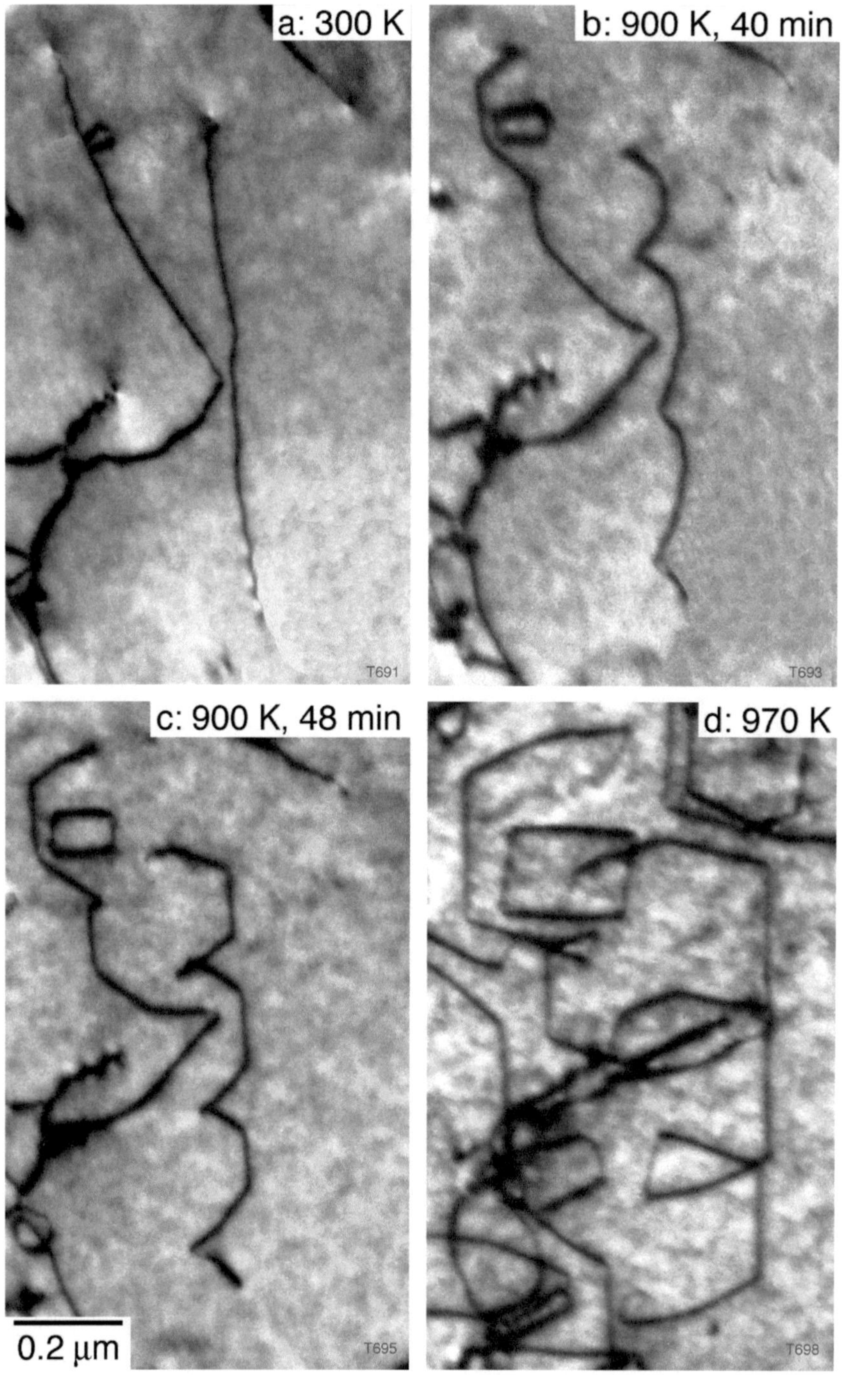

Fig. 11. Climb of ordinary dislocations during in situ heating inside the TEM at an acceleration voltage of 120 kV. Note the formation of a helical dislocation and the growth of prismatic dislocation loops. Cast Ti-48Al-2Cr, pre-deformation at room temperature in compression to ε=3%.

3.3 Dynamic recrystallization

The kinetics of dynamic recrystallization of TiAl alloys depends on several factors; these involve alloy composition, grain size of the starting microstructure, and hot working parameters [41]. The conversion of the coarse-grained lamellar ingot structure is probably the most difficult step in wrought processing. Thus, dynamic recrystallization occurring at this stage will be described as a specific example for TiAl alloys.

There is undisputed evidence that slip transfer through lamellar boundaries is difficult [42]. Plastic strain resulting from a cooperative operation of several deformation modes can be localized between the lamellar boundaries. These facts, combined with the flat plate geometry of the lamellae (Fig. 2), cause a marked plastic anisotropy of lamellar material, which affects the recrystallization behaviour [41]. The recrystallization kinetics is relatively fast if the lamellae orientation is parallel to the deformation axis. Deformation in this orientation apparently involves an element of instability, which is manifested by kinking of the lamellae (Fig. 12) and is reminiscent of buckling of load-carrying structures. In terms of a laminate model [43], the lamellar morphology may be considered as an ensemble of TiAl and Ti_3Al plates. When perfectly aligned with the compression axis, these "columns" are highly stable under compression, as long as the axial load is below a critical value. Above this critical load the equilibrium becomes unstable and the slightest disturbance will cause the structure to buckle. In the lamellar structure an upsetting moment might develop by lateral impinging of the lamellae by dislocation pile-ups or deformation twins. Furthermore, the elastic response of the α_2 and γ phase upon loading is significantly different. Thus, the tendency to instable buckling is expected to increase, if there is an inhomogeneous distribution of α_2 and γ lamellae. The process probably starts with local bending of the lamellae. From the curvature and thickness of the lamellae local strains can be deduced, which are often larger than 10 % and lead to the formation dense dislocation structures and of sub-boundaries (Figs. 13 and 14). Subsequent rotation and coalescence of these sub-grains occurs apparently in such a way that kinking of the lamellae is accomplished. Due to kinking, the lamellae are reoriented with respect to the deformation axis, which may support shearing along the lamellar interfaces. In the regions of highest local bending spheroidization and dissolution of the α_2 phase occur; this suggests that both the non-equilibrium constitution and the local stress provide the driving pressure for the observed phase transformation and recrystallization. All these aspects are manifested in Fig. 15, which shows the initial state of grain nucleation in a kinked α_2 lamella.

Micromechanical modelling [43, 44] has shown that polycrystalline lamellar material deforms very inhomogeneously. Localized stress concentrations have been recognized, which developed within the polycrystal upon straining due to the variations of grain size, shape and orientation. This led the authors to believe that constraints imposed by neighbouring grains are a dominant factor in determining the flow behaviour. Even under compression, very high tensile hydrostatic stresses are generated at the triple points of colony boundaries. There is a strong tendency to develop shear bands, kink bands, lattice rotations, and internal buckling of lamellae. The shear bands consist of extremely fine grains and may traverse the whole work piece, often resulting in gross failure. From a mechanical point of view, buckling failures do not depend on the yield strength of the material but only on the dimensions of the structure and the elastic properties of the material. In lamellar alloys of given lamellar spacing, the tendency to buckling and shear band formation increases with the axial length of plates. Thus, coarse-grained lamellar alloys seem to be prone to deformation instabilities and shear band formation, which turns out to be one of the prime problems in wrought processing of TiAl alloys.

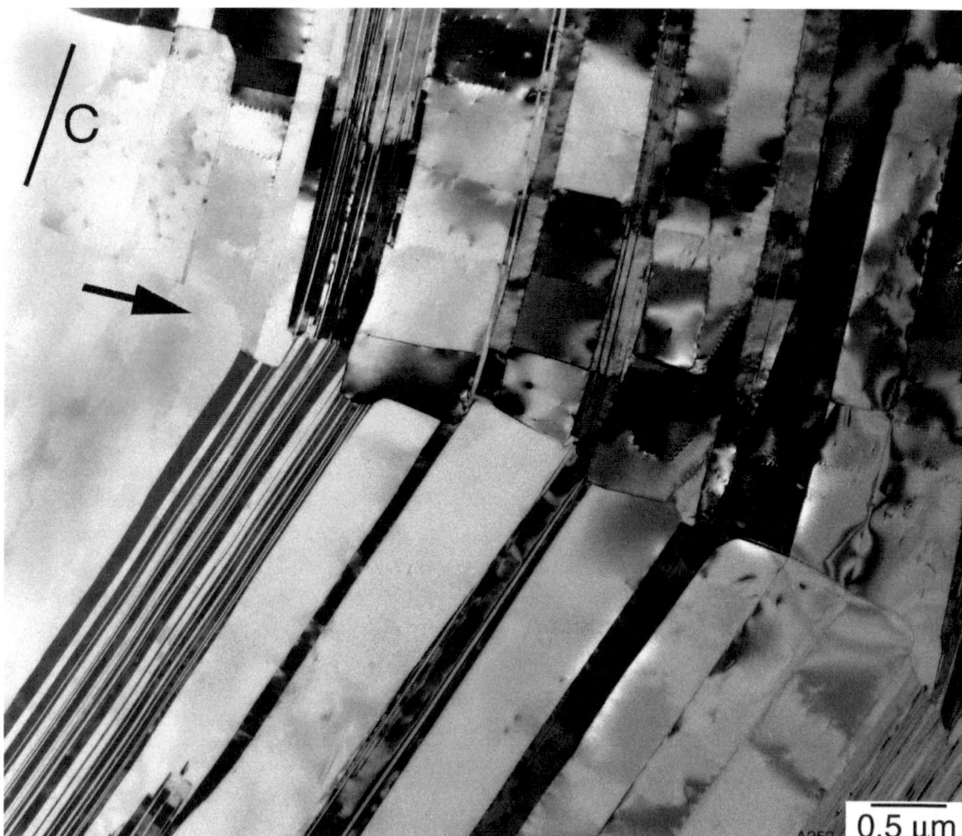

Fig. 12. Kinked lamellae in Ti-47Al-5Nb-0.2B-0.2C. The lamellar ingot material was subject at 1270 °C to a compression stress of σ=(25.4±12.7) MPa fluctuating with 30 Hz. C indicates the orientation of the compression axis, which is parallel to the image plane.

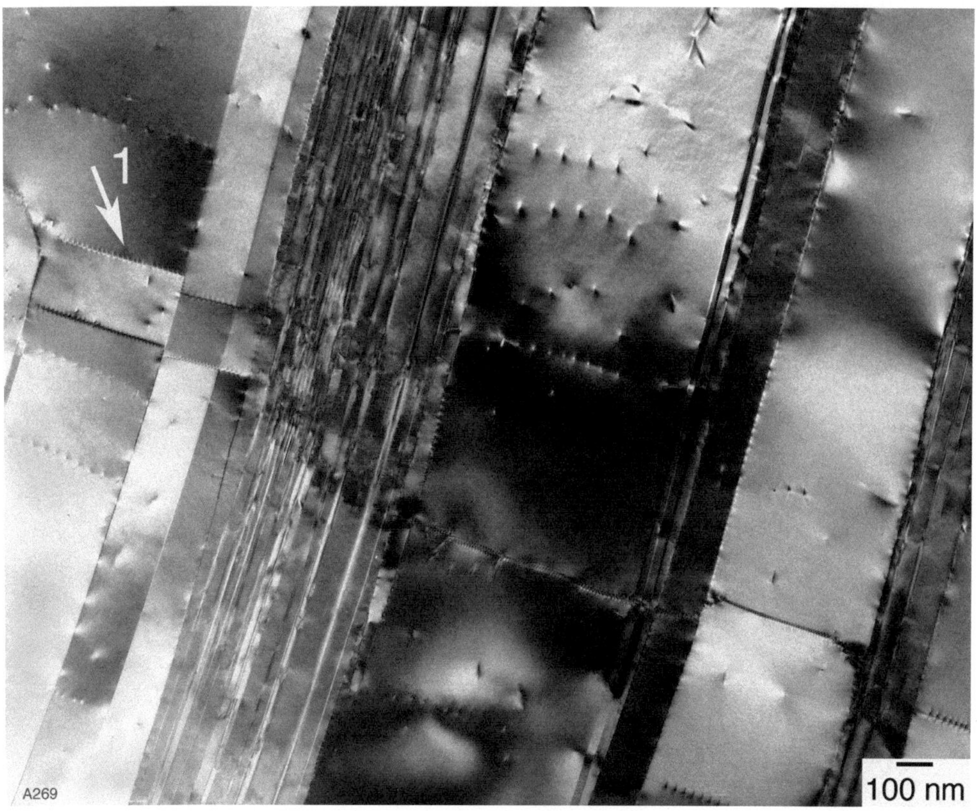

Fig. 13. Sub-boundaries formed at kinked lamellae. Experimental details as for Fig. 12.

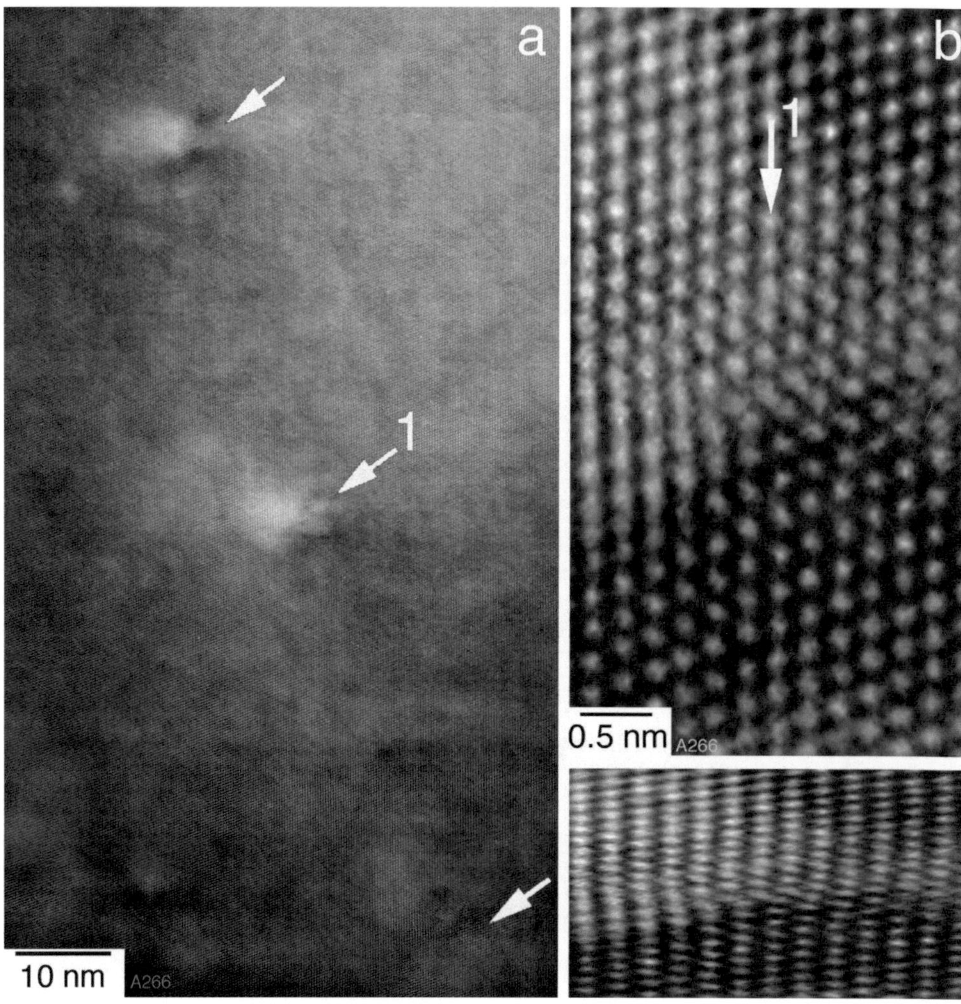

Fig. 14. Atomic structure of a sub-boundary formed at kinked γ lamella; experimental details as for Fig. 12. (a) Mixed ordinary dislocations situated in the sub-boundary (marked with arrow 1 in Fig. 13) and (b) on of the dislocations shown in higher magnification; the compressed image below shows the extra plane of the dislocation more clearly.

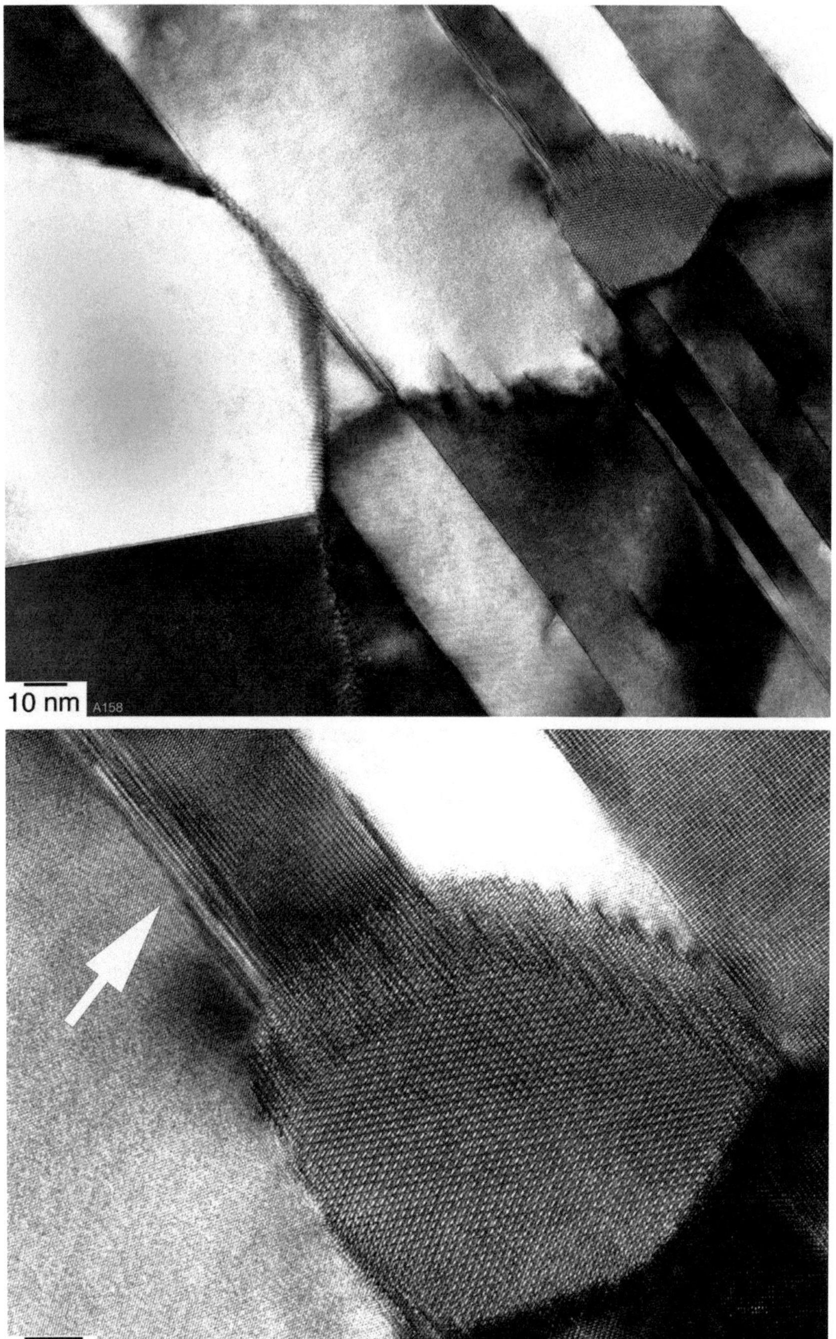

Fig. 15. Initial state of grain nucleation at kinked α_2 lamellae. The image below shows one of the grains in higher magnification. Note the shear processes occurring along the lamellae. Ti-45Al-8Nb-0.2C, sheet material; viewing direction in the transverse direction of the rolling plane.

4. Service-induced phase transformation and recrystallization

4.1 Creep

At temperatures above 650 °C damage of TiAl components may occur due to continuous creep. The relevant mechanisms are numerous and synergistic, depending on the operation conditions; for a review see [45]. Among the various microstructures that can be established in polycrystalline (α_2+γ) alloys, fully lamellar alloys are most creep resistant. However, numerous investigations have demonstrated that the lamellar microstructure degrades upon creep. This structural instability is a serious problem for long-time service of lamellar alloys. In the following, a few examples of TEM analysis will be presented, which demonstrate the complexity of the processes involved in the degradation of the lamellar morphology. Most of these studies have been performed on samples that had been subject to long-term tensile creep at 700 °C under relatively low stresses of 80 to 140 MPa [46].

When compared with the microstructure of undeformed material, the lamellar interfaces in crept samples were highly imperfect. Figure 16a demonstrates the formation of a high interfacial step in a 60° pseudo twin boundary. The interfacial steps had often grown into broad zones, which extended over about 200 nm perpendicular to the interface; Fig. 16b demonstrates an intermediate stage of this growth process. Multiple-height ledges are commonly observed after phase transformation and growth, and several mechanisms have been proposed to explain the phenomenon [47]. Analogous to these models it is speculated that the large ledges observed in the crept TiAl alloy arise from one-plane steps, which moved under diffusional control along the interfaces and were piled up at misfit dislocations. A misfit dislocation with a Burgers vector component out of the interfacial plane is arrowed in Fig. 16a. Once a sharp pile up is formed, the configuration may rearrange into a tilt configuration with a long-range stress field. This would cause further perfect or Shockley partial dislocations to be incorporated into the ledge and would also explain that in all cases the ledges were associated with misfit dislocations. The detailed atomic structure of the macro ledges is not clear. As can be seen in Figs. 16, there is a variation of the contrast in the ledges with a periodicity of three (111) planes. This is reminiscent of the 9R structure, which is a phase that probably has a slightly higher energy than the $L1_0$ ground state. The formation of the 9R structure is a well-known phenomenon in many f.c.c. metals that exhibit twinning. Singh and Howe [48] have recognized the 9R structure in heavily deformed TiAl. It should be noted, however, that similar three-plane structures have been observed in a massively transformed, but undeformed Ti-48.7Al [49]. In this work the contrast phenomena have been interpreted as arising from overlapping twin related γ variants. Nevertheless, the macro-ledges are a characteristic feature in the microstructure of crept samples and represent at least a highly faulted $L1_0$ structure; the question is only whether there is a periodicity in the fault arrangement. When the macro-ledges grow further, it might be energetically favourable to reconstruct the $L1_0$ structure and to nucleate a new γ grain. Figure 17 probably demonstrates an early stage of such a process. The recrystallized grains usually have a certain orientation relationship with respect to the parent γ lamellae; Fig. 18 indicates that the (001) planes of the recrystallized grain are parallel to the $(1\overline{1}\overline{1})$ planes in the parent lamella γ_1. There is a significant mismatch for this orientation relationship, which is manifested by a high density of ledges and dislocations at the $(001)||(1\overline{1}\overline{1})$ interface. Recrystallization of ordered structures has been investigated in several studies [50, 51]. There is a drastic reduction in grain boundary mobility, when compared with disordered metals. Recovery of ordered alloys is also complicated by the fact that the ordered state has to be restored. In this respect it is interesting to note that the small grain shown in Fig. 18 is completely ordered giving the impression that the ordering is immediately established after grain nucleation or that nucleation occurred in the ordered state. This might be a consequence of the fine scale of the lamellar microstructure and the heterogeneous grain nucleation at the interfacial ledges. There are certainly crystallographic constraints exerted by the parent lamellae adjacent to the ledges, which may control nucleation and growth. Clearly, the process needs further investigation.

There is a significant body of evidence in the TiAl literature indicating that dissolution of α_2 lamellae occurs during creep [45, 46]. The phase transformation is probably driven by a non-equilibrium constitution. High-temperature creep is expected to promote phase transformation towards equilibrium constitution; thus, dissolution of α_2 and formation of γ occurs [52]. The high-resolution micrograph shown in Fig. 19 supports this reasoning; there is clear evidence that the density of steps at the α_2/γ interfaces is significantly higher than that at the γ/γ interfaces, meaning that the α_2 lamella dissolves, whereas the γ lamellae are relatively stable. The processes eventually end with the formation of new grains (Fig. 20) and a more or less complete conversion of the lamellar morphology into a fine spheroidized microstructure. The $\alpha_2 \rightarrow \gamma$ phase transformation is often associated with local deformation, as suggested by Fig. 21. The micrograph shows two α_2 terminations that are connected by an interface, thus the α_2 lamella is partially dissolved. Two twins were emitted at one of the terminations and a dislocation with a Burgers vector out of the interface is present (Fig. 22). The α_2 terminations have extremely small principal radii of curvature. The elimination of such structural features reduces the surface energy and provides a driving force towards further coarsening. The interface connecting the two α_2 terminations exhibits a fault translation that corresponds to an intrinsic stacking fault. This is indicated by the stacking sequence *ABC B CAB*. The nature of the interface between two phases is determined by their structural relationships. There is a strong tendency for planes and directions with the highest atomic densities to align across the interface. As suggested by Chalmers and Gleiter [53], a better atomic fit at a boundary could result if atoms were moved away from coincident sites by a rigid-body displacement of one grain relative to the other by a constant displacement vector. Atomic modelling performed in this context (for a review see [33]) has shown that the energy of lamellar interfaces could be minimized by a rigid-body translation along the vectors $f_{APB}=1/2<10\bar{1}]_{\gamma}$, $f_{SISF}=1/6[11\bar{2}]_{\gamma}$ and $f_{CSF}=1/6[\bar{2}11]_{\gamma}$. These translations correspond to the formation of an antiphase boundary (APB), a superlattice intrinsic stacking fault (SISF) and a complex stacking fault (CSF), respectively, at the interface. In the present case the translation vector is of type $f_{SISF}=1/6[11\bar{2}]_{\gamma}$. The observation underlines once again the fine scale of interface processes that may occur during creep of lamellar alloys.

The $\alpha_2 \rightarrow \gamma$ transformation requires a change of both the stacking sequence and the local composition. However, achieving the appropriate composition requires long-range diffusion, which at a creep temperature of 700 °C is very sluggish. Low-temperature diffusion might be supported by the presence of Ti_{Al} antisite defects [6, 32]. In the newly formed γ phase a high density of such defects is certainly formed, in order to accommodate the excess of titanium. Thus, a substantial antistructural disorder occurs, which forms a percolating substructure. Under such conditions, the antisite defects may significantly contribute to diffusion because antistructural bridges (ASB's) are formed. An elementary bridge event involving one vacancy and one antisite defect consists of two nearest neighbour jumps, which result in a nearest neighbour displacement of two atoms of the same species. For the $\alpha_2 \rightarrow \gamma$ transformation the so-called ASB-2 mechanism [32] might be relevant, which requires only low migration energy. Diffusion may also be supported by the mismatch structures present at the interfaces. Dislocations and ledges represent regions where the deviation from the ideal crystalline structure is concentrated. These are paths of easy diffusion, which can effectively support the exchange of Ti and Al atoms. One may expect all these processes to be thermally activated und supported by superimposed external stresses. In this respect, the coherency stresses present at the interfaces are certainly significant because they are comparable to or even higher than the shear stresses applied during creep tests and are often associated with mismatch structures. Thus, given a non-equilibrium constitution, it is understandable that drop in the α_2 volume content also occurs, when a lamellar alloy is subject to the same temperature/time profile without externally applied stress [54].

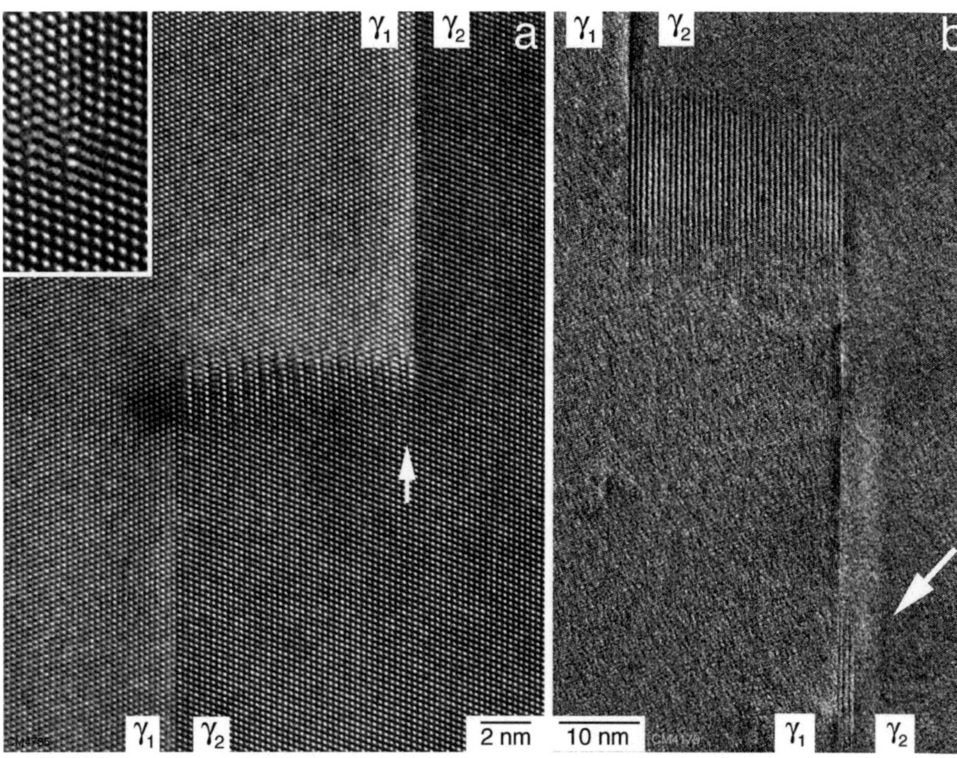

Fig. 16. Degradation of the lamellar structure in Ti-48Al-2Cr under long-term creep at T=700 °C, σ_a=140 MPa, for t=5988 hours to strain ε=0.69 %. (a) Formation of an interfacial step in a 60° pseudo twin boundary. Note the interfacial dislocation (arrowed) that is manifested by an additional {111} plane. (b) A macro-ledge present in a 60° pseudo twin boundary.

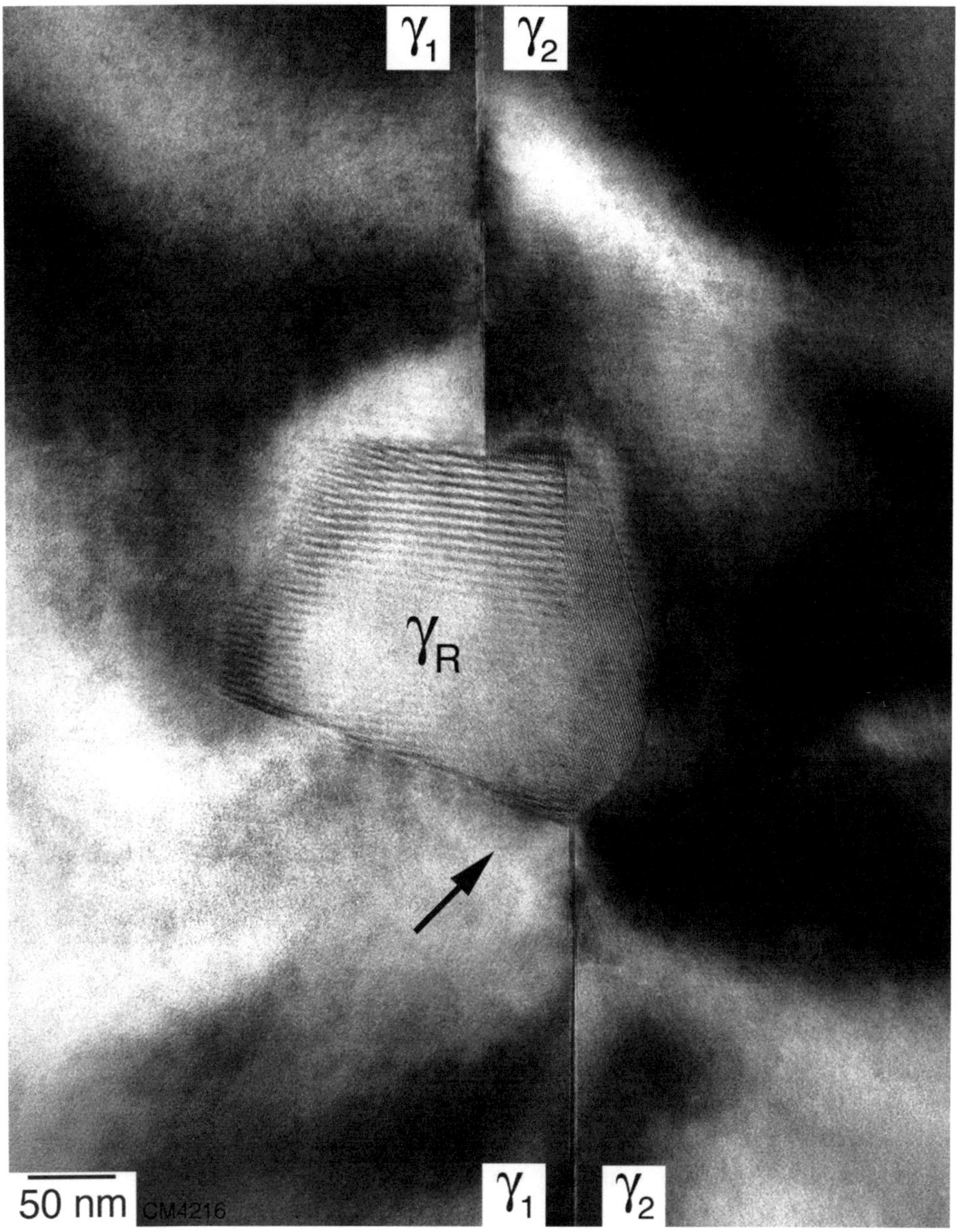

Fig. 17. Recrystallized grain γ_R formed at a ledge in a lamellar interface joining the gamma variants γ_1 and γ_2 with a pseudo twin orientation relationship. Note the step in the interface and the ordered state of the recrystallized grain. Experimental details as for Fig. 16.

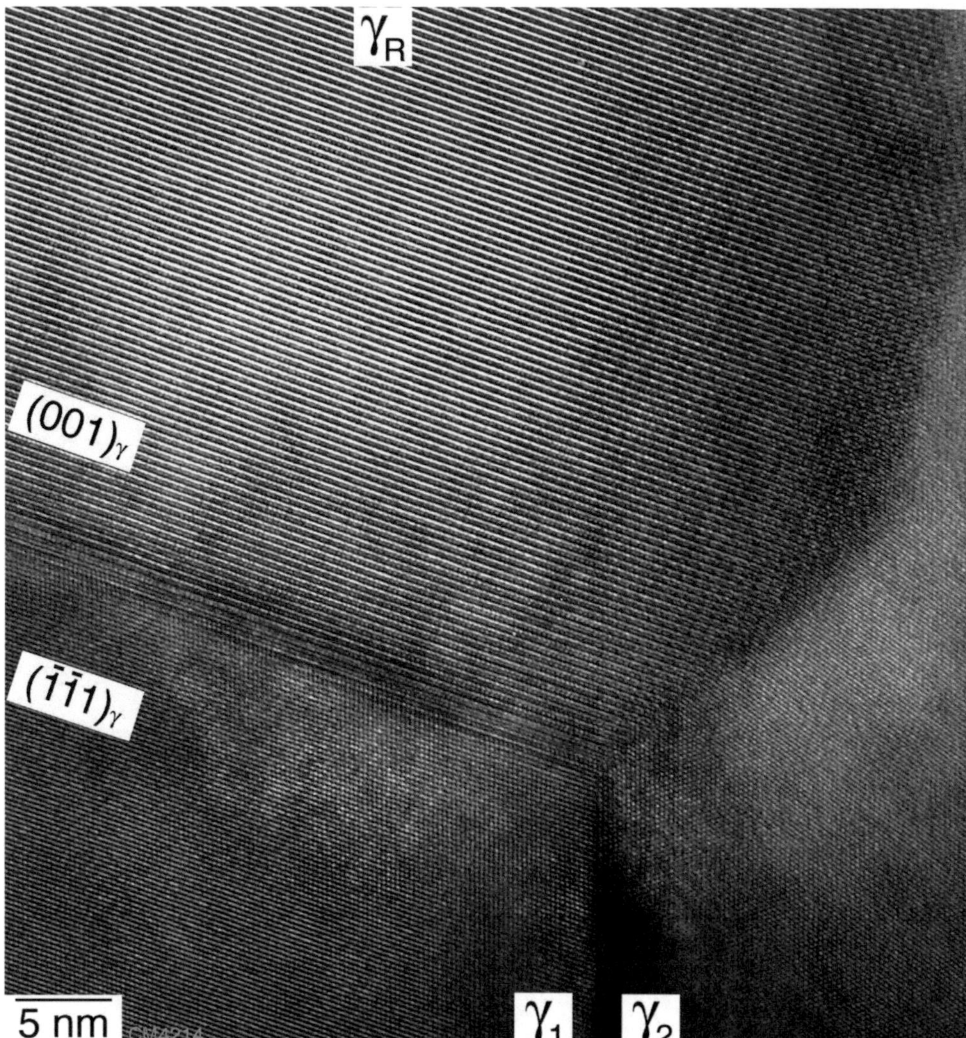

Fig. 18. Higher magnification of the boundary triple-point marked in Fig. 17. Note the orientation relationship (001) | | (111) between the recrystallized grain γ_R and lamella γ_1.

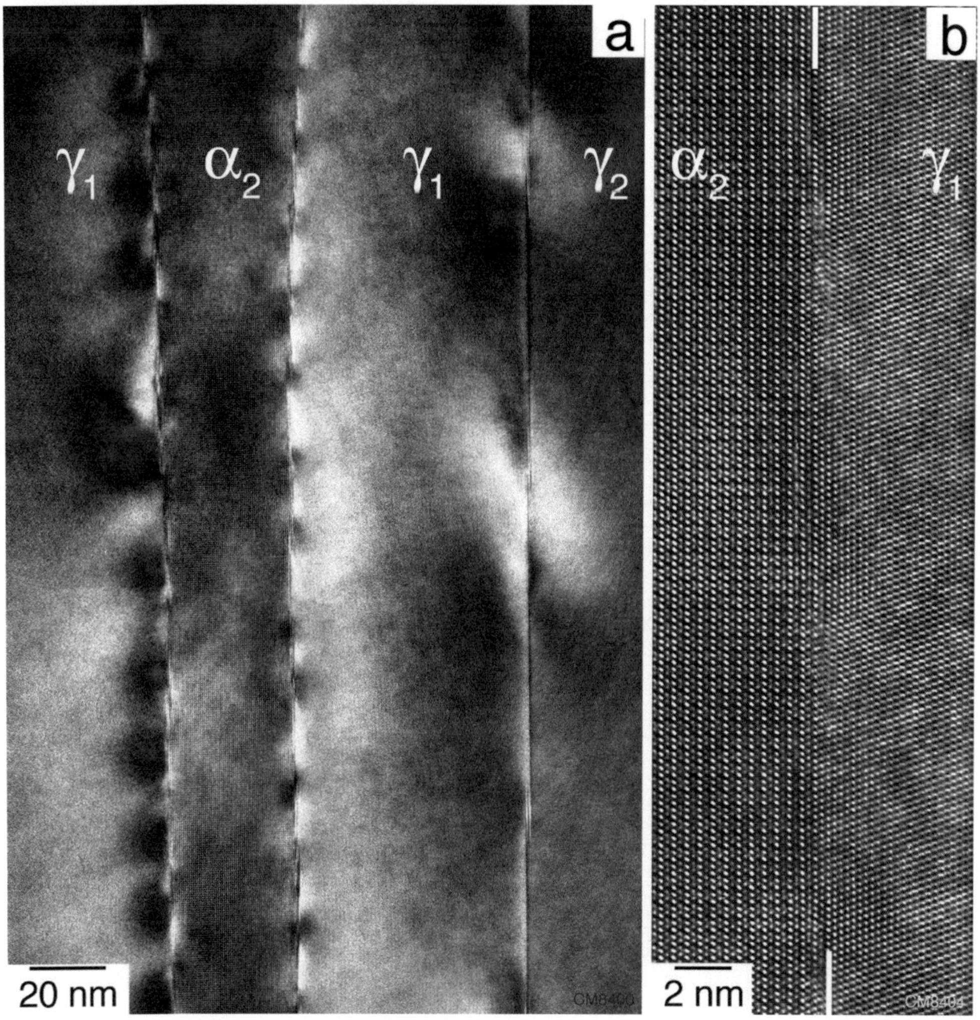

Fig. 19. Initial stage of the $\alpha_2 \rightarrow \gamma$ phase transformation in the lamellar structure of a Ti-48Al-2Cr alloy occurring during long-term creep at T=700 °C, σ_a=110 MPa, t=13400 hours to strain ε=0.46 %. (a) Low-magnification high-resolution image of the lamellar structure. Note the significantly higher density of steps at the α_2/γ interfaces, which indicates dissolution of α_2 phase. (b) Atomic structure of one of the α_2/γ interfaces demonstrating its stepped character.

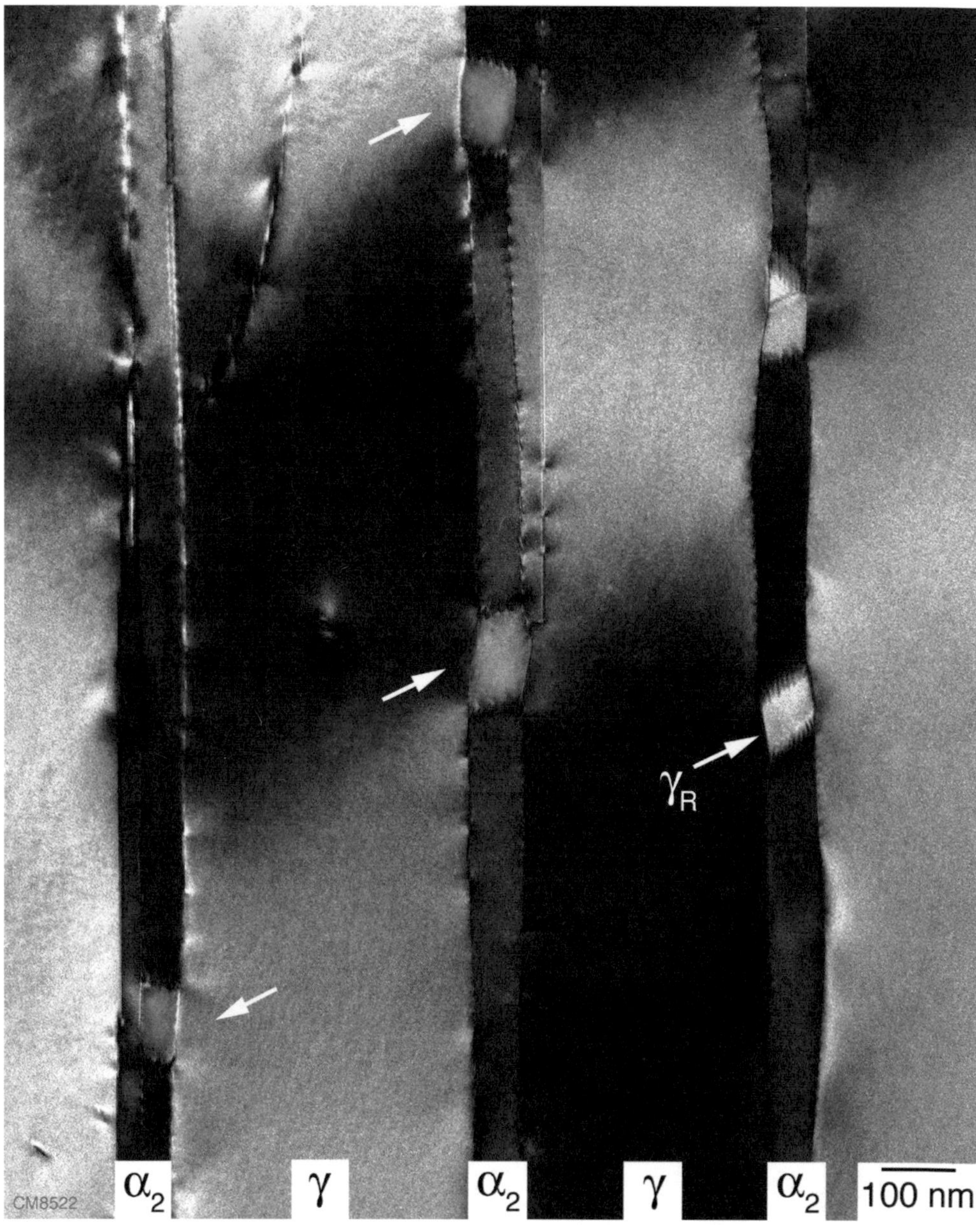

Fig. 20. Spheroidization of α_2 lamellae due to the formation of γ grains (arrowed, designated as γ_R). Experimental details as for Fig. 19.

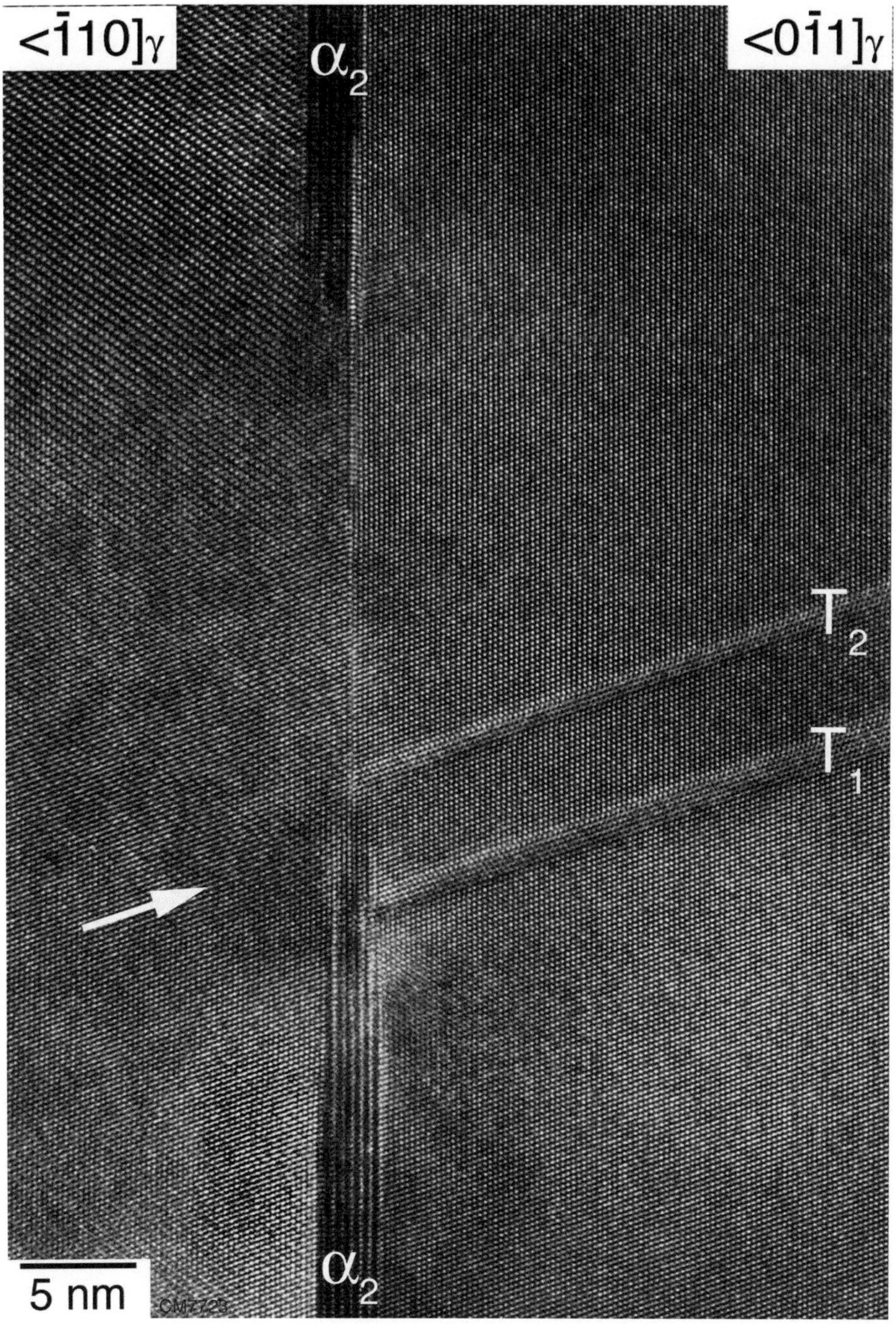

Fig. 21. A partially dissolved α_2 lamella in Ti-46.5Al-4(Cr, Nb, Ta, B) embedded in γ phase. Creep deformation at T=700 °C, σ_a=200 MPa to strain ε=1.35 %. Note the two α_2 terminations that are connected by an interface and the emissions of two twins T_1 and T_2 at one of the terminations. The stacking sequence indicates a rigid body translation of the adjacent γ lamellae.

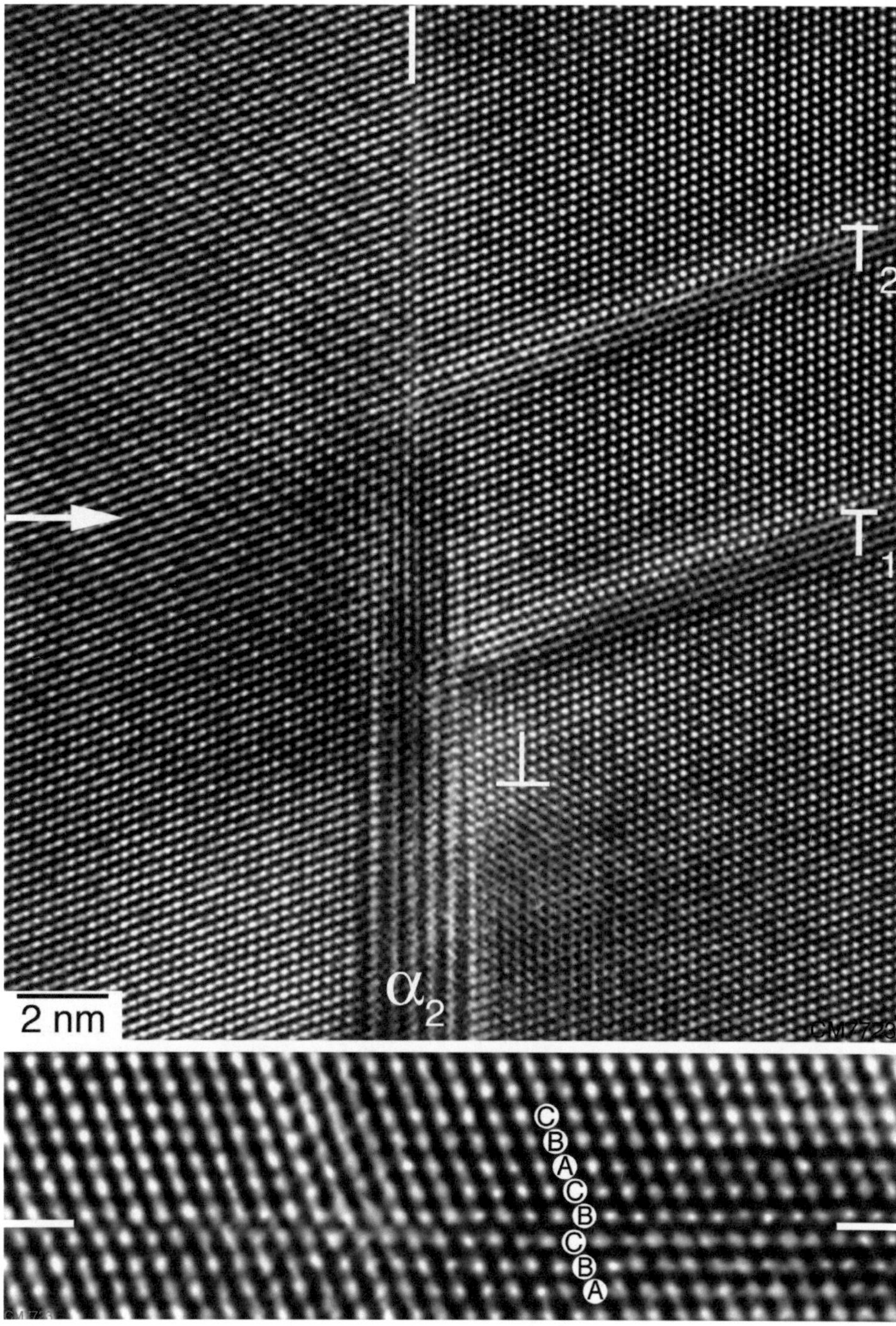

Fig. 22. Higher magnification of the area marked in Fig. 21 showing deformation activity in the vicinity one of the α_2 terminations. Two mechanical twins and a dislocation with a Burgers vector out of the interface plane (marked by symbol) are present. The lower image shows the stacking sequence across the interface (parallel to the arrow) that connects the two α_2 terminations. The stacking sequence indicates the presence of a fault translation.

4.2 Fatigue

By far, the most anticipated engineering applications of TiAl alloys involve components that are subjected to fluctuating or cyclic loading. The capability of TiAl alloys to sustain such loading conditions is inherently limited by the plastic anisotropy at the dislocation level and the lack of independent slip and twinning systems that can operate under reversed plastic straining. Low-cycle fatigue (LCF) is a progressive failure phenomenon brought about by cyclic strains that extend into the plastic range. Thus, the LCF life is largely determined by the amount of inelastic strain in each cycle. While the macroscopic LCF phenomena are well characterized [55], it is only recently that information about the structural degradation occurring upon fatigue has been obtained [56]. A few examples for these processes will be demonstrated in this section.

The fatigue study was performed on an extruded Ti-45Al-8Nb-0.2C alloy (TNB-V2), which contains a significant amount of β/B2 phase and an orthorhombic phase with B19 structure. TEM examination performed after room temperature fatigue has shown that the B19 structure transforms into γ phase. The phase transformation occurs in such a way that extremely fine shear bands are formed (Fig. 23). The result is a lamellar morphology that is comprised of extremely fine γ lamellae adjacent to B19 phase (Fig. 23b). A general observation is that two different γ variants are usually generated adjacent to the B19 phase, as seen for almost all the γ lamellae in Fig. 23b. If, for example, the stacking sequence of the $\{111\}_\gamma$ planes of the variant γ_1 is labelled *ABC*, that of variant γ_2 is *CBA*. This inversion of the stacking sequence is thought to induce strain fields of opposite signs, which eventually reduces the total strain energy. It might be speculated that such a combination of shear processes makes the transformation easier. The transformation often starts at grain boundaries and proceeds through the formation of ledges via distinct atomic shuffle displacements; these details are shown in Fig. 24. It should be noted that the B19→γ transformation has been observed at all the fatigue test temperatures investigated. Due to the difference in lattice constants between the γ and B19 phases the transformation may accommodate local strains and is thus expected to serve as a toughening mechanism. Nevertheless, the life of TiAl testpieces under low cycle fatigue with plastic cyclic strain amplitudes of a few tenth of a percent is limited to a several hundred cycles.

5. Diffusion bonding

Solid-state diffusion bonding provides a means of joining TiAl alloys without melting of the base materials. The principal diffusion bonding parameters, temperature and stress, depend on yield strength, work hardening behaviour and creep resistance of the mating alloy coupons. The bonding conditions should be chosen so that coalescence of contacting surfaces is produced by asperity deformation, but without gross deformation of the component. The effects of bonding temperature and bonding stress are synergistic; at higher temperature less stress is required and vice versa. For more details see [57].

During diffusion bonding of (α_2+γ) alloys a three layer process zone is typically developed that involves a fine grained layer of α_2 phase at the former contact plane of the diffusion couple, a region made up of relatively large recrystallized grains, followed by a region of deformed bulk material (Fig. 25). The bond layer consists of fine stress free grains (Fig. 26), which were identified by EDX and EBSD analysis as α_2 phase. It is well documented in the

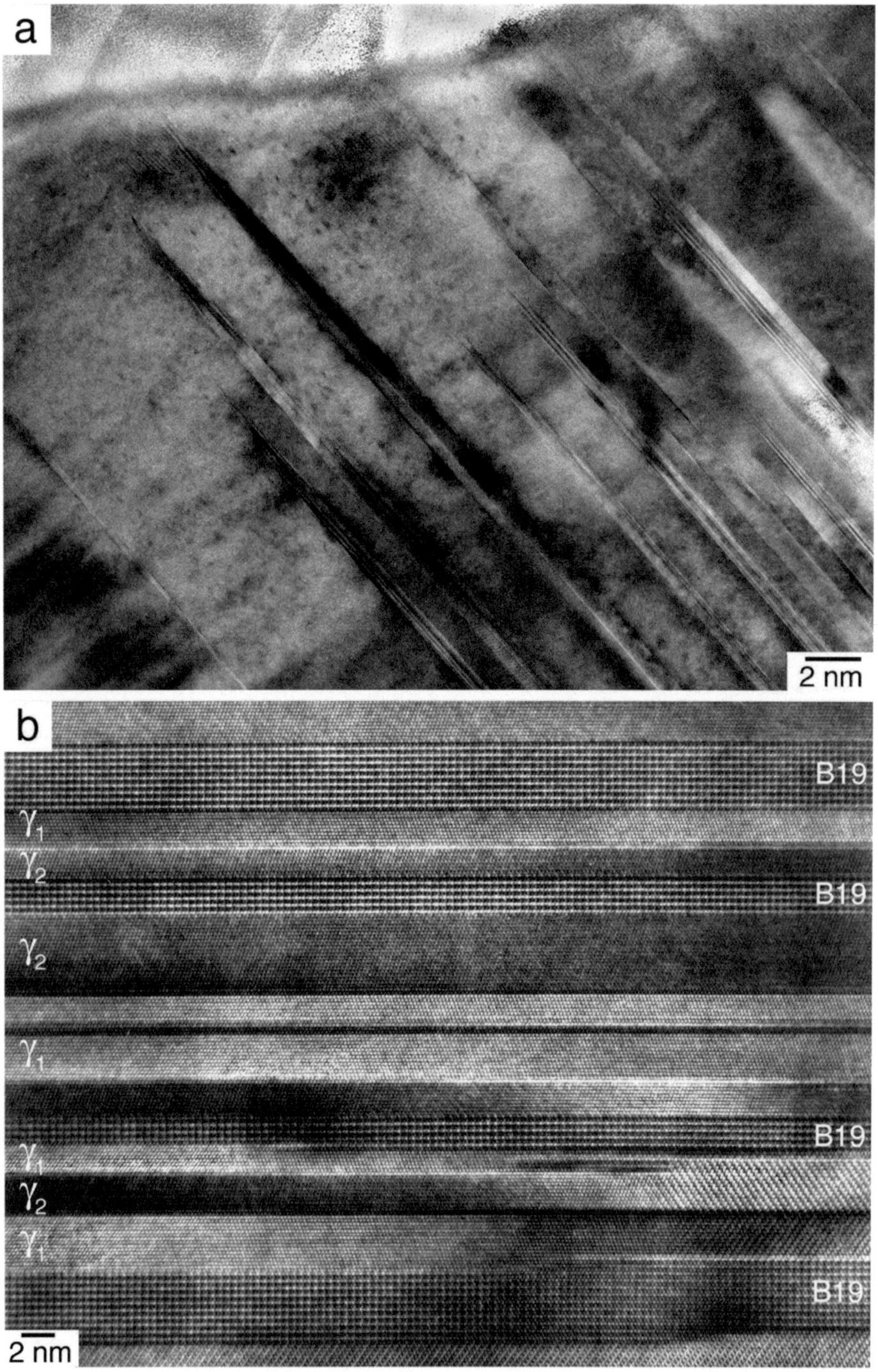

Fig. 23. Phase transformation B19→γ during low cycle fatigue at T=550 °C, $\Delta\varepsilon_t/2$=±0.7 %, N_f=452; nearly lamellar Ti-45Al-8Nb-0.2C. (a) Low-magnification high-resolution TEM micrograph showing fine γ lamellae produced in the B19 phase. (b) Lamellar morphology consisting of fine γ lamellae adjacent to B19 phase.

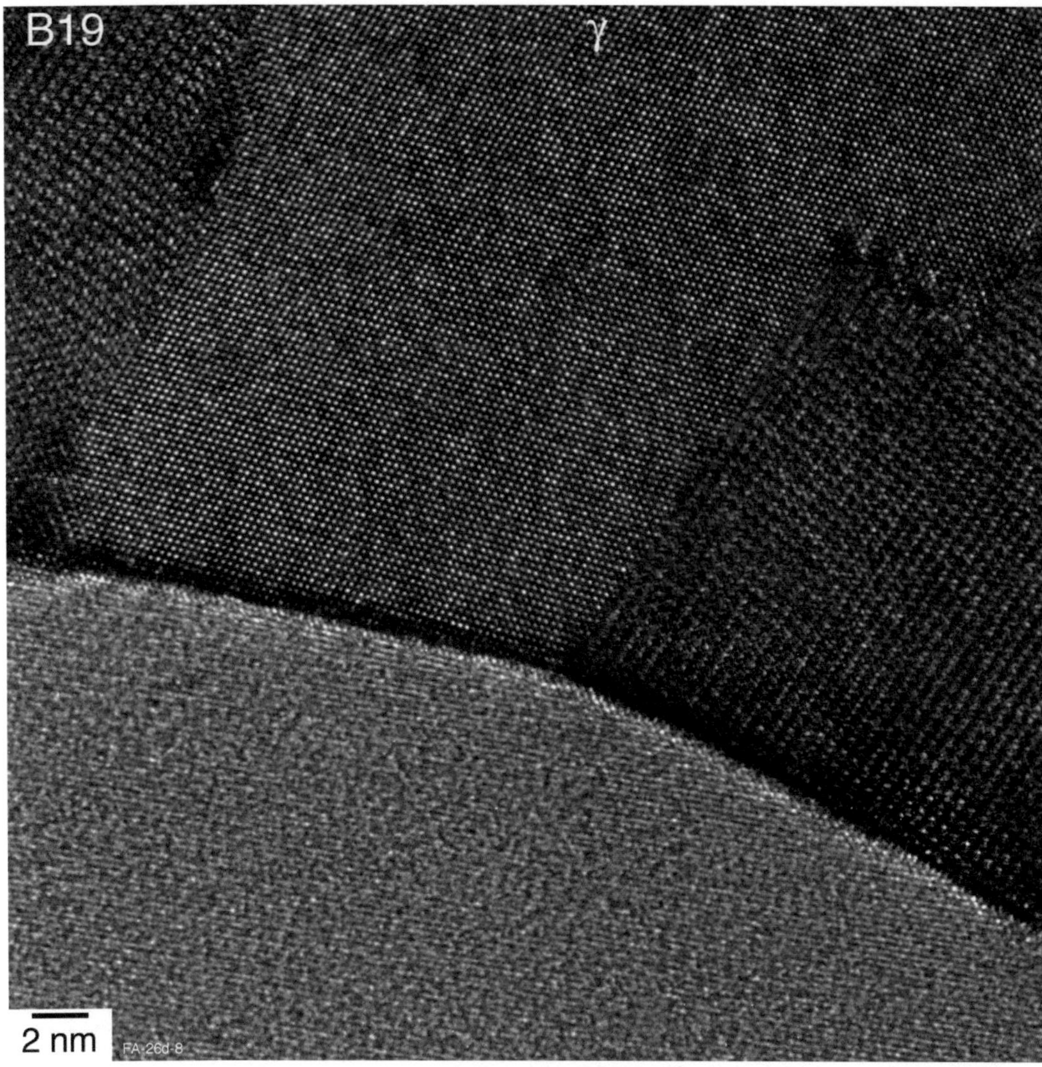

Fig. 24. Transformation B19→γ near to a grain boundary. Nearly lamellar Ti-45Al-8Nb-0.2C, sample fatigued at 25 °C to failure after N=641 cycles with R=-1 and a total strain amplitude $\Delta\varepsilon_t/2=0.7$ %.

literature [58] that a very small amount of oxygen can stabilize the α_2 phase. This finding is consistent with an early investigation of Godfrey et al. [59] performed on diffusion bonded Ti-48Al-2Mn-2Nb. Orientation analysis performed on the α_2 grains has shown that most of the newly formed α_2 grains have an orientation that is suitable for prismatic glide. This data could reflect the well-known plastic anisotropy of the α_2 phase, according to which the activation of prismatic slip along $1/3\langle 11\bar{2}0\rangle\{10\bar{1}0\}$ is by far easiest (Sect. 3.1). It might be speculated that nucleation and growth of new α_2 grains are controlled by the deformation constraints operating during bonding in that the preferred grain orientation ensures strain accommodation on the most favourably slip system.

The α_2 phase at the bond layer is formed at the expense of the Ti content of the adjacent regions, which needs long-range diffusion. The Ti transport could be supported by the anti-structural disorder, as already mentioned in Sect. 4.1. Furthermore, Ti transport could be accomplished by diffusion along the various internal boundaries present in the fine-grained materials. In this respect the lamellar interfaces are probably important because the deviation from the ideal crystal structure occurs and dense arrangements of misfit dislocations are present. Due to the transport processes described, the Ti content of the pre-existing α_2 and β phases situated next to the bonding layer gradually decreases and eventually falls below the critical composition required for their existence; eventually, these phases transform into γ(TiAl). The effect is most pronounced in lamellar colonies that are in contact with the bonding line, presumably because pipe diffusion along the lamellar interfaces is significant. A common observation supporting this mechanism is that the newly formed α_2 grains are connected with pre-existing α_2 lamellae, which in a sense feed the chemically driven generation of α_2 phase at the bonding layer. All in all, the process zone of diffusion bonded TiAl couples reflects the combined influences of chemical driving pressure due to the oxygen contamination at the bonding surfaces and dynamic recrystallization induced by asperity surface deformation.

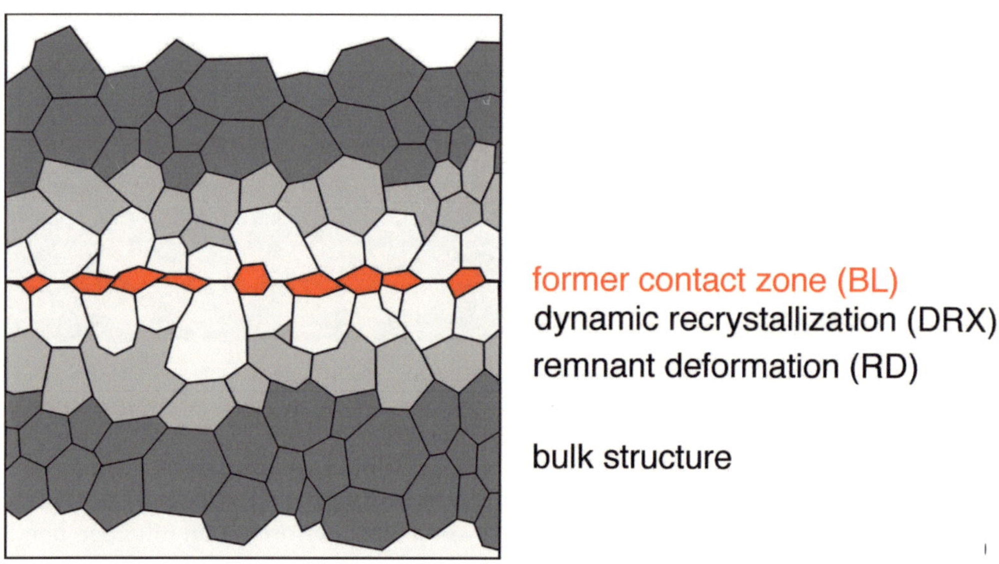

Fig. 25. Schematic illustration of the process zone observed after diffusion bonding at relatively low stresses. BL - fine grained bond layer at the former contact plane of the diffusion couple, DRX - region consisting of relatively large recrystallized grains, RD - initial bulk material but with remnant plastic deformation.

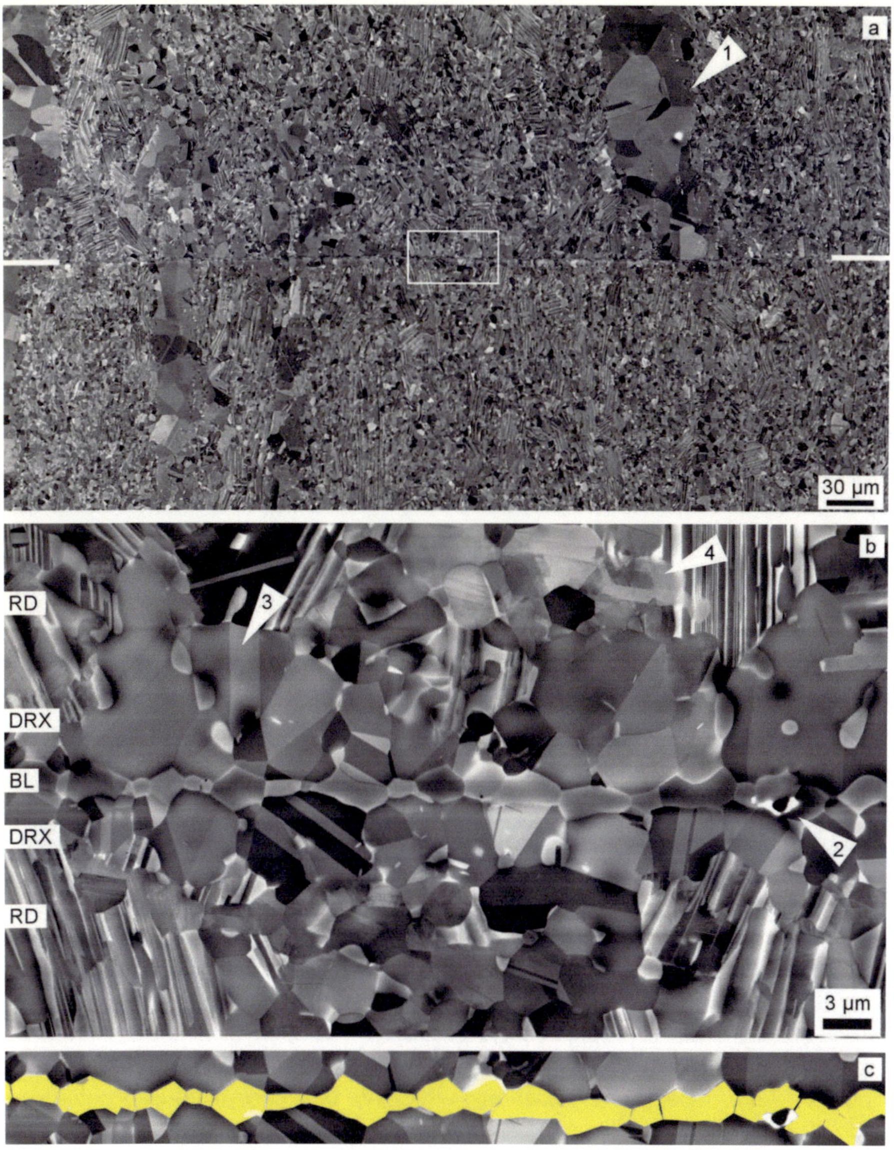

Fig. 26. Cross section of a bond in Ti-46.5Al formed at T= 1273K, σ=20 MPa and t=2h. Scanning electron micrographs taken in the backscattered mode. (a) Low magnification image showing the gross structure of the bond; the horizontally oriented bonding layer is marked by white bars. Note the inhomogeneity of the starting material that is manifested by a banded structure parallel to the extrusion direction (vertical in the micrograph) involving remnant lamellae, fine-grained regions and large γ grains (arrow 1). (b) Higher magnification of the area boxed in (a) showing the bonding layer BL, the recrystallized region DRX and the region with remnant deformation RD in more detail. Note the pore in the bonding layer (arrow 2) and the annealing twins in the γ grains of the DRX region (arrows 3 and 4). (c) Bonding layer consisting of fine-grained α_2 phase as identified by EBSD and EDX analysis [57].

6. Shot peening

Shot peening is a cold working process in which the surface of a component is blasted with small spherical media called shot. A compressive layer is formed by a combination of subsurface compression developed at the Hertzian impression combined with lateral displacement of the surface material around each of the dimples formed. Since fatigue cracks will not initiate nor propagate in compressively stressed regions, shot peening can greatly enhance fatigue life.

In TiAl alloys shot peening produces a heavily deformed surface layer with a thickness of 10 μm to 80 μm depending on the microstructure and yield stress of the substrate alloy [60]. Deformation is characterized by intensive glide and mechanical twinning, involving all potential slip systems available in the major phases $\alpha_2(Ti_3Al)$ and $\gamma(TiAl)$, not only the easy ones. On the mesoscopic scale, buckling and kinking of the lamellae manifest deformation. TEM observations have revealed a remarkable conversion of the microstructure involving dynamic recrystallization and $\alpha_2 \rightarrow \gamma$ phase transformation. The phase transformation starts with a splitting of α_2 lamellae at positions of strong bending or kinking, where the elastic stresses are highest. Another prominent damage mechanism is amorphisation. As shown in Fig. 27a, nano-crystalline grains are embedded in an almost featureless amorphous phase; Fig. 27b demonstrates the gradual loss of crystallinity towards to the adjacent amorphous phase. There seems to be a significant mismatch between the crystalline and amorphous phases, which is indicated by a systematic array of like dislocations (Fig. 28). From the dislocation separation distance it may be concluded that the mismatch is at least 2 % to 5 %.

The observation of an amorphous phase is surprising. However, there are a few arguments that make its existence plausible. Firstly, in the surface layer the material undergoes severe plastic deformation. This introduces various defects, raises the free energy, and creates fresh surfaces due to the formation of slip steps and localized cracking. Secondly, there is certainly a substantial pick up of nitrogen, oxygen, and perhaps hydrogen because the shot peening was performed in air. Thus, several metastable nitride, oxide and hydride phases can be formed. The presence of these interstitial elements in the α_2 and γ phases may favour their amorphisation. Unfortunately, the nature of the crystalline grains embedded into the amorphous phase could not be determined, but it might be speculated that they are oxide, nitride or hydride phases. The formation of one of these crystalline surface phases could be an intermediate state before amorphisation eventually starts. This is suggested by two observations. The crystalline surface phase contains a high density of dislocations, which are often arranged in dipole or multipole configurations. The compressed image reveals significant bending of the lattice planes, which suggests high internal stresses. Another remarkable feature of the crystalline surface phase is antiphase boundaries (APB), as seen in Fig. 29. The APB's provide a local loss of order and are often associated with adjacent amorphous phase. Thus, it is speculated that the formation of APB's represents the initial stage of amorphisation.

Taken together, theses factors apparently make the crystalline surface phase prone to further structural changes, which could be directly observed in the electron microscope. Figure 30 demonstrates the transformation of the crystalline phase by a couple of micrographs. The slurry contrast in the micrograph on the left hand side indicates that the structure was about to transform into the structure shown on the right hand side. The observed conversion of the structure usually involved a very small volume of several ten nanometres and often occurred within a few ten seconds.

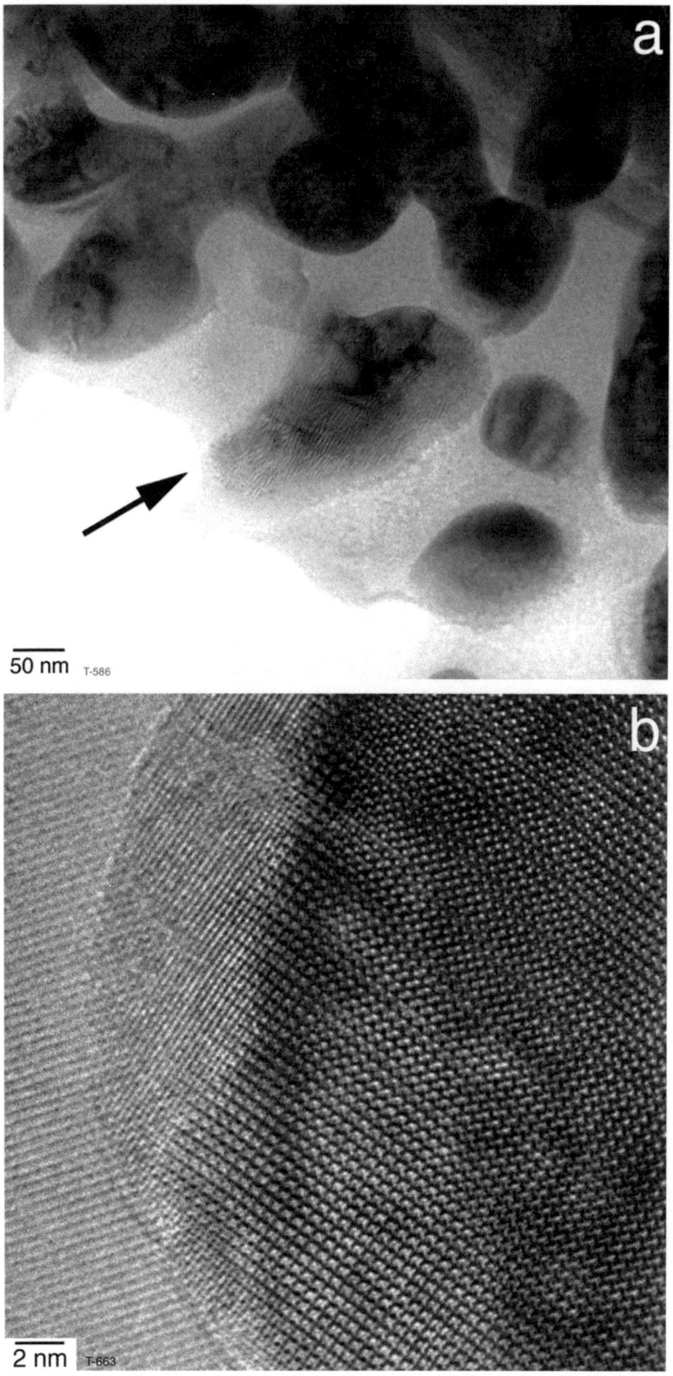

Fig. 27. Partial amorphisation in the shot peened surface layer. Lamellar Ti-45Al-10Nb, shot peened at room temperature with an Almen intensity of 0.4 mm N. (a) Crystalline grains embedded in amorphous phase. (b) Gradual loss of crystallinity of a grain adjacent to the amorphous phase.

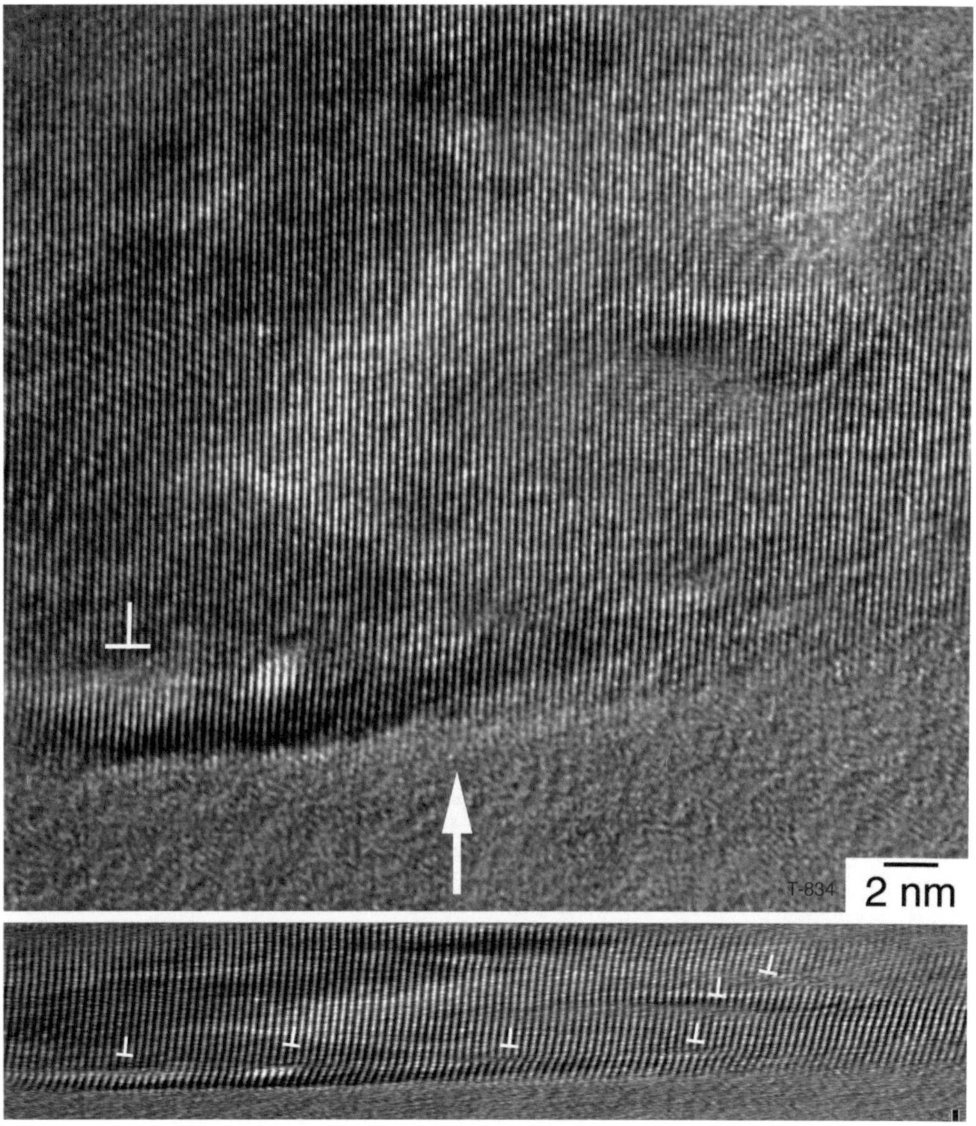

Fig. 28. Misfit between the amorphous and crystalline surface phases produced by shot peening. Note the arrangement of like dislocations at the interface. The compressed image below shows these details more clearly. Experimental conditions as for Fig. 27.

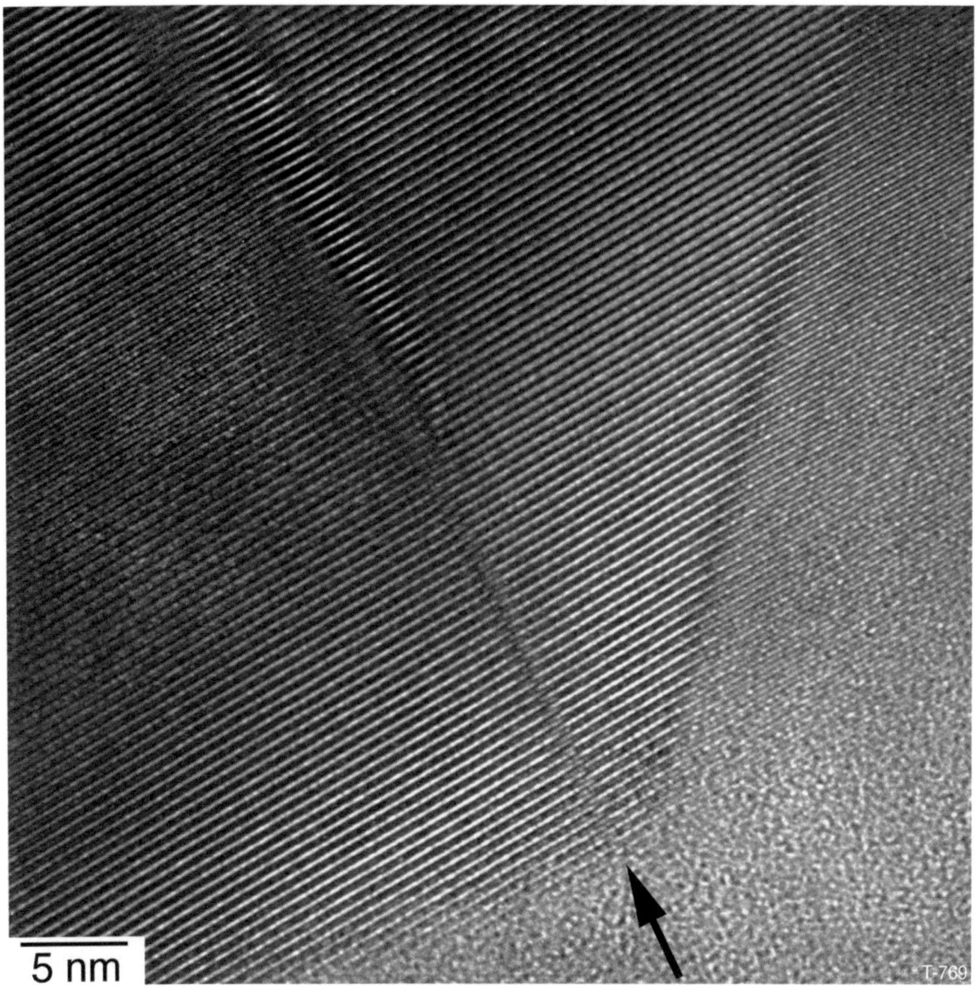

Fig. 29. An antiphase boundary (arrowed) in a crystalline grain embedded into the amorphous surface layer produced by shot peening. Experimental conditions as for Fig. 27.

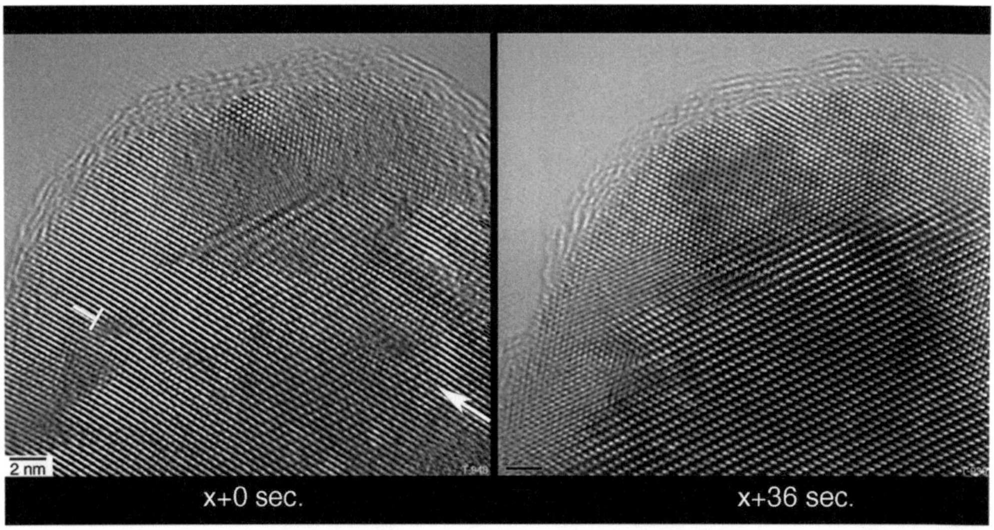

Fig. 30. Recrystallization of the crystalline surface phase observed in situ in a high-resolution transmission microscope (acceleration voltage 300 kV). Experimental conditions as for Fig. 27.

7. Conclusions

As with other metals recrystallization in multiphase titanium aluminide alloys is triggered by heterogeneities in the deformed state. However, there are several specifics that involve the following features.

The deformation heterogeneities are formed by a significant plastic anisotropy of the majority phases $\gamma(TiAl)$ and $\alpha_2(Ti_3Al)$, twin intersections and elastic buckling of lamellar constituents.

The release of strain energy by recovery seems to be relatively easy and probably retards the recrystallization kinetics.

In the evolution of the microstructure concurrent phase transformations occur that are primarily initiated by a non-equilibrium constitution; other factors supporting phase transformations are local mechanical stresses or chemical driving pressure due contamination with gaseous elements. The combination of these effects can give rise to solid state amorphisation.

The complex interactions between deformation, recovery and phase transformation are of great importance for processing and service of TiAl alloys.

8. Acknowledgments

The author acknowledges the continuous support of St. Eggert, D. Herrmann, U. Lorenz, M. Oehring, and J. Paul, from the Helmholtz Zentrum Geesthacht, Germany. Thanks are due to Th. Heckel, A. El-Chaikh and H.-J. Christ from Universität Siegen, Germany, for performing the fatigue experiments. The financial support by the Deutsche Forschungsgemeinschaft (Projects AP 49/5 and AP 49/4-6) is gratefully acknowledged.

8. References

[1] F. Appel, J.D.H. Paul and Michael Oehring, *Gamma Titanium Aluminide Alloys - Science and Technology* (Wiley VCH, Weinheim, 2011).
[2] M. Yamaguchi, H. Inui and K. Ito, Acta Mater. 47, 307 (2000).
[3] C. McCullough, J.J. Valencia, C.G. Levi, and R. Mehrabian, *Acta Metall.* 37, 1321 (1989).
[4] P. M. Hazzledine, Intermetallics 6, 673 (1998).
[5] M.J. Blackburn, in: R.I. Jaffe, N.E. Promisel, eds. *The Science Technology and Applications of Titanium*, Pergamon, Oxford 1970.
[6] U. Fröbel and F. Appel, Acta Mater. 50, 3693 (2002).
[7] H.I. Aaronson, Metall. Trans. A, 24A, 241 (1993).
[8] F.C. Frank and J.H. van der Merwe, Proc. Roy. Soc. Series A - Mathematical and Physical Sciences A 198, 205 (1949).
[9] M.G. Hall, H.I. Aaronson and K.R. Kinsman, Surf. Sci. 31, 257 (1972).
[10] R.C. Pond, in: *Dislocations in Solids, Vol. 8*, ed. F.R.N. Nabarro (North-Holland, Amsterdam, 1989), p.1.
[11] J.M. Howe, R.C. Pond and J.P. Hirth, Progr. Mater. Sci. 54, 792 (2009).
[12] L. Zhao and K. Tangri, Acta Metall. Mater. 39, 2209 (1991)
[13] S.R. Singh and J. M. Howe, Phil. Mag. A 66, 739 (1992).
[14] S. Rao, C. Woodward and P. Hazzledine, in: *Defect Interface Interactions*, Materials Research Society Symposium Proceedings, Vol. 319, eds. E.P. Kvam, A.H. King, M.J. Mills, T.D. Sands, and V. Vitek (MRS, Pittsburgh, PA, 1994), p. 285.
[15] P. Shang, T.T. Cheng and M. Aindow, Phil. Mag. A 79, 2553 (1999).
[16] R.C. Pond, P. Shang, T.T. Cheng, and M. Aindow, Acta Mater. 48, 1047 (2000).
[17] P. Shang, T.T. Cheng and M. Aindow, Phil. Mag. Lett. 80, 1 (2000).
[18] J.P. Hirth and J. Lothe, *Theory of Dislocations* (Krieger, Melbourne, 1992).
[19] P.M. Hazzledine, B.K. Kad, H.L. Fraser, and D.M. Dimiduk, in: *Intermetallic Matrix Composites II*, Mater. Res. Soc. Symp. Proc. Vol. 273, eds. D.B. Miracle, D.L. Anton, J.A. Graves (MRS, Pittsburgh, PA, 1992), p.81.
[20] F. Appel and U. Christoph, Intermetallics 7, 1173 (1999).
[21] B. Shoykhet, M.A. Grinfeld and P.M. Hazzledine, Acta Mater. 46, 3761 (1998).
[22] F. Appel, J.D.H. Paul and M. Oehring, Mater. Sci. Eng. A 493, 232 (2008).
[23] E. Abe, T. Kumagai and M. Nakamura, Intermetallics 4, 327 (1996).
[24] R. Ducher, B. Viguier and J. Lacaze, Scripta Mater. 47, 307 (2002).
[25] A.K. Gogia, T.K. Nandy, D. Banerjee, T. Carisey, J.L. Strudel and J.M. Franchet, Intermetallics 6, 741 (1998).
[26] T. Haibach and W. Steurer, Acta Crystallographica A, 52, 277 (1996).
[27] D. Nguyen-Manh and D.G. Pettifor, in: *Gamma Titanium Aluminides 1999*, eds. Y-W. Kim, D.M. Dimiduk and M.H. Loretto (TMS, Warrendale, PA, 1999), p. 175.
[28] M.H. Yoo and J. Zou, C.L. Fu, Mater. Sci. Eng. A 192-193, 14 (1995).
[29] V. Seetharaman and S.L. Semiatin, Metall. Trans. A, 27A, 1987 (1996).
[30] Y-W. Kim and D.M. Dimiduk, in: *Structural Intermetallics 1997*, eds. M.V. Nathal, R. Darolia, C.T. Liu, P.L. Martin, D.B. Miracle, R. Wagner, and M. Yamaguchi (TMS, Warrendale, PA, 1997), p. 531.
[31] F. Appel and R. Wagner, Mater. Sci. Eng. R22, 187 (1998).
[32] Y. Mishin and Chr. Herzig, Acta Mater. 48, 589 (2000).
[33] M.H. Yoo and C.L. Fu, Metall. Mater. Trans. A, 29A, 49 (1998).

[34] Y. Minonishi, Mater. Sci. Eng. A 192-193, 830 (1995).
[35] Y.Q. Sun, P.M. Hazzledine and J.W. Christian, Phil. Mag. A 68, 471 (1993).
[36] Y.Q. Sun, P.M. Hazzledine and J.W. Christian, Phil. Mag. A 68, 495 (1993).
[37] S. Wardle, I. Phan and G. Hug, Phil. Mag. A 67, 497 (1993).
[38] J.W. Christian and S. Mahajan, Progr. Mater. Sci. 39, 1 (1995).
[39] F. Appel, Phil. Mag. 85, 205 (2005).
[40] F. Appel, U. Sparka and R. Wagner, Intermetallics 7, 325 (1999).
[41] R. M. Imayev, V.M. Imayev, M. Oehring, and F. Appel, Metall. Mater. Trans. A, 36A, 859 (2005).
[42] T. Fujiwara, A. Nakamura, M. Hosomi, S.R. Nishitani, Y. Shirai, and M. Yamaguchi, Phil. Mag. A, 61, 591 (1990).
[43] Th. Schaden, F.D.Fischer, H. Clemens, F. Appel, and A. Bartels, Adv. Eng. Mater. 8, 1109 (2006).
[44] R.A. Brockman, Int. J. Plasticity 19, 1749 (2003).
[45] J. Beddoes, W. Wallace and L. Zhao, Int. Mater. Rev. 40, 197 (1995).
[46] M. Oehring, F. Appel, P.J. Ennis, and R. Wagner, Intermetallics 7, 335 (1999).
[47] T. Furuhara, J.M. Howe and H.J. Anderson, Acta Metall. Mater. 39, 2873 (1991).
[48] S.R. Singh and J.M. Howe, Phil. Mag. Lett. 65, 233 (1992).
[49] E. Abe, S. Kajiwara, T. Kumagai, and N. Nakamura, Phil. Mag. A 75, 975 (1997).
[50] R.W. Cahn, in: *High Temperature Aluminides and Intermetallics*, eds. S.H. Whang, C.T. Liu, D.P. Pope, and J.O. Stiegler (TMS, Warrendale, PA, 1990), p. 245.
[51] F.J. Humphreys and M. Hatherly, *Recrystallization and Related Annealing Phenomena* (Pergamon, Oxford, 1995).
[52] F. Appel, U. Christoph and M. Oehring, Mater. Sci. Eng. A 329-331, 780 (2002).
[53] B. Chalmers and H. Gleiter, Phil. Mag. 23, 1541 (1971).
[54] D. Hu, A.B. Godfrey and M. Loretto, Intermetallics 6, 413 (1998).
[55] G. Hénaff and A.-L. Gloanec, Intermetallics 13, 543 (2005).
[56] F. Appel, Th. Heckel and H.J. Christ, Int. J. Fatigue 32, 792 (2010).
[57] D. Herrmann and F. Appel, Metall. Mater. Trans. A, 40A, 1881 (2009).
[58] U.R. Kattner, J.C. Liu and Y.A. Chang, Metall. Trans. A, 23A, 2081 (1992).
[59] S.P. Godfrey, P.L. Threadgill and M. Strangwood, in: *High-Temperature Ordered Intermetallic Alloys VI*, Materials Research Society Symposia Proceedings, Vol. 364, eds. J.A. Horton, I. Baker, S. Hanada, R.D. Noebe, and D.S. Schwartz (MRS, Pittsburgh, PA, 1995), p. 793.
[60] J. Lindemann, C. Buque and F. Appel, Acta Mater. 54, 1155 (2006).

11

The Effect of Strain Path on the Microstructure and Mechanical Properties in Cu Processed by COT Method

Kinga Rodak
Silesian University of Technology Katowice, Poland

1. Introduction

In recent years a lot of place in research has been devoted to the methods of grain refinement with the aid of large plastic deformations- *Severe Plastic Deformation.* The research of the methods of grain refinement is carried out in parallel to the intensive research of the structural changes occurring in the deformed materials. Some metallic materials which are deformed by means of *Severe Plastic Deformation* are characterized by ultrafine-grained and sometimes even nanograined size. Grain refinement fosters, above all, the increase in the strength of a material. Therefore, a production of nano- and ultafine-grained structures has become one of the most important research issues which are currently the subject of interest in many research facilities. There are many techniques of SPD, for example: *Equal Chanel Angular Pressing* (ECAP), *High Pressure Torsion* (HPT), *Cyclic Extrusion Compression* (CEC), *Hydroextrusion* (HE), the KOBO methods, *Acumulative Roll Bonding* (ARB). The SPD techniques cannot be described as easy ways of obtaining the refined materials mainly because of the insufficient homogeneity of the structure, the low efficiency of the applied methods, and substantial losses of the material [Pakieła, 2009; Olejnik, 2005; Cao 2008]. That is why, the commonly know methods of SPD, excluding some exceptions [Bochniak, 2005], have not become implemented as production technologies, although a lot of time has passed since they were discovered. This is the reason why more and more works concentrate on using SPD techniques to produce a structure that is more homogenous and has higher degree of refinement [Shaarbaf, 2008; Raab, 2004]. There also appear some suggestions to modify SPD techniques [Lugo, 2008; Kulczyk, 2007]. Some intensive research is carried out aiming at preparing new SPD techniques which would have refined a grain to the level of ultrafine-grained or even nano-grained. One of such methods is the Compression with Oscillatory Torsion (COT). The method is regarded as an unconventional way of volume shaping and is developed by the Department Of Materials Technology in the Faculty of Materials Engineering and Metallurgy of the Silesian University of Technology. This method has become recognized mainly as a method that enables deformation of the materials to values of large plastic deformations [Pawlicki, 2007], therefore, it is possible to obtain a refined structure. The benefits from applying the COT method are visible mainly in the aspect of reduction of work hardening effects (lowering of the plastic deformation work) [Grosman, 2006] and formation of a particular type of a spatial configuration of defects in deformed material [Rodak, 2007]. It is, therefore, worth-exploring because of the attractive perspectives of using this method to form ultrafine-grained structures by using a suitable combination of deformation parameters (change of deformation path). It was proven that grain refinement in the COT method happens when the deformation parameters are suitably chosen. Among the parameters there are: torsional frequency *f*, compression velocity *v*, absolute strain Δh, and torsion angle α. An interesting observation coming from the longstanding research on the changes of the structures that accompany the COT torsion, is the fact that the effective deformation ε_f cumulated in a material is not the most important parameter thanks to which obtaining the suitably refined grains is possible.

2. Compression with oscillatory torsion (COT) method and deformation path

Compression with oscillatory torsion is a new method of plastic deformation in which the material is deformed as an effect of a changing deformation path. Fig. 1 is a schematic presentation of the COT set up. The facility for compression with oscillatory torsion consists of upper and lower anvils made from high-strength tool steel. Torsional straining was achieved by rotating the lower anvil, and compression was simultaneously achieved by linear strain from the lower anvil.

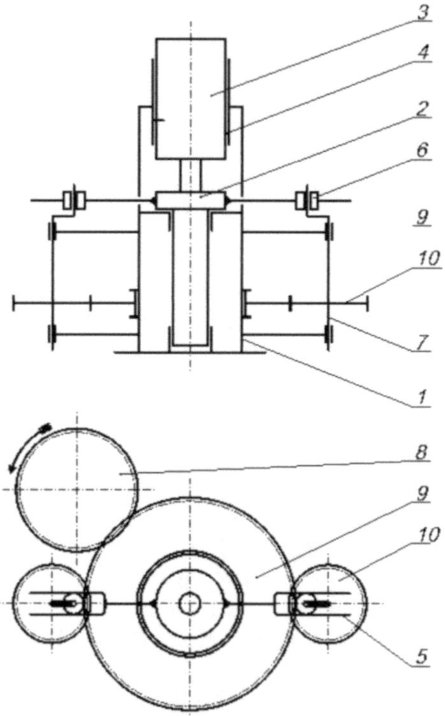

Fig. 1. Schematic illustration of COT process: 1. Frame, 2. Lower punch, 3. Upper punch, 4. Non-rotating slidable bearing, 5. Lower punch arm, 6. Roller, 7. Crankshaft, 8. Driving gear, 9. Ring gear, 10. Gear wheel

The appliance allows for the following parameters to change:

- the compression velocity *v*, (the velocity of the lower punch shift). The maximal value of compression velocity is 0,6 [mm/s]
- the torsional frequency *f*. The frequency of the lower punch oscillation is regulated by the inverter ranging from 0 Hz to 1,8 [Hz].
- the torsional angle amplitude α[°]. The set points of the kinematic magnitudes enable the change of the torsional angle ranging from 0° to 6°
- the absolute strain Δh [mm]
- the compression force F [kN]

The compressive force F and the deformation path are registered by computer. The process can be carried out only at room temperature.

The total cumulated value of deformation equals the sum of deformation dimensions obtained in consecutive cycles of torsion with simultaneous compression, for the following parameters: torsional velocity *v*, reversal torsional frequency *f*, amplitudes of the torsional angle α, the initial height of the test piece h_o, the diameter of a test piece d_o.

The total equivalent deformation ε_f is [Grosman, 2006]:

$$\varepsilon_f = \varepsilon_h + \varepsilon_t \tag{1}$$

where:

ε_h – the effective deformation from compression
ε_t - the effective deformation from torsion

This method regarded as an unconventional way of forming, has become recognized above all as an effective way to lower the force of plastic forming of a material. The force characteristics of the process depend on the component deformations induced by the torsion and compression. The value of the component deformations induced by the torsion is dependent on: torsional angle *a* and torsional frequency *f*. The component deformations induced by the compression are dependent on the compression velocity *v*. The proportions of the particular components describe the deformation path d_ε [Grosman, 2006].

For the constant value of the torsional angle amplitude (*a* = const) the deformation path is proportional to the value of the torsional frequency to compression velocity ratio.

The force characteristics $F = f(\Delta h)$ indicates the raise of the torsional frequency *f*, accompanied by the constant values of the other process parameters: the absolute strain Δh, the compression velocity *v* and the torsional angle *a*, has an impact on the decrease of the axial (compression) force level (Fig.2). The course of curves shows that the double raise of the torsional frequency influences the lowering of the compression force more than 1,5 times and, additionally, reduces the work hardening. Higher torsional frequencies, in connection with the increased compression velocities, have no greater influence on the change of the average unit pressure.

In the method of compression with oscillatory torsion, a various combinations of the parameters can be used in order to obtain comparable values of the effective deformations.

The table 1. shows an example set of the deformation parameters needed to obtain similar values of the total effective deformation ε_f ~5 and ~14. For the constant values of the Δh = 3 mm and the torsional angle α=6°, the proportions between the compression velocity *v* and the torsional frequency *f* were changed. The effective deformation ε_f and the axial forces of compression are close to the test carried on with *f* = 0,1 Hz, *v*=0,015 mm/s and *f* = 0,8 Hz, *v*=0,04 mm/s. Likewise, for the test with *f* = 0,2 Hz, *v*=0,015 mm/s and *f* = 1,6 Hz, *v*=0,04 mm/s similar values of an effective deformation ε_f and the axial forces of compression are obtained. Yet, with such parameters higher effective deformations and lower axial forces of compression are achieved.

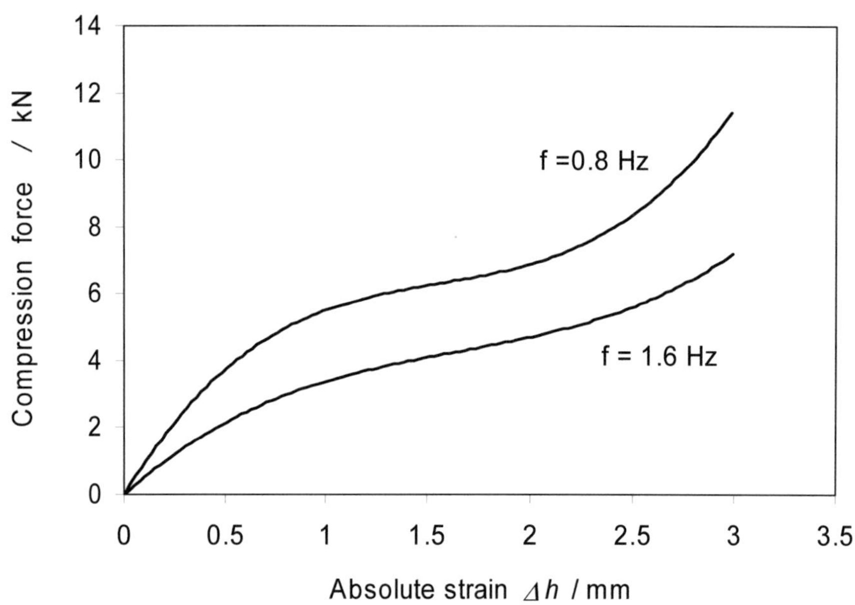

Fig. 2. Dependence of F=f(Δh) for samples 3 and 4

Sample	1	2	3	4
Parameters	Δh = 3 mm α=6° f= 0.1 Hz v=0.015 mm/s	Δh = 3 mm α=6° f= 0.2 Hz v=0.015 mm/s	Δh = 3 mm α=6° f= 0.8 Hz v=0.04 mm/s	Δh = 3 mm α=6° f= 1.6 Hz v=0.04 mm/s
Effective strain, ε_f	~5	~14	~5	~14

Table 1. Deformation parameters of Cu

The structural effects produced by the deformation with such parameters are interesting (Fig.3). Because of the possibility of using the COT method to refine the grain, more effective process is the one of deformation implemented by the use of higher parameters *f* and *v*, even for the comparable values of the effective deformations and the values of the axial forces of compression.

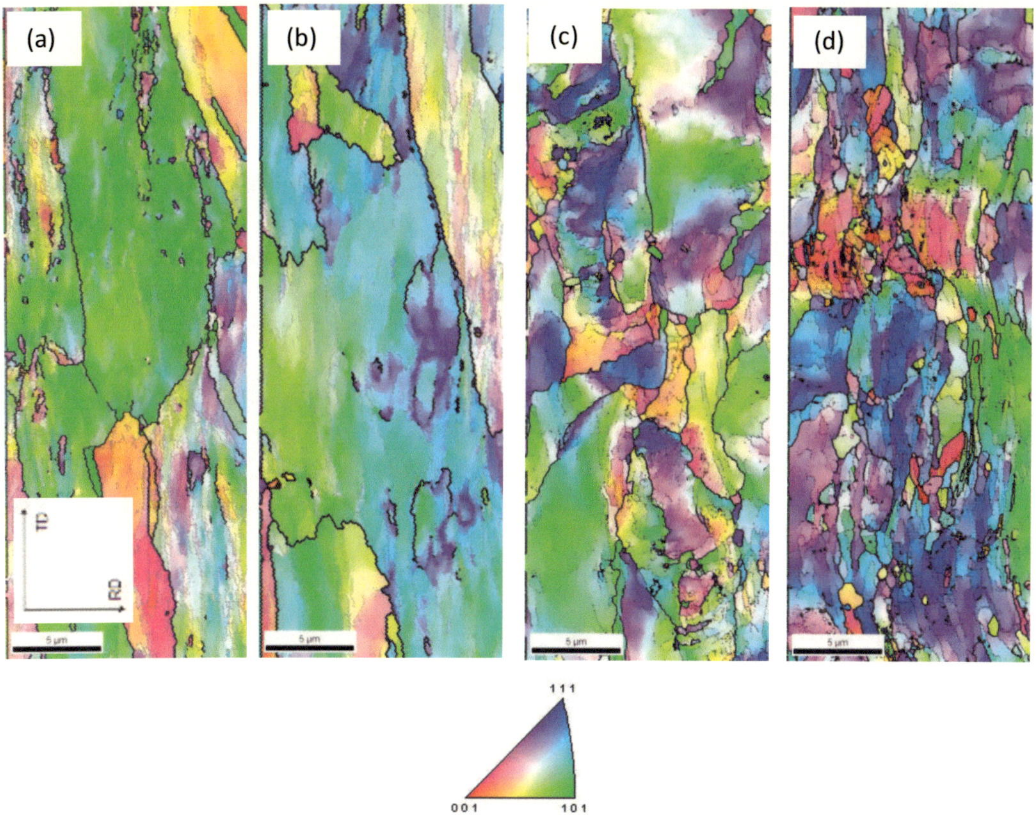

Fig. 3. EBSD maps of Cu after COT processing: a) sample 1, b) sample 2, c) sample 3, and d) sample 4

The deformation realized using high torsional frequencies evoked a decrease of the axial forces of compression in the material (Fig.2.) and the desired structural effect in a form of the grain refining (Fig.3). A similar correctness in a form of structural effects, was not observed in the case of deformable samples with small oscillating frequencies, in spite of the fact that also in this case a comparable decrease of the axial force level was registered. The conclusion might be that the effect of the decrease of the axis force and the increase of the effective deformation is not adequate with the structural effect in a form of grain refining.

In the COT method only the „special" conditions of the process (the proportions of the deformation parameters) can guarantee the effective grain refining. Therefore, the selection of the appropriate deformation parameters guarantee the effective grain refining.

3. Experimental details

The tests were conducted on the samples from copper in the M1E species (chemical composition is shown in the Table.2). This material is eagerly used to refine grains using the SPD techniques therefore, in case of new SPD techniques can constitute a good comparative material especially when it comes to the effectiveness of grain refining as well as to the mechanisms of grain refining process.

Chemical composition [%]						
Cu	Fe	Bi	Pb	Ni	Sn	As
99,8	0,02	0,002	0,008	0,019	0,03	0,001

Table 2. Chemical composition Cu used in experimental

The samples for the tests were taken from the bars having 12 mm in diameter and then, they were exposed to the heat treatment which involves the annealing in temperature of 500°C /2 hours. After this treatment the average diameter of the grain equaled 50 μm. The heating treatment that was carried out, allowed for eliminating structural effects resulting from the previous technological treatments and for obtaining the homogenous grain structure in the whole volume of the material.

The test of compression with oscillatory torsion, required two types of samples. The first series of samples was prepared according to the scheme shown in the Fig. 4, where the ratio of height *h* (6 mm) to the diameter *d* (6 mm) equaled (*h/d*) =1. Whereas, in the second set the height *h* was increased to 9 mm without the change of diameter and this equaled: (*h/d*) =1,5.

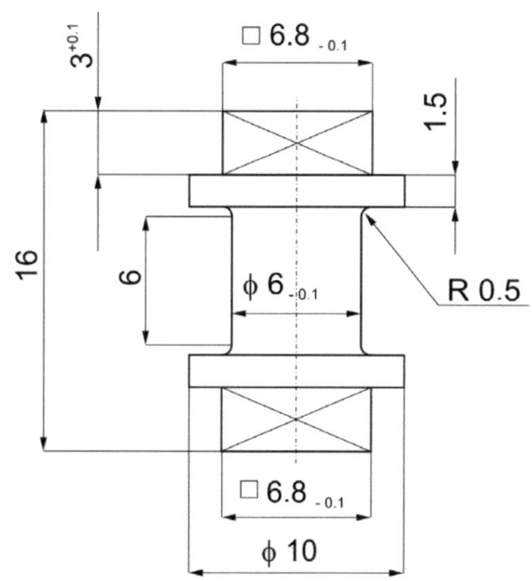

Fig. 4. Geometry and dimensions of samples used in the experiment

The deformation by means of the COT method was conducted using the following parameters:

- the range of the compression velocity; 4 variants of velocity were used: 0,015 mm/s; 0,04 mm/s; 0,1mm/s, 0,6 mm/s.

- the constant values of the torsional angle α=6°.

- the range of torsional frequency; 4 variants of this parameter were used: 0,2 Hz, 0,8 Hz, 1,6 Hz and 1,8 Hz.

- the absolute strain was Δh = 3 mm and Δh = 7 mm correspondingly for (*h/d*) =1 and (*h/d*) =1,5.

- the compression force F equaled 300 kN.

The compression with oscillatory torsion method is regarded as a method characterized by the heterogeneity of deformation. The most intense deformations proceed in places that are the nearest to the lateral surfaces of the material, which is results from the functioning of the torsional moment. The heterogeneity of the plastic deformation in the sample, causes the occurrence of a considerable differentiation of the structure in its sectional view. Because of the heterogeneity of the deformation, the microscope observations and strength testing were carried out in spheres located in a distance of about 0,8 of the sample's ray (Fig.5).

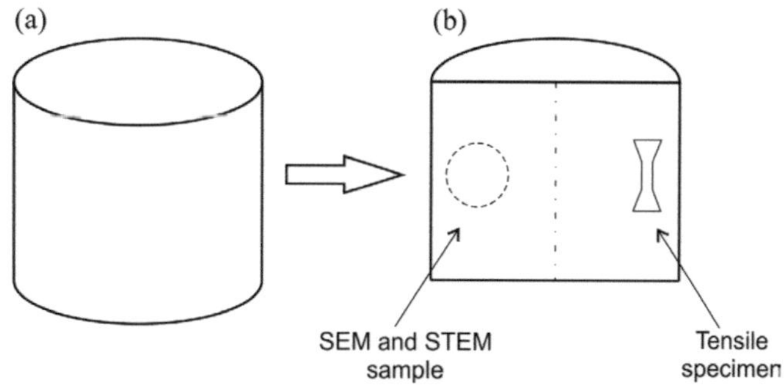

Fig. 5. Simple schematic of a specimen without the collar and handle: (a) sample after processing and (b) procedures for different types of testing: SEM and STEM observations and tensile testing using miniature specimens. The structure and mechanical properties studies were conducted on samples extracted from a distance near 0,8 radius in the longitudinal plane section

The analysis of the dislocating structure was carried out using the Scanning Transmission Electron Microscopy (STEM) technique which was applied thanks to the microscope Hitachi HD 2100A equipped with the FEG gun, working at the accelerating voltage of 200 kV.

With the help of Transmission Electron Microscopy (TEM) Jeol 100B, the orientation of the grains were determined based on the received pictures of Kikuchi lines. For the calculations the KILIN programme was used that was developed on the University of Science and Technology in Cracow.

The detailed quantities research of the ultrafine-grained structures being formed was conducted using Scanning Electron Microscope (SEM) INSPECT F produced by FEI equipped with the gun with cold field emission and the detector of electron back scattering diffraction (EBSD). In order to release the structure of the material by using the SEM/EBSD method, firstly, the mechanical polishing was used and then, electrolytic.

On the basis of the SEM/EBSD method the average equivalent diameter of the subgrains d [μm] and the average equivalent diameter of the grains D [μm] were determined. The boundary between the grain and subgrain was determined on the basis of the misorientation angle measurement. The divisional boundary was an angle equaling 15°. In the measurements of the average diameter of grain/subgrain, the grains located in the periphery of the image were not considered.

Some measurements of misorientation angles between the neighboring subgrains/grains were also taken. Therefore, the fields that had the misorientation degree from 2° were analyzed.

The estimation of the dislocation density in the material that were strongly deformed by means of transmission electron microscopy is almost impossible when these dimensions exceed $5 \cdot 10^{14}$ m^{-2}, and the result have a doubtful statistic value because of the small area of the analysis. Those inconveniences can be avoided by using the X-ray structural analysis.

The analysis of the diffraction line profile is an effective tool needed to characterize the structures of the material which have defects in crystal lattice. The presence in the material of tiny crystallites and the lattice distortions, causes the widening of diffraction reflexes. Both the size of the crystallites and the lattice distortions can be determined in quantity by means of the Williamson-Hall method in which the basis of analysis in this method is the Full Width at Half Maximum (FWHM) of the reflex, and using the Warren-Averbach method based on analyzing the Fourier coefficient [Ungar, 2001].

The difficulties in analysis using these methods occur when the material consists of anisotropy of diffraction line broadening. In such a case, neither the half-value line nor the Fourier coefficient constitute the monotonic function of the diffraction vector $K = (2sin\theta)/\lambda$, where θ - reflection angle, λ - X-ray wavelength. The replacement of K and K^2 in a classic procedure of Williamson-Hall and Warren-Averbach by the expression $K^{1/2}$ and K^2 allows for eliminating the influence of anisotropy of diffraction line broadening.

The measurements of the hardness were taken by the Vickers method using the hardness testing machine FM 700 produced by the Japanese company Future Tech coupled with the metallographical microscope. The test was carried out with the loading of HV0,2.

Applying the hardness measurement, when the small volumes of the material, usually with heterogeneous deformation are obtained in the laboratory tests, is fully justified. However, the method that describes the strength and plastic features is the static tensile test. In the case of a very small dimension of the sample obtained by the SPD methods, the tensile test can be conducted only on the micro samples. Using micro samples is connected with numerous technical inconveniences among which there are: the danger of implementing structural changes resulting from the preparation of the material for the tests, fastening of the samples and the axiality of the meter circuit. In the tests that are described here, the micro samples have the dimensions 1.20 x 0.3 and length 2.20. The cutting was made using the spark wire (the wire was 0,1mm thick) with an intense cooling in distilling water.

The tensile test was carried out by employing the universal testing machine with the screw drive equaling MTS QTest/10. This method is successfully introduced in measuring deformations of small samples by the Faculty of Materials Science and Engineering in the Warsaw University of Technology. The elongation of the samples is measured using the digital image correlation method. The method is based on comparing of the digital recording of the sample image before the deformation (the reference image) with the digital recording of the sample image after the deformation. The computer algorithm compares the images and describes the relocate of small areas within the tested surface. Thanks to this method the local and total deformations on the whole analyzed surface of the sample can be determined. The tensile test was conducted with the initial tension velocity equaling $2*10^{-3}$ [1/sec].

4. The impact of the deformation parameters on the grain refining

4.1 The torsional frequency

Among the several changing parameters of the deformation used in the COT method, the torsional frequency is the factor which, when increased, has a beneficial influence on lowering the value of the compression force (Fig.2.). Moreover, the increase of this parameter causes that the material is deformed to large values of effective deformations. For example, when the torsional frequency is 0,2 Hz and 1,8 Hz with the constant parameters: $\Delta h=7$, $\alpha=6°$, v=0,015mm/s; the effective deformation ε_f equals 15 and 130. The Cu microstructure with the increasing torsional frequency was presented in Fig.6a.

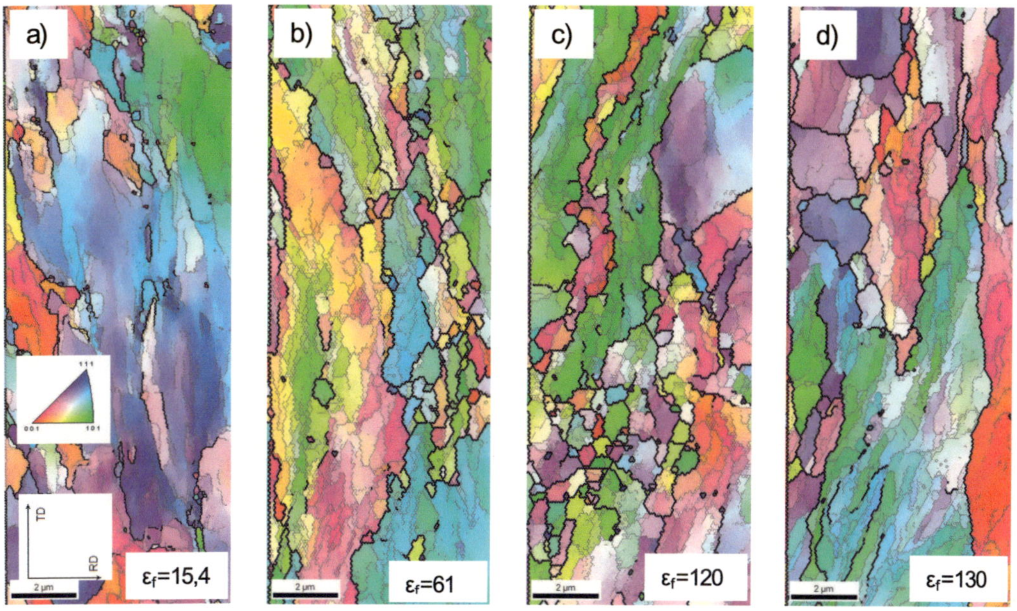

Fig. 6. EBSD maps of Cu after COT processing: a) f=0,2 Hz; b) f=0,8 Hz; c) f=1,6 Hz; d) 1,8 Hz at constant parameters: α =6°, v=0,015 mm/s i Δh=7 mm

The deformed copper with low torsional frequencies - 0,2 Hz is characterized mainly by the boundaries with a small misorientation angle. The boundaries like HABs are seen in the fragmentary outline.The Cu deformation with the increasing torsional frequency to 1,6 Hz, allows for the formation of the equiaxed structures having a great share of wide-angle boundaries (Fig.6b,c.). The maps obtained using the EBSD techniques show that the structures got during the deformation process where the torsional frequencies were - 0,8Hz and 1,6 Hz, are characterized by the considerable grain refining, especially when it comes to the deformation when the torsional frequency was 1,6 Hz. It was observed that the numerous grains had the size no bigger than 1 μm. The neighboring grains are characterized by comparable but yet diversified crystallographic orientation. The great part of the analyzed surfaces are the banding structures isolated by high-angle boundaries and elongated in accordance with the direction of the compression. The example that are shown here prove that even if the values of effective deformations ε_f=130 are high, the microstructure Cu is characterized by heterogeneity.

The deformation of the material when the torsional frequency is f= 1,8 Hz produces an inconvenient phenomenon of developing large grains which often exceed 1 μm.

This is proven by the results of the quantity tests of the grains and subgrains size (Fig.7), the area fraction of the grains with the misorientation above 15° and the size not exceeding 1 μm, (Fig.8) and misorientation angle (Fig.9) that are fulfilling the ultrafine-grained material criteria.

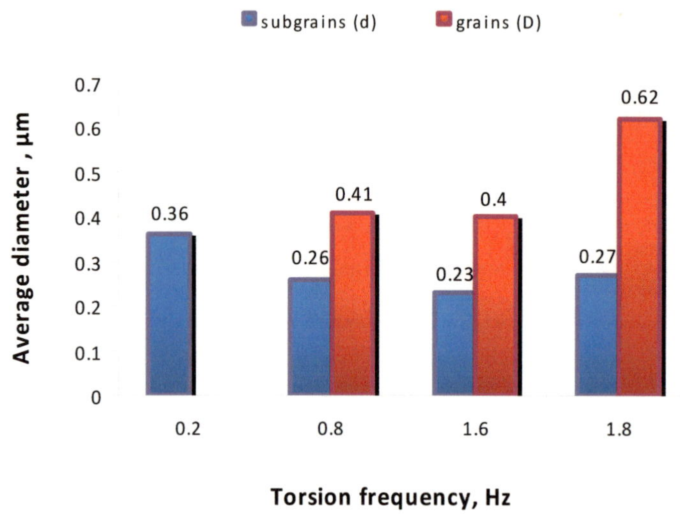

Fig. 7. The variation of the subgrain/grain size in dependence of changes in torsion frequency at constans: α =6°, v = 0,015 mm/s, Δh=7 mm

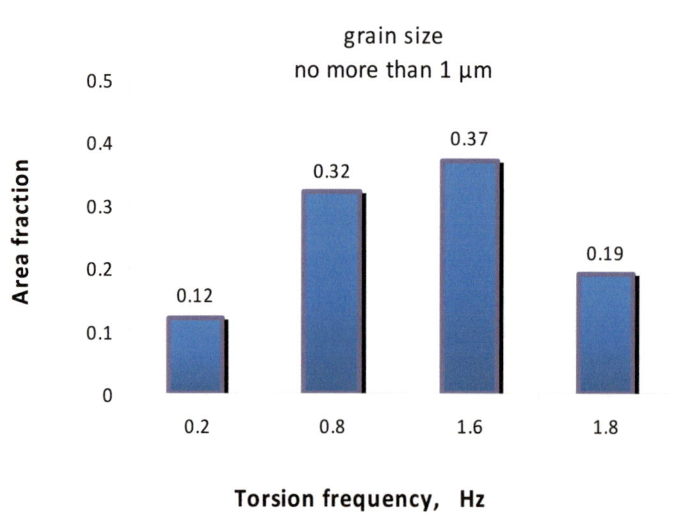

Fig. 8. The variation in the area fraction of ultrafine grain in dependence of changes in torsion frequency at constans: α =6°, v = 0,015 mm/s, Δh=7 mm

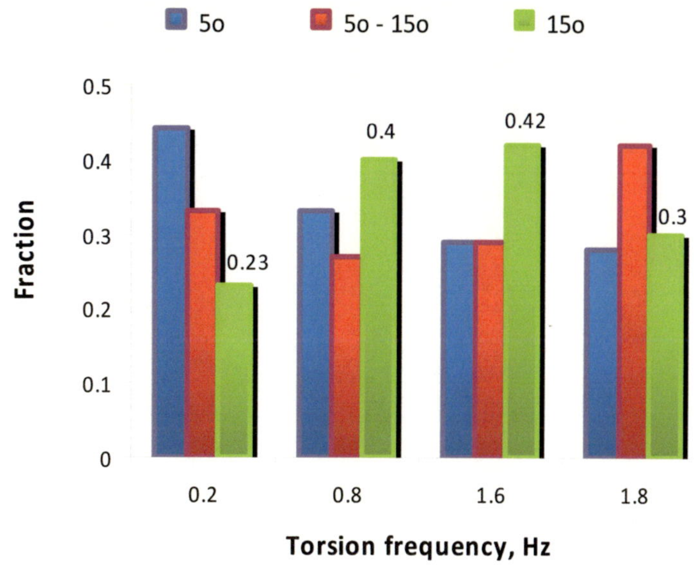

Fig. 9. The variation in the misorientation angle in dependence of changes in torsion frequency at constans: α =6°, v = 0,015 mm/s, Δh=7 mm

When the torsional frequency is 0,8 Hz and 1,6 Hz, the average diameters of the Cu grains have about 400 nm. A distinct increase of the average diameter of the grain is seen when the torsional frequency is increased from 1,6 Hz to 1,8 Hz. The grains have a size of about 600 nm. Roughly, it can be assumed that the subgrains are about 2 times smaller than grains. The area fraction of the grains to the magnitude of 1 μm, also appears to be beneficial for the torsional frequencies 0,8 Hz and 1,6 Hz (especially for 1,6 Hz) (Fig.8.). The Cu grains which have the ultrametric size occupy about 40% of the analyzed surfaces. The application of the high torsional frequencies during the deformation process allows for obtaining about 40% of the high- angle boundaries fractions (Fig.9). It means that more than a half of the deformed structure is dominated by the subgrains.

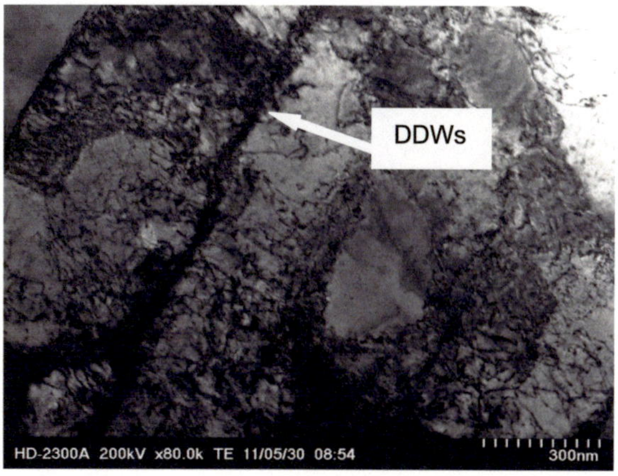

Fig. 10. Microstructure of Cu after COT deformation at: f=0,2 Hz, α =6°, v=0,015 mm/s and Δh=7 mm

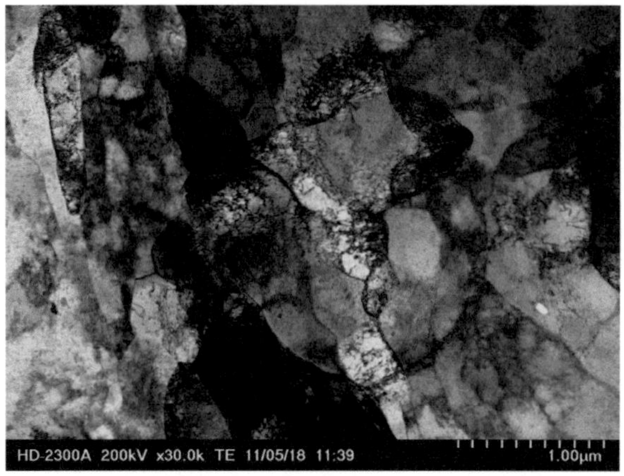

Fig. 11. Microstructure of Cu after COT deformation at: 11. f=0,8 Hz, α =6°, v=0,015 mm/s and Δh= 7 mm

After the COT deformation with the torsional frequency f=0,2 Hz, the dislocation cell structure and the DDWs dislocation walls with a high density of dislocation are observed (Fig.10). The effects of arranging the dislocation structure are seen between the particular DDWs (Fig.10). In general, deformation carried out when the torsional frequencies are small makes the fraction of the narrow-angle boundaries (up to 5°) reaching almost 50% (Fig.9.) and the area fraction of the ultrafine-grains does not exceed 20% (Fig.8.).

It was observed that when the torsional frequency increased, the grain division into smaller volumes happened more often (Fig.11). The result of this is the generation of the greater amount of the LABs dislocation boundaries. A great amount of the tested areas of thin foils, shows that after the deformation with frequencies f= 0,8 Hz and f=1,6 Hz in particular, the dislocation polygonal walls and the grains with the HAB boundaries are created in the structure (Fig.12). The vacancy clusters visible in Cu, are the effects of the dislocation annihilation (Fig.13). The results of measuring the dislocation density (Fig.14) are also the proof of the above mentioned.

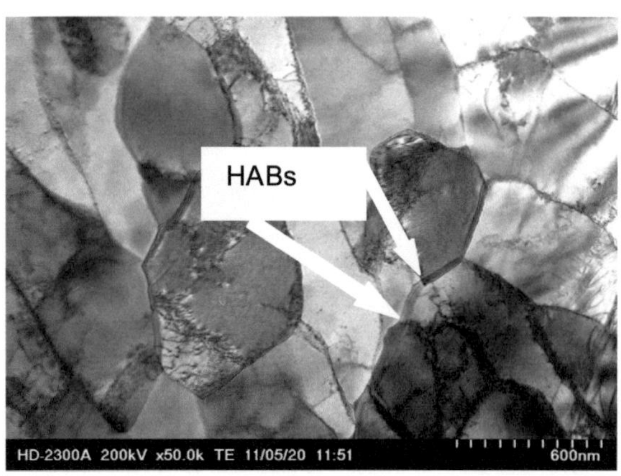

Fig. 12. Microstructure of Cu after COT deformation at: f=1,6 Hz, α =6°, v=0,015 mm/s and Δh= 7 mm

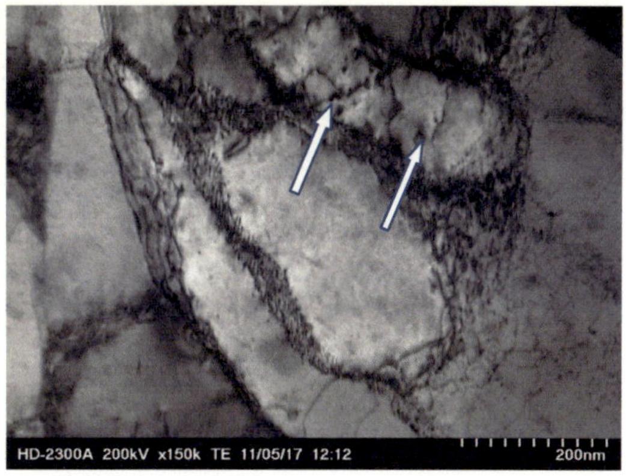

Fig. 13. Microstructure of Cu after COT deformation at: f=1,6 Hz, α =6°, v=0,015 mm/s and Δh=7 mm. The vacancy clusters are arrows marked

Fig. 14. Microstructure of Cu after COT deformation at: f=1,8 Hz, α =6°, v=0,015 mm/s and Δh=7 mm

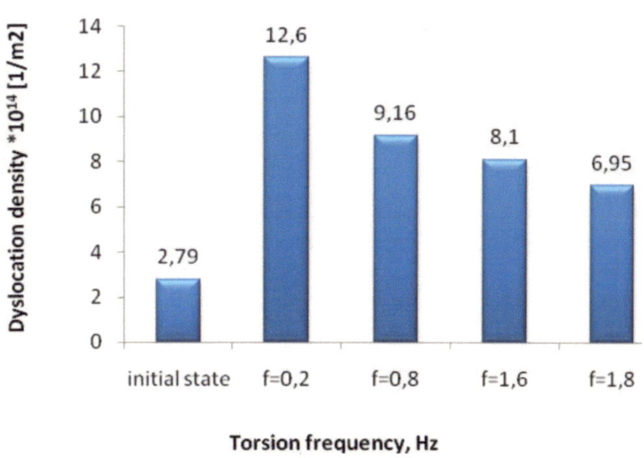

Fig. 15. The variation of the dislocation density in COTes samples with deferent value of torsion frequency and constant: α =6°, v=0,015 mm/s, Δh=7 mm

Together with the increase of the torsional frequency, the decrease of dislocation density is observed (Fig.15). When the frequency is 1,8 Hz the recovery begins to dominate the structure what has a negative influence on the material that is being refined (Fig.14). The transformation of dislocation tangles into polygonal boundaries and the dislocation annihilation, have no impact on raising the amount of the wide-angle boundaries fractions as well as on increasing the area fraction of the ultrafine-grains.

4.2 The compression velocity

In the COT process the compression velocity is the parameter which when increased, has an influence on highering the value of the compression force and on lowering the values of equivalent deformations. For example, the compression velocity 0,015 mm/s and 0,04 mm/s while the rest of the parameters is constant: 1,6 Hz, $\Delta h=7$, $\alpha=6°$; the effective deformation ε_f is 120 and 45. Applying even higher compression velocities i.e. v= 0,1 mm/s and v= 0,6 mm/s causes a significant decrease of effective deformation to the corresponding values $\varepsilon_f=12$ and $\varepsilon_f=2$.

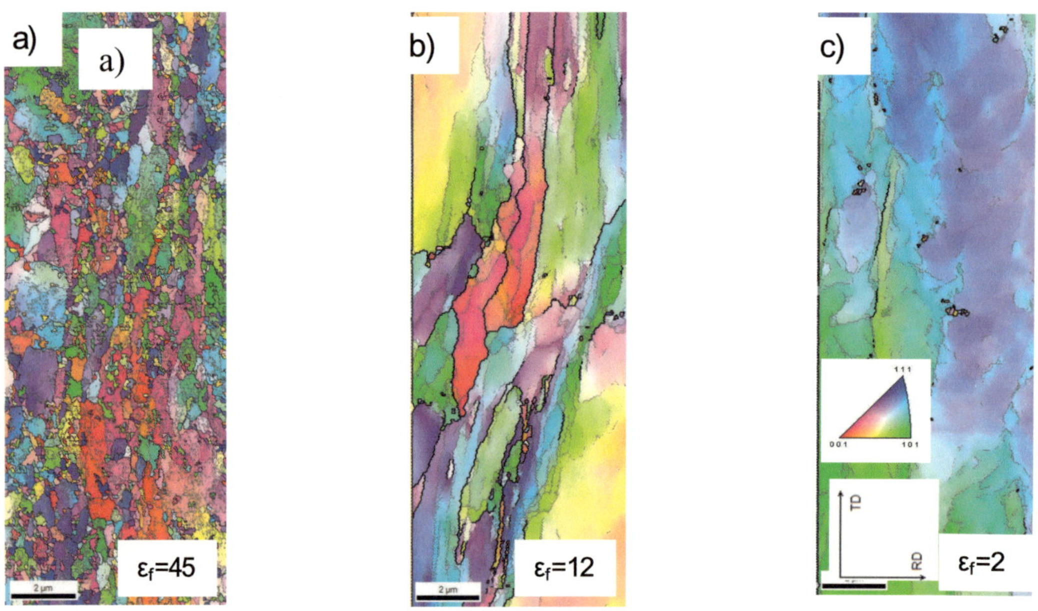

Fig. 16. EBSD maps of Cu after COT processing: a) v=0,04 mm/s; b) v=0,1 mm/s; c) v=0,6 mm/s; at constant parameters: $\alpha=6°$, v=0,015 mm/s i $\Delta h=7$ mm

The deformation realized when the compression velocity is increasing from 0,015 mm/s to 0,04 mm/s while the rest of the parameters is constant: 1,6 Hz, $\Delta h=7$, $\alpha=6°$; has an positive influence on the grain refining as well as on the increase of the homogeneity of the structure (compare Fig.6c and Fig.16a). The structures in the prevailing part of the areas are equiaxed, the grains bigger than 1 μm are not produced, the amount of high-angle boundaries is high. Whereas, the further increase of this parameter delays the reduction of the grain size and the creation of high-angle boundaries (Fig.16 b,c). During the deformation with the velocity of 0,1mm/s the effects of the structure deformation are clearly visible (Fig.16b), but when the compression has the velocity of 0,6 mm/s only the fragments of the near-angle boundaries attest to the insignificant deformation of the material (Fig.16c). A significant reduction of the deformation realized by the increase of the compression velocity, does not lead to the grain refining.

When it comes to the Cu refined-grain, the juxtaposition of the structural changes for two velocities: 0,015 mm/s and 0,04 mm/s as well as of different torsional frequencies is quite interesting (Fig.17). That is why the further structural analysis involved comparing some selected ways of deformation.

Comparing the EBSD images obtained for Cu after the deformation with the compression velocity of v= 0,015 mm/s and v= 0,04 mm/s, the constant torsional frequency 1,6 Hz and the rest of the parameters also being constant (Fig.6c and Fig.16a), it is seen that the deformation happening at higher compression velocity is an effective way of grain-refining. It is also proven by the results of the quantity tests shown in the Figs.18-20. The average diameter of the Cu grain after the increase of the compression velocity, decreases from 400 nm to 300 nm (Fig.18). The average diameter of the subgrains also decreases from 230 nm to 180 nm (Fig.18). The area fraction of the grains up to 1 µm constitute 60% which is advantageous, when the torsional frequency is 1,6Hz and the compression velocity is v= 0,04 mm/s (Fig.19).

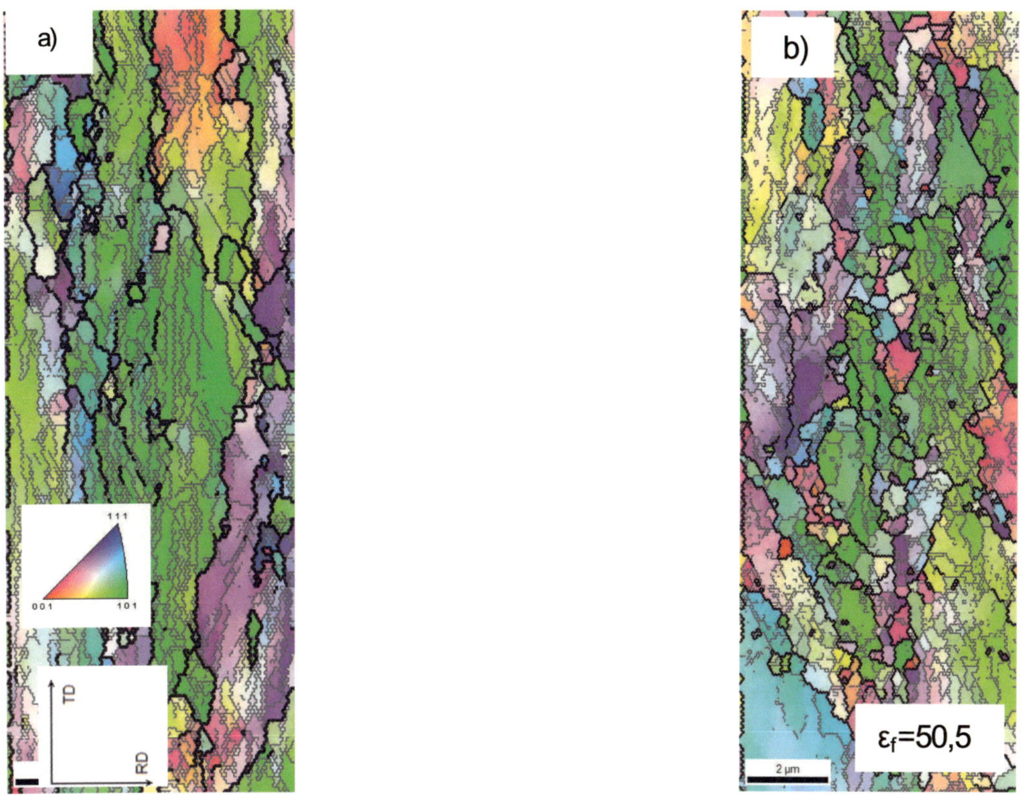

Fig. 17. EBSD maps of Cu after COT processing: a) f=0,8Hz; b) f=1,8Hz; at constant parameters α =6°, v=0,04 mm/s and Δh=7 mm

When the compression velocity is higher, the high-angle boundaries are easily created and they constitute about 50% of the analyzed Cu areas (Fig.20).

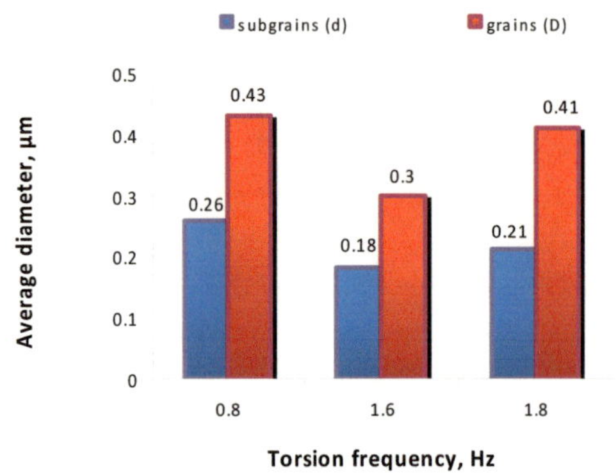

Fig. 18. The variation of the subgrain/grain size in dependence of changes in torsion frequency at constans α =6°, v = 0,04 mm/s, Δh=7 mm

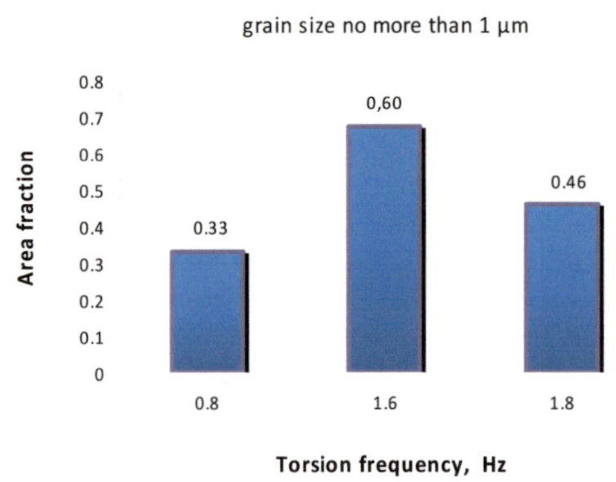

Fig. 19. The variation in the area fraction of ultrafine grain in dependence of changes in torsion frequency at constans: α =6°, v = 0,015 mm/s, Δh=7 mm

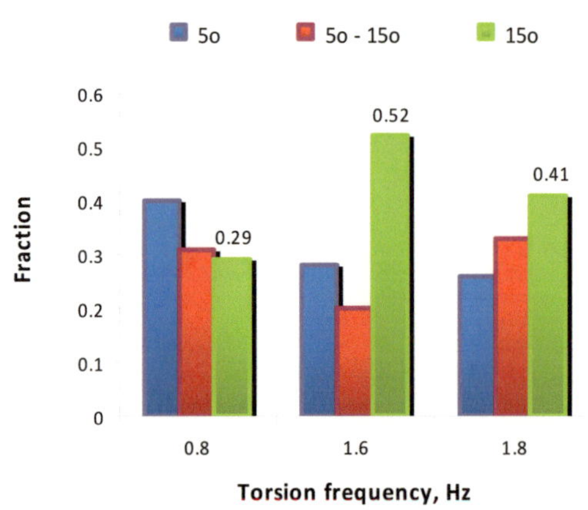

Fig. 20. The variation in the misorientation angle in dependence of changes in torsion frequency at constans: α =6°, v = 0,015 mm/s, Δh=7 mm

Comparing the structure of the deformed samples when the torsional frequency is lower - 0,8 Hz the compression velocities are as mentioned, it is visible that the Cu samples deformed with the higher compression velocity maintain greater fraction of the near-angle boundaries than the samples deformed with the lower compression velocity. The deformation realized when the compression velocity is v= 0,015 mm/s and v= 0,04 mm/s does not cause any significant changes when it comes to the size of the grains and subgrains (Fig.7 and Fig.18). The differences in the area fraction of the grains up to 1 μm were not observed (Fig.8 and Fig.19).

In the case of the deformation when the compression velocity was 0,04 mm/s and the torsional frequency was 1,8 Hz, it was observed that Cu structure was significantly smaller than after the deformation with the compression velocity equaling v= 0,015 mm/s (Fig.6d and Fig.17b). The area fraction of the grains up to 1 μm are also greater (Fig. 19) as well as the high-angle boundaries fractions (Fig.20).

STEM micrographs (Fig.21) evidently demonstrate that deformation at higher value of v parameter leads to generating banded structure with low angle grain boundaries and high value of dislocation denity (Fig.22).

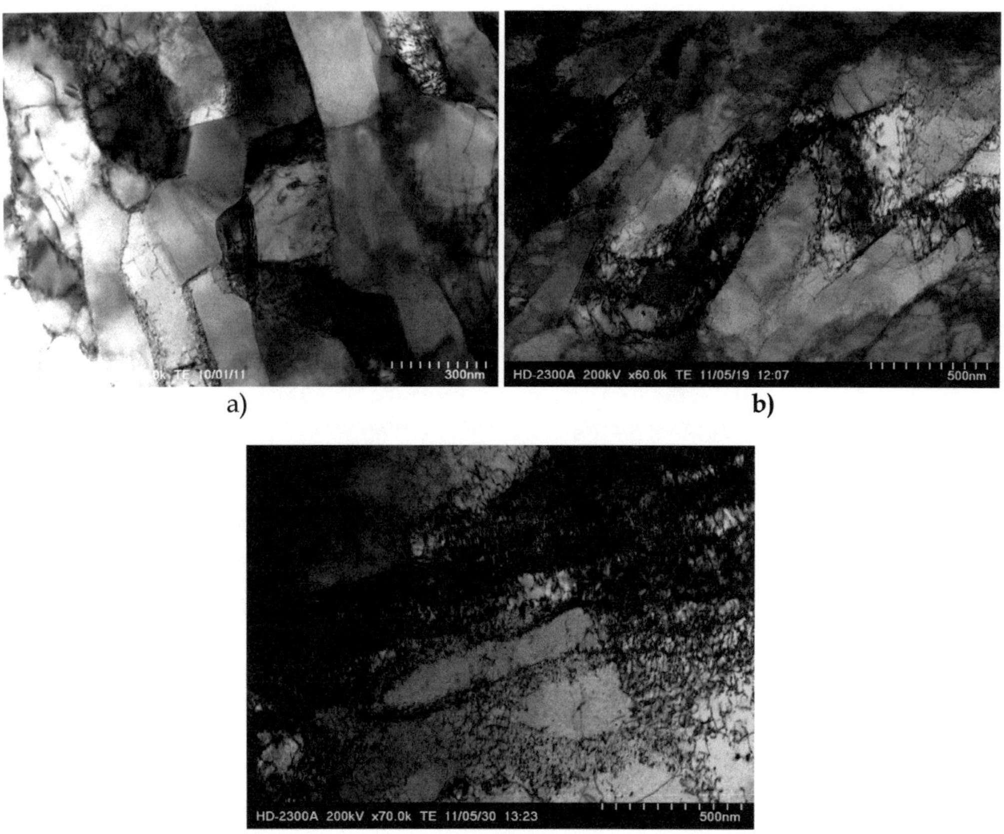

a) b)

c)

Fig. 21. Microstructure of Cu after COT deformation at: a) v=0,04 mm/s, b) 0,1mm/s, c) 0,6 mm/s and constans: f=1,6 Hz, α =6° and Δh=7 mm

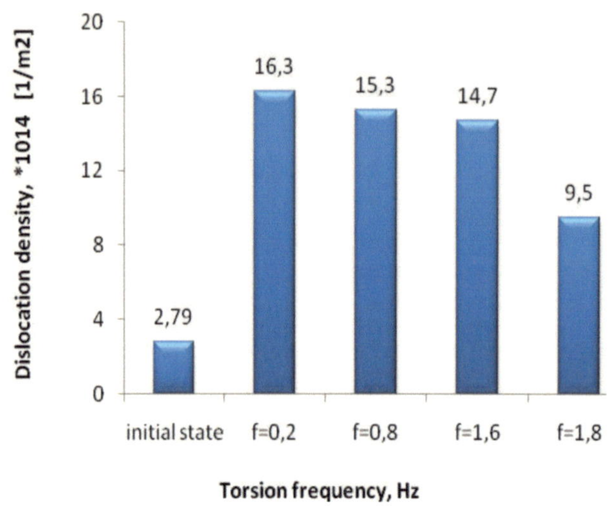

Fig. 22. The variation of the dislocation density In COTes samples with deferent value of torsion frequency and constant α =6°, v=0,04 mm/s and Δh=7 mm

4.3 The absolute deformation

The absolute deformation is the parameter, the increase of which has an impact on the increase of the value of ε_f parameter. Acording with literature, the heterogeneity is connected with too small effective deformation [Kuziak, 2005]. Using larger effective deformation leads to the homogenization of the microstructure and larger grain refining. Analyzing the selected samples of the size distribution of the subgrains (Fig.23), grains (Fig.24), and the shape indicator (Fig.25) using the EBSD technique, it is seen that the microstructure after applying a larger absolute deformation, is characterized by greater homogeneity. The homogenous character of the microstructure is indicated by the narrow spectrum of the size distribution in subgrains/grains and shape indicator. In the samples which are deformed to lower values of the absolute deformation, the great diversity of grain and subgrain size is observed.

The effect connected with the accumulation of the deformation through the increase of Δh is, above all, the grains fragmentation by the generation of the dislocation boundaries. The structures obtained after the absolute strain Δh =3mm, are characterized by the initial stage of creating the DDWs boundaries (Fig.26). In the inside of the areas that are divided by the dislocation boundaries, the cellular dislocation structure can be seen (Fig.26). After applying the deformation Δh =7 mm, the dislocation boundaries indicate a clear contrast proving the accumulating within the dislocation boundaries. A frequent intersection of the dislocation boundaries is observed (Fig.27).

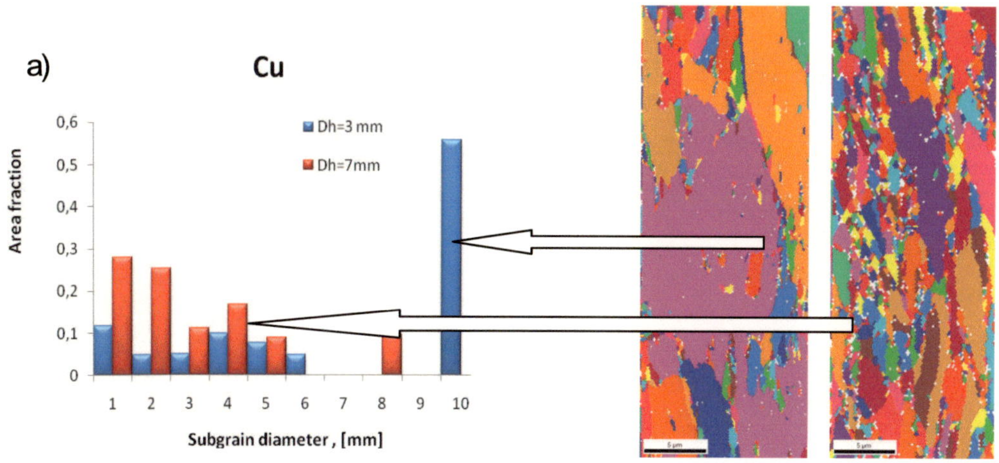

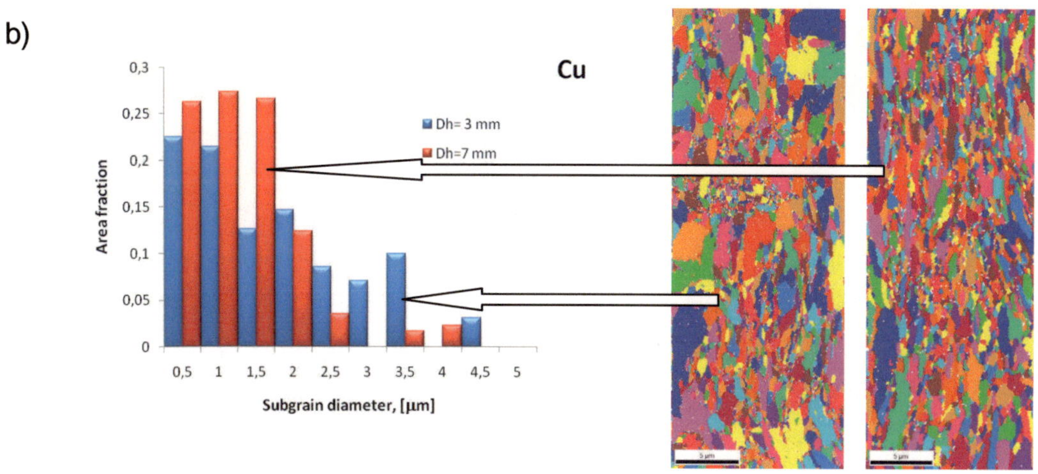

Fig. 23. Area fraction subgrains distributions of COTed Cu samples at: Δh= 3mm, Δh= 7mm; and at constant parameters: a) f=0,8 Hz, α =6°, v=0,04 mm/s , b) f=1,6 Hz, α =6°, v=0,04 mm/s

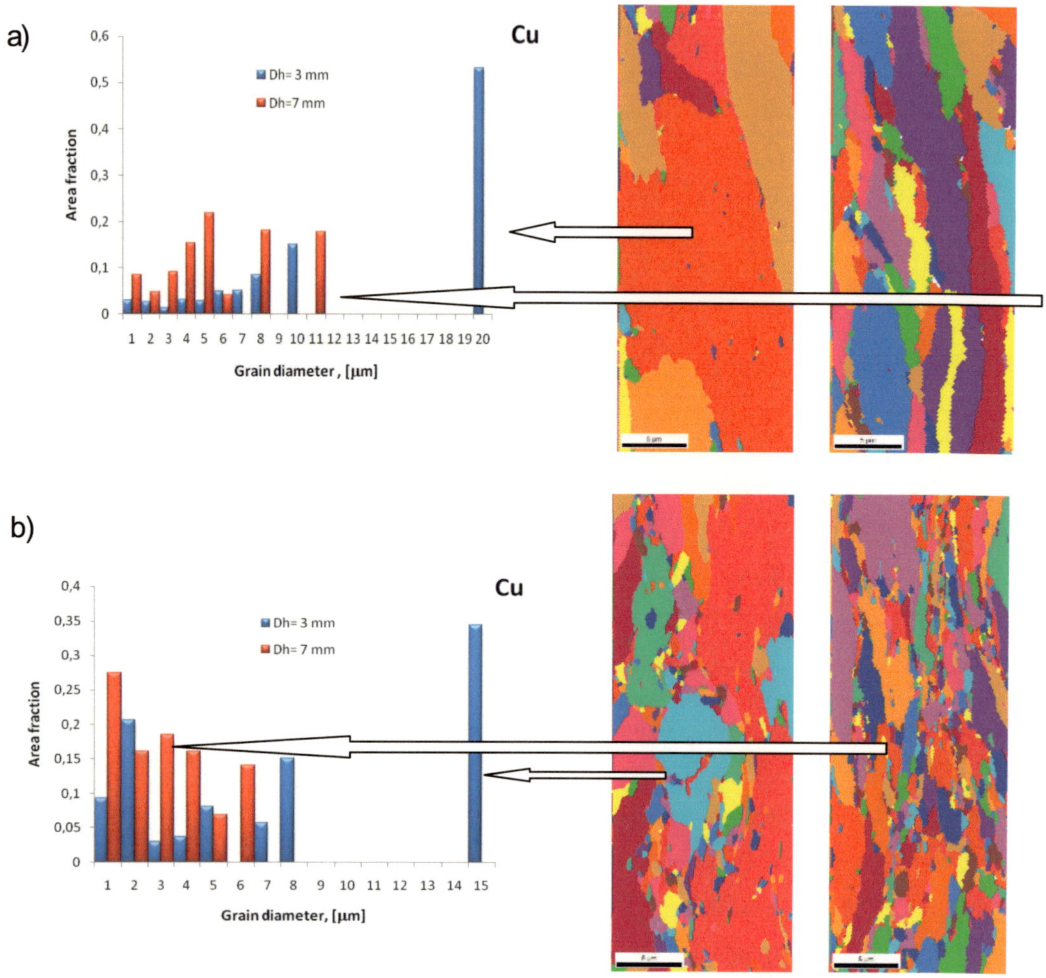

Fig. 24. Area fraction subgrains distributions of COTed Cu samples at: Δh= 3mm, Δh= 7mm; and at constant parameters: a) f=0,8 Hz, α =6°, v=0,04 mm/s , b) f=1,6 Hz, α =6°, v=0,04 mm/s

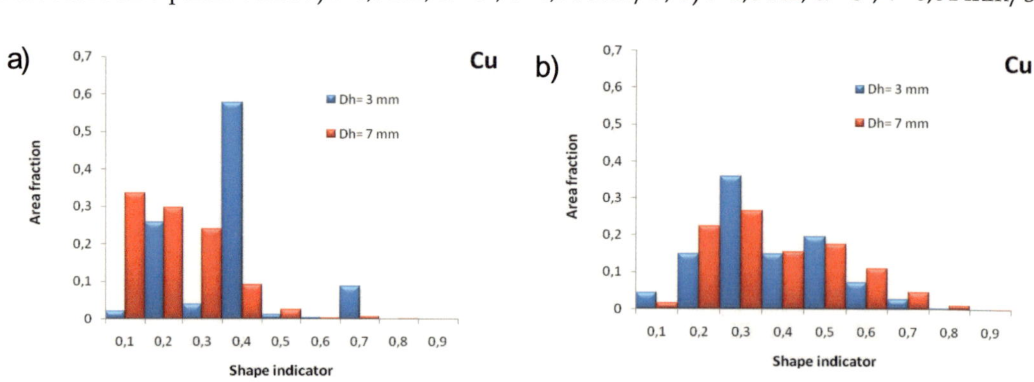

Fig. 25. Shape indicator distributions of COTed Cu samples at: Δh= 3mm, Δh= 7mm; and at constant parameters: a) f=0,8 Hz, α =6°, v=0,04 mm/s , b) f=1,6 Hz, α =6°, v=0,04 mm/s

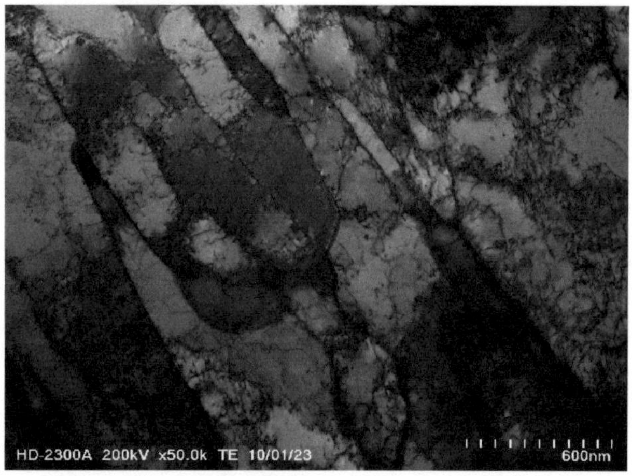

Fig. 26. Microstructure of Cu after COT deformation at: f=1,6 Hz, α =6°, v=0,04 mm/s and Δh=7 mm

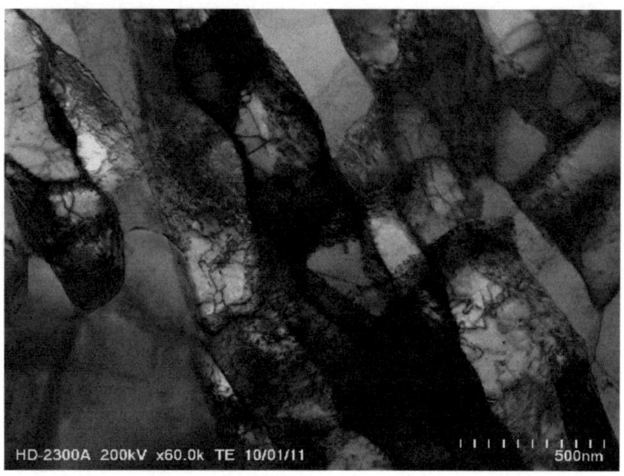

Fig. 27. Microstructure of Cu after COT deformation at: f=1,6 Hz, α =6°, v=0,04 mm/s and Δh=7 mm

The above data shows that for the effective structure refining, it would be beneficial to lead the deformations with:

- the torsional frequency 0,8 Hz and 1,6 Hz. The large equivalent deformations realized by the increase of the torsional frequency ranging from 0,8 Hz to 1,6 Hz, have an impact on the increase of the misorientation value between the created grains (Fig.20), on the increase of the area fraction of the grains with the average diameter up to 1μm (Fig.19), on the creation of the equiaxed grains and subgrains in a range of the ultrametric sizes (Fig.18).
- the high value of the absolute deformation (Δh=7 mm). The increase of the effective deformation by means of the increase of the absolute strain has an influence on the raise of the high-angle boundaries fraction, the increase of the defects, and the generation of the dislocation boundaries which are subject to the mutual intersection (Fig.26,27). The microstructures are characterized by great homogeneity (Figs.23-25).

- the low compression velocity up to approximately 0,04 mm/s. The deformation when the compression velocities are higher than 0,04 mm/s has no positive influence on refining a grain to the ultrametric level (Fig.16b.c). However, the deformation proceeding when the compression velocities are very slow, fosters the structure recovery processes (Fig.6d).

Taking into account the most convenient conditions of the Cu grain refining process it should be stated that:

- the average diameter of the Cu grains and subgrains equals correspondingly about 300 nm and 200 nm.
- the area fraction of the grain which has the ultrafine-grained size, is 60%
- the fraction of high -angle boundaries (HAB) reaches up to about 50%

Despite the majority of the literature data does not discuss the created grain/subgrain separately but rather give the overall values, it should be stated that the obtained are comparable with the literature data [Dobatkin, 2007; Dalla Torre, 2004].

Less attention is paid to determine the fraction of high angle boundaries (HAB) while describing the structural effects. Only the work of Richert [Richert, 2006] and Dobatkin [Dobatkin, 2007] proved that the fraction of high angle boundaries (HABs) in Cu is above 50%. It is difficult to relate the obtained data concerning the area fraction of the ultrametric grains to the literature data because the available literature does not give such results.

The demonstrated decrease of a dislocation density with the increasing deformation, argues for the recovery processes accompanying deformation. The literature also tells about the decrease of the dislocation density after obtaining large deformations [Dalla Torre, 2004].

5. Copper mechanical properties after compression with oscillatory torsion

The measurements of hardness for Cu which was subjected to compression with oscillatory torsion, shows that the changes of hardness depends both on the compression velocity and torsional frequency (Fig.28). The growth of the torsional frequency causes the gradual decrease of hardness. However, the increase of the compression velocity causes the increase of hardness. When the torsional frequency was f= 0,2 Hz, the levels of hardness were the highest - even 130 HV0,2. The material deformed when the values of *f* parameter are low, does not create the grain/subgrain structure but is only characterized by numerous dislocation tangles. Thus, a large dislocation density is responsible for high hardness. The growth of the effective deformations ε_f caused by the increase of the torsional frequency does not contribute to the increase of hardness. A slow decrease of hardness that was observed is connected with the reduction of the dislocation density resulting from the recovering processes that it undergo. The data shown in the Fig.29,30 suggests that the refined grain of Cu leads to 1,5 fold increase of the ultimate tensile strength (UTS) when compared with the initial stage. It was also stated that the highest increase of the mechanical properties accompanies the deformed material when the compression velocities are higher. In general, the increase of deformation realized through the increase of the torsional frequency does not increase the UTS and yeld stress (YS) of the tested materials.

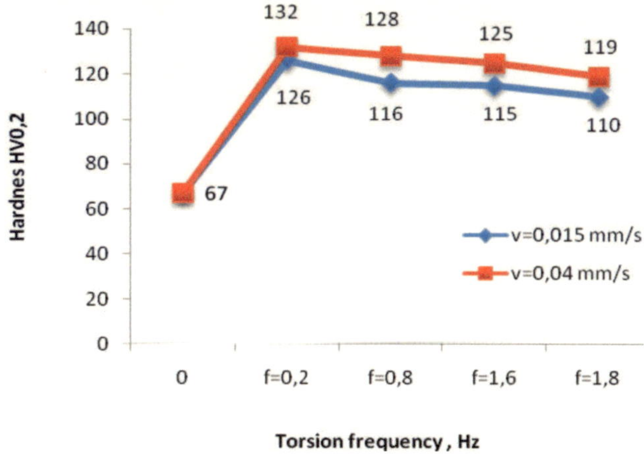

Fig. 28. Hardnes ploted as a function of torsion frequency and compression velocity; constants rest parameters: α =6°, Δh=7 mm

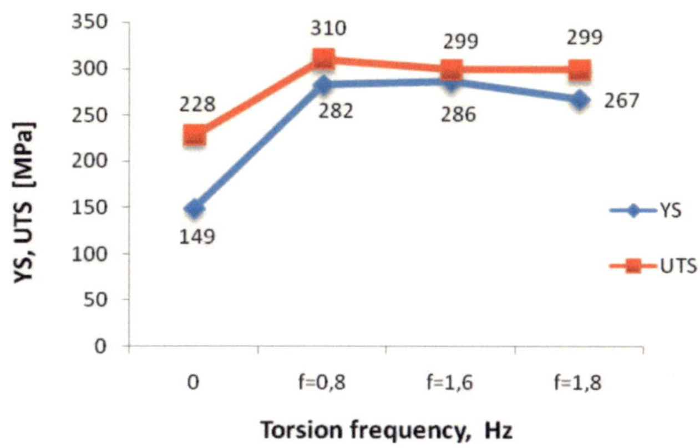

Fig. 29. YS and UTS ploted as a function of torsion frequency; constants rest parameters: v=0,015 mm/s, α =6°, Δh=7 mm

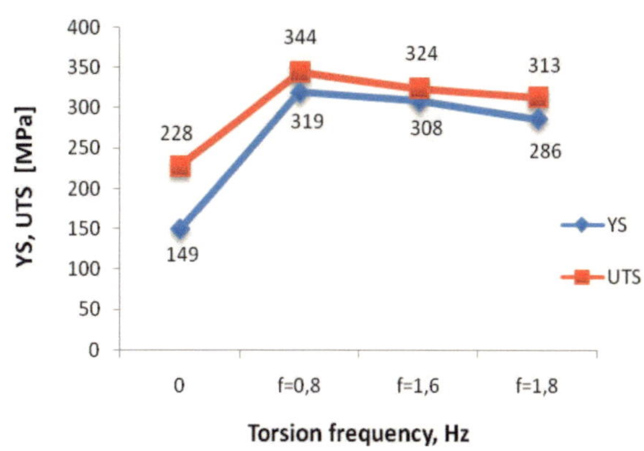

Fig. 30. YS and UTS ploted as a function of torsion frequency; constants rest parameters: v=0,04 mm/s, α =6°, Δh=7 mm

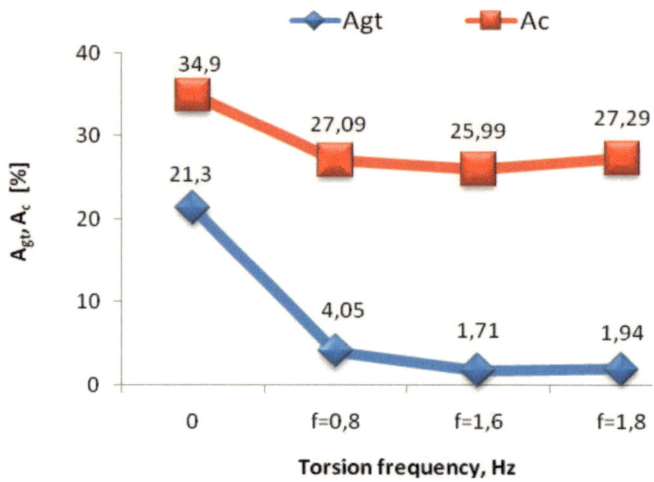

Fig. 31. A_{gt}, A_c ploted as a function of torsion frequency; constants rest parameters: v=0,015 mm/s, α =6°, Δh= 7

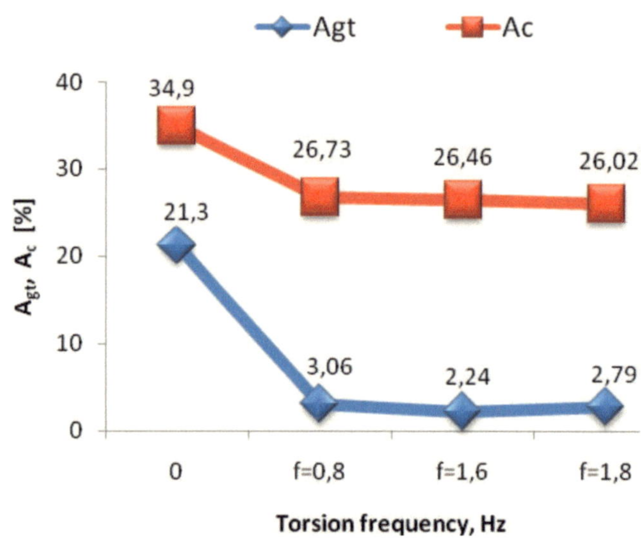

Fig. 32. A_{gt}, A_c ploted as a function of torsion frequency; constants rest parameters: v=0,04 mm/s, α =6°, Δh= 7

The reason of low ductility of the material deformed by the SPD techniques is the localization of deformation. In the case of ultrafine-grained materials with the average diameter of the grain not exceeding 1000 nm, the elongations are usually bigger than in the case of nanograined materials. It is caused by the greater abilities of the material to cumulate the dislocations in grains what leads to higher velocity of the work hardening. The Cu plasticity expressed by the elongation to the rupture A_c is lower in comparison to the initial state and does not change with the variable of deformation parameters (Fig. 31 and Fig. 32). However, the total uniform elongation A_{gt} reaches value of about 2% (Fig. 31 and Fig. 32).

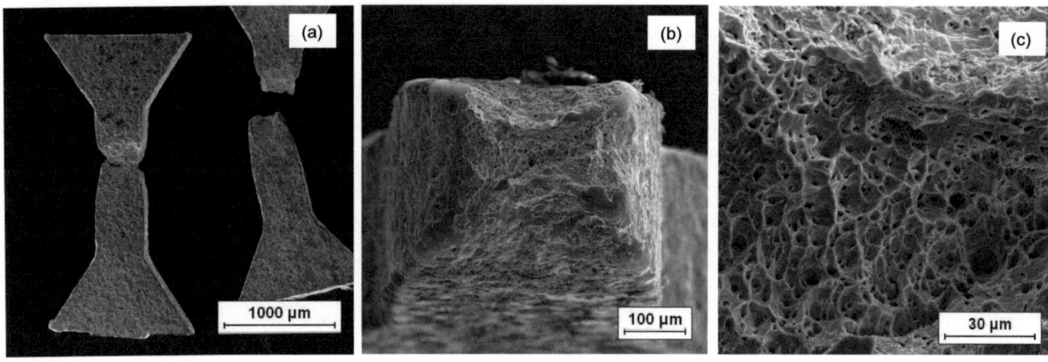

Fig. 33. Example fracture of the specimens 4 subjected to tension, b) necking near fracture section, c) fracture surface

Cracking of the materials after the tensile test happens in the areas of the measured parts (Fig. 33a). The material loses its stability as a result of the deformation localization in form of a neck which concentrates almost whole plastic deformation. In connection with that, the uniform elongation is negligible, however, the (YS) is close to the (UTS). The fracture surface prove the ductile character (Fig. 33b,c).

The mechanical properties of Cu after the deformation ensures the best grain size reduction is when f=1,6 Hz, Δh=7, α=6°, v=0,04 reaches values of about: YS =308 MPa and UTS = 324 MPa. The average UTS of Cu after the ECAP deformation is about 400 MPa [Kurzydłowski, 2004], and sometimes even higher values were noted [Beresterci, 2008]. The plastic properties, especially the uniform elongation for Cu, reaches the values of about 2%. This is the value comparable with the literature data [Kurzydłowski, 2004].

6. The mechanism of producing ultrafine-grained structures after the COT deformation

A lot of place in the structural tests is devoted to the mechanisms of grain-refining with the help of large SPD deformations. This matter is interesting particularly because of the application of different materials and different SPD techniques. The most important conclusions taken from the researches on the mechanisms of grain-refining are as follows:

- The fragmentation of grain takes place when the dislocation boundaries taking different forms are generated.
- The deformation is accompanied by the processes of dynamic recovery or even recrystallization [Kaibyshev, 2005, Wang, 2003]. An example of a material in which the grain-refining is the effect of deformation and of the „extended recovery" is aluminum. Whereas, in copper recrystalized grains are visible in the background of the deformed structure.
- The increase of misorientation happening thanks to the rotation of the grains boundaries and is a diffusion- controlled process. It is based on annihilation and absorption of the dislocation through the grains boundaries.

Hughnes and Hansen [Hughnes, 2000] presented the concept of microstructure evolution for the classic techniques of deformation which is based on the generation of dislocation boundaries which lead to the division of the initial grains into smaller volumes.

The proposed conception of the structure evolution is also characteristic for SPD techniques because many researchers dealing with refining the grain using large plastic deformations, introduce to the description of the structure the terminology in a form of the shear bands or dislocation layers. In general, there are three main mechanisms of the material structure-refining that are known:

- the production of new grains takes place thanks to a gradual increase in misorientation of dislocation boundaries as a result of absorption of new dislocations created during the deformation [Xu, 2005],
- the fragmentation of grains takes place thanks to the generation of the shear bands [Richert, 2006],
- the fragmentation of grains takes place thanks to the production of new grains as a result of the continuous recovery or continuous recrystallization [Kaibyshev, 2005, Wang, 2003] .

The refining of the copper grain after the COT deformation, happens as a result of the generation of dislocation boundaries, which together with the growth of deformation transform themselves into ultrafine-grained structure. The introductory stage of the grain-refining is the creation of DDWs dislocation walls the misorientation of which reaches even a few degrees and they stretch along the considerable fraction of a grain, separating blocks of dislocation cells (CBs) (Fig.34). Within the boundaries a high density of dislocation is accumulated (Fig.34).

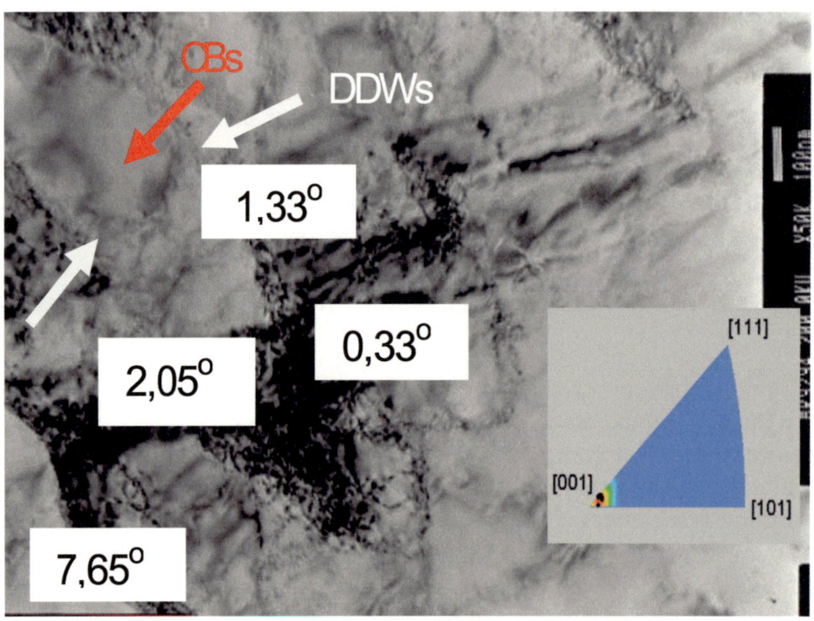

Fig. 34. Microstructure of Cu after COT deformation at: f=0,2 Hz, v=0,015 mm/s, $\alpha=6°$, $\Delta h=7$ mm

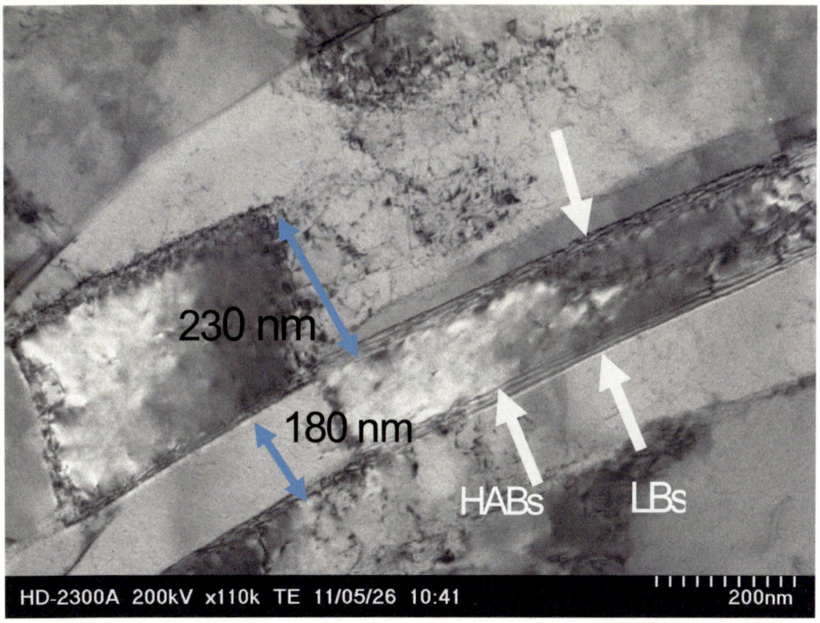

Fig. 35. Microstructure of Cu after COT deformation at: f=0,8 Hz f=0,2 Hz, v=0,015 mm/s, α=6°, Δh=7 mm

The growth of deformation causes the transformation of the DDWs dislocation walls into lamellar boundaries (LBs) which has larger misorientation (sometimes above 15° - HABs) which resembles long subgrains (Fig.35) Inside of the lamellar boundaries, the dislocation structure is regular and singular dislocation cells are usually noticeable. Moreover, the distance between the lamellar areas decreases (Fig.35). It can be assumed that the accumulation of deformation induces not only the creation of new dislocation boundaries and high angle boundaries but above all, it induces the crossing of dislocation boundaries. This phenomenon of intensive boundaries crossing (Fig.36) is a result of activating the subsequent slip systems while the process of deformation. The places where the dislocation boundaries are crossing induce the generation of almost equiaxed subgrains/grains (Fig.37). As a result, the large deformation increases the share of grains/subgrains boundaries and misorientation scattering (Fig.6) This, in turn, leads to the formation of the grains with HABs boundaries (Fig.37). The result of EBSD test shown in the Fig.6 and Fig.17 also constitute the confirmation of the tests. On the basis of EBSD tests it was proved that in a great deal of cases elongated neighboring grains remain in crystallographic compatibility. In the case of fine, equiaxed grains, the orientation was accidental (Fig.6).

The high-angle boundaries (HABs) marked in the Fig.36, 37 have bulges characteristic for the dynamic recrystallization and they can indicate the migration of high-angle boundaries. In many works it was proven that the migration of the grain boundaries created as a result of SPD process as an effect of dynamic recrystallization [Kaibyshev, 2005; Wang, 2003]. The sequence of figures registered during the rotation of the sample at a given angle indicates that the bulges of HAB boundaries are not the effects of the boundary migration but of the mutual superimposing of the boundaries which are in one crystallographic orientation in a given microarea (Fig.38).

This means that the creation of ultrafine-grained structures using the COT method is not determined by the process of recrystallization.

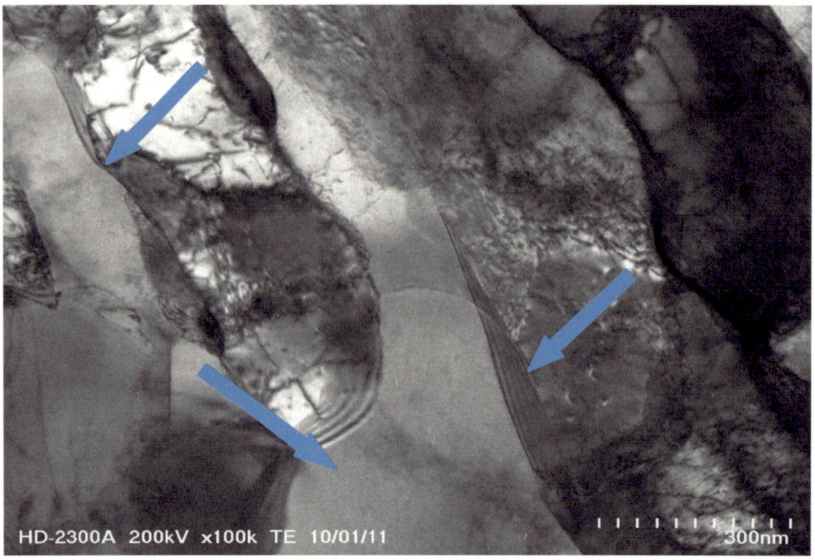

Fig. 36. Microstructure of Cu after COT deformation at: f=0,8Hz, v=0,015 mm/s, α=6°, Δh=7 mm

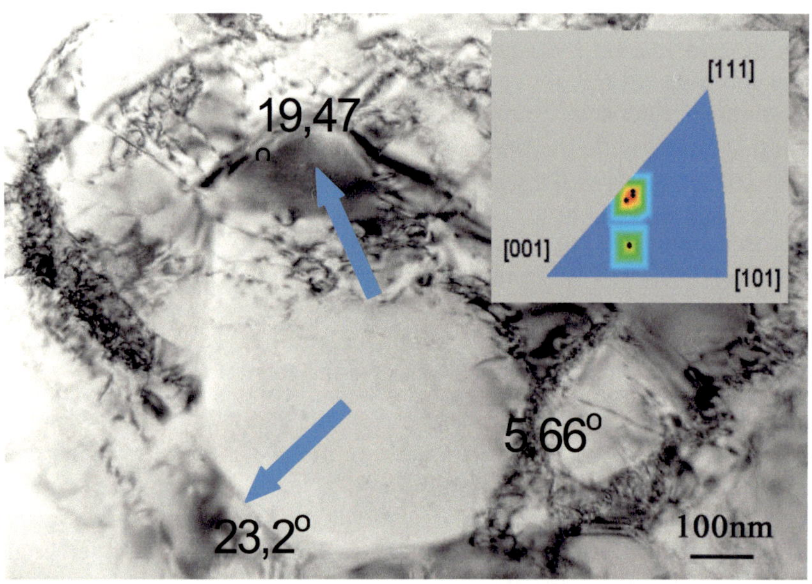

Fig. 37. Microstructure of Cu after COT deformation at: 1,6 Hz, v=0,015 mm/s, α=6°, Δh=7 mm

In order to trace the way in which the low-angle boundaries (LABs) change into HABs boundaries, a series of structural tests was carried out in which the Kikuch diffraction was used. Some examples are presented in the Fig. 39. Defining of the local orientations and crystallographic misorientations of particular areas, allowed to formulate the mechanism in which the high-angle boundaries are formed. The examples presented in Fig.39 suggest that the recovery of Cu caused by EBU happens relatively slow. That is why the dislocations generated during the deformation do not easily annihilation. The cellular type of the dislocation structure remains even if a large deformation is used. The EBSD tests show that the fraction of LABs boundaries is still present even after a large deformation (Fig.9,20).

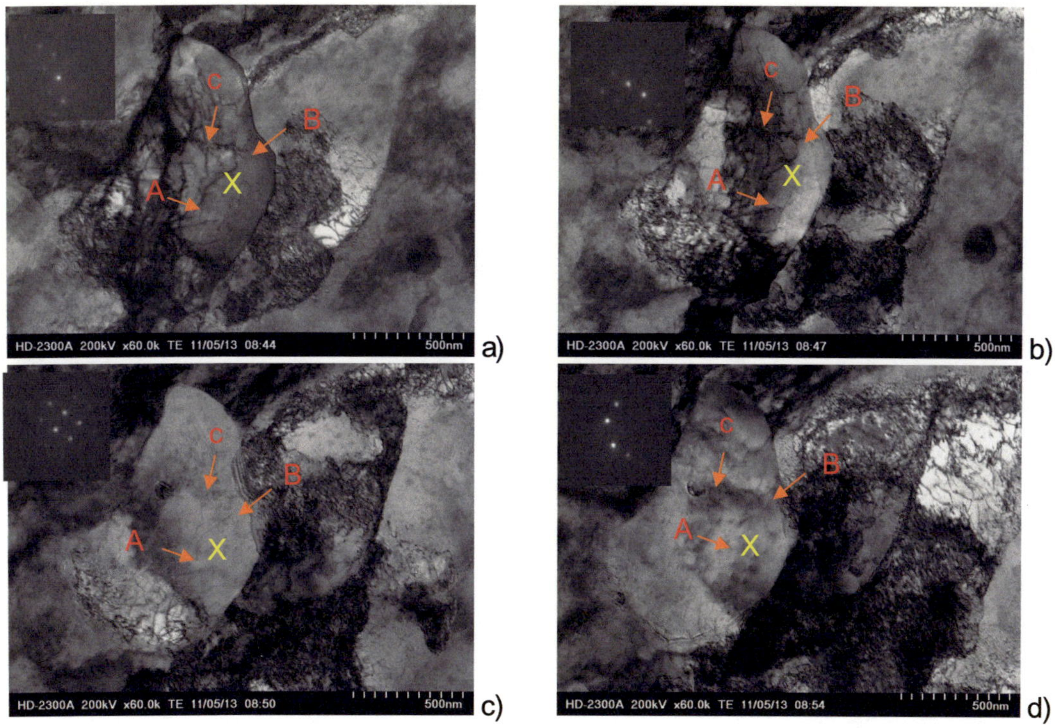

Fig. 38. Microstructure of Cu after COT deformation at: 0,8 Hz, v=0,15, α=6°, Δh=7 mm. ABC marked grains are wisible after changes in sample rotation : a) α= -3,9°, b) α= -6,2° , c) α= -9,6 °, d) α= -10,2° . Diffraction patterns taken from X

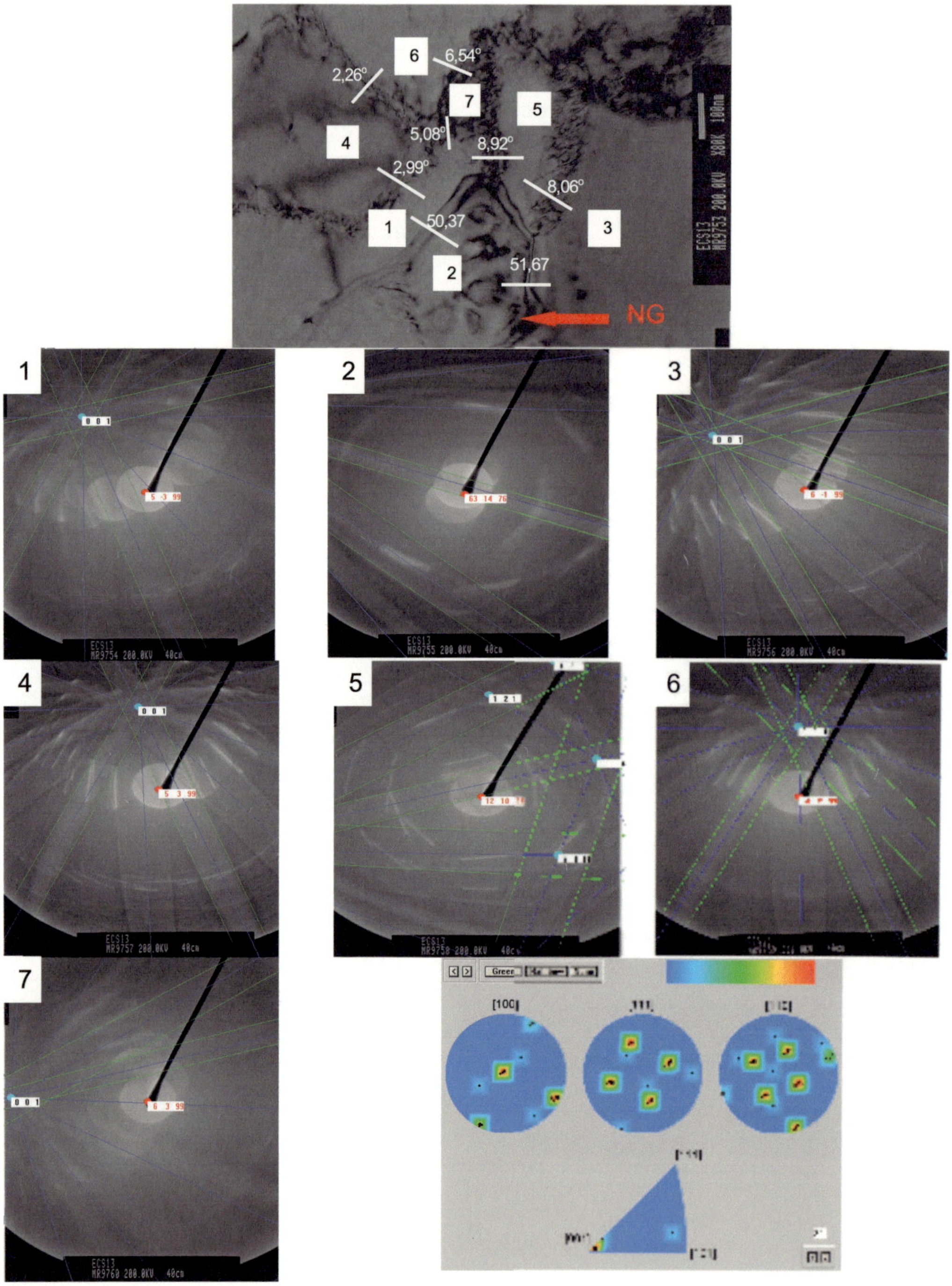

Fig. 39. a) Microstructure of Cu after COT deformation at: f= 1,6 Hz, α°=6, v = 0,04 mm/s. High misorientation of nonequilibrium grain boundary in grain marked 2.; b) Kikuchy difractions with solutions. Numbers 1-7 in Fig. 38a coresponds Kikuchy patterns 1-7 ; c) orientation of analized area

The clusters of dislocation are usually created near the boundaries what causes nonequilibrium state of the grains boundaries. A weak and heterogeneous diffractional contrast inside of the grains indicates a high level of the interior stress in the grain boundaries (Fig.39). Such boundaries are high-angular but almost invisible.

The structural analysis presented here show that the dominating mechanism of forming the high-angle grains in Cu is the growth of misorientation in dislocation boundaries. This, in turn, happens as an effect of absorbing dislocations to the grains boundaries during the deformation process. The nonequilibrium boundaries presented in the Fig.39 are formed from the absorption of numerous dislocations as a result of glide and climb [32].

It was proven that together with the growth of deformation, the fraction of the boundaries increases gradually from 5°-15° (up to approx. 30%) and the fraction of HABs boundaries raises above15° (to about 40%) (Fig.6). Approximately, on a constant level the fraction of narrow-angle boundaries LABs is formed and it does not exceed 30%. When the process parameters are changed (Fig.20) the increase of the HABs fraction to 50% is visible, but also the fraction of boundaries form 5°-15° was reduced to approximately 20%.

The LABs boundaries still constituted 30%. It can mean that the boundaries with misorientation from 5°-15° transform themselves into the grain boundaries. Because of the fact that the boundaries with an average misorientation (5°-15°) are the nonstable boundaries [Cao, 2008], as a result of the deformation they are transformed into high-angle boundaries.

7. Summary

The chapter concentrates on the possibility of a structure forming and on the properties of the metallic materials by means of using the method of compression with oscillatory torsion. The issues that were introduced have a great technical importance because it refers to the formation of ultrafine-grained structure by application of the unfamiliar SPD technique. Because the intensive research on generating new, more economical methods of producing ultrafine-grained materials, the presented matter coincides with the current state of research conducted in many scientific centers.

The aim of the tests in which the COT deformation method was used, was to obtain maximal refining of the grain below 1μm. The structural investigations showed that using the COT method, the grains of the Cu can be refined to ultrafine-grained. The intensity of grain-refining depends on the value of the effective deformation, however, the deformation path (the selection of the deformation parameters) is a decisive factor.

Using the different deformation parameters in process the presence of different phenomena that were controlling the growth of the microstructure. The increase of deformation (realized through the Hz growth) causes a progress in the grain refinement. The deformation conducted when the torsional frequencies were 0,8 Hz and 1,6 Hz is the most beneficial for obtaining the most refined grain. Using a significantly higher torsional frequency -18 Hz during the deformation caused considerable restrictions in the grain-refining because of the intensive recovery which begins to dominate over the deformation process.

It can be noticed that a relatively small change of the compression velocity, had an impact on considerably greater refining of the structure. The acceleration of the deformation process by increasing the compression velocity, causes the delay of the structure recovery process. What is seen through the changes of dislocation density for example. Despite reducing the effective deformation, the progress in the structure refining is observed what denies the results described in the literature [Dalla Tore, 2004; Wang, 2003]. However, the next increase of the compression velocity from 0,1 mm/s and 0,6 mm/s does not foster grain-refining. It should be explained by the fact that the effective deformation is too small.

The main effect of deformation is the increase of the structure's homogeneity. The homogeneity is obtained mainly by the increase of the height parameter Δh.

When the process parameters are as follows: the torsional frequency ranges from 0,8 Hz to 1,6 Hz; Δh=7mm, and the compression velocity ranges from 0,015 to 0,04 mm/s; the maximal refining of the copper grain is obtained but also:

- the average diameter of the grain/subgrain correspondingly 200 nm and 300 nm.
- the share of high-angle boundaries is about 50% and
- the fraction of the ultrafine grains is about 60%. In spite of using large equivalent deformations, narrow-angle boundaries up to 15° constitute a significant fraction in both materials. The fraction of high-angle boundaries does not exceed 70%. This means that the thermal stability of the structures formed in such a way is sufficient [Raab, 2004; Lugo, 2008].

The result of mechanical tests were discussed in details for selected schemes of deformation in which the effects of deformation were visible the most. The influence of grain refining on the mechanical properties of the grain, was determined on the basis of the hardness measurements and the tensile test of the micro samples. The results of the mechanical tests that were carried out, indicate clearly that the changes taking place in the structure and the grain refining had a positive influence on the mechanical properties of the tested materials. However, the uniform elongation is not strongly dependent on the grain refining.

The difference in the properties that were observed, mainly result from using different schemes of deformation.

The results of structural investigations were the basis for determining the mechanisms of grain refining and forming high-angle boundaries. It was proven that the formation of ultrafine grains during the COT deformation is based on the general mechanisms of forming the dislocation boundaries. The process of grain refining proceeds by the generation of the LABs and HABs boundaries. In the introductory stage of deformation the dislocation boundaries are formed which are perpendicular to the direction of the compression force. The formation of the dislocation boundaries which proceed in such a way, suggests that in the initial stage of deformation it is mainly the compression that initiates the process of the grain refining. The non-directional process of deformation (the introduction of an additional torsion causing the change in the direction of loading) leads to the deformation of the material in more and more numerous systems of glides and to the new dislocations having influence on the active system of glides on the previously-generated dislocations. The effect of the introduced, additional loading is the increase in the number of the dislocation boundaries that cross mutually. When the effective deformation in the microstructure increases, the distances between the dislocation boundaries decrease - a new order of LBs dislocation boundaries is created which as a result of the thermal process activation (recovery), transform themselves into subgrains and grains. The process leads to the increase of misorientation of the grain's boundaries.

A significant role in forming the ultrafine-grained structure has the recovery process. Dislocations are rearranged, undergo annihilation and are also absorbed to the grain boundaries. Such a rebuilding of a dislocation structure causes the increase of the misorientation within the grain boundaries.

8. References

Pakieła, Z. (2009). *Mikrostrukturalne uwarunkowania właściwości mechanicznych nanokrystalicznych metali*, Oficyna Wydawnicza Politechniki Warszawskiej

Olejnik, L. et al. (2005). *Bulletin the polish academy of science, Technical sciences*, 53, pp. 413-423

Bochniak, W. et al. (2005). *Journal of Materials Processing Technology*, 169, pp. 44-53

Shaarbaf, M. et al. (2008). *Materials Science Engineering*, A473, pp. 28-33

Raab, G.J. et al. (2004). *Mater. Sci. Eng.*, A 382, pp. 30

Lugo, A. N. et al. (2008). *Materials Science Engineering*, A477, pp. 366-371

Kulczyk, M. et al. (2007). *Materials Science Poland*, 25, pp. 991-999

Pawlicki J.& F. Grosman. (2005). Naprężenia uplastyczniające metali w warunkach złożonych obciążeń, *Materiały XII Konferencji Informatyka w Technologii Metali*, Ustroń 16-19.01

Grosman F. & Pawlicki J. (2004). Processes with forced deformation path. New Forming Technology 2004. *Proceedings of the 1st ICNFT, Harbin Institute of Technology Press*, Harbin, China, Sep. 6–9

Rodak, K. & Goryczka T. (2007). *Solid State Phenom*, 130, pp. 111-113

Ungar, T. et al. (2001). *Materials Science Engineering*, A319-321, pp. 274-278

Kuziak, R.(2005). *Modelowanie zmian struktury i przemian fazowych zachodzących w procesach obróbki cieplno-plastycznej stali*, Instytut Metalurgii Żelaza, Gliwice

Dobatkin, S. V. et al. (2007). *Materials Science Engineering*, A462, pp. 132-138

Dalla Torre, F. (2004), *Acta Materialia*, 52, pp. 4819- 4832

Richert, M. (2006). *Inżynieria nanomateriałów i struktur ultra drobnoziarnistych*, Uczelniane Wydawnictwo Naukowo Techniczne, Kraków

Kurzydłowski, K. J. (2004). *Biuletin of the polish academy of sciences technical sciences*, 52, pp. 301-311

Beresterci, M. et al. (2008). *Metalurgija*, 47, pp. 295-299

Kaibyshev, R. et al. (2005). *Materials Science Engineering*, A398, pp. 341-351

Wang, G. et al. (2003). *Materials Science Engineering*, A346, pp. 83-90

Hughnes, D.A. et al. (2000). *Acta Materialia*, 48, pp. 2985-3004

Xu, Ch. et al. (2005). *Materials Science Engineering*, A398, pp. 66-76
Cao, W.Q. et al. (2008). *Materials Science Engineering*, A492, pp. 74-79

Part 2

Recrystallization of Minerals

12

Zircon Recrystallization History as a Function of the U-Content and Its Geochronologic Implications: Empirical Facts on Zircons from Romanian Carpathians and Dobrogea

Ioan Coriolan Balintoni and Constantin Balica
"Babeş-Bolyai" University, Cluj-Napoca
Romania

1. Introduction

According to Davies et al. (2003), "...the age of Earth and the time scale of pre-human events are central to a civilization's sense of origin and purpose. Therefore, the quest for precise and reliable geochronometers has had a scientific and cultural importance that few other enterprises can match". In this respect, since the beginning of the last century it has been recognized that long-lived radioactive decay systems provide the only valid means of quantifying geologic time.

One of the most reliable Earth's timekeeper has proven to be the mineral zircon, since it records the ages of Earth's earliest evolution stages, the oldest sediments, extinction episodes, mountain-building events and supercontinents' coalescence and dispersal (e.g. Rubatto and Hermann, 2007; Harley et al., 2007). Its widespread use in geochronology is based on the decay of uranium (U) and thorium (Th) to lead (Pb). They provide three distinct radioactive decay series involving the parent isotopes ^{238}U, ^{235}U and ^{232}Th and their daughter isotopes ^{206}Pb, ^{207}Pb and ^{208}Pb, respectively. Through the incorporation of U and Th at the time of growth, every zircon grain hosts three different clocks. In an ideal closed system, the three estimates would agree with each other within the analytical errors of measurements. However, in natural systems the zircon grains are not equally closed for Th and U with respect to post-crystallization effects. The usual approach in zircon geochronology is to consider the U-Pb system alone, as there is no natural non-nuclear ways of fractionating ^{235}U from ^{238}U. As the modern day-ratio of $^{235}U/^{238}U$ is well known (1/137.88) the need to actually analyze very low abundances of ^{235}U is obviated. Besides U and Th, zircon can incorporate some other incompatible elements such as P, Sc, Nb, Hf, Ti, and REE in trace (up to thousands of ppm) or minor (up to 3% wt) amounts. The primary control factor on the substitutions is the ionic radii of the substituting cations compared with Zr^{4+} and Si^{4+} cations. Substitutions that minimize strain effects on either or both sites will be favored. The crystal-chemical limitations are that Zr^{4+} in 8-fold coordination has an ionic radius of $84*10^{-3}$ nm and Si^{4+}, in tetrahedral coordination, has an ionic radius of $26*10^{-3}$ nm. U^{4+} (ionic radius of $10*10^{-2}$ nm in 8-fold coordination) and Th ($105*10^{-3}$ nm in 8-fold coordination) can be accommodated in the Zr^{4+} sites. Uranium concentrations are usually less than 5000 ppm and Th concentration less than 1000 ppm. Because of its ionic radius of $129*10^{-3}$ nm (8-fold coordination), Pb^{2+} is highly incompatible with growing zircon crystal lattice and therefore is not incorporated more than ppb levels, which is crucial in geochronology. Because of the same reason, Pb^{2+} can easily escape from zircon lattice when some conditions are fulfilled.

Based on idea of concordant ages between $^{235}U/^{207}Pb$ and $^{238}U/^{206}Pb$, Wetherill (1956) has developed the Concordia diagram. Quite soon though it has been shown (Tilton et al., 1957) that the concordance situations are rather rare and usually zircon shows evidence of discordance (disagreement between $^{235}U/^{237}Pb$ and $^{238}U/^{206}Pb$ ages) due to Pb loss caused by some post-crystallization geologic events.

Since then, the attempts made for understanding the causes of discordance had become the main preoccupation of the U-Pb geochronologists over the next quarter of century. Finally, it has been understood that discordance can be attributed to two major causes: (1) - mixing, in the analyzed sample volume, of discrete zones of different ages from within the same zircon grain; (2) - partial loss of radiogenic Pb by the entire zircon grain or by fractions of it. While the former cause can be bypassed by in situ dating, the second still hinders unequivocal results.

Continuously growing database of published in situ age data provided further insights into the behavior of U-Pb system in zircons: (1) - radiogenic Pb can be entirely lost and even the concordant data do not indicate the initial crystallization age of zircon; (2) – frequently, Pb loss is accompanied by Th loss; (3) – occasionally, even the U can be lost.

The disturbance of the isotope systematics in zircon is related to: (1) - amorphization; (2) - alteration; (3) - recrystallization.

The focus of the present contribution is on the discussion of the above three processes exemplified with their geochronological consequences on zircons from the Romanian Carpathians and Dobrogea. All data presented here were obtained by in situ dating through Laser Ablation-Inductively Coupled Plasma-Mass Specterometry (LA ICP-MS).

1.1 Amorphization process

This process is due to α-decay events associated with U and Th radioactive disintegration. Based on X-ray diffraction and High Resolution Transmited Electron Microscopy (HRTEM) analysis, Murakami et al. (1991) suggested three stages of damage accumulation in Sri Lankan zircon: in stage I (at $<3 \times 10^{18}$ α-decay events/g) the accumulation of isolated point defects predominate. These defects have the potential to recover through geologic time; stage II (at $3x10^{18}$ to $8x10^{18}$ α-decay events/g) is evidenced by crystalline regions with point defects and amorphous tracks caused by overlapped α-recoil nuclei; during stage III (at $> 8x10^{18}$ α-decay events/g) only aperiodic material can be screened by X-ray and electron diffraction.

Salje et al. (1999) suggested a two-phase transition during increasing amorphization process. The first phase is related to the percolation of amorphous material into the crystalline matrix and the second one to the percolation of crystalline material into the amorphous matrix.

According to Nasdala et al. (2001) the α-decay events in the decay chains of U and Th cause the zircon amorphization. An α-particle generates about 120-130 Frenkel type defect pairs along penetration distances from 9.9 to 29.5μm. Recoils of heavy daughter nuclei are only a few hundred Å in length but the recoil damage clusters include 600-1200 Frenkel defect pairs. Spontaneous fission fragments produce heavy damage locally, but their contribution to the overall radiation damage is of minor importance because of their relative rarity. In the absence of recovery, radiation damage is stored in zircon, causing transformation from the crystalline to the metamict state. These authors propose the following stages of radiation damage accumulation.

1. Scattered nano-regions with high defect concentration. The amorphous component is still insignificant.
2. Moderately radiation damaged zircon in which amorphous nano-regions form a domain structure. Amorphous domains in a crystalline zircon or crystalline remnants in an amorphous matrix can be observed.
3. Entirely aperiodic zircon.

Ewing et al. (2003) speak about α-particles ionization processes over a range of 16 to 22 microns that produce several hundreds isolated atomic displacements. The associated α-recoils lose their energy during elastic collisions over 30 to 40 nm, producing localized collision cascades of 1000 to 2000 displacements. However, with increasing temperature the amorphization dose increases. This can be due either, by a decrease in the average cascade size caused by thermal relaxation or by a reduction in the surviving amorphous volume as a result of thermal recovery of irradiation induced defects.

According to Geisler et al. (2007), the amorphization of a crystalline zircon is characterized by transformation from an initial stage, where isolated amorphous domains are surrounded by slightly disordered crystalline material, to a more advanced stage of damage where few isolated, disordered, nano-crystalline islands occur in an amorphous matrix. These authors refer to Salje et al. (1999) when describing the crystalline- to-amorphous transformation as a geometrical phase transition. During the first phase transition, the amorphous domains form clusters that percolate over macroscopic length scales, whereas during the second transition, the crystalline domains cease to be interconnected. The first percolation point appears at an amorphous fraction of ~30% while the second percolation point appears when the amorphous fraction reaches ~ 70%. In other words, less than 30% amorphous fraction and less than 30% crystalline fraction can not be interconnected.

1.2 Alteration processes

According to Geisler et al. (2002), alteration is the interaction between the metamict zircon and fluids, including the meteoric ones (weathering processes), characterized by distinct chemical and structural changes. Geisler et al. (2003) describe two anomalous stages in the alteration rate with increasing degree of amorphization. The first stage takes place when the amorphous domains form interconnected clusters within the zircon structure, namely at the first percolation point suggested by Salje et al. (1999). At this point, the percolation interfaces that represent low-density areas between crystalline and amorphous domains open high diffusivity pathways. A new dramatic increase in alteration rate is observed at the second percolation point of Salje et al. (1999). Around this point, a network of nanometer-size regions of depleted matter interconnects discrete amorphous domains without crystalline dams.

According to Geisler et al. (2007), the structural changes that take place in a zircon grain are defined by the fact that above the 200° C limit the thermal recrystallization front generated by epitaxial reordering moves inward the zircon crystal. Bellow the 200° C threshold, the thermal recrystallization involves enhanced defect diffusion only along the reordering front. The above model hypothesizes an increase in the effective diffusion of any species within the zircon lattice with increasing α-decay dose or of amorphization degree.

1.3 Recrystallization

There are several types of recrystallization processes that take place in zircons and several different opinions in terms of its significance exist on this matter.

According to Nasdala et al. (2001), *recrystallization* is a "re-growth" process in the crystallographic sense. During recrystallization a new zircon lattice forms along a crystallization front, typically replacing a more disordered and polluted zircon. Recrystallization leads to healing of the radiation damage and partial or complete resetting of the U-Pb isotopic system. According to the same authors, *annealing* requires only re-formation of disrupted bonds by re-ordering of nearest neighboring atoms, thus annealing of zircon structure is not necessarily associated with any Pb loss. As defined in the literature, *recrystallization* presumes an epitaxial migration of an interface between an ordered region in zircon and a metamict vicinity, therefore it could be described as a defect annihilation or point defect diffusion process. In a general view, Nasdala et al. (2001) enumerate the following thermal recovery mechanisms: (1) - point defect diffusion in the crystalline and amorphous phase; (2) - epitaxial growth of crystalline residuals; (3) - random nucleation in the amorphous phase.

Ewing et al. (2003) put forward a more comprehensive description of the concept of recrystallization, while offering the foundation for a clear distinction between *Type I* and *TypeII* recrystallization. Type I recrystallization is purely thermal and occurs on time scales longer than cascade quench time. This behavior is due to point defect diffusion and epitaxial migration of the crystalline residuals toward the amorphous domains. It can be seen that Type I recrystallization, as described by these authors, covers both the recrystallization and annealing processes of Nasdala et al. (2001). Type I recrystallization prevails over longer periods of time or at higher temperatures and becomes particularly important in natural specimens stored at ambient conditions for geological periods. In moderately damaged zircons, two stages of recovery process have been described. The first stage is defined by the recovery of short-range order and point defect recombination. It occurs bellow ~ 600° C (Farges, 1994) or ~727° C (Geisler et al., 2001). The second stage occurs at higher temperatures and is caused by epitaxial recrystallization along the internal crystalline-amorphous boundaries. It is worth noting that an initially moderately damaged zircon grain consists of distorted crystalline phases embedded in an aperiodic matrix at the completion of the first phase. In contrast to the moderately damaged specimens, heavily damaged or amorphous zircon segregates into its constituent oxides at higher temperatures and recrystallization takes place in three stages: (1) - decomposition of amorphous zircon into tetragonal ZrO_2 crystallites and amorphous SiO_2; (2) - tetragonal to monoclinic phase transformation in the ZrO_2 crystallites; (3) - formation of coarse-grained (several hundred μm) randomly oriented, polycrystalline zircon. This process is highly improbable to take place in natural zircon and it excludes the idea of new zircon nucleation within amorphous phase as postulated by Nasdala et al. (2001).

Type II recrystallization occurs as a nearly instantaneous process during irradiation and can be divided into two distinct phases. The first, Type *IIa* recrystallization, is due to increased mobility of interstitials and other point defects during irradiation. The irradiation enhanced diffusion leads to a greater degree of point defect recombination and annihilation. Point defect annihilation is most effective at structural boundaries between amorphous and damaged, but still crystalline regions. Type *IIb* recrystallization occurs within individual displacement cascades. Disordered and highly energetic material can epitaxially recrystallize at the cascade peripheries along with the cooling of the displacement cascade to the ambient temperature. During irradiation, both *Type I* and *Type II* recrystallization processes contribute to the dynamic recovery of zircon, but some mechanisms can prevail over others in certain temperature regimes. Obviously, at the Earth's surface conditions, both Type I and Type II recrystallization processes are less effective than the irradiation damage over an α-dose range, allowing amorphization accumulation in time.

Geisler et al. (2007) proposed two more kinds of zircon recrystallization, fluid or melt assisted. The Hf, U and Th content of zircon can be also distributed in solid solutions between zircon and hafnon ($HfSiO_4$), coffinite ($USiO_4$), thorite ($ThSiO_4$). Because solid-state exsolution structures have not been yet reported in zircon, it is argued that such solid solutions are metastable after cooling and characterized by structural strain. Such structural strain enhances surface reactivity and thus the dissolution rate. We note that the effects of the structural strain are supplementary added to irradiation damage, increasing the reactivity between the zircon and fluids. It was shown (Geisler et al., 2003) that the treatment of radiation-damaged zircon crystals in various aqueous solutions produces inward-penetrating, irregular, and curved reaction domains that resemble those found in natural zircon. Recrystallization of zircon on the expense of the amorphous phase inside the reacted domains occur at experimental temperatures above 200° C. Recrystallization of amorphous zircon dramatically reduces the molar volume of the reacted areas inducing a strain that is partially released by fracturing. Porous structure at nanometer-scale level is also very likely to occur. The nanoporosity created between the crystallites provides pathways for chemical exchange between reaction front and the fluids. This is the diffusion-reaction process in which a moving recrystallization front follows at some distance behind the percolation-controlled, inward diffusion of a hydrous species. According to Geisler et al. (2007), a prerequisite for the diffusion-reaction process is the presence of more than 30 % amorphous fraction, which is of an interconnected amorphous network.

The coupled dissolution-reprecipitation is a process by which the breaking of the bonds and dissolution is accompanied by contemporaneous nucleation and precipitation of new zircon. The coupled dissolution-reprecipitation process is independent of the absolute solubility of zircon in natural aqueous fluids, which is very low (Tromans, 2006). This can result in a complete replacement of one zircon crystal by a new one within the same space, without losing the external shape or crystal morphology (Putnis, 2002; Putnis et al., 2005). The chemical exchange between the dissolution-reprecipitation front and external fluid is maintained by the formation of porosity. This porosity results from the higher solubility and the higher molar volume of the dissolved parent zircon, as compared with the more pure zircon, chemically reprecipitated . Since the dissolution of a metastable zircon solid solution is kinetically favored by structural strain, the radiation damage may also enhance kinetically a coupled dissolution-reprecipitation process. Considering the two proposed mechanisms, the reaction of zircon with fluids and melts provides an effective way of its re-equilibration.

1.4 Temperature conditions of zircon recrystallization

According to Mezger and Krogstad (1997), the 600-650° C temperature interval is a good estimate for recrystallization of damaged zircon. Consequently, lattice damage through α-decay and spontaneous fission may accumulate bellow this temperature. The α-recoil tracks in minerals recrystallize at similar temperature as tracks formed by spontaneous fission fragments (e. g. Murakami et al., 1991). And because fission tracks are retained in zircon up to 200-250° C (Tagami and Shimada, 1996), the α-events damage can also be recovered immediately over 200-250° C (Nasdala et al., 2001). According to Ewing et al. (2003), "temperatures as low as 100-200° C seem to be sufficient to produce measurable recovery over extremely long time scales". Such statements are valid knowing that the quantity of radioactive elements is continuously decreasing in zircon through fission process. The same authors suggest that the critical amorphization temperature for zircon should have an upper limit of 460 K, or 187° C. This temperature is close to the interval indicated by Tagami and Shimada (1996) for recovery initiation. However, Ewing et al. (2003) say that "depending on the mass of the incident ions, the critical amorphization temperature for zircon is between 527 and 750° C". This statement can be understood in the context of external ionic bombardment. Commenting the data of Meldrum et al. (1998), Cherniak and Watson (2003) say that "the critical amorphization temperature for zircons with 1,000 ppm U is about 360° C, and only about 20° C higher for zircons with as much as 10,000 ppm U; and it varies as a function of zircon age by less than a degree per billion years for a given U content. Zircon exposed to temperatures bellow the critical amorphization temperature can accumulate radiation damage, but only over long time scales".

1.5 Isotope systems resetting

According to Mezger and Krogstad (1997) Pb-loss in zircon may occur in four distinct ways:(1)-diffusion in metamict zircon;(2)-diffusion in pristine zircon;(3)-leaching from metamict zircon;(4)-recrystallization of metamict zircon. Pb-loss can be also accompanied by U and Th loss. As mentioned, Geisler et al. (2007) discussing re-equilibration of zircon in aqueous fluids and melts describe two more processes in such environments: (5) - diffusion-reaction process and (6) - coupled dissolution-re-precipitation process. Possibly, leaching from metamict zircon is always accompanied by diffusion-reaction.

According to Cherniak and Watson (2003), cation diffusion in pristine zircon appears to be exceedingly slow under normal crustal conditions. They mention that the closure temperature for Pb in zircon of 100 μm effective diffusion radius for a cooling rate of 10° C/Ma is 991° C. Field based studies showed that Pb-diffusion in the pristine zircon lattice is insignificant around 950-1000° C (e.g. Black et al., 1986; Williams, 1992). Thus, we do not further explore the possibility of isotope systems resetting in pristine zircon.

Regarding the diffusion in metamict zircon, Davies and Paces (1990) and Heaman et al. (1992) observed that even metamict zircon can retain Pb if the temperature is low enough to inhibit recrystallization or there has been no chemical attack of the metamict parts.In conclusion, resetting of U, Th, and Pb istope systems is strongly dependent on leaching and recrystallization of metamict zircon.

Accumulation of radiation damage in zircon is a competition between the α-dose induced disorder (as a function of its content in U and Th) and recrystallization processes. With increasing temperature the irradiation defects become gradually neutralized by instantaneous reordering of lattice, which is equivalent with *Type II* recrystallization. When no new radiation defects accumulate, the critical amorphization temperature is reached. However, we should stress out that most of the previously amorphizated lattice volume is preserved at the critical amorphization temperature. As mentioned before, there is little consensus with respect to the critical amorphization temperature, which is also a function of the α-dose. If the data of Meldrum et al. (1998) are applicable for natural zircon, then the accumulation of amorphization in grains containing 10,000 ppm U is possible bellow 380° C only. In addition to that, if zircon contains 100 ppm U or less, the amorphization becomes improbable (e.g. Mezger and Krogstad, 1997), as a consequence of the low α-dose. Above the critical amorphization temperature and especially above 600-650° C, *Type I* recrystallization will recover the zircon's structure (e.g. Mezger and Krogstad, 1997). During this process, Pb, Th, and U will be variably lost from zircon lattice. If the metamict zircon interacts with fluids, Pb, Th, and U can be lost at temperatures lower than 200° C due to structural recovery front of low temperature *Type I* recrystallization, and over 200° C due to a moving recovery front of high temperature *Type I* recrystallization (Geisler et al., 2007). The diffusion-reaction process causes only partial loss of radiogenic Pb. According to Geisler et al. (2007) an unambiguous chemical indication of the alteration of radiation-damaged zircon by a diffusion-reaction process is the enrichment in Ca, Al, and Fe, and also in common Pb. Enrichment in common Pb can be easily recognized in the recent geochronologic data sets. Briefly, to arrive to a more or less metamict state, which is a function of its U and Th content, and to allow to its isotope systematics to be disturbed, zircon should have: (1) - remained, in geological time terms, bellow the critical amorphization temperature; (2) –reheated subsequently over the critical amorphization temperature; (3) – interacted with aqueous fluids independently of temperature and/or with melts.

2. U, Th, and Pb isotope systems in granites

Dating granites is not always an easy enterprise. They can have a lot of inherited zircons, some of these could have recrystallized in granitic magma and the true magmatic zircons could have been isotopically destabilized at various degrees, depending on their U content and/or on the interaction with fluids. Different aspects of these questions will be exemplified with the Carpathian granites and migmatites.

1. *Variscan granites in the Danubian Domain of Romanian South Carpathians.* Peri-Amazonian basement of the Alpine Danubian Domain of South Carpathians (Balintoni et al., 2011) was massively intruded by granite bodies and their accompanying dyke-swarms (Berza and Seghedi, 1983). There are Cadomian granites with their migmatic escort (Grünenfelder et al., 1983; Liégeois et al., 1996; Balintoni et al., 2011) and Variscan bodies accompanied by cross-cutting dykes (Balica et al., 2007; Balintoni et al., 2011).

From the Variscan intrusions we will further consider the *Buta* pluton (sample 266)* for its low U and isotopically undisturbed zircons in comparison with *Cherbelezu* and *Sfârdinu* plutons defined by high U and strongly isotopically disturbed zircons.

The grains in sample 266 are characterized by: (1) - an U-content in zircon dominantly less than 400 ppm; (2) - a high $^{206}Pb/^{204}Pb$ ratio; (3) - a small U/Th ratio; (4) - nearly concordant ages (Fig. 1).

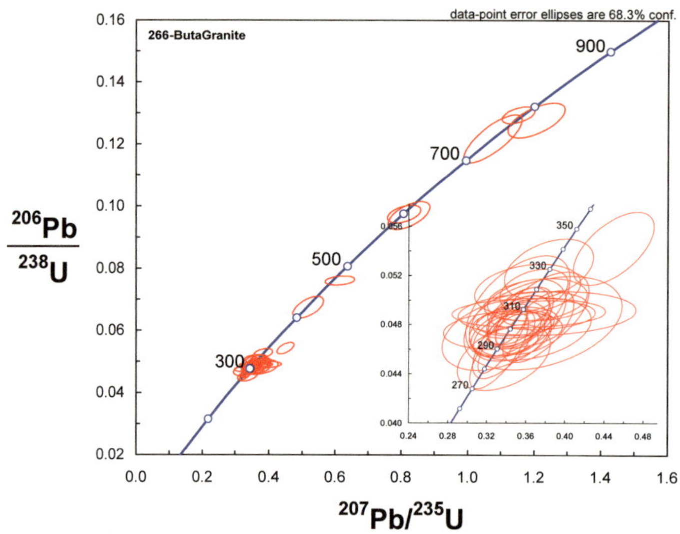

Fig. 1. Concordia projection for sample 266. Good concordances and great concentration around 300 Ma can be noticed. Inset: distribution of ages around 300 Ma.

The vast majority of these zircons represent originally Variscan magmatic grains. Twenty seven $^{206}Pb/^{238}U$ apparent ages ranging between 288.9±7.7 Ma and 314.8±4.3 Ma yielded a crystallization weighted mean age of 303.7±2.4 Ma and a Concordia age of 303.8±0.85 Ma (Fig. 2a, b).

Cherbelezu puton data presented further on were yielded by sample 227. The grains from sample 227 are characterized by two data sets with different parameters: older than 315.2±3.3 Ma and younger than this age. The younger grains with ages ranging between 295.2±3.1 Ma and 122.5±15.1 Ma show: (1) - higher U-content; (2) - higher content in ^{204}Pb; (3) - much greater discordances than the older ones (Table 1 and Fig. 3).

In our interpretation, the older ages suggest inherited recrystallized grains in the Variscan granitic magma. Their isotope systems were completely reset by high temperature *Type I* recrystallization without fluid intervention. The zircons lost all their radiogenic Pb (i.e., good concordance of the data) and partially Th. The available data doesn't allow us to draw some conclusion with respect to any potential U loss. The younger grains most likely represent true Variscan magmatic zircons. Due to their high U content they were amorphizated post 300 Ma and partially lost radiogenic Pb (i.e., poor concordance of the data) and Th during the Alpine thermotectonic events, while gaining ^{204}Pb. The isotope

* Analytical data are available upon request from the authors

systems were probably disturbed in the presence of fluids and amorphization grade reached the first percolation point of Salje et al. (1999). The data suggest that a diffusion-reaction process was active due to structural recovery front of low temperature *Type I* recrystallization, bellow 200° C. The true zircons' crystallization time was probably less than 315.2±3.3 Ma and more than 295.2±3.1 Ma. Thus we suggest for Cherbelezu pluton a crystallization age closed to that of Buta pluton.

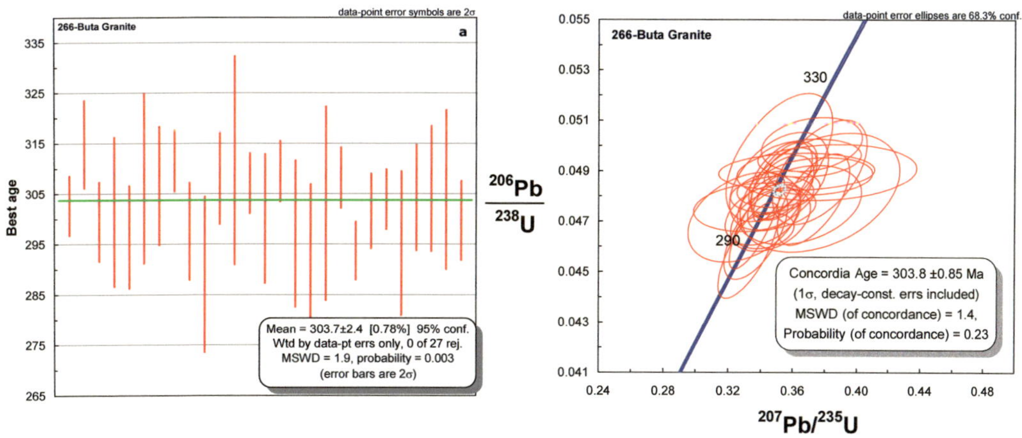

Fig. 2a, b. Weighted mean age and Concordia age for sample 266.

	Older grains	Younger grains
U average content (ppm)	3990	13623
204Pb average content (ppm)	3.46	87.9
U/Th average ratio	8.5	6.3

Table 1. Comparative isotope parameters between the two sets of grains in sample 227.

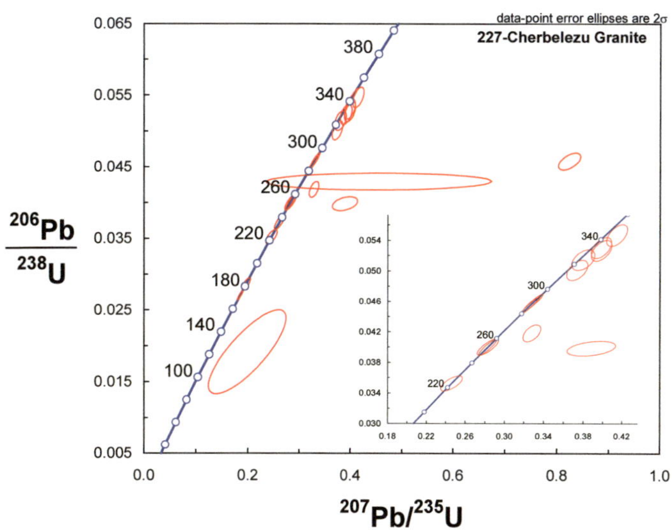

Fig. 3. Concordia projection for sample 227.

The data presented for *Sfârdinu* pluton were provided by sample 229. The grains pertaining to this sample are characterized by: (1) - high U-content (4413 ppm average); (2) - high ^{204}Pb-content; (3) - great discordances for most of the ages, except for 2 out of 50 (Fig. 4); (4) - variable U/Th ratios, generally over the normal ratios in undisturbed magmatic zircon.

Considering the above observations, we infer that all the analyzed crystals represent inherited zircons disturbed initially during the Variscan orogeny, followed by the Alpine thermotectonic events. In their present state all the grains show isotopic disturbance assisted by fluids. Amorphization grade reached the first percolation point of Salje et al. (1999) and grains probably remained bellow 200° C along their entire post Variscan history. Clearly, also in this case, the grains with more than 5000 ppm U gained much more ^{204}Pb than the grains with less than 5000 ppm U. The crystallization age of Sfârdinu pluton is difficult to ascertain. If we consider the most concordant age sets (between 305.4 and 318.6 and between 292.1 and 305.6 Ma) we get a weighted mean age of 301.5±6 Ma (Fig.5).

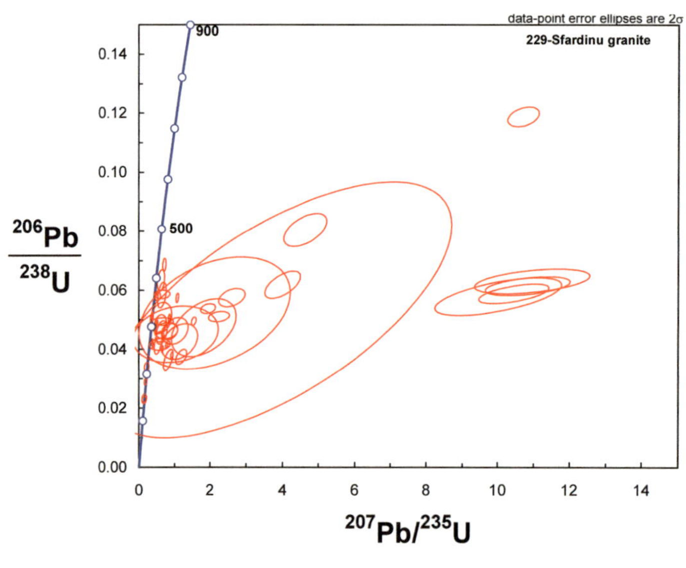

Fig. 4. Concordia projection for sample 229.

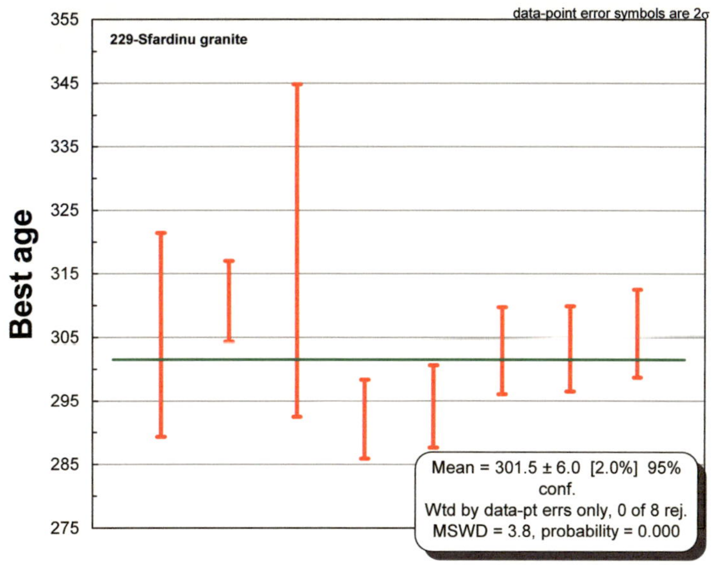

Fig. 5. Weighted mean age for sample 229.

2. *Cadomian granites and migmatites.* Grünenfelder et al. (1983), Liégeois et al. (1996), and Balintoni et al. (2011) dated the Tismana, Şuşiţa, Novaci, and Olteţ plutons and their results bracketed the ages around 600 Ma. These plutons cross-cut a dense swarm of migmatic dykes characterized by black K-feldspar grains (Berza and Seghedi, 1983). Out of the four Cadomian plutons we exemplify the relationship between U-content and isotope disturbance in Şuşiţa pluton. Two samples (277A and 277B) collected several km apart each other show quite different isotope data.

Using the data from sample 227A, Balintoni et al. (2011) obtained a crystallization weighted mean age of 591.0±3.5 Ma and a Concordia age of 591.6±1.8 Ma (Fig. 6a, b).

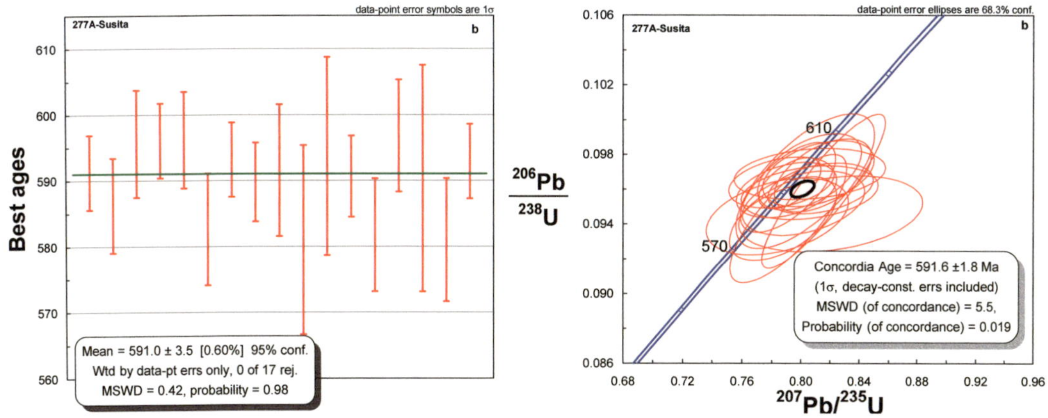

Fig. 6a, b. Weighted mean age and Concordia age for sample 277A.

In comparison with the grains from sample 277A, the grains from sample 277B show: (1) - higher U-content; (2) - higher U/Th ratio; (3) - lower $^{206}Pb/^{204}Pb$ ratio; (4) - greater discordances (Fig.7); (5) - complete lack of protolith ages. Judging from the age discordances, is safe to assume that all the ages were partially reset. The comparative parameters are presented in Table 2.

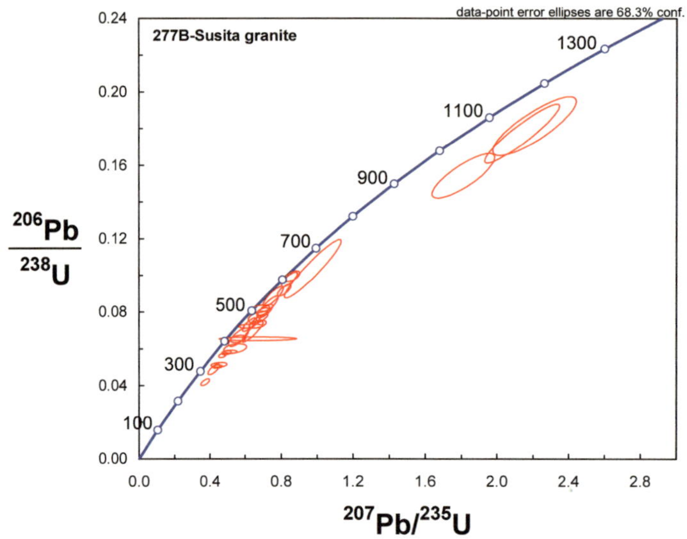

Fig. 7. Spread of ages along Concordia for sample 277B. No ages around 600 Ma.

	Sample 277A	Sample 277B
U average content (ppm)	392	1412
U/Th average ratio	3.2	22.7
$^{206}Pb/^{204}Pb$ average ratio	71374	42429

Table 2. Comparative isotope parameters between the grains in samples 277A and 277B.

Zircons from sample 277B lost radiogenic Pb and Th, and gained ^{204}Pb. There is a direct corellation between U-content, ^{204}Pb gain, age rejuvenation, and concordance deterioration. Therefore, for ages bellow 400 Ma the U average content is of 1883 ppm and $^{206}Pb/^{204}Pb$ average ratio is 27351. For ages between 400 and 500 Ma, the average content of U is of 1235 ppm and the $^{206}Pb/^{204}Pb$ average ratio is 36560. For sample 277B we infer an amorphization grade of up to 40 % and *Type I* recrystallization by point defect diffusion under moderate influence of fluids, as the main recrystallization mechanism. According to Geisler et al. (2007), we assume a diffusion-reaction process at temperatures bellow 200° C. The inferences made for the sample 277B are strongly sustained by the data from the samples 334 (Fig. 8) and 10-406 (Fig. 9) representing Cadomian migmatic material.

The main conclusion that can be drawn is that by increasing the U-content, the isotopic parameters become gradually deteriorated (e.g., discordances in Figs. 8 and 9), while making a distinction between the Variscan and Alpine reworking virtually impossible. At higher U contents the radiogenic Pb and Th are lost in greater quantities and concomitantly more common Pb is added as portrayed in Table 3.

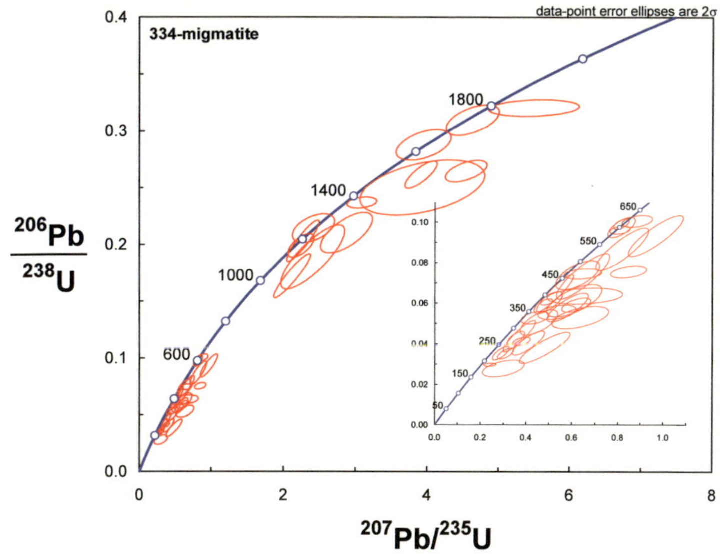

Fig. 8. Spread of ages along Concordia for sample 334. All the magamatic ages have been reset.

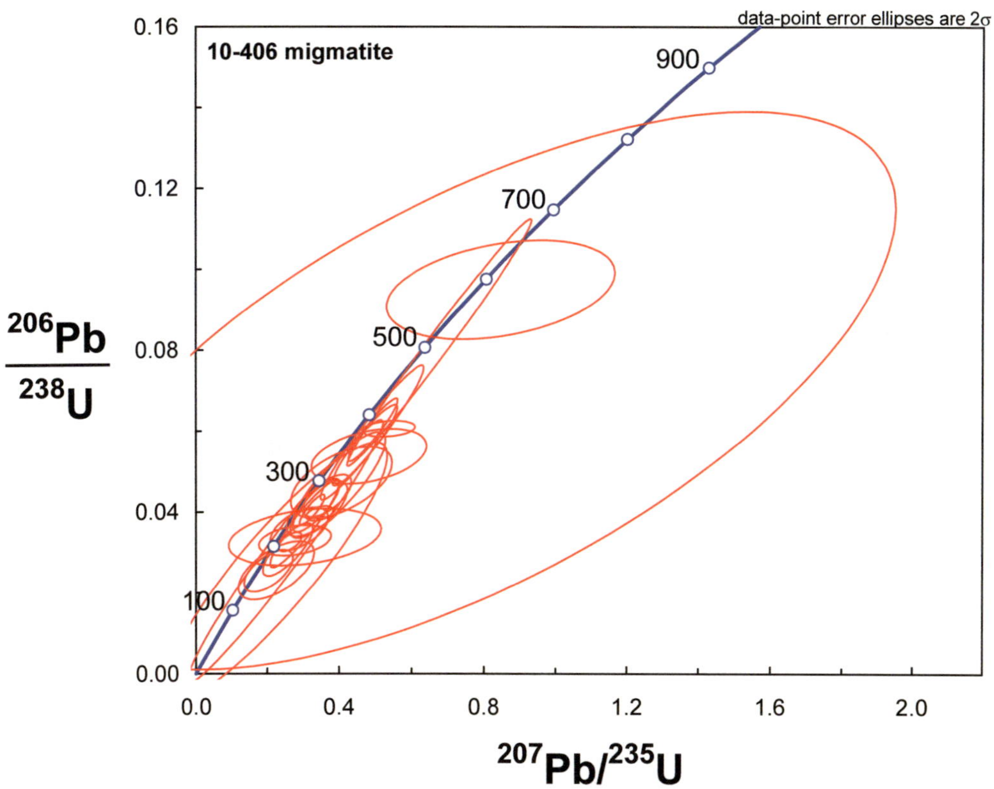

Fig. 9. Spread of ages along Concordia for sample 10-406.Again all the magmatic ages have been reset.

Sample 334		
Ages (Ma)	U-average content (ppm)	U/Th average ratio
>617	385	2.6
617-383	1236	23.9
<383	2049	55.5
Sample 10-406	4465	115.7

Table 3. Comparative isotope parameters between sets of grains from sample 334 and the same parameters for the grains from sample 10-406.

No age was completely reset and the involvement of fluids in these processes is highly suspected. We postulate a poor zircon lattice recovery by defect diffusion processes but the amorphizated material did not recrystallized. During post Cadomian events the grains remained bellow the amorphization critical temperature, very likely bellow 200° C. In the case of sample 10-406 the massive loss of Th and gain of common Pb can be interpreted as an amorphization grade above the second percolation point of Salje et al. (1999), that is more than 70 % amorphization.

3. U, Th, and Pb isotope systems in orthogneisses, metamorphosed under medium grade conditions

The igneous protoliths of orthogneisses are isotopically stabilized during metamorphism. However, as a function of the U-content, temperature history, and fluids intervention, the isotope systems can potentially be reset. Several examples from Apuseni Mountains and East Carpathians will illustrate the behavior of zircon in orthogneisses during Ordovician initial metamorphism and during later Variscan thermotectonic events.

1. *Orthogneisses from the basement of Someş pre-Alpine terrane Apuseni Mountains.* The geochronology of the Apuseni Mountains pre-Alpine terranes has been detailed by Balintoni et al. (2010b). The Someş terrane basement of Ordovician age was intruded by Variscan granites when it functioned as an upper plate. We will further discuss the data from the samples 166 and 167, representing the same orthogneiss body.

The grains from the sample 166 show good concordances (Fig.10), a low ^{204}Pb, small U/Th ratios, and variable U-content. Only three grains show strong reset of U/Pb isotopic system during the Variscan thermotectonic events, with an U average-content of 1642 ppm. The majority of zircon grains preserve Ordovician ages, yielding a weighted mean age of 452.3±5.2 Ma and a Concordia age of 452.4± 3.7 Ma (Fig. 11a, b). The data from sample 166 suggest a Type I recrystallization process of higher temperature of the amorphizated grains, without fluids intervention.

At an U average-content of 2002 ppm, all the grains from the sample 167 lost the radiogenic Pb while the youngest of them gained ^{204}Pb. Rejuvenation lead to the deterioration of the concordances (Fig. 12) but the U/Th ratio remained relatively constant. This is further proof that amorphization grade and resetting of the isotopic systems are a function of the U-content. In the case of sample 167, we advocate for a *Type I* recrystallization at a temperature over the critical amorphization threshold.

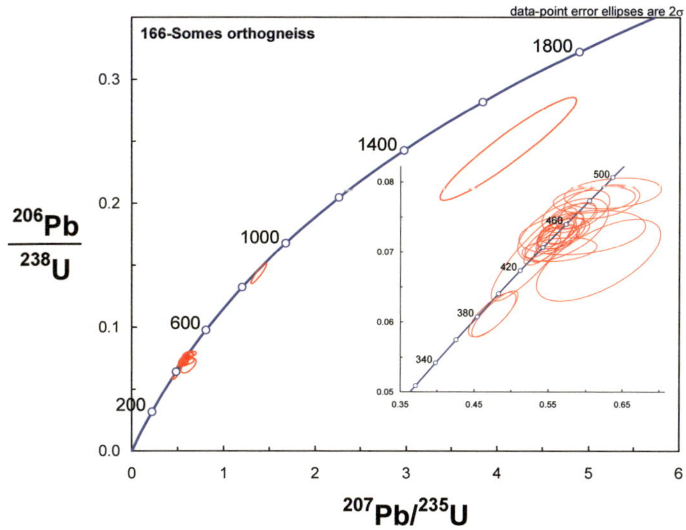

Fig. 10. Concordia projection for sample 166.Concordant ages around 450 Ma.

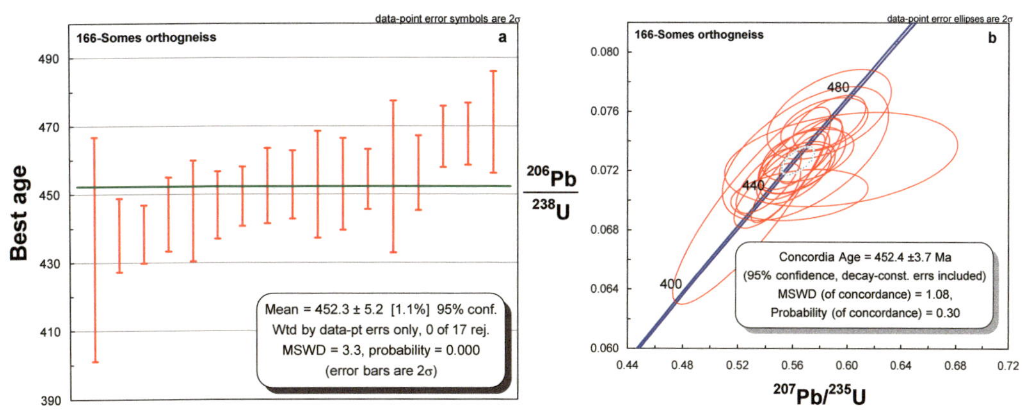

Fig. 11a, b. Weighted mean age and Concordia age for sample 166.

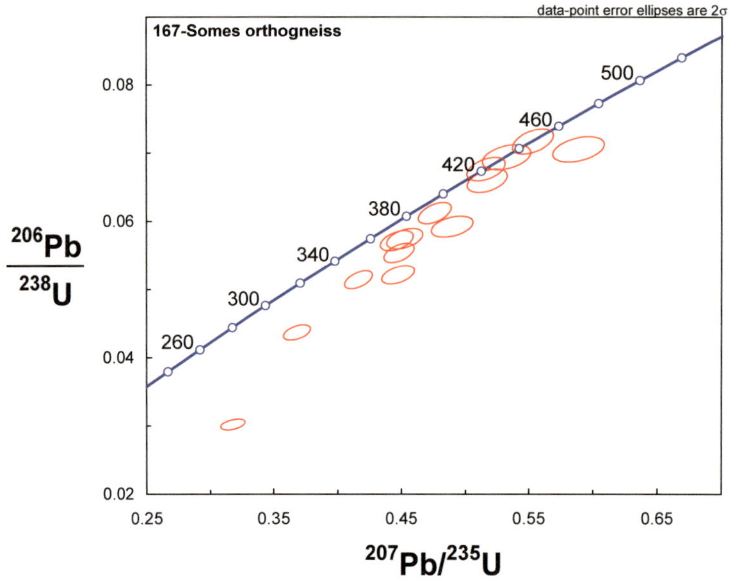

Fig. 12. Spread of ages along Concordia for sample 167.Reset and discordant ages at higher U-content.

2. *Orthogneisses from the basement of the East Carpathians pre-Alpine terranes.* Preliminary geochronologic data on the basement of the pre-Alpine terranes from East Carpathians were presented by Balintoni et al. (2009). In this contribution we will focus mainly on the data from the basement of the Tulgheş terrane and from the basement of the Rebra terrane represented by Negrişoara metramorphic unit. These terranes composed the median part of the Variscan nappe pile and were affected by retrogression down to the chlorite grade temperature (Balintoni, 1997) when the K/Ar ages were reset around 300 Ma (Kräutner et al., 1976).

Tulgheş orthogneiss, sample 10-476. Except a single anomalous age, the data in sample 10-476 show: (1) - good concordances (Fig 13); (2) - low U/Th ratio; (3) - high $^{206}Pb/^{204}Pb$ ratio; (4) - low U-content; (5) - no age younger than the Ordovician. Considering all the Ordovician ages, these can be divided in two distinct groups: (i) between 449.6 and 469.0 Ma; (ii) between 474.4 and 488.8Ma. The younger set corresponds to an U average - content of 203 ppm and to an U/Th average ratio of 3.7., while the older set corresponds to an U average-content of 185 ppm and to an U/Th average ratio of 3.0. The above data suggest a slight increase in the U-content and a decrease in the Th content toward younger ages. Because there is little evidence for any disturbance in the isotopic systems we interpret the age range as an evolution of the magmatic system from its source to the crystallization time. The two data sets yielded a younger $^{206}Pb/^{238}U$ weighted mean age of 462.6±3.1 Ma and an older one of 478.3±5.5 Ma (Fig. 14a, b). We consider the first age to be a better candidate for the protolith crystallization age. There is no doubt that the dated grains constituted isotopically closed systems post protolith crystallization. Therefore, we conclude that bellow the concentration of 250 ppm U, the zircon lattice is prone to recovering even at low temperatures, by *Type* II recrystallization, while the radiation damage and the amorphization can not accumulate.

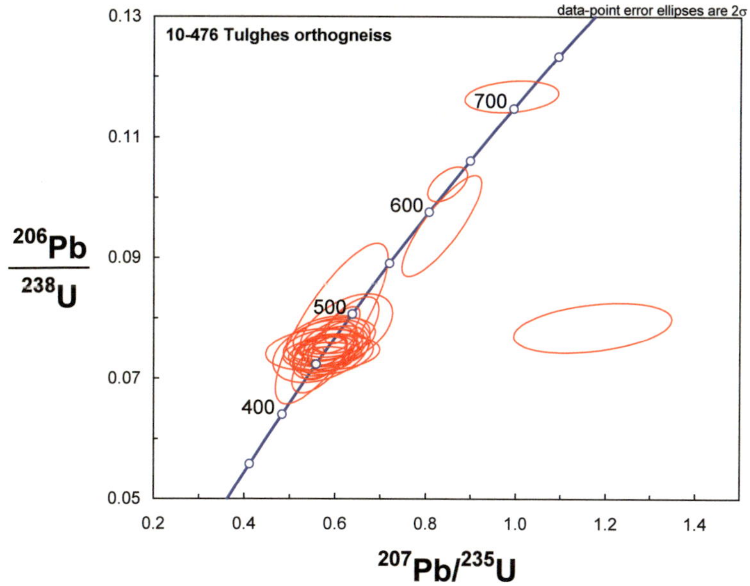

Fig. 13. Concordia projection for sample 10-476.No one age was reset.

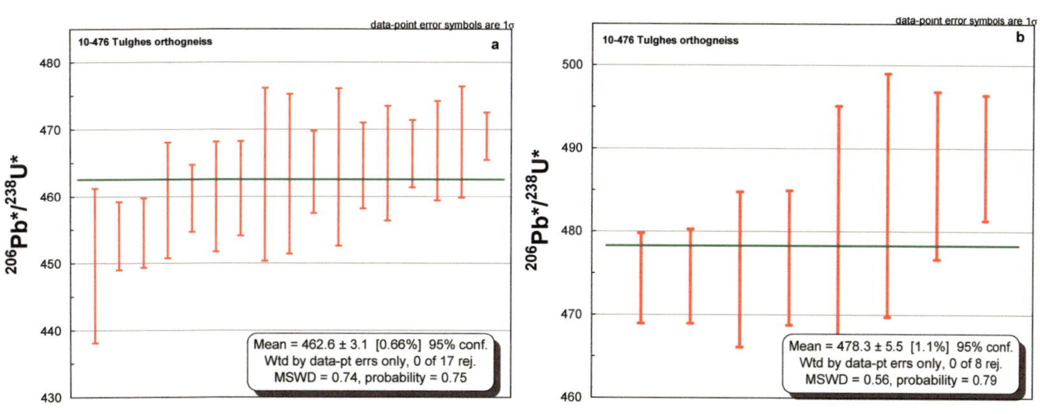

Fig. 14a, b. Weighted mean ages for the two sets of ages from sample 10-476.

Pietrosu Bistriței orthogneiss, Negrişoara metamorphic unit, sample 10-475.

The data from this sample are similar to data from sample 10-476.The younger grains show 226 ppm U in average and an U/Th average ratio of 8.4, while the older grains have an average concentration of 226 ppm U and an U/Th average ratio of 5.5. The younger data set yielded a $^{206}Pb/^{238}U$ weighted mean age of 461.5±5.2 Ma and the older one yielded 477.8±4.2 Ma (Fig. 15a, b). We interpret these ages identically to ages from sample 10-476. As in the case of the previous sample, there is no evidence for zircon damage due to radiation at an U-content bellow 300 ppm.

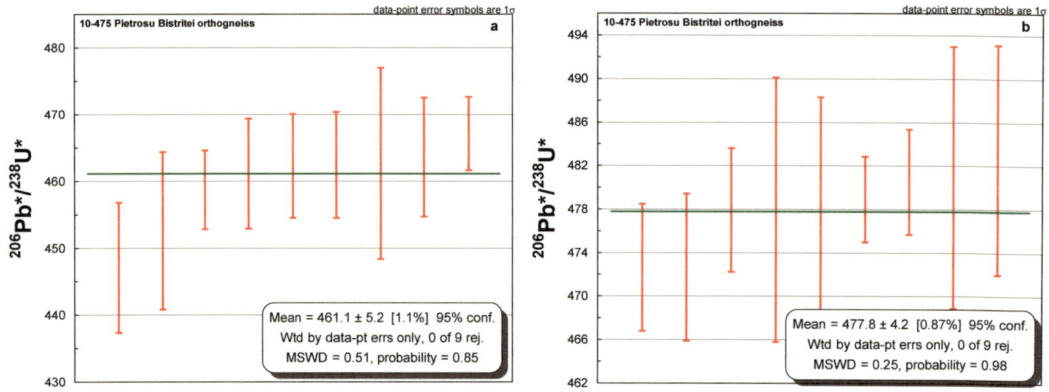

Fig. 15a, b. Weighted mean ages for the two sets of ages from sample 10-475.

4. U, Th and Pb isotope systems in orthogneiss zircons affected by Variscan eclogite-facies metamorphism

Medaris et al. (2003) argued for a Variscan age of the eclogite-grade metamorphism known from the basement of the Sebeş-Lotru pre-Alpine terrane (Iancu et al., 1998; Săbău and Massonne, 2003). Balintoni et al. (2010c) published geochronologic data that revealed the composite nature of the Sebeş-Lotru terrane basement, which consists of Cadomian and Ordovician (Caledonian) igneous protoliths. In the following paragraphs the Cadomian *Frumosu* orthogneiss and *Tău* Ordovician orthogneiss will be discussed.

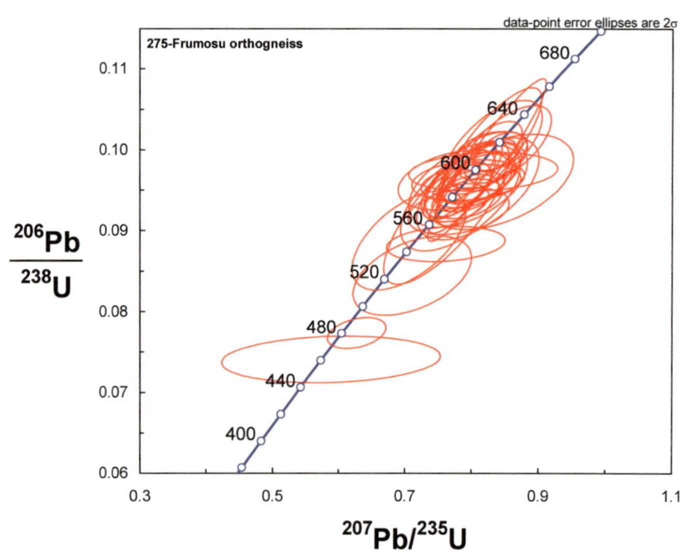

Fig. 16. Concordia projection for sample 275

Frumosu orthogneiss .The grains from sample 275 are characterized by: (1) - very good concordances (Fig 16); (2) - low U/Th ratios; (3) - high $^{206}Pb/^{204}Pb$ ratios; (4) - low U-contents. The U average-content for all grains is around 300 ppm. From 36 dated zircon grains, only 5 of them show Pb loss and 3 of them exhibit Th loss in various degrees. The ages tend to cluster on

Concordia diagram and can be divided in two sets as in East Carpathians orthogneisses, between 568 and 593 Ma and between 597 and 618 Ma. The first data set yielded a $^{206}Pb/^{238}U$ weighted mean age of 584.8±3.6 Ma and the second set yielded a mean age of 606.6±4.4 Ma (Fig. 17a, b). We consider the younger age closer to the crystallization time of igneous protolith while the slightly older mean age is interpreted as an early crystallization event during melt genesis. The main conclusion is that even under the eclogite metamorphic facies conditions, the zircons with less than 300 ppm U do not show isotope systems resetting.

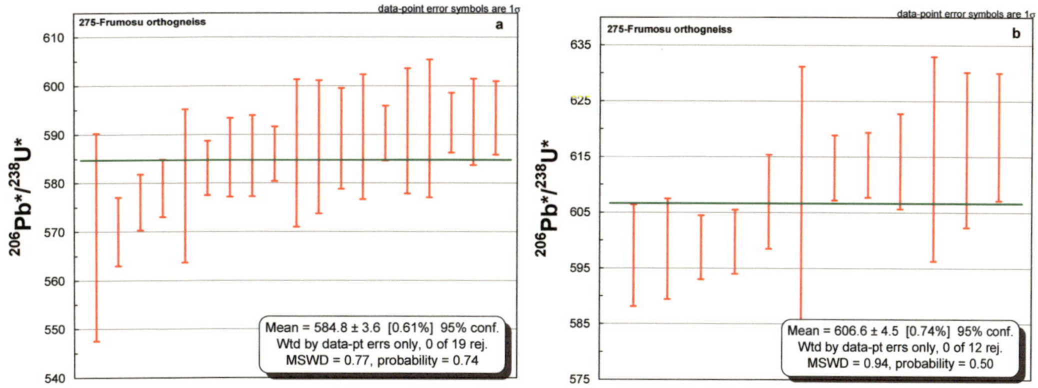

Fig. 17a, b. Weighted mean ages for the two sets of ages from sample 275.

Tău orthogneiss crops out in the median part of the Sebeş valley as a component of the Cumpăna Ordovician metamorfic unit.

The grains from sample 272 are characterized by: (1) - over 1000 ppm U in all the grains; (2) - a deterioration of concordances, yet acceptable for many grains (Fig.18); (3) - a single grain from 36 preserving a protolith age; (4) - Th loss toward the younger ages; (5) - ^{204}Pb gain is in variable quantities and not recorded by all grains.

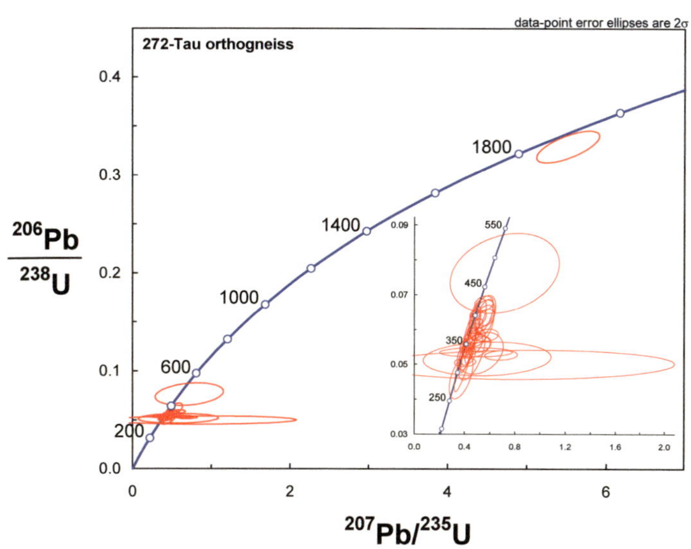

Fig. 18. Concordia projection for sample 272.

It is only obvious that at such high U-content all the grains suffered amorphization prior to the Variscan eclogite event and that during this event they underwent the *Type I* high temperature recrystallization process. The incomplete radiogenic Pb loss can be explained by fluid intervention proved by ^{204}Pb gain.

5. U, Th and Pb isotope systems in detrital zircons from medium grade metaquartzites and paragneisses

Most frequently the detrital zircons remain stable under crustal thermodinamic conditions because they were well selected with respect to their U-content during weathering, transport, and sedimentation. This observation is generally valid for zircons from quartzitic rocks that underwent long and possible repeated sedimentary cycles. In paragneisses, however, the material is often poorly sorted and can originate from proximal sources. The above situations will be exemplified by samples from the pre-Alpine Orliga terrane in North Dobrogea, involved in a Variscan suture as a lower plate (Balintoni et al., 2010a). The basement of the Orliga terrane has been intensely migmatized during Variscan orogenic event, fact that suggests minimum temperatures around 650-700° C. We will begin by scrutinizing the detrital zircons from the sample 336, a metaquartzite.

In this particular case, the U-content in grains is quite low (out of 72 measured grains, 52 have less than 200 ppm U). Furthermore, the U/Th ratio is typical for magmatic zircons (generally smaller than 3.5), the $^{206}Pb/^{204}Pb$ ratio indicates low ^{204}Pb content, and the concordances are surprisingly good even for early Proterozoic or Archean ages (Fig. 19). These observations confirm the lack of the isotopic disturbances in all the grains. The age data are interpreted to represent original crystallization ages in the zircon sources with no signs of the Variscan thermotectonic events recorded by zircons. The metasediment deposition age is not older than the late Cambrian.

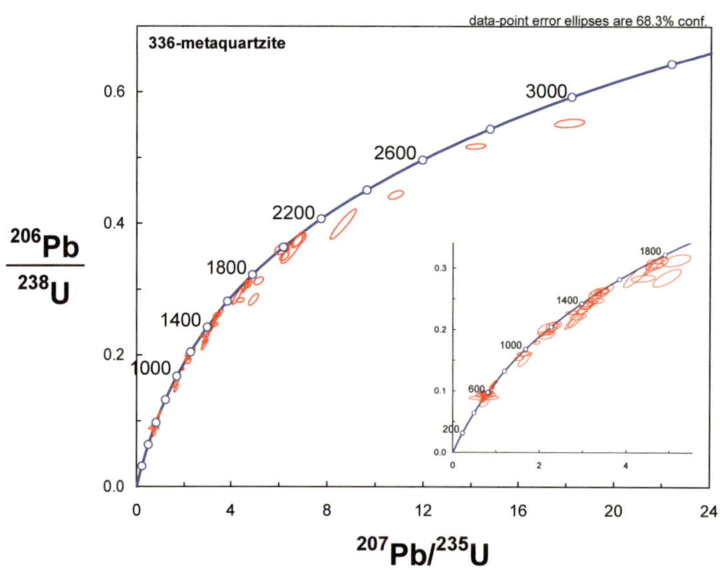

Fig. 19. Concordia projection for sample 336

The zircon grains from sample 167GPS (a paragneiss) are characterized by the followings: (1) - numerous ages clustered around 300 Ma; (2) - variable U-content in zircon grains; (3) - strongly modified U/Th ratio in comparison with the sample 336; (4) - high $^{206}Pb/^{204}Pb$ in all the grains; (5) - good concordances (Fig.20).

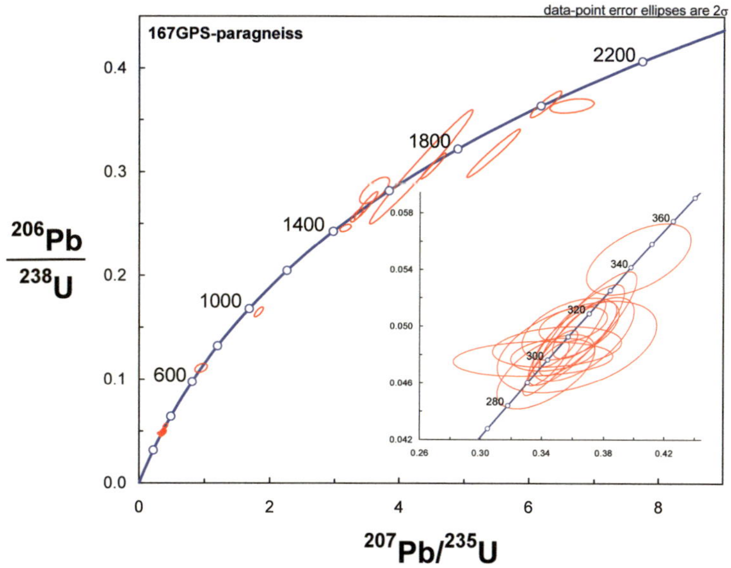

Fig. 20. Concordia projection for sample 167GPS.

The isotopic parameters for the young as well as for the old grains are presented in Table 4.

	Ages between 343.5 -294.8 Ma	Ages older than 600 Ma
U average content (ppm)	686.7	209.9
U/Th average ratio	25.8	2.4

Table 4. Comparative isotopic parameters for different sets of grains from sample 167GPS.

Considering all the data, several conclusions can be drawn. From 26 measured grains, 15 grains recrystallized during Variscan incipient melting. All these grains have an U content usually exceeding 300 ppm U. They lost all the previous radiogenic Pb, a great part of Th, and possible some U. In the same time, the recrystallized grains did not gain any ^{204}Pb and their concordances are remarkable good. These observations suggest a *Type I* recrystallization process by epitaxial migration of the interfaces between the crystalline and amorphizated parts of the grains. The grain lattices have recovered completely without fluid intervention and in presence of a melt. Apparently, ca. 300 ppm U was the boundary between the damaged and undamaged zircons.

A more complex situation is depicted by the data of sample 335.

In great lines the processes are similar to those in sample 167GPS. However, two parameters are more deteriorated in the zircons of sample 335 than in sample 167GPS: many grains gained ^{204}Pb and the ages moved away of Concordia (Fig. 21).

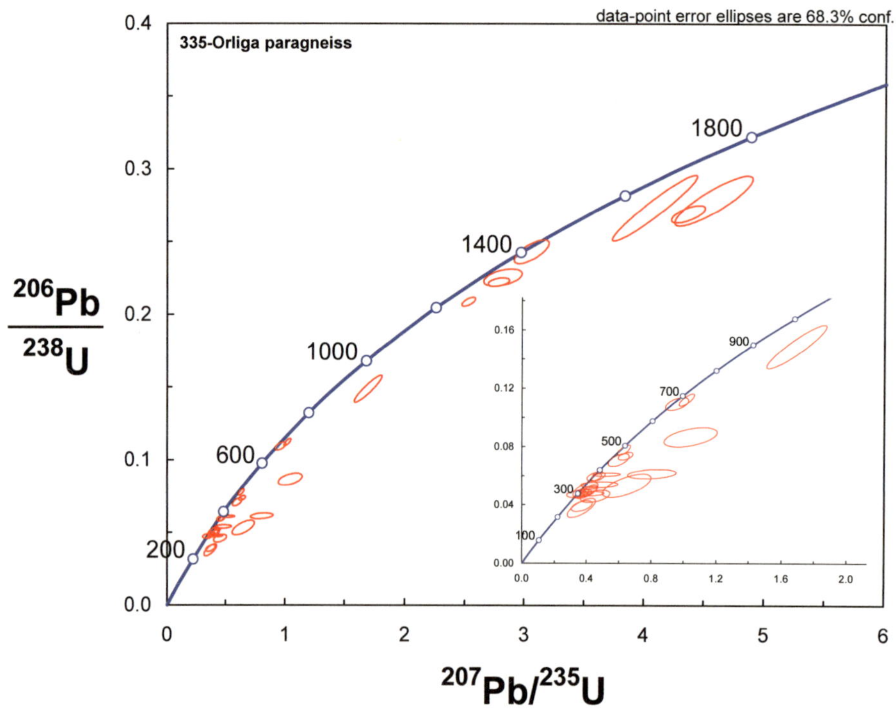

Fig. 21. Concordia projection for sample 335.

These two facts suggest that fluids were extensively involved in the recrystallization processes. Several zircon grains exhibiting ages bellow 300 Ma indicate also later disturbing events.

To see again the role of the U content in zircons history we exemplify by the sample 168GPS.

With the exception of 3 grains (out of 34 analyzed grains), all the other have their U content less than 200 ppm. None of the analyzed grain was isotopically disturbed, clear evidence that at such low U content the effects of the amorphization process are indiscernible and the lattice damage do not accumulate even in geological time. All the grains show the original ages and no sign of the Variscan thermotectonic event is evident (Fig.22).

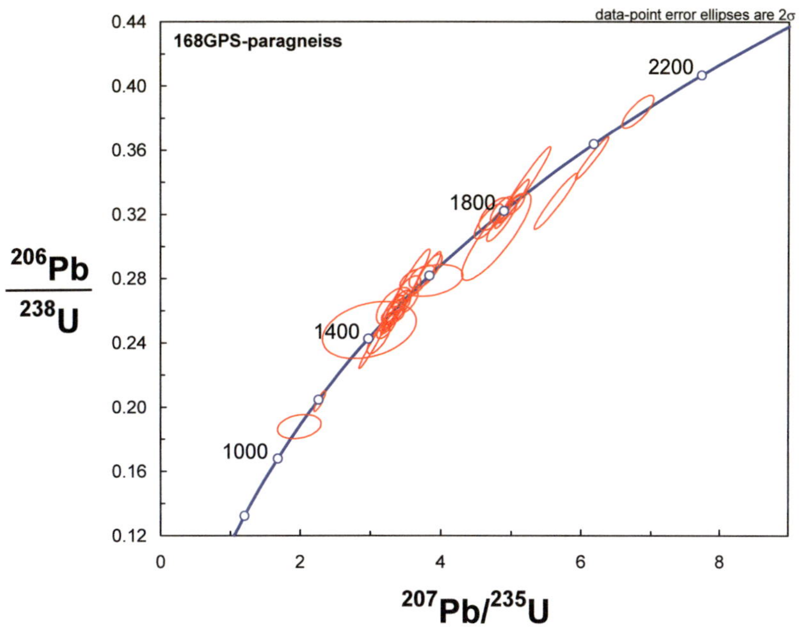

Fig. 22. Concordia projection for sample 168GPS.

6. Conclusions

According to the presented data, the boundary between accumulation of lattice damage and continuous recovery, below the critical amorphization temperature can be set at around 300 ppm U concentration. The undamaged lattices with less than 300 ppm U, sometimes show along the Concordia a spread of ages of ca. 40 Ma from which two valid weighted mean of Concordia ages can be obtained. Usually, the grains display a slight increase in the U-content and U/Th ratio toward the younger ages. The younger age is probably closer to the real crystallization age of the rock.

The migration of ages with respect to the Concordia, when the zircon grains contain over 300 ppm U, can not be explained easily in many situations. This is at least partly because some of the grains appear to lose all the radiogenic Pb at intermediate stages between well known major thermotectonic events. However, the high U-content grains can be useful in deciphering the thermotectonic history of the rocks whenever they reset dominantly around younger ages. If all the grains from a sample contain more than 300 ppm U, the initial age can be totally reset by the subsequent thermotectonic events.

Both, the magmatic and detrital zircons with over 300 ppm U record the metamorphic events, yet the low-U detrital zircon from metaquartzitic rocks do not reset throughout the crustal thermotectonic history.

Alteration in the presence of fluids strongly promotes the resetting of isotope systems as well as recrystallization even at temperatures bellow the crytical amorphization threshold. The amorphization grade is proportional with the U content, while there is little evidence for the role played by the pressure during recrystallization.

Most frequently, Th is lost together with radiogenic Pb, but U loss is by far less a common phenomenon.

7. Acknowledgements

This work was supported by a grant of the Romanian National Authority for Scientific Research, CNCS - UEFISCDI, project number PN-II-ID-PCE-2011-3-0100

8. References

Balica C., Hann H.P., Chen F., Balintoni I. & Zaharia L. 2007: The Age of the intra-Danubian Suture (Southern Carpathians, Romania). *Eos Trans. AGU* 88 (52), Abstract T31B-0476.

Balintoni I. 1997: Geotectonics of Romanian metamorphic terrains (*in Romanian*): Cluj Napoca, Ed. Carpatica, 176 p.

Balintoni I., Balica C., Ducea M.N., Chen F.K., Hann H.P. & Şabliovschi V. 2009: Late Cambrian-Early Ordovician Gondwanan terranes in the Romanian Carpathians: A zircon U-Pb provenance study. *Gondwana Research* 16, 1, 119-133.

Balintoni I., Balica C., Ducea M.N. & Stremţan C. 2011: Peri-Amazonian, Avalonian-type and Ganderian-type terranes in the South Carpathians, Romania: The Danubian domain basement. *Gondwana Research* 19, 4, 945-957.

Balintoni I., Balica C., Seghedi A. & Ducea M.N. 2010a: Avalonian and Cadomian terranes in North Dobrogea, Romania. *Precambrian Research* 182, 3, 217-229.

Balintoni I., Balica C., Ducea M.N., Zaharia L., Chen F.K., Cliveţi M., Hann H.P., Li L.Q. & Ghergari L. 2010b: Late Cambrian-Ordovician northeastern Gondwanan terranes in the basement of the Apuseni Mountains, Romania. *Journal of the Geological Society* 167, 6, 1131-1145.

Balintoni I., Balica C., Ducea M.N., Hann H.P. & Şabliovschi V. 2010c: The anatomy of a Gondwanan terrane: The Neoproterozoic–Ordovician basement of the pre-Alpine Sebeş–Lotru composite terrane (South Carpathians, Romania). *Gondwana Research* 17, 561-572.

Berza T. & Seghedi A. 1983: The crystalline basement of the Dnubian units in the Central South Carpathians: Constitution and metamorphic history. *Anuarul Instititului de Geologie şi Geofizică* LXI, 15-22.

Black L.P., Williams I.S. & Compston W. 1986: Four zircon ages from one rock: the history of a 3939 Ma-old granulite from Mt. Stones, Enderby Land, Antarctica. Contrib. Mineral. Petrology 94, 427-437.

Cherniak D.J. & Watson B.E. 2000: Difusion in zircon. In: Hanchar W.M., Hoskin P.W.O. (Eds) Zircon. *MSA Reviews in Mineralogy and Geochemistry* 53, 113-143.

Davis D.W., Williams I.S. & Krogh T. 2003: Historical development of zircon Geochronology. In: Hanchar W.M., Hoskin P.W.O. (Eds) Zircon. *MSA Reviews in Mineralogy and Geochemistry* 53, 145-181.

Davis D.W. & Paces J.B. 1990: Time Resolution of Geologic Events on the Keweenaw Peninsula and Implications for Development of the Midcontinent Rift System. *Earth and Planetary Science Letters* 97, 1-2, 54-64.

Ewing R.C., Meldrum A., Wang L.M., Weber W.J. & Corrales L.R. 2003: Radiation effects in zircon. In: Hanchar W.M., Hoskin P.W.O. (Eds) Zircon. *MSA Reviews in Mineralogy and Geochemistry* 53, 387-425.

Farges F. 1994: The structure of metamict zircon: A temperature dependent EXAFS study. *Physics and Chemistry of Minerals* 20, 504-514.

Geisler T., Ulonska M., Schleicher H., Pidgeon R.T. & van Bronswijk W. 2001: Leaching and differential recrystallization of metamict zircon under hydrotermal conditions. *Contributions to Mineralogy and Petrology* 141, 53-65.

Geisler T., Pidgeon R.T., van Bronswijk W. & Kurtz R. 2002: Transport of uranium, thorium and lead in metamict zircon under low temperature hydrothermal conditions. *Chemical Geology* 191, 141-154.

Geisler T., Pidgeon R.T., Kurtz R. & van Bronswijk W. 2003: Experimental hydrothermal alteration of partially metamorphic terranes. *American Mineralogist* 88, 1496-1513.

Geisler T., Schaltteger U. & Tomascheck F. 2007: Re-equilibration of zircon in aequous fluids and melts. *Elements* 3 (1), 43-50.

Gruenenfelder M., Popescu G., Soroiu M., Arsenescu V. & Berza T. 1983: K-Ar and U-Pb Dating of the Metamorphic Formations and the Associated Igneous Bodies of the Central South Carpathians. *Anuarul Institutului de Geologie şi Geofizică Bucureşti* LXI, 37-46.

Harley S.L. & Kelly N.M. 2007: Zircon, tiny but timely. *Elements* 3, 13-18.

Heaman L.M., Lecheminant A.N. & Rainbird R.H. 1992: Nature and Timing of Franklin Igneous Events, Canada - Implications for a Late Proterozoic Mantle Plume and the Break-up of Laurentia. *Earth and Planetary Science Letters* 109, 1-2, 117-131.

Iancu V., Mărunţiu M., Johan V. & Ledru P. 1998: High-grade metamorphic rocks in the pre-Alpine nappe stack of the Getic-Supragetic basement (Median Dacides, South Carpathians, Romania). *Mineralogy and Petrology* 63, 173-198.

Kräutner H.G., Kräutner F., Tănăsescu A. & Neacşu V. 1976: Interpretation des ages radiometique K/Ar pour les roches metamorphiques régénérées. Un exemple - les Carpathes Orientales. *Anuarul Institutului de Geologie şi Geofizică Bucureşti* L, 167-229.

Liégeois J.-P., Berza T., Tatu M. & Duchesne J.C. 1996: The Neoproterozoic Pan-African basement from the Alpine Lower Danubian nappe system (South Carpathians, Romania). *Precambrian Research* 80, 281-301.

Medaris G., Ducea M., Ghent E. & Iancu V. 2003: Conditions and timing of high-pressure Variscan metamorphism in the South Carpathians, Romania. *Lithos* 70, 141-161.

Meldrum A., Boatner L.A., Weber W.J. & Ewing R.C. 1998: Radiation damage in zircon and monazite. *Geochimica et Cosmochimica Acta* 62, 14, 2509-2520.

Mezger K. & Krogstad J.E. 1997: Interpretation of discordant U-Pb zircon ages: An evaluation. *Journal of Metamorphic Geology* 15, 127-140.

Murakami T., Chakoumakos B.C., Ewing R.C., Lumpkin G.R. & Weber W.J. 1991: Alpha decay event damage in zircon. *American Mineralogist* 76, 1510-1532.

Nasdala L., Wenzel M., Vavra G., Irmer G., Wenzel T. & Kober B. 2001: Metamictisaton of natural zircon: accumulation versus thermal annealing of radioactivity-induced damage. *Contributions to Mineralogy and Petrology* 141, 125-144.

Putnis A. 2002: Mineral replacement reactions: from macroscopic observations to microscopic mechanisms. *Mineralogical Magazine* 66, 5, 689-708.

Putnis C.V., Tsukamoto K. & Nishimura Y. 2005: Direct observations of pseudomorphism: compositional and textural evolution at a fluid-solid interface. *American Mineralogist* 90, 11-12, 1909-1912.

Rubatto D. & Hermann J. 2007: Zircon behaviour in deeply subducted rocks. *Elements* 3, 1, 31-36.

Salje E.K.H., Chrosch J. & Ewing R.C. 1999: Is "metamictization" of zircon a phase transition? *American Mineralogist* 84, 1107-1116.

Săbău G. & Massone H.J. 2003: Relationships among eclogite bodies and host rocks in the Lotru metamorphic suite (South Carpathians, Romania): Petrological evidence for multistage tectonic emplacement of eclogites in a medium-pressure terrain. *International Geology Review* 45, 1-38, 225-262.

Tagami T. & Shimada C. 1996: Natural long-term annealing of the zircon fission track system around a granitic pluton. *Journal of Geophysical Research-Solid Earth* 101, B4, 8245-8255.

Tilton G.R., Davies G.L., Wetherill, G.W. & Aldrich L.T. 1957: Isotopic ages of zircon from granites and pegmatites. *Transactions - American Geophysical Union* 38, 360-371.

Tromans D. 2006: Solubility of crystalline and metamict zircons: A thermodynamic analysis. *Journals of Nuclear Materials* 357, 221-233.

Wetherill, G.W. 1956: Discordant Uranium-Lead ages, I. *Transactions - American Geophysical Union* 37, 320-326.

Williams I.S. 1992: Some observations on the use of zircon U-Pb geochronology in the study of granitic rocks. *Transactions of the Royal Society of Edinburgh* 83, 447-458.

13

Recrystallization of Fe_2O_3 During the Induration of Iron Ore Oxidation Pellets

Guanghui Li, Tao Jiang, Yuanbo Zhang and Zhaokun Tang
Department of Ferrous Metallurgy, Central South University Changsha, Hunan, China

1. Introduction

Magnetite and hematite concentrates are the two main raw materials for oxidized pellet production. Magnetite concentrates have more predominance due to the heat release by oxidation during roasting and may reduce energy consumption (Jiang et al., 2008; Li et al., 2009). However, with the continuous consumption of iron resources driven by the rapid development of iron and steel industry, magnetite resources are becoming scarce and so the development of pellet production is restricted to some extent. Thus, it is imperative to make better use of hematite resources to meet the raw material supply for pellet production (Xu, 2001).

In the pelletizing process, the firing of hematite materials leads to the development of pellet strength by oxide crystal bridging, recrystallization of the Fe_2O_3, as well as the formation of a small quantity of slag (Ball et al., 1973).

As for the firing of hematite materials, more heat should be supplied from external sources due to the absence of the exothermic reaction of oxidation of magnetite. So the energy consumption of hematite pellet production is greater than that of magnetite pellets (Jiang et al., 2008). Moreover, it has been shown that hematite pellet has poor roasting properties and do not achieve adequate physical strength until the roasting temperature is higher than 1300°C. Findings show that the hematite particles and pellet structure keep their original shapes if the temperature is below 1200°C. Thus, the size of hematite particles are not enlarged, nor the Fe_2O_3 crystal lattice defects are eliminated until the temperature is higher than 1300°C. At high temperatures, initial connecting bridges are formed between crystal grains and recrystallization of Fe_2O_3 is observed. However, if the roasting temperature is too high (>1350°C), something detrimental would happen as Fe_2O_3 decomposes to Fe_3O_4 expressed as reaction (1), which adversely results in the loss of pellet quality :

$$6Fe_2O_3 \rightarrow 4Fe_3O_4 + O_2 \quad \begin{array}{l} \Delta G^{\theta} = 140380 - 81.38T(J) \\ \ln Po_2 = -\dfrac{70649.22}{T} + 40.96 \end{array} \qquad (1)$$

From the thermodynamic equation of reaction (1), it can be seen that decomposition temperature of Fe_2O_3 increases with increasing oxygen partial pressure. Therefore, excessively high firing temperature and low oxygen partial pressure should be avoided to restrain the decomposition of Fe_2O_3. Thus, it is necessary to maintain at higher roasting temperature for hematite pellet as well as narrower firing temperature range, which makes the operation of firing equipments difficult.

To enhance the induration of hematite pellets, both magnetite-addition and carbon-burdened methods are found to be the favourable techniques in practice. In this chapter, the induration mechanisms of hematite pellet with addition of magnetite concentrate and anthracite powder are revealed by characterization of recrystallization rules of Fe_2O_3 during the oxidization roasting.

2. Fe_2O_3 recrystallization during the firing of mixed hematite/magnetite concentrates pellet

2.1 Materials and methods

2.1.1 Materials

The chemical compositions of iron ore materials and bentonite are shown in Table 1. The size distribution of iron concentrates is shown in Table 2.

Materials	Fe_{total}	FeO	SiO_2	CaO	MgO	Al_2O_3
Hematite	67.60	0.72	1.55	0.15	0.17	1.16
Magnetite	69.31	27.88	1.32	0.23	0.55	1.02
Bentonite	4.38	/	59.05	0.68	1.73	18.72

Table 1. Chemical compositions of materials / %

Materials	+0.075mm	0.0375-0.075mm	-0.0375mm
Hematite	13.50	16.35	70.15
Magnetite	5.93	15.35	78.72

Table 2. Size distribution of iron ore materials / %

2.1.2 Methods

The experimental procedure includes ball preparation, preheating and roasting tests, strength measurement and mineralogical analysis.

For each trial, 5 kg mixed concentrates at the given hematite/magnetite (H/M) ratio was blended with 8% moisture and 0.5% bentonite was used as binder. The green balls were prepared in a disc pelletizer with a diameter of 1000 mm. The green balls of 10-12 mm in diameter were statically dried at 105°C in an electrical furnace for 4 hrs.

Preheating and roasting tests were carried out in an electrically heated horizontal tube furnace with an internal and external diameter of 50 mm and 70 mm respectively. Firstly, the dry balls were put into a corundum crucible and pushed into the preheating zone of the furnace step by step, preheated at the given temperature for a given period. Then the preheated pellets were taken out of the furnace and cooled in the air, or directly pushed forwards into a higher temperature zone for roasting. Finally, the roasted pellets were taken out and naturally cooled in the air.

The compression strength of cooled pellets was measured with an LJ-1000 material experimental machine. An average value of 20 pellets is expressed as the compression strength for each test.

2.2 Effects of H/M ratio on the compression strength of pellet

According to the orthogonal experimental results, the relationship between H/M ratio in pellet and the compression strength of preheated and roasted pellet was investigated and shown in Fig. 1. Experimental conditions are preheating temperature of 900°C and preheating time of 10 min for Fig. 1a, and 900°C preheating temperature, 10 min preheating time, 1275°C roasting temperature and 15 min roasting time for Fig. 1b.

It can be seen from Fig. 1a, that the preheated pellet strength is continuously improved from 190 N/P to 1132 N/P when magnetite ratio is increased from 0 to 100%. The main reason is that magnetite has fine particle size and great specific surface area, and magnetite particles are rapidly oxidized into Fe_2O_3 grains during the preheated stage; moreover, the atoms on the newborn Fe_2O_3 grain surface have greater migrating capability than those on the original hematite grains, and consequently the Fe_2O_3 crystallites are easily formed between particles. It also can be found that the compression strength of roasted pellet increases with magnetite ratio below 70%; above 70%, the strength will be decreased with the addition of magnetite. This may be due to the fact that some of magnetite particles are residual in the core of pellet during the preheating stage, which cannot be completely oxidized in the high temperature roasting stage; the magnetite particles are recrystallized and bonded with the formation of slag phases.

2.3 Firing properties of mixed H/M pellet

2.3.1 Preheating characteristics

2.3.1.1 Compression strength of preheated pellet

The effects of preheating parameters on the compression strength of preheated pellets with different H/M ratios are shown in Fig. 2. It can be seen from Fig. 2a that the strength of pellet with H/M=70:30 reaches 559 N/P when preheated at 900°C for 10 min, while the strength of pellet with H/M=50:50 exceeds 560 N/P when the time is more than 5 min. Fig. 2b shows that the preheated pellet strength keeps increasing with the preheating temperature varying from 800 to 1000°C. The strength of the pellets with H/M=50:50 reaches 470 N/P when the temperature is 800°C. However, the compression strength of the preheated pellets with H/M=70:30 is less than 400 N/P until the temperature is over 850°C. Thus, the results show that the preheating time can be shortened and preheating temperature can be decreased with the increase of magnetite ratio in the pellet.

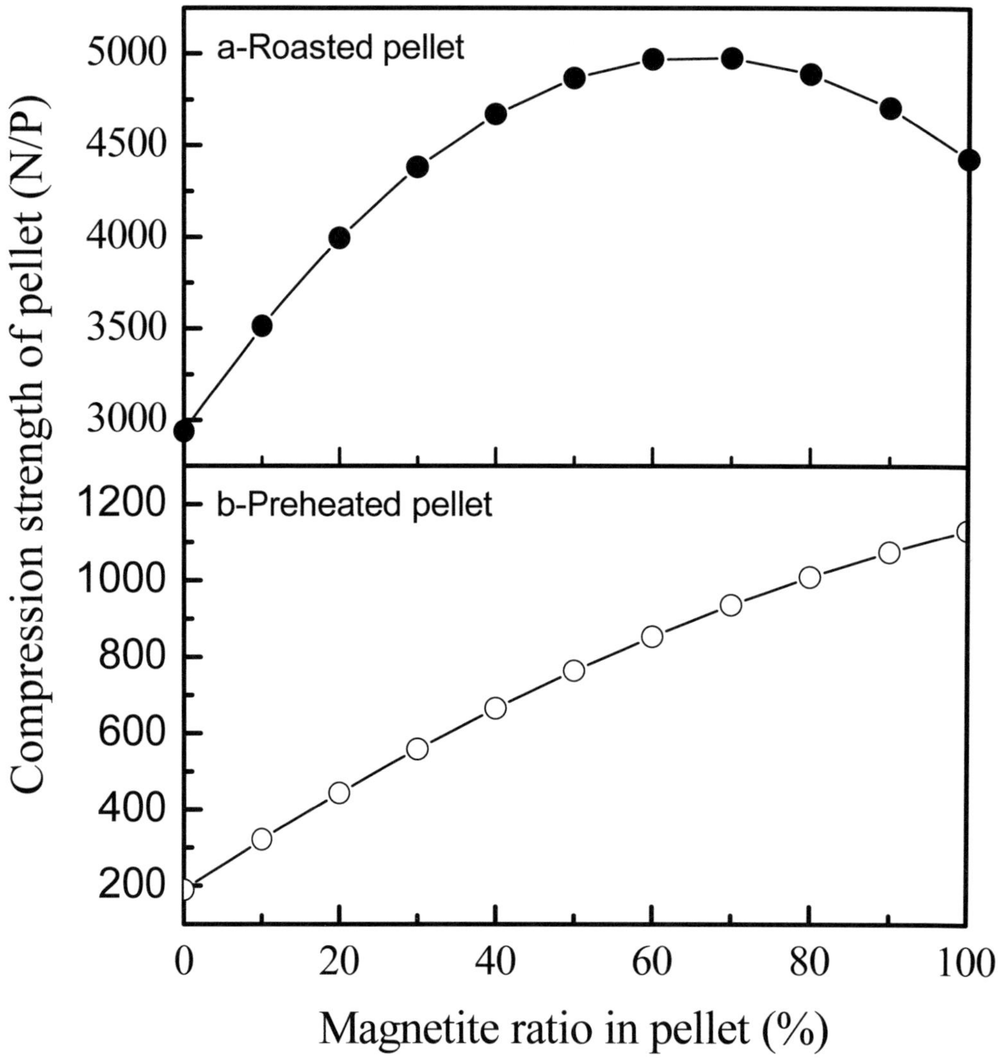

Fig. 1. Effects of magnetite ratio on the compression strength of pellet

2.3.1.2 FeO content of preheated pellet

Whether the magnetite in pellet is oxidized completely or not has an important effect on the roasted pellet performance (compression strength, microstructure, mineral composition, etc.). Generally speaking, residual FeO content in the preheated pellet is required to be less than 3% in operation. The FeO content in pellets with H/M=70:30 under different preheating conditions were analyzed and are shown in Fig. 3. FeO content falls gradually with the increase in preheating temperature and time. FeO content is less than 3% only if the preheating time is more than 11 min at 850°C; however, at 900°C, FeO content goes down to 2.97% when the preheating time is 8 min. In consideration of the compression strength and FeO content together, the suitable preheating parameters for the pellet with H/M=70:30 should be 8~11 min at 850°C~900°C.

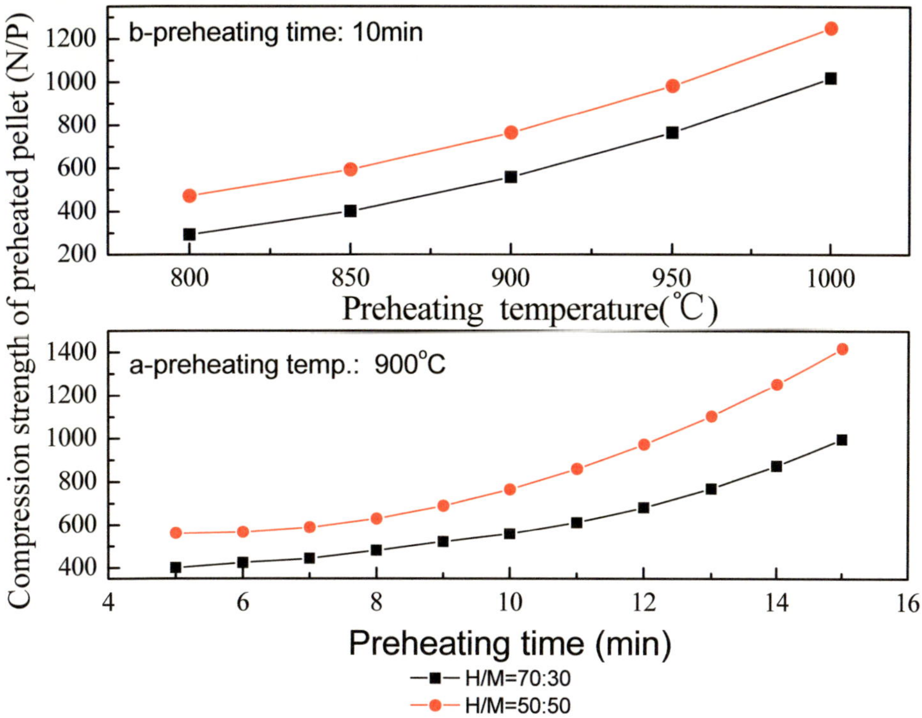

Fig. 2. Effects of preheating on the compression strength of the preheated pellets with different H/M ratios

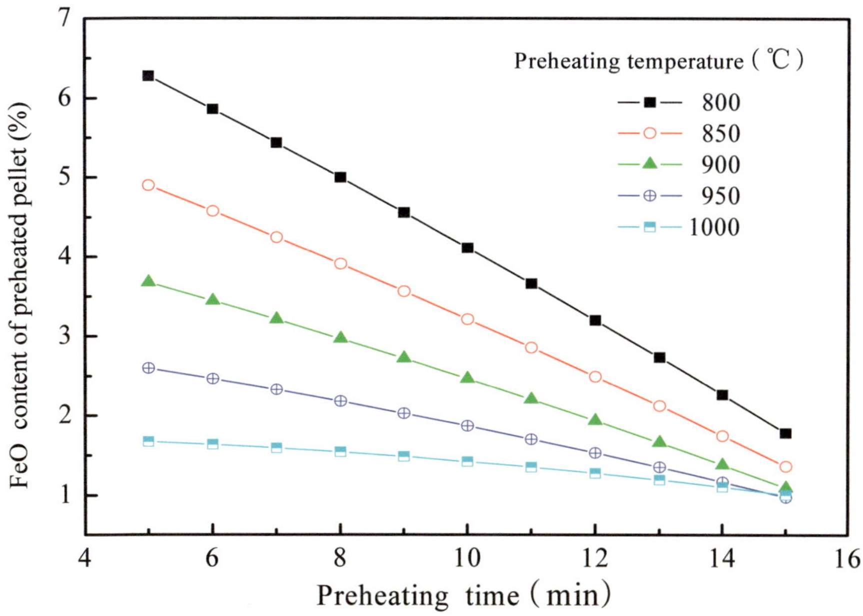

Fig. 3. Effects of preheating on FeO content of the preheated pellets with H/M=70:30

2.3.2 Roasting characteristics

The effects of preheating and roasting parameters on the compression strength of the finished pellets with different H/M ratios are presented in Fig. 4.

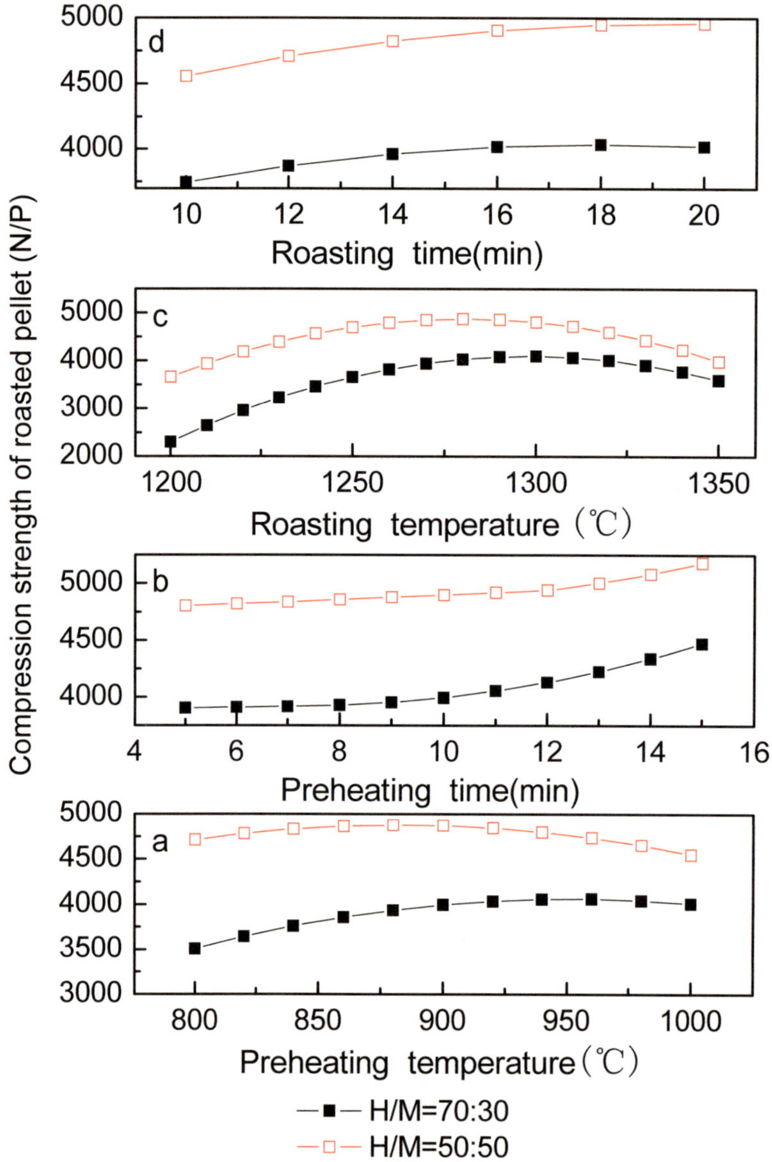

a-preheating time 10 min, roasting temp. 1275°C, roasting time 15 min;
b-preheating temp. 900°C, roasting temp. 1275°C, roasting time 15 min;
c-preheating temp. 900°C, preheating time 10 min, roasting time 15 min;
d-preheating temp. 900°C, preheating time 10 min, roasting temp. 1275°C

Fig. 4. Effects of preheating and roasting parameters on the compression strength of the finished pellet with different H/M ratios

As shown in Fig. 4a, the compression strength of roasted pellets increases quickly with the preheating temperatures varying from 800 to 900°C and remains almost unchanged when temperature is between 900 and 950°C. While the temperature is further increased, the pellet strength decreases slightly. The main reason is that a compact layer is quickly formed and the residual magnetite in the core of pellet is difficult to be oxidized completely. If pellet is roasted in poor oxidative atmosphere at relatively high temperature, low melting point slag phases (such as $2FeO \cdot SiO_2$) are formed, and result in deteriorating the compression strength of roasted pellet. From Fig. 4b, the strength of the roasted pellets increased gradually with increasing preheating time, the reason for which is that magnetite particles are gradually oxidized into Fe_2O_3 crystallites and they are recrystallized more completely with the prolonged preheating time.

As shown in Fig. 4c, the pellet strength increases steadily with the increase of roasting temperature below 1300°C. The possible reason is that the Fe_3O_4 is gradually oxidized into Fe_2O_3, which leads to Fe^{3+} diffusion, rearrangement of Fe_2O_3 crystal lattices and a compact microstructure formed. When the roasting temperature is over 1300°C, however, the pellet strength is reduced with rising temperature, the main reason for which is that: on one hand, it is too late for the residual magnetite in the core of preheated pellet to be oxidized adequately at high temperature, so that low melting point slag phases are formed priorly; and on the other hand, part of Fe_2O_3 crystal grains decomposes at high temperature reversely, so that the structure of the pellet is destroyed to some degree. As far as the roasting time is concerned, as given in Fig. 4d, the pellet strength is significantly improved with the roasting time in the range of 10–15 min and remains constant after 15 min.

The results mentioned above show that adding a proper proportion of magnetite concentrate into the hematite pellet is able to improve the compression strength of both preheated and roasted pellets.

2.4 Induration mechanisms of mixed H/M pellet

Mineral composition, microstructure and Fe_2O_3 crystallization of fired pellets at various preheating and roasting temperatures were studied by using Leica DMRXE microscope with an automatic image analyzer.

2.4.1 Crystallization behavior of Fe_2O_3 during preheating

Oxidation of magnetite into hematite is the main reaction during preheating. Oxygen readily diffuses into the interior of the porous pellet and reacts with magnetite particles during preheating. Oxidation always occurs firstly on the surface of particles and cracks, and usually a few small spotty or lamellae hematite are formed (Figs. 5, 6). The oxidation process will advance towards the core with increasing temperature.

It is shown that the hematite in the pellets involves two types: the one is original hematite (OH), which is the unreacted hematite from the raw hematite concentrate (particle A), and the other, namely secondary hematite (SH), is formed from the oxidation of magnetite concentrate (particle B).

Obvious differences in colour and shape can be observed between the OH and SH grains. For the OH grains, their colours are bright and white, and their shapes are original and with discernible angularity. The strip, triangular and rectangular grains can be distinctly observed, and the surface of crystal grain is smooth and the compact inner structure is unchanged. Moreover, the distance between the two close OH particles is large; there is no crystallitic bond formation between them.

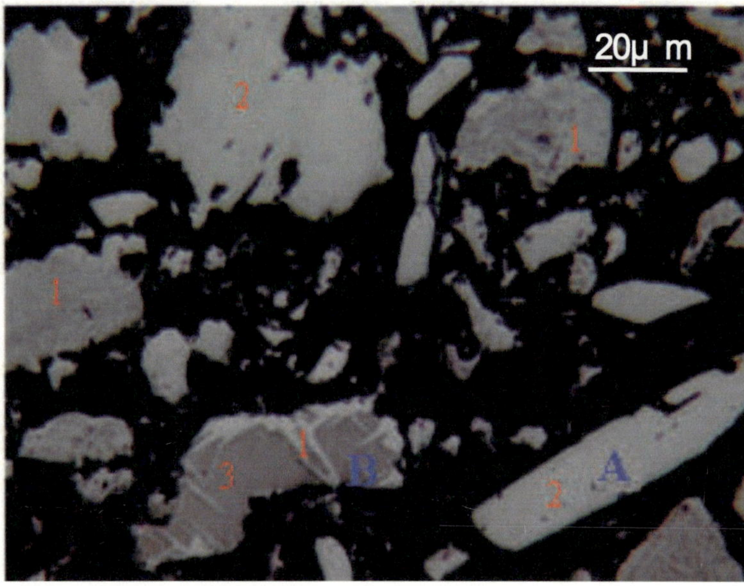

1-SH (columnar); 2-OH; 3-residual magnetite (irregular, tabular)

Fig. 5. Shapes of SH and OH grains in the pellet preheated at 950°C

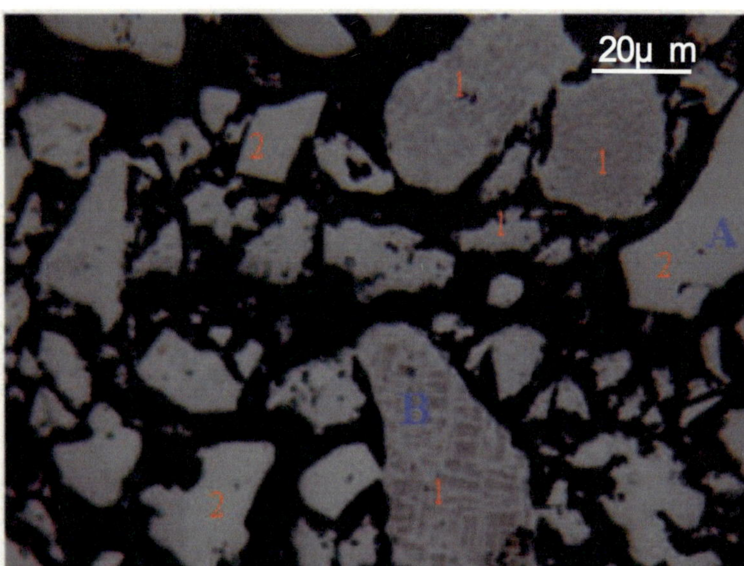

1-SH (reticular); 2-OH; 3-residual magnetite (vein)

Fig. 6. Shapes of SH and OH grains in the pellet preheated at 1000°C

By contrast, for the SH grains, they are relatively heavy-coloured, and some residual magnetite domains can be observed. Tabular or massive grains are the main morphology of residual Fe_3O_4 preheated at 950°C (Fig.5), but the grains change into reticular and vein shape at 1000°C (Fig. 6).

In comparison with OH grains, the angularities of the SH particles disappear or become unclear during oxidation, and they transform into massive, zonal or columnar-shaped particles. The formation of Fe_2O_3 microcrystalline junctions between the close SH particles is significantly different from the OH particles at this stage.

The above results suggest that Fe_2O_3 in SH particles is more active than that in the OH particles. Fe_2O_3 microcrystalline junctions between SH particles are formed when the pellet is preheated, which is able to improve the strength of the preheated pellet. The OH particles keep their original shapes, and no Fe_2O_3 microcrystalline junction can be observed at the preheating stage, thus OH has little contribution to the strength of preheated pellets. Therefore, the improvement of oxidative atmosphere during preheating is able to enhance Fe_2O_3 microcrystalline junctions between the SH particles, the strength of preheated pellet will be improved accordingly.

2.4.2 Crystallization behaviour of Fe_2O_3 during roasting

The crystal morphology of Fe_2O_3 and microstructure of pellets roasted at different temperatures are shown in Figs. 7–9.

1-SH particle; 2-OH particle

Fig. 7. OH grains enclosed by or connected with SH grains in the pellet roasted at 1150°C

As shown in Fig. 7, a large number of junctions between particles are formed by recrystallization of Fe_2O_3 at 1150°C and the strength of the roasted pellet highly increases. However, the inner structure and shape of the OH particles remain visible, which indicates that none of Fe_2O_3 recrystallization takes place in the OH particles at 1150°C. The junctions between particles come from Fe_2O_3 crystallization of the SH particles.

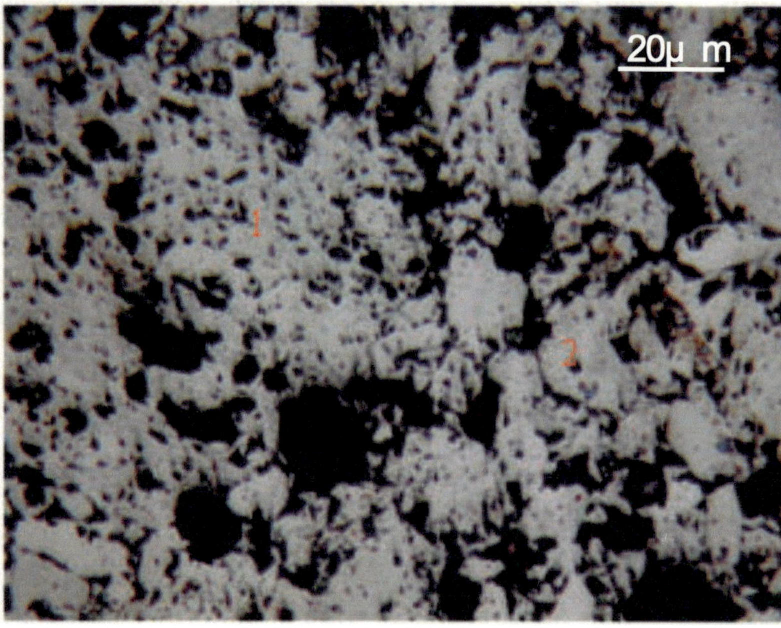

1-SH particle; 2-OH particle

Fig. 8. Recrystallization of Fe_2O_3 in the pellet roasted at 1230°C

Fig. 9. Developed Fe_2O_3 recrystallization in the pellet roasted at 1280°C

At 1230°C (Fig. 8), Fe_2O_3 recrystallization is further developed and the interconnection between particles has been enhanced, and individual particles are scarce in pellet. The OH particles have been connected with or even enclosed by SH particles, and the strength of pellet is further increased. However, the inner structure of the OH particles are still compact, with marked difference from the SH particles, which suggests that recrystallization of Fe_2O_3 in the OH particles is still undeveloped at 1230°C, and the junctions between particles are from the Fe_2O_3 recrystallization of SH particles mainly.

At 1280°C, it can be seen from Fig. 9 that Fe_2O_3 recrystallization in the OH particles develops satisfactorily; all of hematite particles become porous, the profile of the OH particles almost disappears, and all the particles (both SH and OH) are connected with each other through Fe_2O_3 recrystallization and form a whole crystal structure; therefore, the strength of pellet is further improved.

In summary, the results indicate that Fe_2O_3 from SH and OH particles have different activities during roasting, which results in the difference of strength formation mechanisms of the mixed H/M concentrates pellets at various roasting temperatures. At lower temperature (1150°C), the pellet strength is mainly provided by Fe_2O_3 recrystallization of the SH particles. However, OH particles can be connected with the SH particles through crystal bridges formed by high activity Fe_2O_3 of SH particles, or even enclosed by the SH particles. So, OH particles also contribute to the pellet strength to a certain degree.

When the temperature goes up to 1250°C, Fe_2O_3 recrystallization in the SH particles is further developed, and the crystal junctions between the particles become stronger. However, Fe_2O_3 recrystallization within OH particles does not obviously take place, and few joint is formed between the close OH particles. Pellet strength is mainly provided by Fe_2O_3 recrystallization junctions of SH particles.

At 1280°C, Fe_2O_3 recrystallization in the SH and OH particles simultaneously occurs, and the joints between the close particles, including OH and SH, are well developed. Particles are connected with each other and the roasted pellet forms a whole crystal structure. The Fe_2O_3 recrystallization of SH and OH particles plays a crucial role in improving the roasted pellet strength.

The results indicate that, because the activity of Fe_2O_3 in the SH grains is higher than that in the OH grains, Fe_2O_3 recrystallization bonds between particles can be enhanced by the SH grains during preheating and roasting, thus the newborn SH is able to improve the pellet strength and decrease the roasting temperature of mixed H/M concentrates pellet. It is the reason that adding a certain proportion of magnetite concentrate is for the enhanced roasting performance of hematite pellet.

3. Fe_2O_3 recrystallization during the firing of carbon-burdened hematite pellet

3.1 Materials and methods

3.1.1 Materials

The hematite concentrate is characterized by high total iron grade (67.2% TFe) and less impurities (Table 3). The particle size is 92% undersize 0.074mm and the specific surface area is 1629.5cm^2/g (Table 4).

Anthracite powder was used as the material of burdened carbon, of which the specific surface area reaches 6599 cm^2/g after grinding.

Materials	Fe_{total}	FeO	SiO_2	CaO	MgO	Al_2O_3	LOI*
Hematite	67.22	0.55	2.17	0.01	0.05	0.55	0.59
Anthracite	1.48	/	6.86	1.51	0.24	5.88	84.08
Bentonite	7.07	/	60.61	0.94	2.2	17.98	10.41

*LOI-Loss on ignition

Table 3. Chemical compositions of the materials / %

+0.074mm/ %	0.074-0.045mm/ %	-0.045mm/ %	specific surface area/ $cm^2.g^{-1}$
7.79	28.99	63.22	1629.5

Table 4. Size distribution and specific surface area of hematite concentrate

3.1.2 Methods

For each test, 5 kg of hematite concentrate was blended with the given proportion anthracite powder, using 1.25% bentonite as binder. The green balls were prepared in a disc pelletizer with a diameter of 1000 mm, and the green balls with 9~15 mm in diameter were statically dried at 105°C in a drying oven for 4 hours.

Firing teats were carried out in an electrically heated shaft furnace. To simulate firing atmosphere, mixed gas of N_2/O_2 at the given oxygen partial pressure (volume fraction) was pumped into the shaft furnace at a certain flow-rate. The dry balls were charged into heat-resistant pot, and then the pot was pushed downwards into the furnace step by step. The pellets were fired at the given temperature for a given period. Afterwards, the roasted pellets were taken out and naturally cooled to ambient temperature. The compression strength of cooled pellets was measured with an LJ-1000 material experimental machine. An average value of 10 pellets is expressed as the compression strength for each test.

3.2 Firing characteristics of carbon-burdened hematite pellet

3.2.1 Effects of anthracite dosage on compression strength

The effects of anthracite dosage on the compression strength of the roasted pellets are shown in Fig. 10.

As shown in Fig. 10, the pellet strength with 0.5% anthracite is a little lower than that with no anthracite. When anthracite dosage reaches 1.0%~1.25%, the compression strength goes up to the maximum, however, the strength decreases greatly if the anthracite further increases from 1.5% to 4%. The results indicate appropriate anthracite amount of 1.0%~1.25% may improve the strength of hematite pellet.

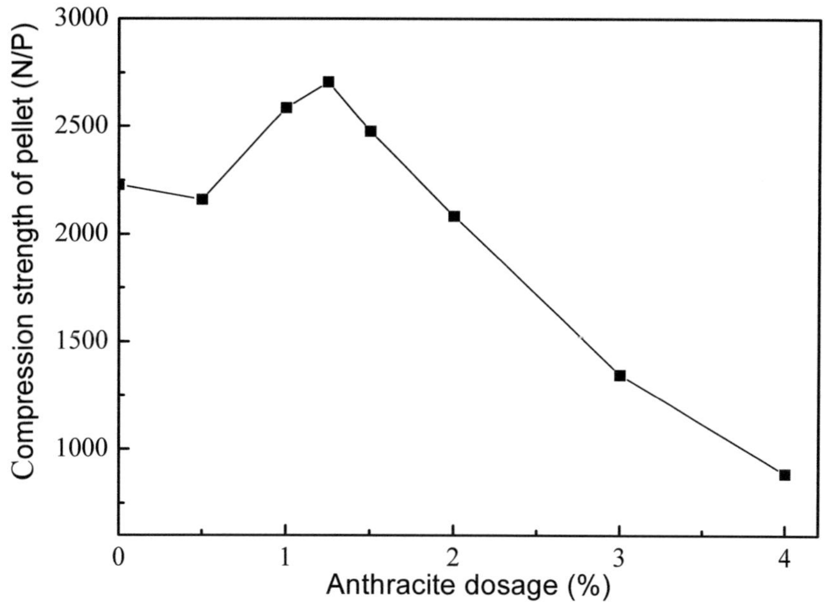

(Oxygen partial pressure: 20%, airflow: 6L/min, roasting temperature: 1280°C, roasting time: 20min)

Fig. 10. Effects of anthracite dosage on the compression strength of pellet

3.2.2 Effects of roasting temperature on compression strength

The effects of roasting temperature on compression strength of pellet are shown in Fig. 11.

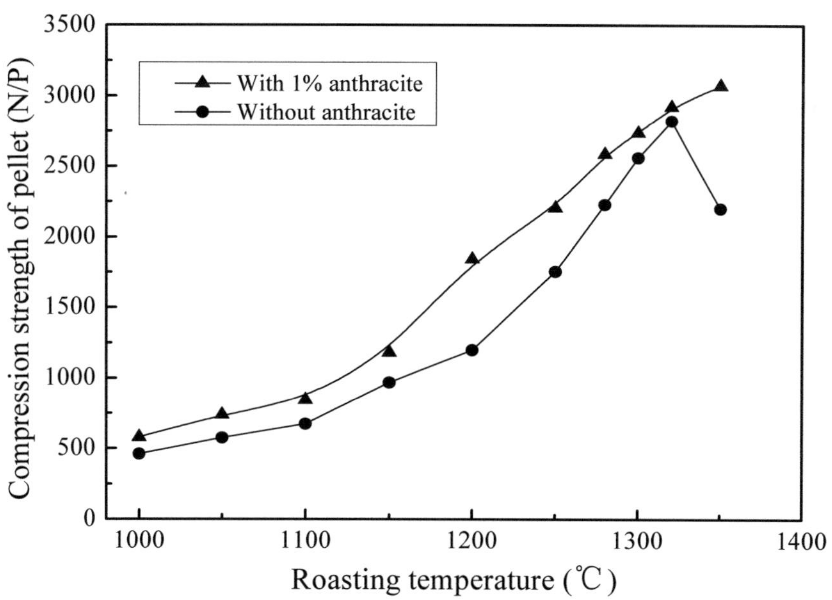

(Oxygen partial pressure 20%, airflow: 6L/min, roasting time: 20min)

Fig. 11. Effects of roasting temperature on the compression strength of pellet

As shown in Fig.11, the strength of carbon-burdened pellet is always higher than that of pellet without carbon when roasted at the same temperature; moreover, for carbon-burdened hematite pellet, the compression strength increases constantly with the temperature. However, the strength of pellet in the absence of carbon not only doesn't increase markedly until 1250°C, but also decreases over 1320°C due to the decomposition of Fe_2O_3. The results imply that the roasting temperature can be decreased and the firing temperature range is enlarged by adding an appropriate amount of anthracite into hematite pellet.

3.2.3 Effects of roasting time on compression strength

The compression strength of roasted pellet increases gradually with the roasting time prolonging and a maximum strength is obtained at 25 min (in Fig.12), hereafter, the strength almost remains constant.

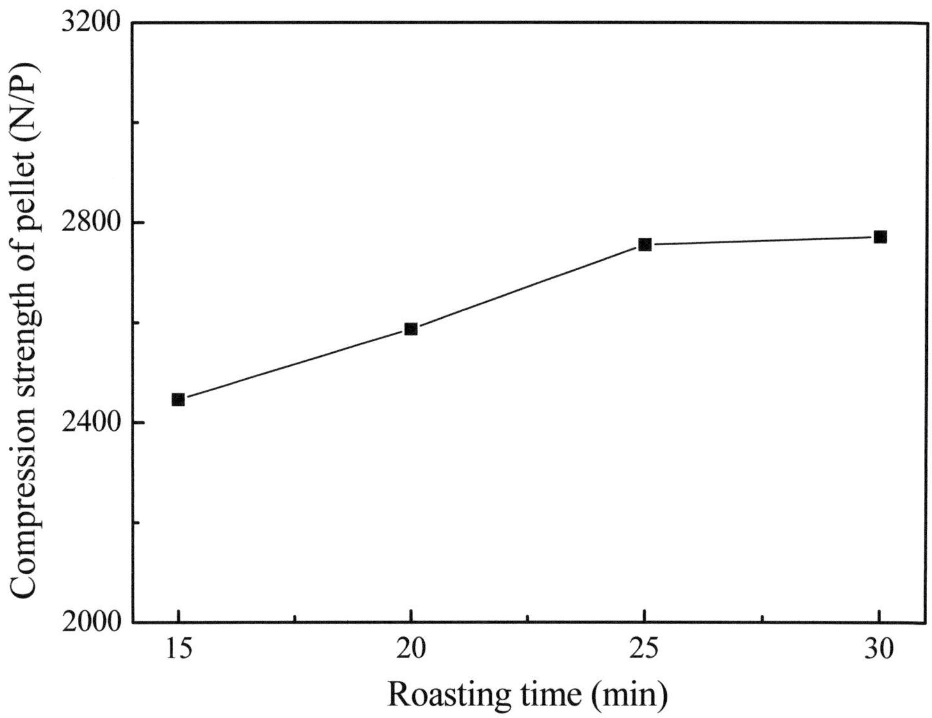

(Oxygen partial pressure: 20%, airflow: 6L/min, roasting temperature: 1280°C, anthracite: 1.0%)

Fig. 12. Effects of roasting time on the compression strength of carbon-burdened pellet

3.2.4 Effects of oxygen partial pressure on compression strength

The compression strength of pellet is susceptible to the change of oxygen partial pressure as shown in Fig.13. The strength of the pellet roasted in N_2, that is, in the inert atmosphere, is slightly higher than that in 10% O_2. The strength nearly reaches the maximum at 20% O_2, and then decreases gradually with increasing the oxygen partial pressure.

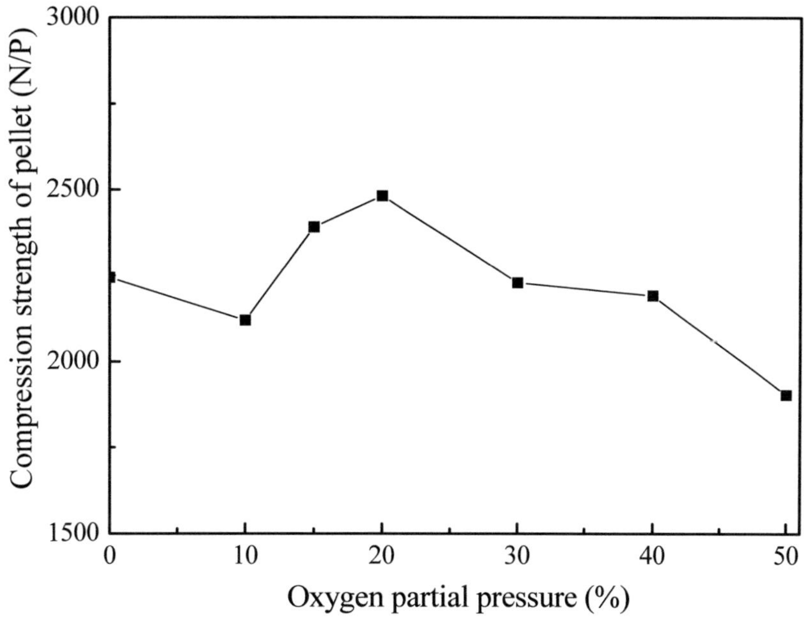

(Airflow: 6L/min, roasting temperature: 1280°C, roasting time: 20min, anthracite: 1.0%)

Fig. 13. Effects of oxygen partial pressure on the compression strength of carbon-burdened pellet

3.3 Roles of burdened carbon during the roasting

To shed light on the effects of the burdened carbon on the induration of hematite pellet, a test, as shown in Fig. 14, was designed to identify the reduction and decomposition of hematite in carbon-burdened pellet during roasting.

As shown in Fig. 14, a cylinder was made by briquetting hematite concentrate firstly, the cylinder bottom is closed, and its inner diameter is 20 mm, the outer diameter is 30 mm. To allow the gas upward injecting into the inner cylinder and penetrating through the anthracite powder layer, many ventages with 0.1 mm diameter were drilled through the cylinder bottom.

Dry pellets with 2~3 mm in diameter were prepared from hematite concentrate in advance, and then were charged into the surface layer of inner cylinder. The cylinder bottom, anthracite powder layer and pellet layer were separated by inert material of Al_2O_3 powder to avoid their contact with each other.

At the beginning of trail, the sample prepared according to Fig. 14 was placed in an electrically heated shaft furnace, and 6L/min N_2 with 99.99% purity was pumped into the shaft furnace from the bottom. The sample was taken out and immersed into water immediately after roasted at 1280°C for 20 minutes. Subsequently, the FeO content of the cylinder bottom and pellet was measured respectively.

It is shown that FeO content of the cylinder bottom and pellet is 4.35% and 28.69% respectively. FeO content of the pellet is obviously higher than that of the cylinder bottom.

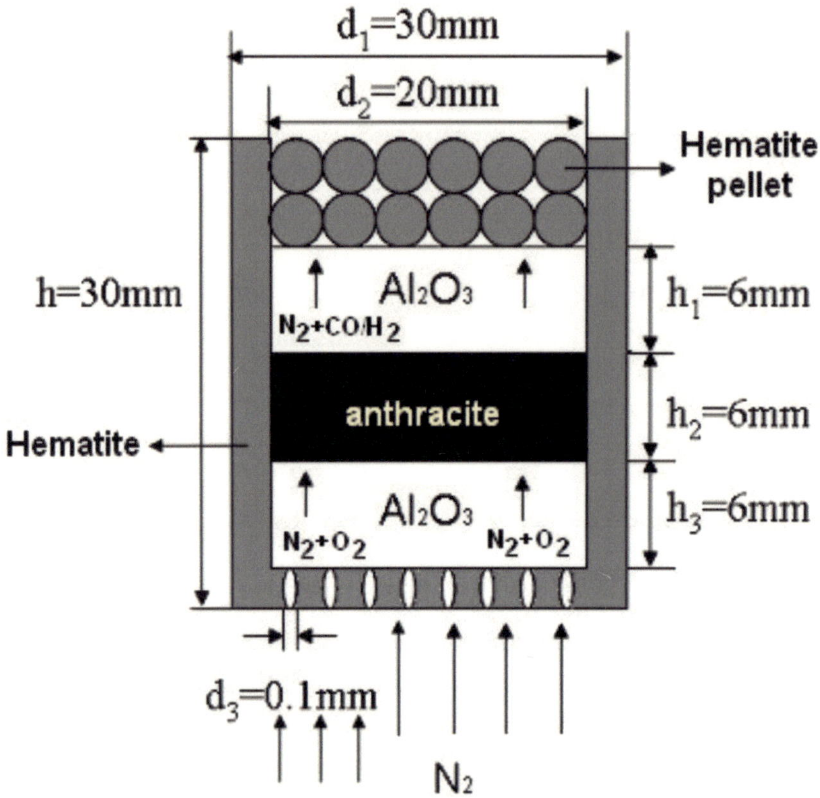

Fig. 14. Schematic diagram of the test on the role of burdened carbon during roasting

In N_2 atmosphere, hematite may decompose into magnetite and release O_2 according to formula (1), and FeO content increases accordingly. It is the reaction that the hematite within the cylinder bottom occurs. However, because of being separated by Al_2O_3 powder, the hematite within the cylinder bottom can't be reduced by anthracite or upwards flowing reductive gases CO/H_2, which are produced by gasification of anthracite. Therefore, the increase of FeO content in the cylinder bottom is only caused by the decomposition of hematite.

However, as regards as the hematite within the pellet, on one hand, it may be decomposed into magnetite in N_2 gas as same as the hematite within the cylinder bottom; on the other hand, it can be also reduced into magnetite by upwards flowing CO/H_2 produced by gasification of anthracite. Therefore, the increase of FeO content of the pellet is caused by both the decomposition and the reduction of hematite, and the latter is more crucial.

The above results identify that the burdened carbon plays the role of reductant during the roasting, and a large number of newborn magnetite are created due to the reductive reaction of hematite by CO/H_2, the products of gasification of anthracite.

Of cause, the burdened carbon can release heat via combustion and heat up the inner pellet, which is advantageous for the induration of hematite pellet. The heating function of the burdened carbon resembles the heat release by oxidation of the magnetite concentrate added into hematite pellet.

3.4 Induration mechanisms of carbon-burdened hematite pellet

3.4.1 Changes of FeO content during the roasting of carbon-burdened pellet

3.4.1.1 Effects of anthracite dosage

FeO content of the roasted pellets with different dosage of anthracite is measured by chemical analysis and the results are shown in Fig. 15.

As shown in Fig.15, the FeO content of the roasted pellets increases gradually with the increasing the dosage of anthracite. This suggests that the anthracite in pellet benefits the reduction of hematite during roasting. Because of the relatively strong reductive atmosphere under the condition of high dosage of anthracite, a large amount of Fe_2O_3 is reduced into Fe_3O_4.

The above results that FeO content of the pellet varies with the anthracite dosage are in accord with the conclusions obtained from the tests as shown in Fig. 14

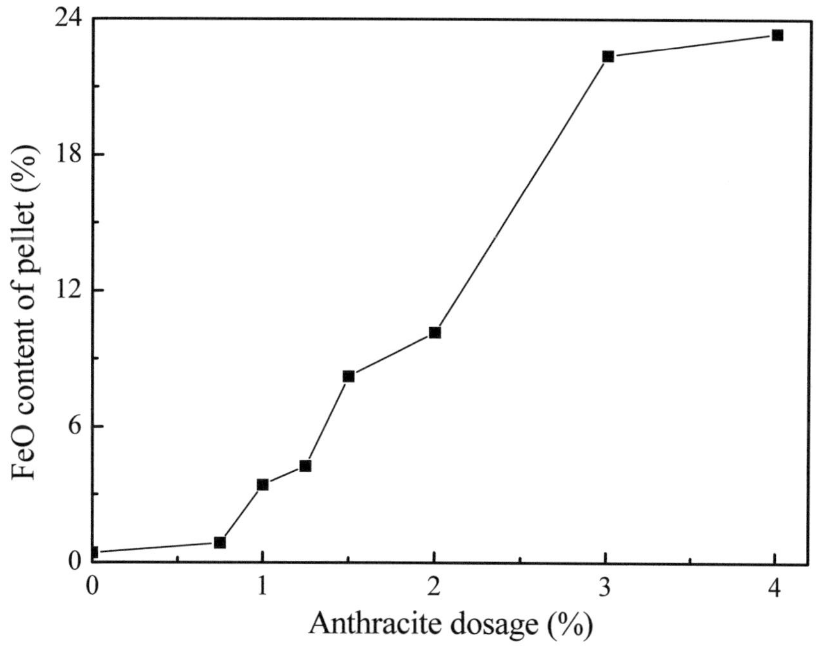

(Oxygen partial pressure: 20%, airflow: 6L/min, roasting temperature: 1280°C, roasting time: 6min)

Fig. 15. Effects of anthracite dosage on FeO content of the carbon-burdened pellet

3.4.1.2 Effects of oxygen partial pressure

FeO content of the pellets roasted at different oxygen contents is measured and the results are shown in Fig. 16.

As shown in Fig.16, the reduction of hematite mainly occurs at the initial roasting stage. The time, when the maximum FeO content attains to, is shortened with increasing the oxygen partial pressure, which indicates that the increase of oxygen partial pressure has favourable effect on oxidization rate of newborn magnetite.

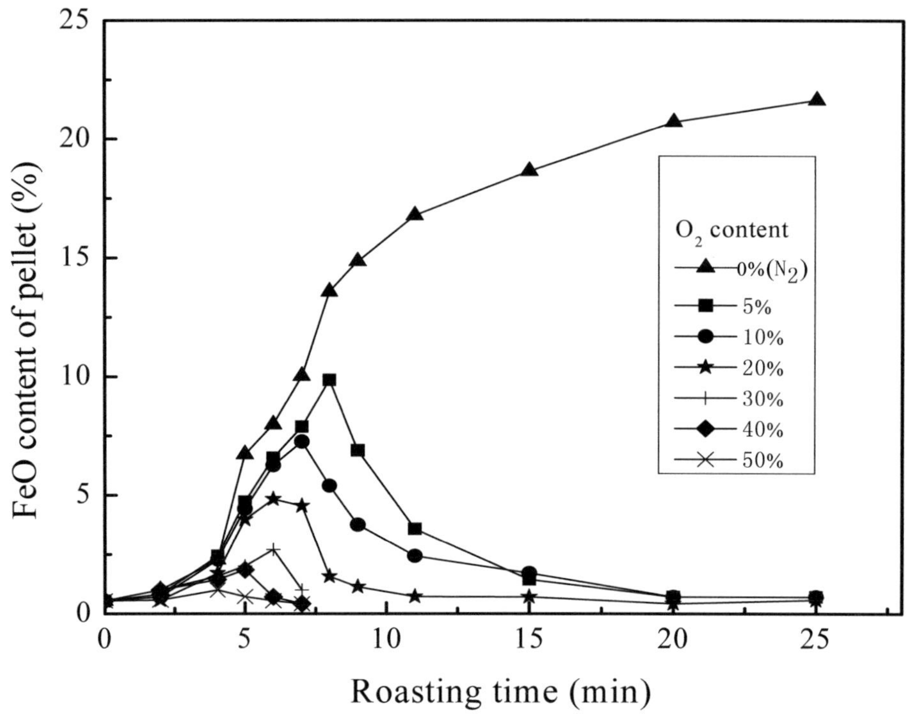

(Anthracite: 1.0%, airflow: 6L/min, roasting temperature: 1280°C)

Fig. 16. Change of FeO content of the pellets roasted at different oxygen partial pressures for different period

The maximum FeO content decreases when the carbon-burdened hematite pellet is roasted at high oxygen partial pressure. The result shows that the lower the oxygen partial pressure during roasting, the stronger the reductive atmosphere within the pellet; and the reduction of Fe_2O_3 can be enhanced. At high oxygen partial pressure, the burdened carbon in pellet combusts completely, and more carbon comes into being not CO but CO_2, thus reductive atmosphere is weakened, which prevents the hematite from reducing into magnetite.

In oxidative atmosphere, FeO content of the pellet increases firstly and then decreases during roasting, the reason for which is that, at the initial roasting stage, the reduction rate of hematite into magnetite is higher than the oxidation rate of newborn magnetite into hematite, however, accompanied with the consumption of burdened carbon, the reduction rate decreases while the oxidation rate increases, the maximum FeO content attains when the reduction rate equals to the oxidation rate. Subsequently, FeO content decreases alone with the oxidation of magnetite untill the oxidation is complete.

It indicates that some original hematite (OH) can be reduced to magnetite by the burdened carbon firstly, and then the newborn magnetite is oxidized into secondary hematite (SH) again during the roasting. The transformation of OH to magnetite and newborn magnetite to SH changes the route of Fe_2O_3 recrystallization during the induration of carbon-burdened hematite pellet.

3.4.2 Fe_2O_3 recrystallization behaviours of carbon-burdened hematite pellet

To shed light on the Fe_2O_3 recrystallization behaviour, the microstructure of carbon-burdened hematite pellets at different roasting stage is investigated and the results are presented in Fig. 17.

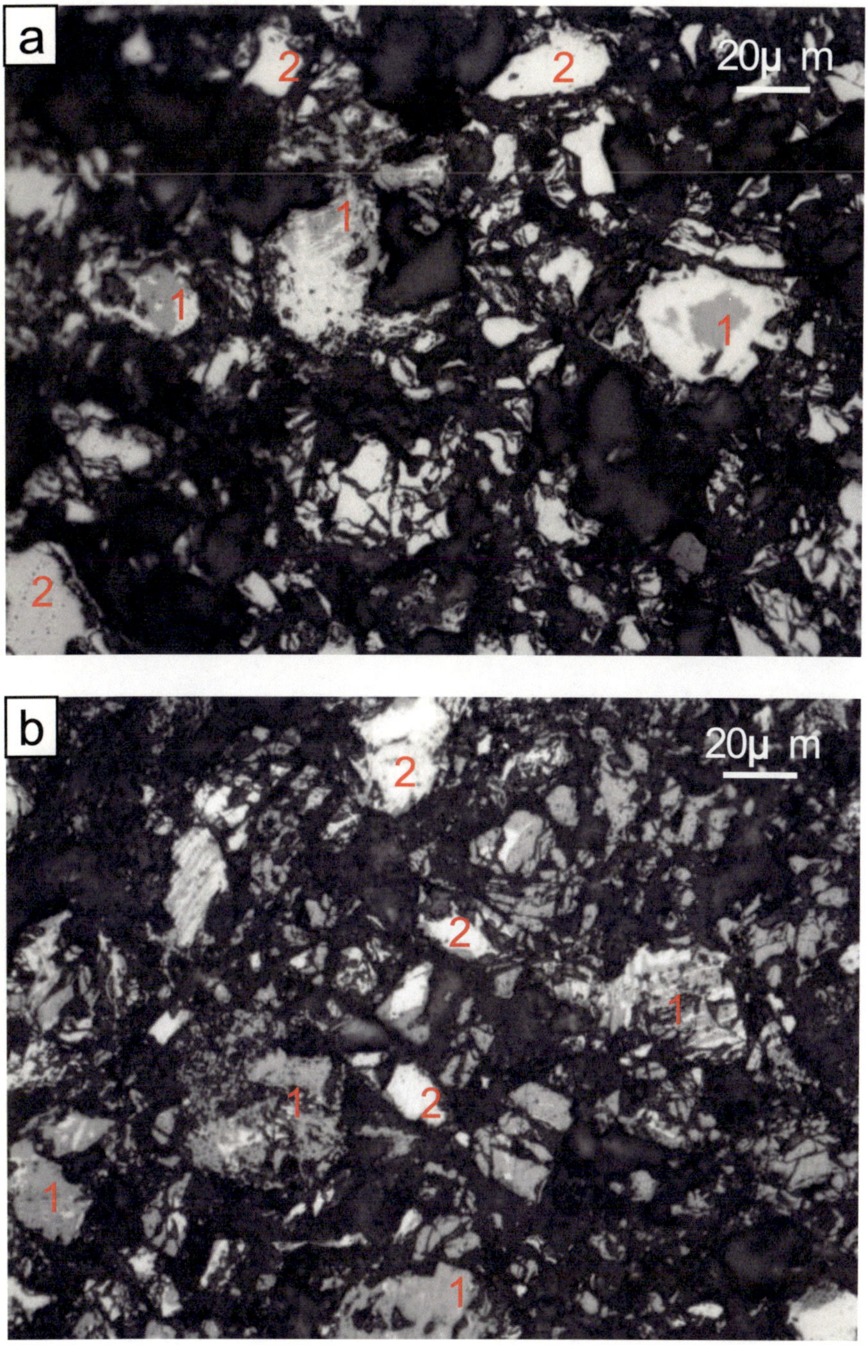

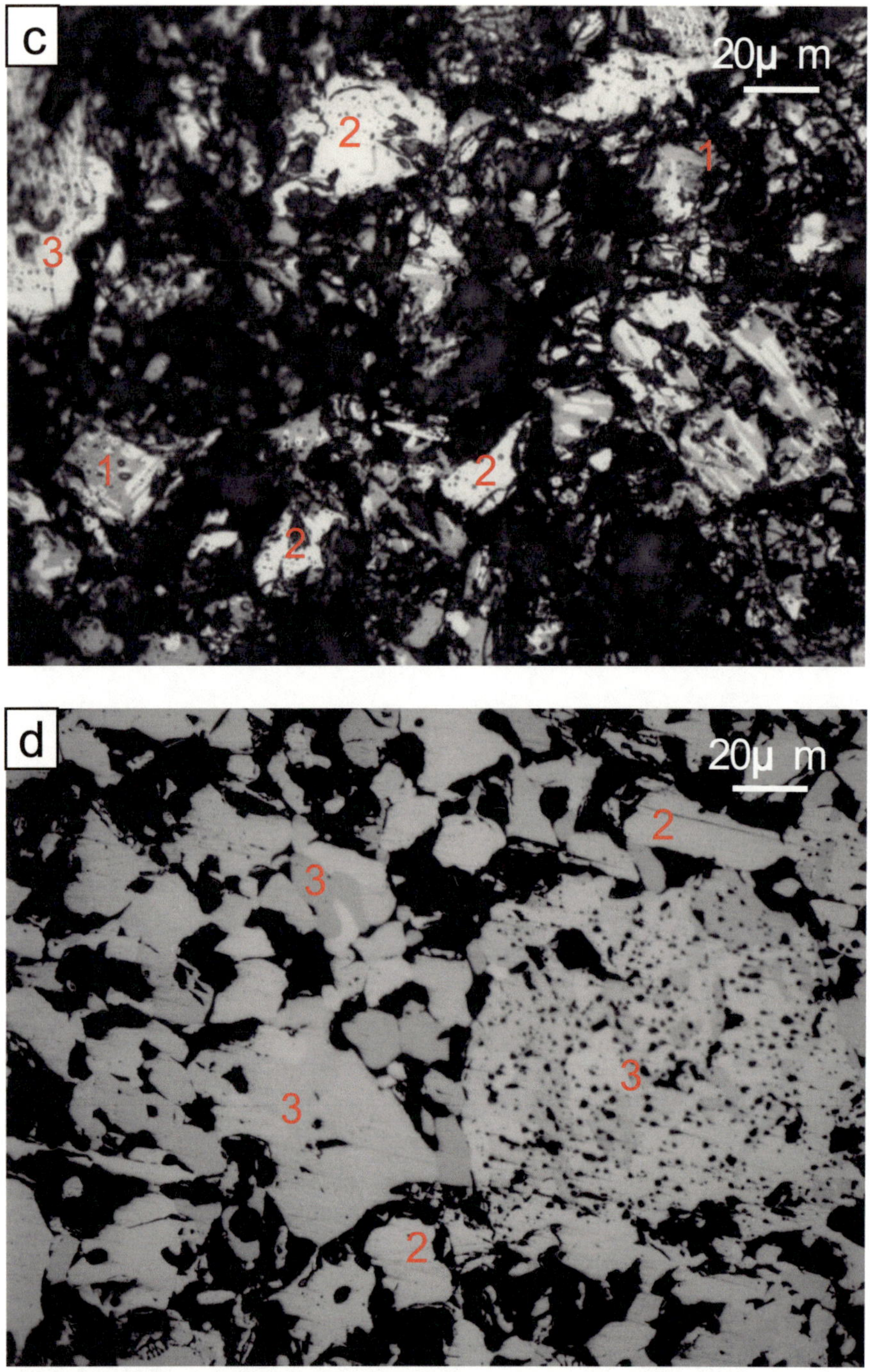

Roasting time: a-4min; b-6min; c-8min; d-20min
1-Magnetite; 2-OH ; 3- SH
(Anthracite: 1.0%, Oxygen partial pressure: 20%, airflow: 6L/min)

Fig. 17. Microstructure of the carbon-burdened pellets roasted at 1280°C

When the volatile matter in anthracite is pyrolyzed and gasified during initial roasting stage, the reduction atmosphere is gradually enhanced within the pellet, and a few magnetite grains, shown in Fig. 17a, are formed by the reduction of OH. An apparent colour difference can be observed between the hematite grains and magnetite grains. The hematite presents bright and white colour, while the magnetite is grey. However, the OH particles keep their original shape and discernible angularity. The inner crystal structure of OH is compact and there is no crystal bond formed between OH particles.

It can be seen from Fig. 17b, a large amount of newly created magnetite is observed as the roasting process progresses; however, the OH particles keep their original shapes, and there is no crystal junction observed still.

As shown in Fig. 17c, the content of newborn magnetite decreases, and a few SH, which comes from the oxidation of the newborn magnetite, are formed. In comparison with OH grains, the angularities of the SH particles become unclear, and a few crystal junctions can also be observed between them. In this stage, with the consumption of anthracite, reduction atmosphere is weakened and oxidation rate of SH is accelerated.

At the end of roasting, the newborn magnetite has been completely oxidized into SH, and a large number of crystal junctions between particles are formed (as shown in Fig. 17d). Pellet strength is mainly provided by Fe_2O_3 recrystallization junctions between SH particles and OH particles.

It can be concluded from the results mentioned above that partial OH particles can be reduced firstly and turn into magnetite particles by the anthracite powder dispersed in the carbon-burdened hematite pellet, however, the newborn magnetite can be subsequently oxidized into SH particles with higher activity, and the route of Fe_2O_3 recrystallization is changed from the recrystallization of OH particles to the recrystallization bonds among SH particles and OH particles. Therefore, the formation of SH during the roasting of carbon-burdened hematite pellet is able to improve the pellet strength and decrease the roasting temperature. It is the reason why adding a certain proportion anthracite is also an effective way to enhance the roasting performance of hematite pellet.

4. Conclusions

Both magnetite and burdened carbon are found to be the favourable techniques for enhancing the induration of hematite pellet. The induration mechanisms of hematite pellet with addition of magnetite concentrate and anthracite powder are revealed by characterization of Fe_2O_3 recrystallization rules during the oxidization roasting.

The crystallization behaviours of Fe_2O_3 during preheating and roasting of pellets made from mixed hematite/magnetite (H/M) concentrates have been revealed. The results indicate that the strength of pellet is mainly provided by the crystalline connections between the particles during preheating. This occurs because the activity of Fe_2O_3 from secondary hematite (oxidized from magnetite concentrate) is higher than that from original hematite (from the raw hematite concentrate). In the roasting process, when temperature is lower than 1250°C, the strength is mainly provided by the development, connection and growth of Fe_2O_3 crystalline grains from secondary hematite. Only if the temperature exceeds 1280°C does Fe_2O_3 recrystallization in original hematite grains develop very well.

Because the activity of Fe_2O_3 in the secondary hematite grains is higher than that in the original hematite grains, Fe_2O_3 recrystallization bonds between particles can be enhanced by the secondary hematite grains during preheating and roasting, and the secondary hematite in H/M concentrate pellet is able to improve the strength and decrease the roasting temperature of hematite pellet. Thus, adding a certain proportion of magnetite concentrate is an effective way to improve the roasting performance of hematite pellet.

The effects of anthracite on oxidation roasting behaviour for hematite pellet have been elucidated. Anthracite in pellet has two functions: the one lies on that a part of heat needed in roasting process can be supplied by the carbon combustion, and the other is that the reduction of partial hematite by the carbon dispersed in pellet, as well as the partial decomposition of hematite at relatively low oxygen partial pressure, leads to the transformation of hematite into magnetite during the roasting.

Based on microstructure analysis, it can be founded that the new-born magnetite, produced from the reduction and the decomposition of original hematite (OH), is oxidized into the secondary hematite (SH) during roasting. Thus, Fe_2O_3 recrystallization bonds between particles can be consolidated by the secondary hematite grains at lower roasting temperature. The strength of carbon-burdened hematite pellet is enhanced and the roasting temperature is decreased due to the formation of secondary hematite. Therefore, adding a certain proportion of anthracite is also an effective way to enhance the roasting performance of hematite pellet.

5. Acknowledgments

The authors want to express their thanks to National Science Fund for Distinguished Young Scholars (50725416), National Natural Science Foundation of China (50604015 and 50804059) and Fundamental Research Funds for the Central Universities for financial supports of this research. Dr. Mingjun Rao is appreciated for his helpful remarks on spelling and expression.

6. References

Jiang, T.; Zhang, Y. & Huang, Z. (2008). Preheating and Roasting Characteristics of Hematite–Magnetite (H–M) Concentrate Pellets. *Ironmaking Steelmaking*, Vol. 35, No. 1, pp. (21-26), ISSN 0301-9233

Li, G.; Li, X. & Zhang, Y. (2009). Induration Mechanisms of Oxidised Pellets Prepared from Mixed Magnetite–Haematite Concentrates. *Ironmaking Steelmaking*, Vol. 36, No. 5, pp. (393-396), ISSN 0301-9233

Xu, M. (2001). Deve1opment of BF Burden Structure in China. *Ironmaking*, Vol. 20, No. 2, pp. (24-27), ISSN 1001-1471

Ball, D.; Dartnell, J. & Davison, J. (1973). *Agglomeration of Iron Ores*, Heinemann Educational, ISBN 0435720104, London

APbill, J. (1985). Carbonaceous Additives for Pelletizing Production, *4TH International Symposium on Agglomeration*, ISBN 0-932897-00-2, Toronto, Canada, June, 1985

Clout, J.; Manuel J. (2003). Fundamental Investigations of Differences in Bonding Mechanisms in Iron Ore Sinter Formed from Magnetite Concentrates and Hematite Ores. *Powder Technology*, Vol. 130, pp. (393-399), ISSN 0921-8831

Yang, X.; Guo, Z. & Wang, D. (1995). Research on the Reduction Mechanism of Iron Ore Pellets Containing Graphite. *Engineering Chemistry and Metallurgy*, Vol. 16, No. 2, pp. (118-126), ISSN 1001-2052

14

Ion-Beam-Induced Epitaxial Recrystallization Method and Its Recent Applications

Rossano Lang[1,4], Alan de Menezes[1], Adenilson dos Santos[2], Shay Reboh[3,4], Eliermes Meneses[1], Livio Amaral[4] and Lisandro Cardoso[1]
[1]*Instituto de Física Gleb Wataghin - UNICAMP, Campinas, SP*
[2]*CCSST, Universidade Federal do Maranhão, Imperatriz, MA*
[3]*Groupe nMat, CEMES-CNRS, Toulouse*
[4]*Instituto de Física - UFRGS, Porto Alegre, RS*
[1,2,4]*Brazil*
[3]*France*

1. Introduction

The transition from amorphous Silicon to crystalline Silicon is a process of great technological importance and has raised an enormous interest also from a purely scientific perspective. Ion irradiation through an amorphous/crystalline interface may stimulate recrystallization or layer-by-layer amorphization depending on the sample temperature and ion beam parameters. In this chapter, we address some key features of this recrystallization phenomenon. The recrystallization/amorphization process will be discussed in relation to its dependence on the energy deposited during the ion beam irradiation, the sample temperature, and the presence of impurities, such as iron atoms dissolved within an amorphous silicon layer. Also it will be discussed the specific experimental condition under which a metastable phase of the $FeSi_2$ binary compound is trapped within a region of the recrystallized Silicon, in the form of nanoparticles. Theses nanoparticles, with different orientations and morphologies, are shown to cause interesting distortions in the surrounding Si crystal lattice. This effect reflects an important application of the ion-beam induced recrystallization process, as a method that could be used to synthesize ordered nanoparticles within a Silicon matrix.

2. Epitaxial crystallization

The amorphous Silicon (a-Si) is a Si phase with well-defined thermodynamic properties and that presents Gibbs free energy of ~ 0.12 eV/at. higher than that of the crystalline phase (c-Si) (Donovan et al., 1985, 1989; Roorda et al., 1989). This implies the existence of a driving force for the transition from the amorphous to the crystalline phase to occur. In other words, there is a natural tendency for the (a-Si) → (c-Si) transformation. At room temperature, the a-Si phase is metastable and it is transformed into c-Si only when submitted to high temperatures, typically higher than 450 °C (Olson & Roth, 1988; Williams, 1983). For the case of an amorphized layer on top of a Si single-crystal substrate, the transition occurs by a planar motion of the crystal-amorphous (c-a) interface, from the interior towards the surface, and hence decreasing the thickness of the existing amorphous layer with time, as schematically illustrated in Fig. 1. This effect is called epitaxial crystallization. For a pure thermal recrystallization process - SPEG (*Solid Phase Epitaxial Growth*), the growth rate is strongly dependent on the temperature and it presents an Arrhenius-like behavior with a unique activation energy of (2.68 ± 0.05) eV over a growth rate range of more than six orders of magnitude (Olson & Roth, 1988). For instance, at 470 °C the c-a interface displacement velocity is ~ 1 Å/min while at lower temperatures, this value decreases considerably and the amorphous to crystal transition becomes kinetically inhibited. However, epitaxial recrystallization of a-Si layers can also be achieved at lower temperatures (~ 200 - 320 °C) by ion-beam irradiation. This phenomenon represents a typical example of a dynamic annealing process and it is usually referred to as *Ion-Beam-Induced Epitaxial Crystallization* - IBIEC (Priolo & Rimini, 1990).

It is well known that both the implantation and the irradiation processes (depending on the dose and the beam ion mass) have the amorphization of the host matrix, as one of their main effects. For the specific case of Si, when the deposited energy (energy loss of ion beam mainly by nuclear collisions) handed by the projectile exceeds the threshold value of ~ 13 eV/at. a buried amorphous Si layer is formed (Narayan et al., 1984). However, if the Si is irradiated at a temperature above ~ 150 °C, there is a competition between both phenomena: amorphization and recrystallization.

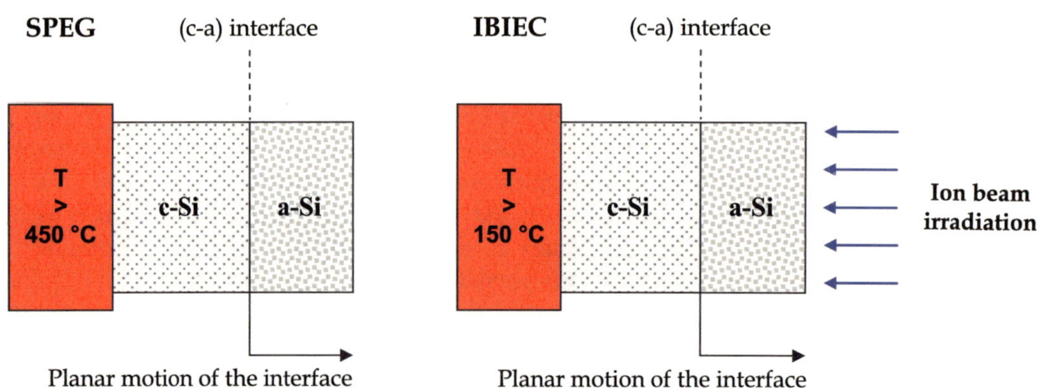

Fig. 1. Schematic illustration of two recrystallization processes in Silicon. SPEG (Solid Phase Epitaxial Growth) is a purely thermal process while, IBIEC is an ion beam induced epitaxial recrystallization process.

In fact, the temperature rise of the Si substrate is not solely responsible for this trend of the crystalline order recovery, since the recrystallization rate is greater than that obtained by a purely thermal process at the same temperature. A more complex mechanism involving a dynamic reordering stimulated by the ion beam is responsible for this effect. The recrystallization under ion-beam irradiation is the result of the prevalence of dynamic annealing over defect generation rate. Both processes depend on the substrate temperature and on the parameters of the irradiating beam, such as ion species, energy, ion dose and ion flux - also known as dose rate.

2.1 Crystal-amorphous interface displacement velocity

Rutherford backscattering spectrometry in combination with the ion channeling technique (RBS/C) is commonly used to directly monitor the c-a interface motion and infer the kinetics of the IBIEC process. The following RBS/C results were carried out with He^+ beam at 1 MeV in a 170° scattering geometry. Figure 2 presents c-a interface position measurements of [100] oriented Si substrate previously implanted at room temperature with Fe^+ ions at 100 keV energy. Channeled implantation was avoided by tilting the sample 7° normal with respect to the incident beam direction. Subsequently, the recrystallization of the Fe-implanted surface amorphous Si layer (filled circles in the figure) was induced by Si^+ irradiation at high energy - 600 keV (whose projected range is well beyond the original c-a interface). In the IBIEC experiments the substrate temperature was fixed and controlled. This allows discrimination between the effects due to the heating of the sample holder and those due to ion-beam irradiation. In order to avoid beam heating effects low current density (≈ 1 μA/cm²) was therefore used with ion flux of ~ 6.2×10^{12} ions/cm²s. The substrate temperature was maintained at 350 °C. Essentially, one observes that the increase of the Si^+ irradiated dose, leads to a decrease in the distance between the c-a interface and the surface. That is, there is a recrystallization towards the surface.

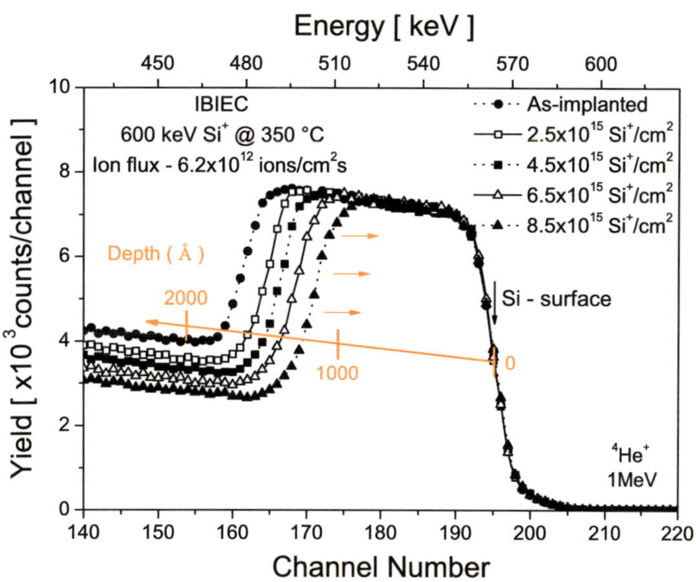

Fig. 2. Determination of the thickness of an amorphous layer using the ion channeling technique. This particular case corresponds to the recrystallization of ~ 176 nm a-Si under 600 keV Si^+ beam irradiation keeping the substrate at 350 °C.

In implantation processes (Dearnaley et al., 1973), the implanted dose Φ is the integral of the ion flux in time, which corresponds to the total number of ions that were focused on the sample per unit area

$$\Phi\ (\mathrm{ion/cm^2}) = \int_0^T \frac{I}{nq_e} dt \quad ; \quad T(\mathrm{s}) = \frac{\Phi \cdot nq_e}{\mathrm{A / cm^2}}, \tag{1}$$

where: I is the beam current in ampere per unit area of the sample (A/cm^2), q_e is the electron charge, n = 1 for once ionized ions, n = 2 for doubly ionized species, and so on. The term I/q_e is designated as the ion flux ϕ (ion/cm^{2}s) and T(s) is the implantation time in seconds. For the IBIEC, one designates a certain irradiation dose Φ and, consequently, the ion flux for such irradiation by $\phi = d\Phi/dt$. When the ion flux is kept constant at a given crystallization procedure, the irradiation dose may be interpreted as a measure of the processing time described by

$$\frac{\mathrm{d}X_{c-a}}{\mathrm{d}t} = \frac{dX_{c-a}}{d\Phi}\frac{d\Phi}{dt} = \phi\frac{dX_{c-a}}{d\Phi}. \qquad (2)$$

Several reports define a c-a interface velocity R as the derivative of the recrystallized thickness (nm) *versus* dose (at/cm^2) curve. Thus, the unit of such "velocity" is expressed in nm/(at/cm^2). In the present study, figure 3 is an example that enables us to extract the interface displacement velocity or recrystallization rate.

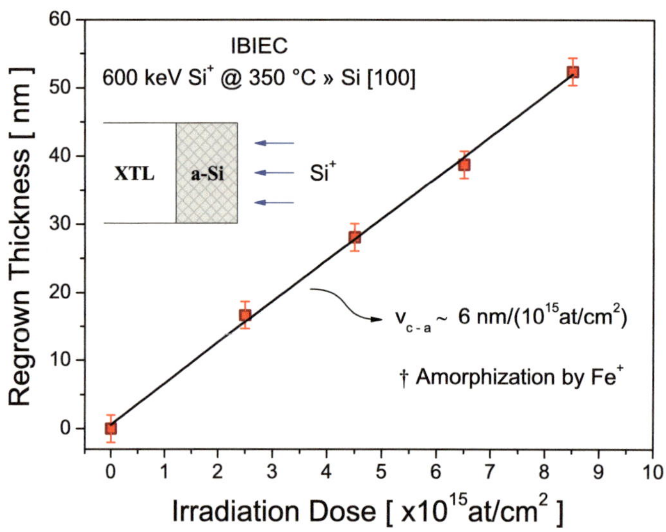

Fig. 3. Regrown thickness for [100] Si amorphized by Fe^+ ions and subsequently irradiated at 350 °C by 600 keV Si^+ ions. The interfacial displacement velocity R can be extracted from the curve.

It should be noted that a comparison between the thermal crystallization rate $V \equiv (\mathrm{d}X_{c-a}/\mathrm{d}t)$ and the recrystallization rate defined for the IBIEC $R \equiv (dX_{c-a}/d\Phi)$ is always possible, if one knows the ion flux ϕ of the process as shown in equation (2).

2.2 Influence of the beam parameters in the IBIEC process

Several experimental results have indicated that the beam parameters such as ion species (by the factor of nuclear energy loss S_n) and the ion flux ϕ have direct influence on the recrystallization rate. With regards to the energy loss by projectile-target elastic collisions, the following result is quite interesting. By using the recrystallized thickness *versus* irradiation dose plot (Fig. 3), the recrystallization rate R as a function of depth was determined as shown in figure 4. In the same figure, one observes the number of vacancies produced by irradiation of Si^{+} ions as a function of depth. This evaluation corresponds to a calculation obtained by the SRIM algorithm with a displacement energy of 15 eV and a lattice binding energy of 2 eV) (Ziegler, 2011). The depth dependence of the regrowth experimental rates seems to follow the profile of defect generation.

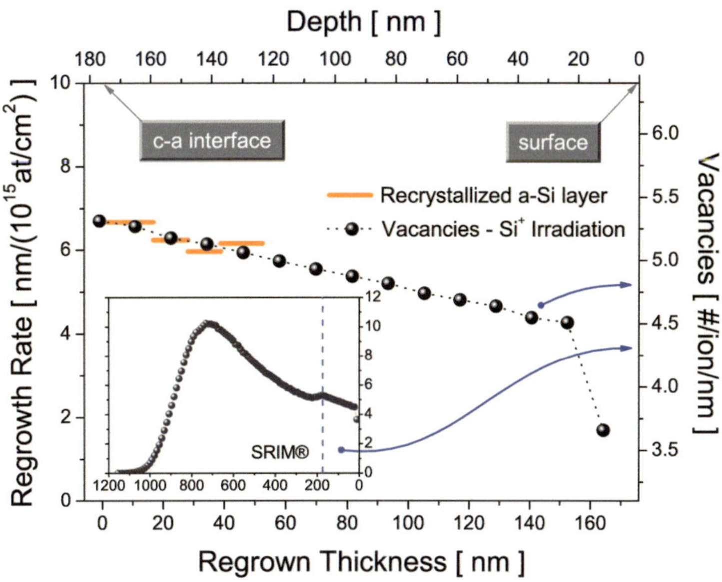

Fig. 4. Regrowth rate versus depth for a Si layer amorphized by Fe^{+} ions and recrystallized at 350 °C by 600 keV Si^{+} ions. The inset shows the number of vacancies generated by Si irradiation calculated by the SRIM code (closed circles).

This result suggests that the ion-induced recrystallization rate is associated with the production of point defects, or in a more general way, with the energy loss into elastic collisions at the c-a interface. Therefore, only those defects generated nearby or directly at the c-a interface are available for the recrystallization process. However, it should be mentioned that the diffusion or vacancy migration is not the promoter mechanism of the IBIEC. This issue will be addressed later. The abovementioned conclusions were also confirmed by other experimental results (Holmén et al., 1984; Linnros et al., 1984, 1985; Miyao et al., 1986; Williams et al., 1985). For instance, Linnros (Linnros et al., 1985) reported a clear experimental evidence of a linear dependence of the ion-induced recrystallization rate on the nuclear energy loss. In their experiments, ion beams of different masses (He, N, Ne, Si, As, Kr) were used to stimulate IBIEC where the recrystallization rate was observed to increase with increasing ion mass. This demonstrates that nuclear energy loss is the mechanism which produces the defects responsible for the IBIEC. Furthermore, the effects produced by both kinds of energy loss (nuclear and electronic) were discriminated through the dependence of the recrystallization rate on the beam energy. It was inferred that the electronic excitations and ionizations play practically no role in the recrystallization process (Elliman et al., 1985; Williams et al., 1985). In other experiments, in which a-Si layers were irradiated by the electron-beam, the recrystallization was observed only for energies above a threshold for atomic displacement of ~ 150 keV while below this threshold no epitaxial regrowth was produced, even after irradiation at very high electron doses (Lulli et al., 1987; Miyao et al., 1986; Washburn et al., 1983). Therefore, the observed epitaxy is associated with elastic collisions that transfer sufficient momentum to displace target atoms from their lattice site.

As mentioned earlier in this section, there is another parameter that directly influences the process - the ion flux. In general, it is observed that the lower the flux ϕ, the higher is the effective velocity $R \equiv (dX_{c\text{-}a}/d\Phi)$ defined by equation 2. This parameter is so important that high fluxes almost inhibit the process, especially for heavier ions (Linnros et al., 1985).

2.3 Temperature dependence

The sample temperature during irradiation is a fundamental variable in the IBIEC process (Elliman et al., 1985; Linnros et al., 1984; Priolo et al., 1988, 1989; Williams et al., 1985). Figure 5 shows, in an Arrhenius plot, the ion-induced growth rate (or recrystallization) as a function of the reciprocal temperature of an a-Si/Si(100) layer recrystallized by 600 keV Kr^{2+} ion irradiation (dose: $1x10^{15}$ Kr/cm^2 and ion flux: $1x10^{12}$ Kr/cm^2s). It also shows the recrystallization rate that represents the thermal contribution (SPEG) with an activation energy of (2.68 ± 0.05) eV. The data were extracted from references (Olson & Roth, 1988; Priolo et al., 1988) and reported in the figure. The growth rate is reported in Å/s (left-hand side) and in $Å^4$/eV (right-hand side). The latter scale represents the growth rate in the form $\Delta X/\Phi \nu(E)$, ΔX being the recrystallized thickness, Φ the dose and $\nu(E)$ the total energy deposited responsible for the displacement production at the c-a interface.

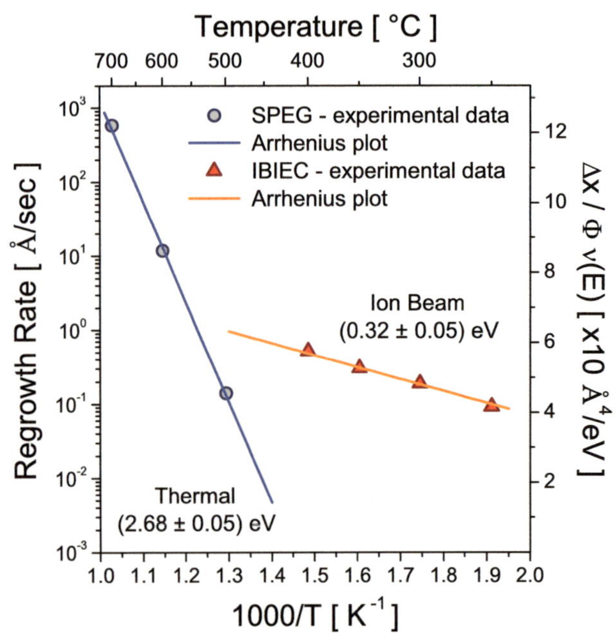

Fig. 5. Ion-induced growth rate versus reciprocal temperature for a-Si layers produced by Si^+ implantation and recrystallized by 600 keV Kr^{2+} ion irradiation. The thermal contribution to the growth rate is also displayed. Data extracted from references (Olson & Roth, 1988; Priolo et al., 1988) and reported in the figure.

It should be noted that recrystallization due to ion-beam irradiation occurs in a temperature range for which the thermal process is kinetically inhibited. Therefore, ion-beam irradiation strongly enhances the kinetics of recrystallization. For instance, at 250 °C the ion-induced growth rate is 0.07 Å/s while, an extrapolation of the thermal data gives a rate of only 10^{-10} Å/s. Furthermore, in the temperature range shown, the IBIEC presents an Arrhenius-like temperature dependence with an apparent activation energy of (0.32 ± 0.05) eV as also demonstrated by several other experiments (Elliman et al., 1985; Linnros et al., 1984; Williams et al., 1985).

It has been proposed that the activation energy for thermal recrystallization of a-Si layers is composed of two terms: one for defect generation and other for defect migration. Ion-beam irradiation clearly removes the main activated process usually limiting conventional thermal regrowth. In fact, during ion-beam irradiation defects are not thermally generated but rather being produced by means of atomic collisions. Jackson (Jackson, 1988) proposed that the activation energy inferred from the IBIEC is not associated with any activated process and therefore considered just as an apparent activation energy. The linear dependence of the regrowth rate as a function of reciprocal temperature, in a logarithmic plot, comes instead, from a balance between different effects. In addition, Jackson (Jackson, 1988) in his intracascade model suggested the dangling bond in the amorphous phase as the promoter for IBIEC. Dangling bonds are structural defects which by moving nearby the c-a interfacial region should produce a rearrangement of the bonds leading to recrystallization. In the mid-1990s, Priolo (Priolo et al., 1990) proposed a phenomenological model of IBIEC which combines the approach of the Jackson model (Jackson, 1988) with the structural and electronic features of models proposed for conventional thermal regrowth (Williams & Elliman, 1986). This model has explained all the experimental results so far.

2.4 Planar amorphization

In the previous sections, it was shown how a combination of thermal energy and energy deposited by ballistic effects can produce a non-equilibrium epitaxial recrystallization. However, IBIEC is a reversible process, where the increase in the ion flux of the irradiating beam, and /or the decrease in the target temperature can cause a planar layer-by-layer amorphization instead of an epitaxial recrystallization (Elliman et al., 1987; Linnros et al., 1986, 1988). Both processes are schematically illustrated in figure 6. At a constant flux, there is a substrate temperature T_R (reversal temperature) such that, when $T > T_R$, the ion irradiation produces epitaxial regrowth whereas, when $T < T_R$ the irradiation produces a layer-by-layer amorphization. The remarkable point is that the amorphization occurs just from the pre-existing amorphous seed and not from regions below the c-a interface, despite the fact that the energy loss by elastic collisions increases with the increasing depth. In this context, the net velocity of the interface motion can be described by two terms: a crystallization and an amorphization, as

$$R = \frac{dX_{c-a}}{d\Phi} = \frac{\mathrm{d}X_{c-a}}{\mathrm{d}t} / \phi = \left[\frac{\mathrm{d}X_c}{\mathrm{d}t} - \frac{\mathrm{d}X_a}{\mathrm{d}t}\right] / \phi . \tag{3}$$

When R is positive, the crystallization regime is prevalent and when R is negative the system is in the amorphization regime.

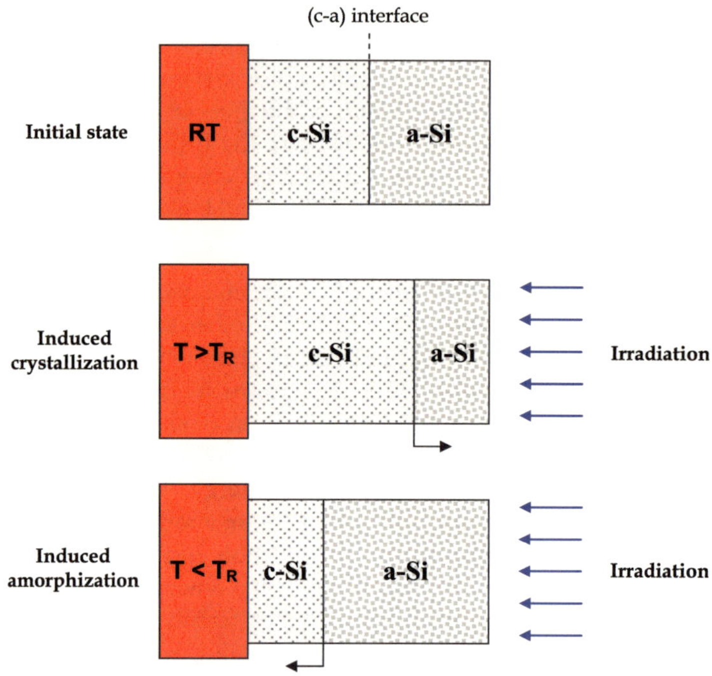

Fig. 6. Schematic representation of the ion-induced crystallization and amorphization. At temperatures below T_R the amorphous layer grows (amorphization regime), whereas at temperatures above T_R, it shrinks (crystallization regime).

An example of planar amorphization is shown in figure 7. In this example, a ~ 90 nm thick layer onto a Si(100) substrate was amorphized at room temperature by 40 keV Fe^+ ions at a dose of $1x10^{16}$ /cm^2. The irradiation was performed by using a 380 KeV Ne^+ beam at a dose of $1x10^{17}$ /cm^2 and ion flux of ~ $1.5x10^{13}$/cm^2s. The substrate temperature was fixed at 100 °C. In the figure two RBS spectra are displayed in channeling condition for the a-Si layer before (open circles) and after (closed circles) Ne irradiation. A random spectrum is also reported. Ne irradiation clearly produces a great amount of damage beyond the original c-a interface. The a-Si surface layer is clearly seen to enlarge under a planar motion towards the sample interior. The amorphous layer has become ~ 50 nm thicker.

2.5 Impurity effects

Besides temperature and beam parameters, the kind of impurity dissolved in the sample plays an important role in the IBIEC. The presence of impurities dispersed within the a-Si layer can dramatically affect the recrystallization process. Depending on their behavior, they can be divided in two major categories: fast and slow diffusers. Fast diffusers comprehend species like Cu, Ag and Au which, at typical temperatures of the IBIEC process (~ 300 °C) have diffusivities of the order of 10^{-12} - 10^{-15} cm^2/s and low solid solubility. These impurities have therefore enough mobility to be redistributed at the advancing c-a interface during recrystallization, modifying the impurity initial profile through the segregation towards the surface imposed by the planar advance. On the other hand, the slow diffusers such as B, P, As, do not present the effect of segregation - are immobile in the time-temperature windows used during the IBIEC experiments, since the interface displacement velocity is much higher than the values of their mobilities in amorphous silicon. In this case, the initial concentration profile of these impurities remains unchanged after recrystallization. This allows one to produce non-equilibrium structures with impurities trapped in the c-Si at concentrations well above their solid solubility. However, the presence of impurities at the c-a interface can modify the recrystallization rate which is observed to increase or decrease, according to the particular species and to its concentration.

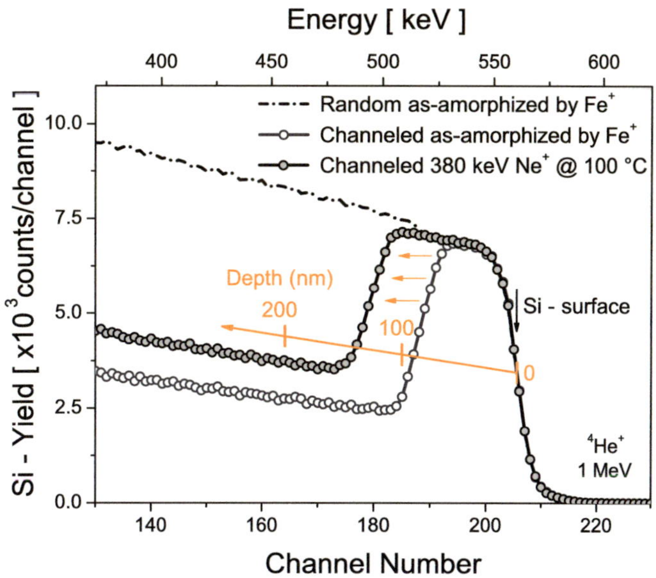

Fig. 7. RBS channeling spectra along the [100] Si axis for a ~ 90 nm thick amorphous layer before (open circles) and after (closed circles) 380 keV Ne^+ irradiation at 100 °C and dose of $1x10^{17}/cm^2$. Note the planar growth of the amorphous phase.

From this brief discussion, it is noticed that the behavior of crucial parameters in the IBIEC process such as temperature and impurities is very similar to that in solid-phase epitaxial growth case. Despite the fact that the impact of these parameters is quite different for each phenomenon, there is a strong indication that similar microscopic processes occur in both cases. More precisely, the same interfacial defect responsible for the thermal recrystallization is considered to be active in IBIEC, being the ion beam the precursor of the increase in the average concentration of these defects.

3. Nanoparticles synthesis by IBIEC

The IBIEC technique has been used as a method to synthesize nanoparticles in Silicon matrix [Lang et al., 2010a, 2010b]. Specifically, the structural properties of the $FeSi_2$ nanoparticles synthesized in Fe^+ low dose implanted Si(100) substrates were investigated. Particularly in this experimental observation, the Fe proved to be a fast diffuser for IBIEC, despite of its action as a retardant of the process, whose recrystallization rate was dependent on the implanted Fe^+ concentration. Nevertheless, only the main results are reported here. The remarkable results which are presented show that the shape of the synthesized material (observed by high resolution transmission electron microscopy - HRTEM) singularly affects the surrounding Si lattice. The lattice strain shape-dependent distribution in both directions: out-of-plane and in-plane was tailored by X-ray Bragg-Surface Diffraction technique.

3.1 Fe^{+} ion implantation in Si(100) and recrystallization process

A Si(100) *n*-type Czochralski wafer (thickness 500 μm, resistivity 10 - 20 Ω cm) was used as host matrix. Fe^{+} ions at 40 keV were implanted at room temperature at an ion dose of $5x10^{15}$ cm^{-2}. Channeling effects were avoided by tilting the sample 7° normal with respect to the incident beam direction. The typical iron beam current density during implantation was about 150 nA/cm^2. Subsequently, ion-beam recrystallization experiments were performed at 350 °C using a 600 keV Si^{+} beam (current density ≈ 1 $\mu A/cm^2$) to a total dose of $6x10^{16}$ $ions/cm^2$. The dose rate resulting from the Si beam current was $6.2x10^{12}$ $ions/cm^2s$. It is worth noting that Si^{+} ions at 600 keV energy have a projected range (average depth) of ~ 770 nm with a straggle (standard deviation) of ~ 150 nm. These values ensure that the irradiation exceeds the pre-existing amorphous layer (well beyond the original c-a interface).

The structures obtained into the as-implanted and recrystallized samples were analyzed and characterized by Rutherford Backscattering Spectrometry combined with ion channeling technique (RBS/C - with He^{+} beam at 1 MeV in a 170° scattering geometry) and also by transmission electron microscopy (TEM - JEOL 2010 operating at 200 kV). High-resolution rocking curves, as well as the reflection mappings of the Bragg-Surface Diffraction reflections, i.e., at the exact multiple diffraction condition, were carried out using the Huber multiaxis diffractometer of the XRD1 beam-line (Brazilian Synchrotron Radiation Facility - LNLS), with an incident beam wavelength of λ = 1.5495(5) Å, as defined by using a Si(111) channel-cut monochromator.

3.1.1 RBS and TEM – Results and discussion

Figure 8 shows RBS/C spectra obtained at random and [100]-channeled direction from the samples before (as-implanted) and after (recrystallized) irradiation. As observed in the aligned as-implanted spectrum, the implantation has produced an amorphous layer over ~ 90 nm while no channeling in the Fe signal was observed. However, expressive reduction of dechanneling yield in the Si profile is detected after IBIEC process (aligned IBIEC spectra relative to aligned as-implanted spectra). This decrease reflects the crystalline order recovery. The entire amorphous Si layer was recrystallized (at an average rate of ~ 0.04 nm/s) with a minimum backscattering yield on the subsurface region being $\chi_{min} \approx 6.3$ %. The RBS concentration-depth profile (not shown here) has indicated an implanted Fe peak concentration of ~ 2.8 at.% at 40 nm from the surface. However, the recrystallization process caused a slight narrowing of the Fe peak and a small segregation towards the surface. A significant degree of channeling ($\chi_{Fe} \approx 46$ %) was also observed in the Fe spectrum.

Figure 9 shows bright-field TEM cross-section images (taken at $[110]_{Si}$ zone axis) of the as-implanted sample. One clearly observes (Fig. 9a) the amorphous Si layer of ~ 90 nm produced by 40 keV Fe^{+} implantation and the crystal-amorphous interface. Figure 9b shows in detail the c-a interface, where one can note a large amount of defects called "end-of-range defects" generated by the implantation process. This defective intermediate zone between the two distinct regions (crystalline and amorphous) is mainly composed of dangling bonds, and this particular kind of structural defect is responsible for the IBIEC process. Upon IBIEC conditions (temperature + irradiation) there is a dynamic rearrangement of these dangling bonds with annihilation in pairs which promotes the layer-by-layer planar recrystallization toward the surface.

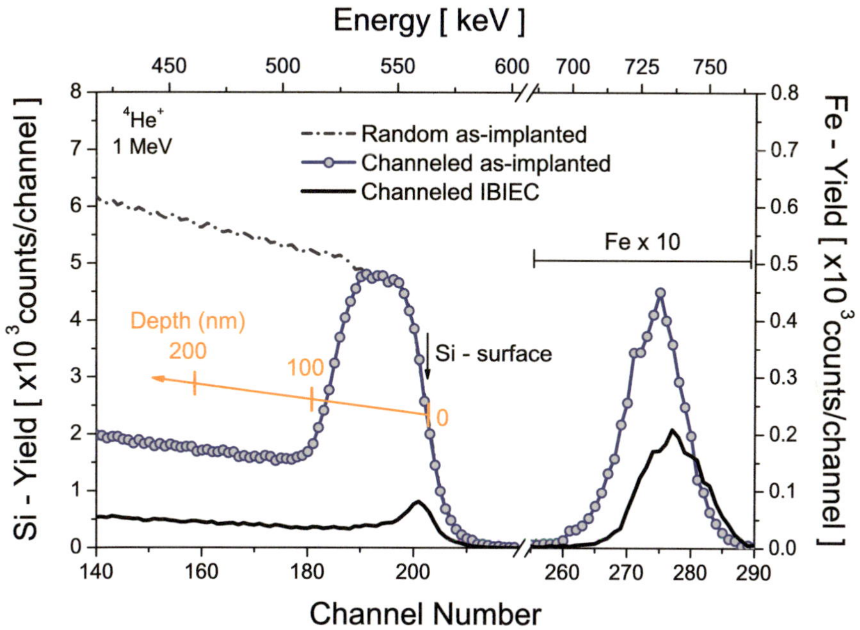

Fig. 8. 1 MeV He^+ RBS spectra in channeling along the [100] direction from a 90 nm a-Si layer containing Fe before (as-implanted sample) and after (recrystallized sample) irradiation with 600 keV Si^+ to a dose of $6x10^{16}$ ions/cm^2 at 350 °C.

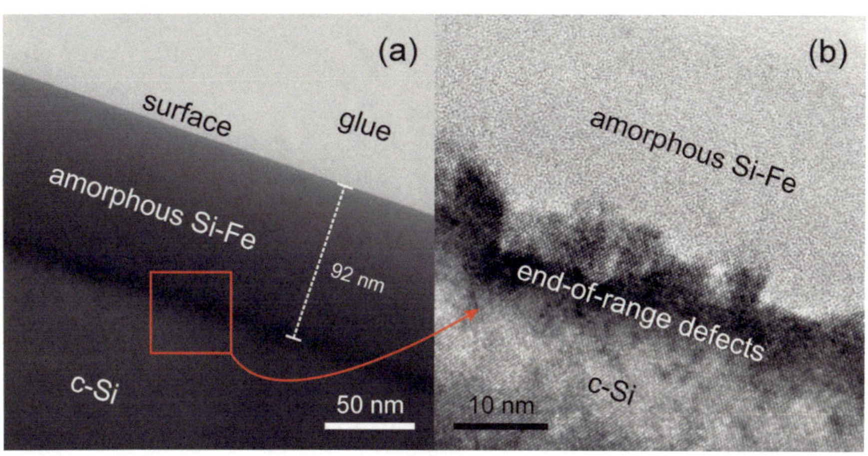

Fig. 9. Bright-field TEM cross-sectional images of the as-implanted sample. **(a)** Overview of the 90 nm thick a-Si layer containing Fe. **(b)** High-resolution of the selected region in **(a)** showing in detail the c-a interface region.

The complete recrystallization of the amorphous Si layer was also confirmed by TEM analyses. It should be noted that for a conventional thermal annealing at the same temperature (350 °C) and time (irradiation time ~ 160 minutes) would have produced a regrowth of only 7.7×10^{-3} nm which is a negligible amount. Therefore, ion-beam irradiation strongly enhances the kinetics of recrystallization. Figure 10 exhibits representative TEM micrographs of the recrystallized sample. The cross-section image, such as Fig. 10a taken along the $[110]_{Si}$ pole and slightly tilted on the zone axis, revealed the efficient a-Si regrowth and an impurity redistribution - nanoparticles formation after the IBIEC process. Iron was completely swept by the moving c-a interface and retained within the precipitated narrowing layer. Despite the amount of Fe present in the recrystallized region, the quality of the recovered crystal appears to be very good, as demonstrated by the TEM image. Three regions regarding the nanoparticles distribution are observed: a thin region a few nanometers thick which is closer to the surface (R_1); a Si region about 5 nm with a small concentration of nanoparticles (R_2); and a layer (≈ 40 nm wide) with a higher concentration of nanoparticles (R_3).

High-resolution cross-sectional images (HRTEM) of the R_1 and R_3 regions are shown in the insets 10b, 10c and 10d. In inset 10b, it is possible to identify small irregular shaped nanoparticles at Si subsurface R_1. In the deeper layers (R_2 and R_3), two morphological variants of the metastable γ-$FeSi_2$ phase were observed and recognized: spherical-like nanoparticles epitaxially formed in the substrate with a fully aligned orientation regarding the Si matrix (Fig. 10c) and plate-like nanoparticles rotated with respect to the Si matrix (Fig. 10d) as previously reported (Lin et al., 1994). The spherical-like nanoparticles form coherent interfaces with the Si matrix, while the plate-like ones are elongated along Si$\langle \bar{1}12 \rangle$ directions.

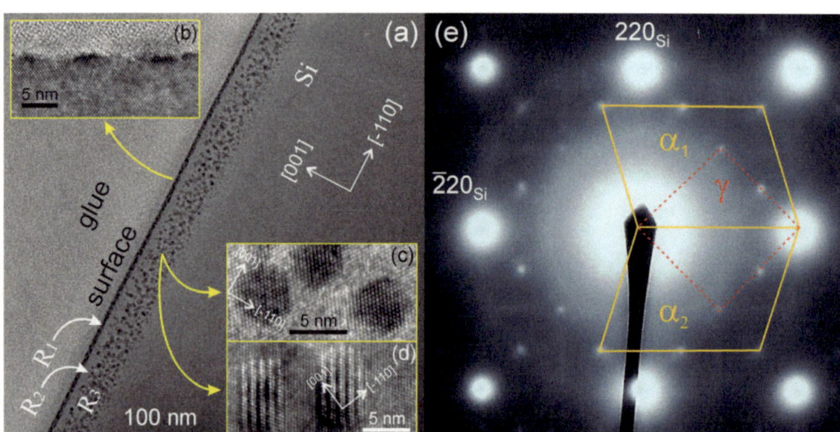

Fig. 10. TEM images of the recrystallized sample. **(a)** $[110]_{Si}$ cross-sectional revealing three nanoparticles regions in depth. HRTEM showing: **(b)** irregular shaped α-$FeSi_2$ nanoparticles at Si subsurface, **(c)** γ-$FeSi_2$ spherical-like and **(d)** γ-$FeSi_2$ plate-like nanoparticles in a deeper region. **(e)** $[001]_{Si}$ SAED pattern indicating the concomitant presence of both α- and γ-$FeSi_2$ phases.

Selected-area electron diffraction (SAED) pattern obtained from a plan-view specimen is presented in figure 10e. In addition to strong Bragg reflections of Si, extra spots due to the nanoparticles are apparent in the diffraction pattern. The extra spots show symmetric net patterns suggesting that there is a certain orientation relationship between the nanoparticles and the substrate lattice. From the analysis of the extra spots symmetry and lattice spacing, the diffraction pattern of Figure 10e can be explained as the overlap of cubic γ-$FeSi_2$ and tetragonal α-$FeSi_2$ phases. The reflections are consistent with $[\bar{1}12]_{\alpha 1}$, $[1\bar{1}\bar{2}]_{\alpha 2}$ (straight lines) and $[100]_{\gamma}$ (dotted line) net patterns (Behar et al., 1996; Vouroutzis et al., 2008). As the SAED measurements have identified two crystalline phases, the near-surface precipitated layer should contain α-$FeSi_2$ nanoparticles.

3.2 X-ray multiple diffraction

Bragg-Surface Diffraction (BSD) (Chang, 2004) is a special diffraction case of the X-ray multiple diffraction (XRMD) technique which has become a very useful and high resolution probe to study in-plane effects in single crystals in general, and also, with several interesting contributions to semiconductor epitaxial systems (dos Santos et al., 2009; Morelhão & Cardoso, 1993; Morelhão et al., 1998; Orloski et al., 2005; Lang et al., 2010b). For a more complete understanding of the experimental results that will follow in this chapter, we briefly discuss the physical aspects of the XRMD technique.

The XRMD phenomenon is systematically generated by aligning the primary planes of a single crystal - generally parallel to the sample surface, to diffract the incident beam and, by rotating it around the normal to the primary (h_p, k_p, l_p) planes while the diffracted beam is monitored by a detector. Under rotation (ϕ-axis), several other secondary (h_s, k_s, l_s) planes which are inclined with respect to the surface can enter into diffraction condition simultaneously with the primary ones. A closer observation of the diffraction geometry shows that other diffraction planes, the so-called coupling (h_c, k_c, l_c), also interact with the secondary diffracted beams to re-scatter them towards the detector. The obtained XRMD pattern, called Renninger scanning (RS) (Renninger, 1937), shows a series of peaks distributed according to the symmetry of the chosen primary vector and also to the symmetry plane established by rotation of the several reciprocal space secondary points when entering and leaving the Ewald sphere. Therefore, one can clearly observe in a RS, these two types of symmetry mirrors whose position and intensity distributions are essential for most of the applications of this technique. When a peak in the RS represents an interaction of the incident, the primary and one secondary beam, it shows up as a three-beam peak (or three-beam case). However, one can have two or three secondary beams simultaneously interacting to provide four or five-beam cases (or even cases for $n > 5$ interacting beams) being these secondary beams either Bragg (reflected) or Laue (transmitted) cases.

A special three-beam XRMD case, called Bragg-Surface Diffraction (BSD), appears under adequate conditions, that is, when the secondary diffracted beam propagates along the crystal surface under an extreme asymmetric geometry. A schematic diagram of the multiple scattering for the BSD case occurring inside the crystal, can be seen in Figure 11, where $\mathbf{H}_{ij}$ are the reciprocal lattice vectors corresponding to the primary planes ($\mathbf{H}_{01}$), secondary planes ($\mathbf{H}_{02}$) and coupling plane ($\mathbf{H}_{21}$). These BSD beams carry information on the sample surface which are useful for studying surface impurity incorporation effects (Lai et al., 2005), or even on the interface (layer/substrate), if such an interface is present, as in the semiconductor epitaxial structures. The technique has provided significant information when successfully applied as a special high-resolution 3D probe for the study of epitaxial layered heterostructures. Here, the layer and substrate lattices may be separately studied just by selecting one adequate layer or substrate peak. The technique is also a method capable of measuring strain fields at the interfaces in epilayer/substrate systems with depth penetration of 2 Å resolution and with enough sensitivity to detect lattices distortions in the range of 72 Å around the epilayer/substrate interface (Sun et al., 2006).

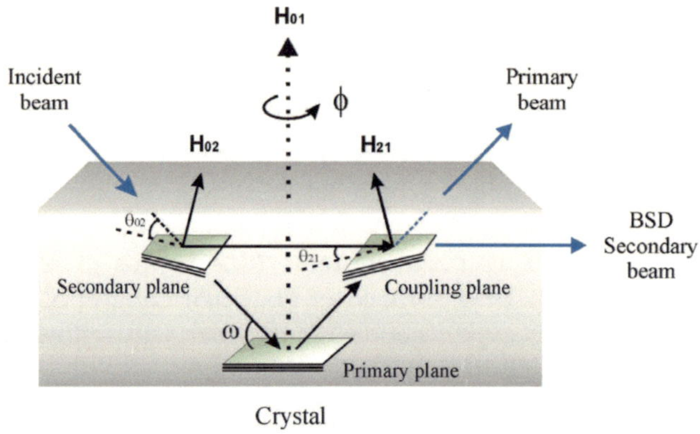

Fig. 11. BSD scheme using its consecutive scattering model with $\mathbf{H}_{01}$ (primary), $\mathbf{H}_{02}$ (secondary) and $\mathbf{H}_{21}$ (coupling) vectors. The coupling planes re-scatter the secondary beam towards the primary direction.

Besides the standard RS, another XRMD scanning methodology can give information on the crystalline quality obtained from an analysis of the ω:ϕ mapping scans (Morelhão & Cardoso, 1996). By using this method, the ϕ rotation is performed for a range of ω angles each targeting an exact angular position of the multiple-beam Bragg condition. This approach results in a three dimensional plot of the primary intensity versus ω and ϕ in a coupled way from which, through analysis of the iso-intensity contours of such plots, one can obtain information on the lattice coherence along the beam path and hence, on the crystalline quality. It has been shown that when the FWHM (full width at half maximum) of the peak in the ϕ scan is larger in comparison to the one in the ω scan, there is almost no loss of coherence, i.e. confirming that the crystal is perfect or nearly perfect.

3.2.1 BSD reflections – Results and discussion

Measurements of the (004) symmetrical high resolution rocking curves (HRRC) are shown in figure 12 at two perpendicular orientations on the sample surface: $\phi = 0°$ (Fig. 12a) and 90° (Fig. 12b). Both patterns present practically the same result with two distinct peaks corresponding to R_2 and R_3 regions, clearly seen in each pattern, with smaller perpendicular lattice parameters in comparison to the matrix peak (stronger). Also, (002) HRRC were measured at ($\bar{1}11$) and (111) BSD reflections then, at two azimuth angles on the recrystallized sample ($\phi = -6.04°$ and 83.96°). The results are shown in figures 12c and 12d.

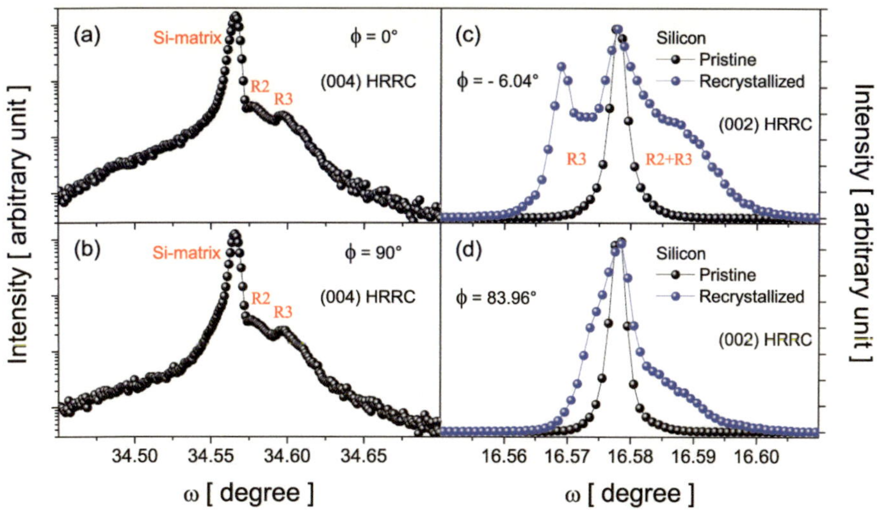

Fig. 12. High resolution rocking curves of the recrystallized sample for (004) reflection at $\phi = 0^o$ **(a)** and $\phi = 90^o$ **(b)** and (002) reflection at two BSD peaks: – 6.04° **(c)** and 83.96° **(d)**. Si pristine is also added for comparison purposes.

As (002) is a forbidden reflection of the Si space group, no primary intensity can be observed out of the BSD secondary peaks. The rocking obtained at $\phi = -6.04°$ shows three different contributions: the stronger peak due to the Si matrix; the broad peak to the right (higher angles), due to the R_2 and R_3 convoluted peaks; and a distinct peak to the left (lower angles), probably associated only with the R_3 region. It should be noticed that the rocking at ϕ = 83.96° exhibits a meaningful profile difference, that is, the peak to the left (lower angles) does not appear as discriminated as in the previous measurement ($\phi = -6.04°$), it means, a noticeable confirmation of the anisotropic behavior. This anisotropy, observed between the $[\bar{1}10]$ and [110] in-plane directions, could be associated with the plate-like nanoparticles (Fig. 10d) since the shape and orientation of these ordered nanoparticles should introduce different strains in both perpendicular directions.

Figure 13 shows the measured ω:ϕ mappings for a Si matrix (pristine) and a recrystallized sample for comparison purposes to provide a better visualization and characterization of the detected anisotropy. The exact BSD reflection is tailored in both ω and ϕ-directions for each of the two above mentioned BSD secondary reflections: $(\bar{1}11)$ Si pristine in Fig. 13a and recrystallized sample in 13b and, for (111) Si pristine in 13c and recrystallized one in 13d. In fact, these mappings give a more complete 3D view of the BSD reflection condition which complements the 2D analysis obtained from the HRRC in Figs. 12c and 12d. Furthermore, the mappings allow for the lattice parameters and 2D-strain determination of the distorted regions (R_2 and R_3). The mappings obtained for the Si matrix (pristine) along the two in-plane perpendicular directions as depicted in 13a and 13c, exhibit only the BSD matrix peak at ω = 16.578° and, as expected, no difference is observed. In turn, the recrystallization process induces $FeSi_2$ nanoparticles nucleation within the implanted matrix and then, a huge broadening as well as a striking difference is clearly observed in Figs. 13b and 13d mappings. Besides the BSD matrix peak in Fig. 13b, two other peaks are also detected: one upper-side (ω ~ 16.59°) and one lower-side (ω ~ 16.57°) with respect to the matrix peak whereas, in Fig. 13d just the matrix and the upper-side peak are clearly seen since the lower-side peak appears as a shoulder of the matrix one. This result confirms the one obtained in Figs. 12c and 12d.

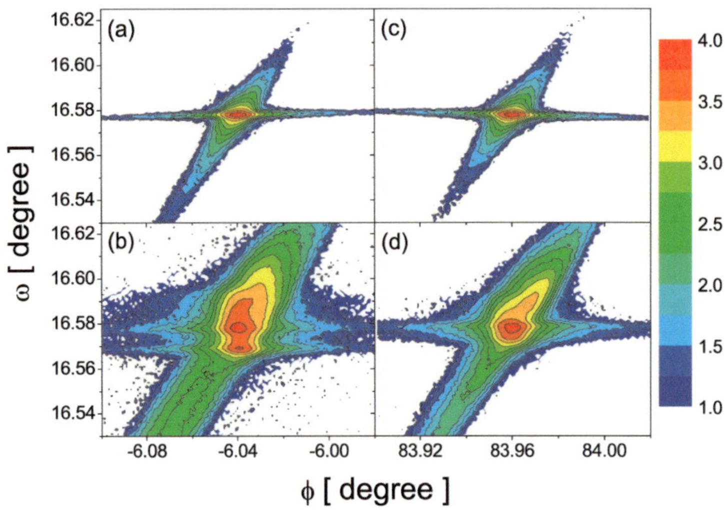

Fig. 13. Strain anisotropy in MBSD: Si pristine **(a)** and recrystallized sample **(b)** at ($\bar{1}11$) BSD ($\phi = -6.04°$) and Si pristine **(c)** and recrystallized **(d)** at (111) BSD ($\phi = 83.96°$).

Perpendicular and in-plane lattice parameters as well as strain values were obtained from the IBIEC sample ω:ϕ mappings for $\phi = -6.04°$ ($\bar{1}11$) and 83.96° (111). The perpendicular strain obtained with the upper-side peaks of the BSD ($\bar{1}11$) and (111) ω:ϕ mappings are $\varepsilon_\perp = -5.8(6)\text{x}10^{-4}$ whereas, the in-plane strains are $\varepsilon_{||} = 0$. Then, as no anisotropy strain is detected from this upper-side peak (ω = 16.588°) one can assume that most of this result can be assigned to the spherical-like nanoparticles rather than to the plate-like ones. In turn, for the lower-side peak, one observes that $\varepsilon_\perp$ and $\varepsilon_{||}$ values are distinct: $\varepsilon_\perp = 5.4(6)\text{x}10^{-4}$ and $\varepsilon_{||} = 3.1(7)\text{x}10^{-4}$ for ($\bar{1}11$) BSD peak and $\varepsilon_\perp = 2.4(8)\text{x}10^{-4}$ and $\varepsilon_{||} = 1.3(9)\text{x}10^{-4}$ for (111) and then, an analogous behavior happens for $a_\perp$ and $a_{||}$. Thus, one concludes there is anisotropy in the lattice parameters and strain in both sample directions: out-of-plane and in-plane. This anisotropy is attributed to the shape and distortion along the ($\bar{1}11$) and (111) crystallographic planes of the plate-like nanoparticles.

4. Conclusion

The conditions for the occurrence of the recrystallization and amorphization phenomena of a Si thin layer have been discussed in terms of the energy deposited by an ion beam as well as the sample temperature during an irradiation process. Also, the dependence of the recrystallization in relation to some impurity species, in particular, iron atoms dissolved into an amorphous Si layer has been discussed. Furthermore, it was shown how the ion-beam-induced epitaxial crystallization (IBIEC) process can be used as a method to synthesize nanoparticles within a Si matrix. For a specific case of Fe^+ low dose implanted in Si(100) substrate, nanoparticles with different orientations and morphologies were observed after the IBIEC process. These nanoparticles have caused interesting distortions in the surrounding Si lattice. In order to structurally characterize these distortions the Bragg Surface Diffraction was used as a 3D high-resolution tool. This non-conventional X-ray diffraction technique was able to discriminate between the out-of-plane and in-plane strain effects and to provide the direct observation of an important in-plane strain anisotropy.

5. References

Behar, M.; Bernas, H.; Desimoni, J.; Lin, X. W. & Maltez, R. L. (1996). Sequential phase formation by ion-induced epitaxy in Fe-implanted Si(001). Study of their properties and thermal behavior. *Journal of Applied Physics*, Vol. 79, No. 2, (January 1996), pp. 752-762, ISSN 1089-7550

Chang, S. L. (2004). *X-ray multiple-wave diffraction: theory and applications*, Springer Series in Solid-State Sciences, Vol. 143, Springer-Verlag, ISBN 3-540-21196-9, Berlin, Germany

Dearnaley, G.; Freeman, J. H.; Nelson, R. S. & Stephen, J. (1973). *Ion Implantation*, Noth-Holland Publishing Company, ISBN 0-7204-1758-9, Amsterdam, The Netherlands

Donovan, E. P.; Spaepen, F.; Turnbull, D.; Poate, J. M. & Jacobson, D. C. (1985). Calorimetric studies of crystallization and relaxation of amorphous Si and Ge prepared by ion implantation. *Journal of Applied Physics*, Vol. 57, No. 6, (March 1985), pp. 1795-1804, ISSN 1089-7550

Donovan, E. P.; Spaepen, F.; Poate, J. M. & Jacobson, D. C. (1989). Homogeneous and interfacial heat releases in amorphous Silicon. *Applied Physics Letters*, Vol. 55, No. 15, (October 1989), pp. 1516-1518, ISSN 1077-3118

dos Santos, A. O.; Lang, R.; de Menezes, A. S.; Meneses, E. A.; Amaral, L.; Reboh, S. & Cardoso, L. P. (2009). Synchrotron x-ray multiple diffraction in the study of Fe^{+} ion implantation in Si(0 0 1). *Journal of Physics D: Applied Physics*, Vol. 42, No. 19, (October 2009), pp. 195401-195407, ISSN 1361-6463

Elliman, R. G.; William, J. S.; Maher, D. M. & Brown, W. L. (1985). Kinetics, microstructure and mechanisms of ion beam induced epitaxial crystallization of semiconductors. *Materials Research Society Symposium Proceedings*, Vol. 51, (January 1985), pp. 319-328, ISSN 0272-9172

Elliman, R. G.; Williams, J. S.; Brown, W. L.; Leiberich, A.; Maher, D. M. & Knoell, R. V. (1987). Ion-beam-induced crystallization and amorphization of silicon. *Nuclear Instruments and Methods in Physics Research Section B Beam Interactions with Materials and Atoms*, Vol. 19-20, No. Part 2, (January 1987), pp. 435-442, ISSN 0168-583X

Holmén, G.; Linnros, J. & Svensson, B. (1984). Influence of energy transfer in nuclear collisions on the ion beam annealing of amorphous layers in silicon. *Applied Physics Letters*, Vol. 45, No. 10, (November 1984), pp. 1116-1118, ISSN 1077-3118

Jackson, K. A. (1988). A defect model for ion-induced crystallization and amorphization. *Journal of Materials Research*, Vol. 3, No. 6, (November 1988), pp. 1218-1226, ISSN 2044-5326

Lai, X.; Roberts, K. J.; Bedzyk, M. J.; Lyman, P. F.; Cardoso, L. P. & Sasaki, J. M. (2005). Structure of habit-modifying trivalent transition metal cations (Mn^{3+}, Cr^{3+}) in nearly perfect single crystals of potassium dihydrogenphosphate as examined by X-ray standing waves, X-ray absorption spectroscopy, and molecular modeling. *Chemistry of Materials*, Vol. 17, No. 16, (August 2005), pp. 4053-4061, ISSN 1520-5002

Lang, R.; Amaral, L. & Meneses, E. A. (2010a). Indirect optical absorption and origin of the emission from β-$FeSi_2$ nanoparticles: Bound exciton (0.809 eV) and band to acceptor impurity (0.795 eV) transitions. *Journal of Applied Physics*, Vol. 107, No. 10, (May 2010), pp. 103508/1-103508/7, ISSN 1089-7550

Lang, R.; de Menezes, A. S.; dos Santos, A. O.; Reboh, S.; Meneses, E. A.; Amaral, L. & Cardoso, L. P. (2010b). X-ray Bragg-Surface Diffraction: a tool to study in-plane strain anisotropy due to ion-beam-induced epitaxial crystallization in Fe^+-implanted Si(001). *Crystal Growth and Design*, Vol. 10, No. 10, (August 2010), pp. 4363-4369, ISSN 1528-7505

Lin, X. W.; Washburn, J.; Liliental-Weber, Z. & Bernas, H. (1994). Coarsening and phase transition of $FeSi_2$ precipitates in Si. *Journal of Applied Physics*, Vol. 75, No. 9, (May 1994), pp. 4686-4694, ISSN 1089-7550

Linnros, J.; Svensson, B. & Holmén, G. (1984). Ion-beam-induced epitaxial regrowth of amorphous layers in silicon on sapphire. *Physical Review B*, Vol. 30, No. 7, (October 1984), pp. 3629-3638, ISSN 1550-235X

Linnros, J.; Holmén, G. & Svensson, B. (1985). Proportionality between ion-beam-induced epitaxial regrowth in silicon and nuclear energy deposition. *Physical Review B*, Vol. 32, No. 5, (September 1985), pp. 2770-2777, ISSN 1550-235X

Linnros, J.; Elliman, R. G. & Brown, W. L. (1986). The competition between ion beam induced epitaxial crystallization and amorphization in Silicon: The role of the divacancy. *Materials Research Society Symposium Proceedings*, Vol. 74, (January 1986), pp. 477-480, ISSN 0272-9172

Linnros, J.; Elliman, R. G. & Brown, W. L. (1988). Divacancy control of the balance between ion-beam-induced epitaxial cyrstallization and amorphization in silicon. *Journal of Materials Research*, Vol. 3, No. 6, (November 1988), pp. 1208-1211, ISSN 2044-5326.

Lulli, G.; Merli, P. G. & Antisari, M. V. (1987). Solid-phase epitaxy of amorphous silicon induced by electron irradiation at room temperature. *Physical Review B*, Vol. 36, No. 15, (November 1987), pp. 8038-8042, ISSN 1550-235X

Miyao, M.; Polman, A.; Sinke, W.; Saris, F. W. & van Kemp, R. (1986). Electron irradiation-activated low-temperature annealing of phosphorus-implanted silicon. *Applied Physics Letters*, Vol. 48, No. 17, (April 1986), pp. 1132-1134, ISSN 1077-3118

Morelhão, S. L. & Cardoso, L. P. (1993). Analysis of interfacial misfit dislocation by X-ray multiple diffraction. *Solid State Communications*, Vol. 88, No. 6, (November 1993), pp. 465-469, ISSN 0038-1098

Morelhão, S. L. & Cardoso, L. P. (1996). X-ray multiple diffraction phenomenon in the evaluation of semiconductor crystalline perfection. *Journal of Applied Crystallography*, Vol. 29, No. 4, (August 1996), pp. 446-456, ISSN 1600-5767

Morelhão, S. L.; Avanci, L. H.; Hayashi, M. A.; Cardoso, L. P. & Collins, S. P. (1998). Observation of coherent hybrid reflection with synchrotron radiation. *Applied Physics Letters*, Vol. 73, No. 15, (October 1998), pp. 2194-2196, ISSN 1077-3118

Narayan, J.; Fathy, D.; Oen, O. S. & Holland, O. W. (1984). High-resolution imaging of ion-implantation damage and mechanism of amortization in semiconductors. *Materials Letters*, Vol. 2, No. 3, (February 1984), pp. 211-218, ISSN 0167-577X

Olson, G. L. & Roth, J. A. (1988). Kinetics of solid phase crystallization in amorphous silicon. *Materials Science Reports*, Vol. 3, No. 1, (May 1988), pp. 1-77, ISSN: 0920-2307

Orloski, R. V.; Pudenzi, M. A. A.; Hayashi, M. A.; Swart, J. W. & Cardoso, L. P. (2005). X-ray multiple diffraction on the shallow junction of B in Si(0 0 1). *Journal of Molecular Catalysis A: Chemical*, Vol. 228, No. 1-2, (March 2005), pp. 177-182, ISSN 1381-1169

Priolo, F.; La Ferla, A. & Rimini, E. (1988). Ion-beam-assisted growth of doped Si layers. *Journal of Materials Research*, Vol. 3, No. 6, (November 1988), pp. 1212-1217, ISSN 2044-5326

Priolo, F.; Spinella, C.; La Ferla, A.; Rimini, E. & Ferla, G. (1989). Ion-assisted recrystallization of amorphous silicon. *Applied Surface Science*, Vol. 43, No. 1-4, (December 1989), pp. 178-186, ISSN 0169-4332

Priolo, F. & Rimini, E. (1990). Ion-beam-induced epitaxial crystallization and amorphization in silicon. *Materials Science Reports*, Vol. 5, No. 6, (June 1990), pp. 319-379, ISSN: 0920-2307

Priolo, F.; Spinella, C. & Rimini, E. (1990). Phenomenological description of ion-beam-induced epitaxial crystallization of amorphous silicon. *Physical Review B*, Vol. 41, No. 8, (March 1990), pp. 5235-5242, ISSN 1550-235X

Renninger, M. (1937). Umweganregung, eine bisher unbeachtete Wechselwirkungserscheinung bei Raumgitterinterferenzen. *Zeitschrift für Physik A Hadrons and Nuclei*, Vol. 106, No. 3-4, (July 1937), pp. 141-176

Roorda, S.; Doorn. S.; Sinke, W. C.; Scholte, P. M. L. O. & Van Loenen, E. (1989). Calorimetric Evidence for Structural Relaxation in Amorphous Silicon. *Physical Review Letters*, Vol. 62, No. 16, (April 1989), pp. 1880-1883, ISSN 1079-7114

Sun, W. C.; Chang, H. C.; Wu, B. K.; Chen, Y. R.; Chu, C. H.; Chang, S. L.; Hong, M.; Tang, M. T. & Stetsko, Y. P. (2006). Measuring interface strains at the atomic resolution in depth using x-ray Bragg-surface diffraction. *Applied Physics Letters*, Vol. 89, No. 9, (August 2006), pp. 091915/1-091915/3, ISSN 1077-3118

Vouroutzis, N.; Zorba, T. T.; Dimitriadis, C. A.; Paraskevopoulos, K. M.; Dózsa, L. & Molnár, G. (2008). Growth of β-$FeSi_2$ particles on silicon by reactive deposition epitaxy. *Journal of Alloys and Compounds*, Vol. 448, No. 1-2, (January 2008), pp. 202-205, ISSN 0925-8388

Washburn, J.; Murty, C. S.; Sadana, D.; Byrne, P.; Gronsky, R.; Cheung, N. & Kilaas, R. (1983). The crystalline to amorphous transformation in silicon. *Nuclear Instruments and Methods in Physics Research*, Vol. 209-210, No. Part 1, (May 1983), pp. 345-350, ISSN 0167-5087

Williams, J. S. (1983). Solid phase recrystallisation process in Silicon, In: Surface modification and alloying by laser, ion and electron beams, Poate, J. M.; Foti, G.; Jacobson, D. C., pp. 133, Plenum Press, ISBN 0306413736, New York, United States of America

Williams, J. S.; Elliman, R. G.; Brown, W. L. & Seidel, T. E. (1985). Dominant Influence of beam-induced interface rearrangement on solid-phase epitaxial crystallization of amorphous silicon. *Physical Review Letters*, Vol. 55, No. 14, (September 1985), pp. 1482-1485, ISSN 1079-7114

Williams, J. S. & Elliman, R. G. (1986). Role of electronic processes in epitaxial recrystallization of amorphous semicontluctors. *Physical Review Letters*, Vol. 51, No. 12, (September 1983), pp. 1069-1072, ISSN 1079-7114

Ziegler, J. F. (2011). *Stopping and range of ions in matter*, SRIM-2011. Available from http://www.srim.org/

15

Steady-State Grain Size in Dynamic Recrystallization of Minerals

Ichiko Shimizu
Department of Earth and Planetary Science, University of Tokyo, Tokyo Japan

1. Introduction

Dynamic recrystallization (DRX) is a strain restoration and grain refinement mechanism that occurs in high-temperature dislocation creep of metals and minerals (Humphreys & Hatherly, 2004). Microstructures indicative of DRX are commonly observed in rock-forming minerals that have been subjected to natural deformation in the Earth's crust and mantle (Fig. 1).

Laboratory studies have revealed that the average size d of recrystallized grains approaches a steady-state value, which is determined by the applied stress and is independent of the initial grain size. Twiss (1977) proposed a stress–grain size relation of the following form:

$$\frac{d}{b} = K\left(\frac{\sigma}{\mu}\right)^{-p} \tag{1}$$

where σ is the flow stress, μ is the shear modulus, b is the length of the Burgers vector, and K is a non-dimensional constant. The grain size exponent p ranges between 1 and 1.5 for most materials. Empirically determined σ–d relations of minerals have been used to estimate the stress states in the Earth's interior. However, detailed studies of a Mg alloy (De Bresser et al., 1998) and NaCl (Ter Heege et al., 2005) revealed that K has a weak dependence on temperature. Derby & Ashby (1987) modeled the DRX processes of metals and predicted the temperature dependence of the recrystallized grain size, but they failed to account for the observed range of exponent p (Derby, 1992; Shimizu, 2011).

In this chapter, we focus on deformation and recrystallization processes in minerals and examine the effects of stress and temperature on the steady-state grain size.

2. Recrystallization mechanisms in minerals

DRX was first observed in hot deformation of cubic metals such as Cu, Ni, and austenitic iron. A simplified description of DRX in these metals is as follows. Strain-free new grains are usually formed by bulging of pre-existing grain boundaries and they grow at the expense of old grains to reduce the dislocation energy of the material (Sakai, 1989; Sakai & Jones, 1984). As the dislocation density of the new grains increases, they cease to grow and new nucleation events occur at their margins. These processes repeat cyclically during dislocation creep.

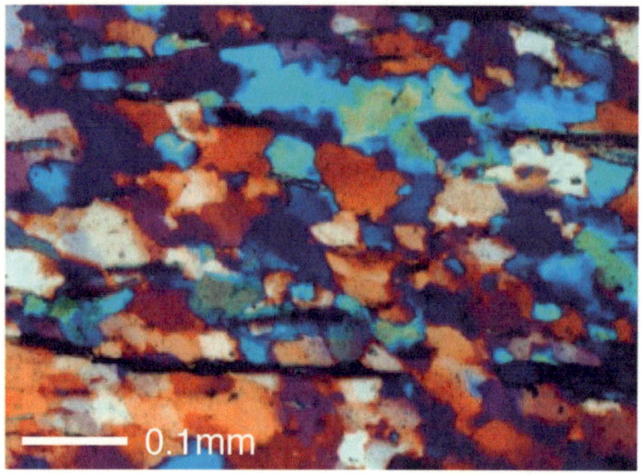

Fig. 1. Optical micrograph of a thin section of a quartz schist (Sanbagawa metamorphic belt, Japan) under polarized transmitted light with a sensitive color plate. Blue and red represent the orientation of the crystallographic *c*-axis of quartz grains. New small grains form at the margins and interiors of larger grains.

In contrast to the classical view of DRX described above, syndeformational recrystallization of minerals such as quartz, calcite, and olivine proceeds with progressive misorientation of subgrain boundaries (Poirier, 1985). Subgrain rotation (SGR) recrystallization also occurs in some metals such as Mg and Al alloys and is termed continuous DRX, whereas DRX in the original sense is currently referred to as discontinuous DRX (Humphreys & Hatherly, 2004). At low temperatures (T) and high strain rates ($\dot{\varepsilon}$), SGR is localized at grain margins (Hirth & Tullis, 1992; Schmid et al., 1980); however, intracrystalline SGR becomes more important and grain boundary migration (GBM) occurs at high T and low $\dot{\varepsilon}$ (Hirth & Tullis, 1992; Rutter, 1995) (Fig. 2). Consequently, the recrystallized grain size is much larger than the subgrain size (Guillopé & Poirier, 1979; Karato et al., 1980).

For both discontinuous and continuous DRX, grain size reduction occurs at nucleation events, whereas strain-induced GBM leads to overall coarsening. The steady-state grain size is determined by the dynamic balance between nucleation and grain growth (Derby & Ashby, 1987).

3. Grain size distribution

In the σ–d relation (Eq. 1), the steady-state microstructure is represented by a single value of the 'average' grain size d, but dynamically recrystallized materials generally have wide grain size distributions. As a simplified model of DRX, Shimizu (1998a; 1999; 2003) considered following nucleation and growth processes and analyzed the evolution of the grain size distribution:

1. Nucleation occurs at a constant rate I per unit volume.
2. Nucleation sites are randomly distributed.
3. Each grain grows with a radial growth rate $\dot{R}$.

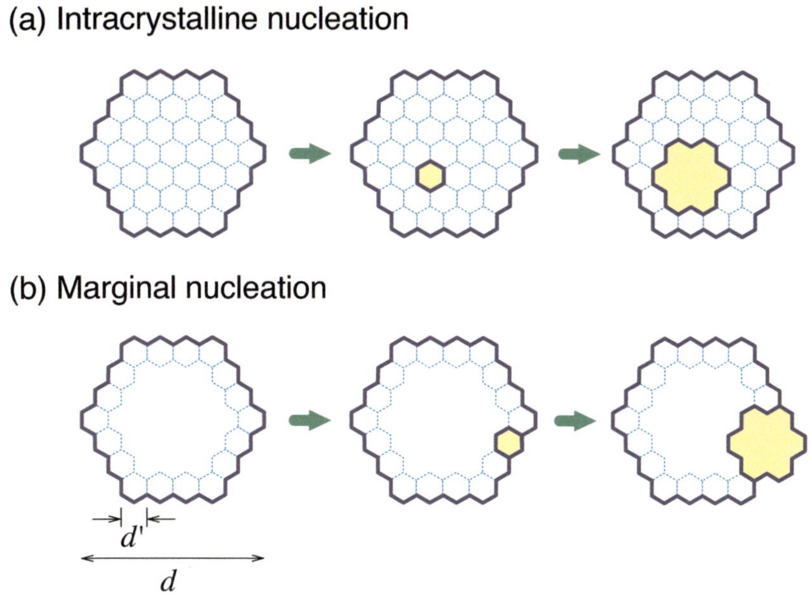

Fig. 2. Nucleation and growth in continuous DRX. Solid lines represent grain boundaries and thin dotted lines represent subgrain boundaries. Nucleated grains (yellow) are formed by SGR and grow in the deformed matrix.

4. Newly crystallized grains replace older grains.

In the steady state, the grain size has a nearly a log-normal distribution and many newly crystallized grains coexist with a few old grains in a certain population balance. The average grain size satisfies

$$d = a\left(\frac{\dot{R}}{I}\right)^{\frac{1}{4}} \quad (2)$$

where a is a scaling factor; $a = 1.14$ for a 3D distribution and $a = 1.12$ for a distribution measured in a 2D section. Shimizu (1998b; 2008; 2011) considered strain-induced grain growth for $\dot{R}$ (Sec. 4) and SGR nucleation for I (Sec. 5) and derived the σ–d relation for continuous DRX (Sec. 6). In Sec. 7, we revise the theoretical model to incorporate the influence of the surface-energy drag.

4. Strain-induced grain growth

4.1 Dislocation energy

During high-T dislocation creep of minerals, dynamic recovery cooperates with continuous DRX and assists subgrain formation. Unrecovered microstructures such as tangled dislocations are rarely observed in recrystallized grains (Hirth & Tullis, 1992). Hence, the strain energy (E_{strain}) is given by a sum of the energies of isolated dislocations and sub-boundaries (E_{disl} and E_{sub}, respectively):

$$E_{strain} = E_{disl} + E_{sub} \quad (3)$$

The free dislocation energy per unit volume is

$$E_{disl} = \rho\zeta \tag{4}$$

where ρ is the dislocation density and ζ is the dislocation line tension. When the internal stress around dislocations is equilibrated with the applied stress σ, the following equation holds (Nabarro, 1987):

$$\sigma = \alpha\mu b\rho^{\frac{1}{2}} \tag{5}$$

where α is a constant that depends on the configuration of the dislocation arrays. Hence,

$$\rho = \left(\frac{\sigma}{\alpha\mu b}\right)^2 \tag{6}$$

The dislocation line tension is given by (Hirth & Lothe, 1982)

$$\zeta = \frac{\mu b^2 \chi}{4\pi} \ln\left(\frac{\beta r}{b}\right) \tag{7}$$

where r is the characteristic radius of the elastic field around the dislocation core and the constant β is typically in the range 3–4. The parameter χ depends on the dislocation configuration:

$$\begin{cases} \chi = 1; & \text{for a screw dislocation} \\ \chi = \dfrac{1}{1-\nu}; & \text{for an edge dislocation} \end{cases} \tag{8}$$

where ν is Poisson's ratio. For a first-order approximation, we assume that all dislocations are edge dislocations. Considering that the elastic field around a dislocation is canceled by other dislocations at half the distance between them, r is scaled as

$$r = \frac{1}{2}\rho^{-\frac{1}{2}} \tag{9}$$

Substituting Eqs. (6) and (9) into Eq. (7) yields

$$\zeta = \frac{\mu b^2}{4\pi(1-\nu)} \ln\left(\frac{\beta\alpha\mu}{2\sigma}\right) \tag{10}$$

Substituting Eq. (10) into Eq. (4) and using Eq. (6) again, we have

$$E_{disl} = \frac{\sigma^2}{4\pi\alpha^2\mu(1-\nu)} \ln\left(\frac{\beta\alpha\mu}{2\sigma}\right) \tag{11}$$

4.2 Sub-boundary energy

Consider nearly spherical subgrains with a diameter d' that occupy a deformed matrix (Fig. 2a). The number density of subgrains is

$$N = \frac{6}{\pi d'^3} \tag{12}$$

and the area of subgrain boundaries per unit volume is

$$A = N \cdot \pi d'^2 \cdot \frac{1}{2} = \frac{3}{d'} \tag{13}$$

The factor 1/2 is included because the area of each subgrain wall is counted twice. The energy of sub-boundaries in a unit volume of the material can thus be written as

$$E_{sub} = \frac{3\gamma}{d'} \tag{14}$$

where γ is the sub-boundary energy per unit area.

The theory of dislocations gives

$$\gamma = \frac{\mu b^2}{4\pi(1-\nu)h}\xi(\eta) \tag{15}$$

$$\xi(\eta) \equiv \eta \coth \eta - \ln(2 \sinh \eta) \tag{16}$$

$$\eta \equiv \frac{\pi b}{\beta h} \tag{17}$$

where h is the mean dislocation spacing (Hirth & Lothe, 1982).

For a tilt boundary (Fig. 3), h and the misorientation angle θ are related by (Poirier, 1985)

$$\frac{b}{h} = 2 \tan\left(\frac{\theta}{2}\right) \simeq \theta \tag{18}$$

The last approximation is justified for low-angle boundaries. Then, Eq. (17) becomes

$$\eta = \frac{\pi}{\beta}\theta \ll 1 \tag{19}$$

Hence, the following approximations can be applied to Eq. (16):

$$\coth \eta \simeq \frac{1}{\eta}, \quad \sinh \eta \simeq \eta \tag{20}$$

Then, Eq. (15) becomes

$$\gamma = \frac{\lambda}{2}\mu b\theta \tag{21}$$

where

$$\lambda \equiv \frac{1}{2\pi(1-\nu)}\left[1 - \ln\left(\frac{2\pi\theta}{\beta}\right)\right] \tag{22}$$

The subgrain size is empirically expressed as (Takeuch & Argon, 1976; Twiss, 1977)

$$\frac{d'}{b} = K'\left(\frac{\sigma}{\mu}\right)^{-1} \tag{23}$$

where K' is a constant. A theoretical expression for K' is given below. Substituting Eqs. (21), (22), and (23) into Eq. (14), we have

$$E_{sub} = \frac{3\lambda\theta\sigma}{2K'} \tag{24}$$

4.3 Subgrain size

We consider a recovery process in which free dislocations with a dislocation density ρ rearrange into sub-boundaries. Conservation of the total dislocation length during subgrain formation requires

$$\rho = \frac{A}{h} \tag{25}$$

The right-hand side represents the length of dislocations in sub-boundaries. Using Eqs. (13) and (18), the above expression is modified to become

$$\rho = \frac{3\theta}{d'b} \tag{26}$$

Subgrains are formed if the total sub-boundary energy is smaller than the free dislocation energy (Twiss, 1977):

$$E_{disl} \geq E_{sub} \tag{27}$$

The equality represents the critical state for the initiation of subgrain formation. From Eqs. (4) and (14), this condition can be written as

$$\rho\zeta \geq \frac{3\gamma}{d'} \tag{28}$$

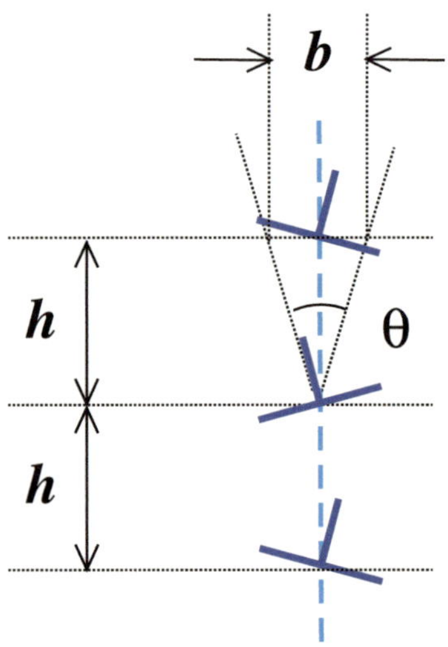

Fig. 3. Schematic illustration of a tilt boundary with a misorientation angle θ, dislocation spacing h, and Burgers vector b.

Substituting Eqs. (6), (7), and (21)–(22) into the above expression for ρ, ζ, and γ, respectively, Eq. (28) becomes

$$\left(\frac{\sigma}{\alpha\mu b}\right)^2 b\ln\left(\frac{\beta\alpha\mu}{2\sigma}\right) \geq \frac{3}{d'}\theta\left[1-\ln\left(\frac{2\pi\theta}{\beta}\right)\right] \tag{29}$$

Equating Eqs. (26) and (6), we have

$$\left(\frac{\sigma}{\alpha\mu b}\right)^2 = \frac{3\theta}{d'b} \tag{30}$$

Then, Eq. (29) reduces to

$$\ln\left(\frac{\beta\alpha\mu}{2\sigma}\right) \geq \left[1-\ln\left(\frac{2\pi\theta}{\beta}\right)\right] \tag{31}$$

The stability limit of θ is derived as

$$\theta \geq \frac{e}{\pi\alpha}\left(\frac{\sigma}{\mu}\right) \tag{32}$$

where e is the Napierian base. The equality gives the initial misorientation angle θ_i:

$$\theta_i = \frac{e}{\pi\alpha}\left(\frac{\sigma}{\mu}\right) \tag{33}$$

Applying θ_i to θ of Eq. (30), the initial subgrain size d_i' is obtained as

$$\frac{d_i'}{b} = \frac{3e\alpha}{\pi}\left(\frac{\sigma}{\mu}\right)^{-1} \tag{34}$$

Once the subgrain boundary is established, it functions as a dislocation sink because progressive subgrain misorientation is an energetically favorable process. We thus assume that the subgrain size is maintained during the subsequent misorientation. Substituting $d' = d_i'$ into Eq. (34), we obtain Eq. (23), where

$$K' = \frac{3e\alpha}{\pi} \tag{35}$$

Using Eqs. (22) and (35), the full expression of Eq. (24) is obtained as

$$E_{sub} = \frac{\pi\lambda\theta\sigma}{2e\alpha} = \frac{\theta}{4e\alpha(1-\nu)}\left[1-\ln\left(\frac{2\pi\theta}{\beta}\right)\right]\sigma \tag{36}$$

4.4 Growth kinetics

The kinetic law of grain growth is generally written as

$$\dot{R} = MF \tag{37}$$

where M is the mobility of the grain boundary and F is the driving force. M depends on T as

$$M = \frac{bwD_{gb}}{kT} \tag{38}$$

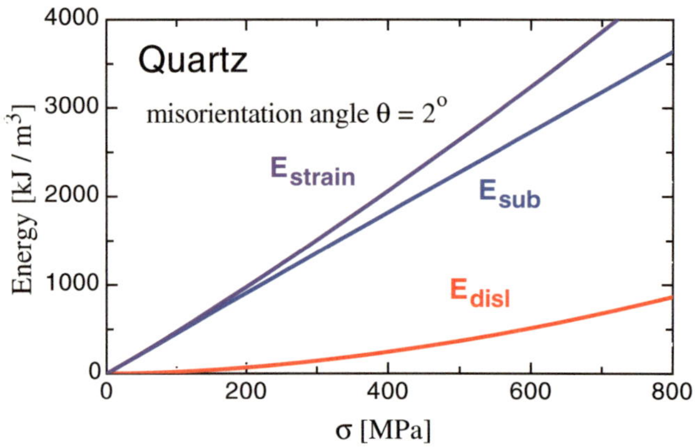

Fig. 4. Strain energy of quartz calculated using Eqs. (11) and (36). The physical parameters of quartz are given as (Shimizu, 2008) $\alpha = 3$ (Kohlstedt & Weathers, 1980), $\mu = 4.2 \times 10^4$ MPa, and $\nu = 0.15$ (Twiss, 1977). Because no data are available for β of quartz, we apply $\beta = 3$ of ionic crystals (Hirth & Lothe, 1982).

$$D_{gb} = D^{\circ}_{gb} \exp\left(-\frac{Q_{gb}}{RT}\right) \tag{39}$$

where w is the boundary width, k is the Boltzmann constant, D_{gb} is the diffusion coefficient at the grain boundary, D°_{gb} is a constant, R is the gas constant, and Q_{gb} is the activation energy for grain boundary diffusion.

In a single-phase material, grain growth occurs to reduce the bulk strain energy and the energy of grain surfaces. Hence, Eq. (37) is written as

$$\dot{R} = M(F_{strain} + F_{surf}) \tag{40}$$

where F_{strain} and F_{surf} represent the driving forces due to strain energy and surface energy (grain boundary energy), respectively. The strain energy in dynamically recrystallized materials is not homogeneous. The strain energy of deformed grains is given by the sum of E_{disl} in Eq. (11) and E_{sub} in Eq. (36), whereas newly recrystallized grains are almost strain free. This difference in strain energy drives grain growth. Hence,

$$F_{strain} = E_{strain} \tag{41}$$

With increasing strain, free dislocations multiply and excess dislocations rearrange into sub-boundaries. Then, θ increases and the sub-boundary energy exceeds the free dislocation energy. Fig. 4 shows the calculations for quartz. When the average misorientation angle reaches several degrees, the following approximation can be used instead of Eq. (3):

$$E_{strain} \simeq E_{sub} \tag{42}$$

5. Nucleation rate

In SGR nucleation, the nuclei are approximately the same size as the original subgrains. Thus, the number of potential nucleation sites per unit volume of crystals is given by Eq. (12) for intracrystalline nucleation and

$$N = \frac{6}{\pi d d'^2} \tag{43}$$

for nucleation at grain margins (Fig. 2b). The nucleation rate is scaled as

$$I = \frac{N}{\tau_c} \tag{44}$$

where τ_c is the interval of nucleation events.

The subgrain becomes a nucleus when the misorientation angle θ exceeds a critical value θ_c. The flux of dislocations that move toward the sub-boundary is given by ρu, where u is the climb velocity. The time required for dislocations to accumulate at the sub-boundary is equal to the nucleation cycle τ_c. From Eq. (18), a critical nucleus has a dislocation spacing of $h_c = b/\theta_c$; hence, the number of dislocations per unit area of the boundary is $1/h_c = \theta_c/b$. Dividing this value by the flux ρu, the nucleation cycle is evaluated as

$$\tau_c \simeq \frac{\theta_c}{b\rho u} \tag{45}$$

The climb velocity of dislocations is given by (Hirth & Lothe, 1982)

$$u = \frac{\sigma \Omega D_v}{lkT} \tag{46}$$

where Ω is the atomic volume, D_v is the self-diffusion coefficient, and l is a length scale given by

$$l \equiv \frac{b}{2\pi} \ln\left(\frac{r}{b}\right) \tag{47}$$

Using Eqs. (6) and (9), Eq. (47) can be rewritten as

$$l = \frac{b}{2\pi} \ln\left(\frac{\alpha\mu}{2\sigma}\right) \tag{48}$$

The temperature dependence of D_v is expressed as

$$D_v = D_v^\circ \exp\left(-\frac{Q_v}{RT}\right) \tag{49}$$

where D_v° is a constant and Q_v is the activation energy for volume diffusion.

Combining Eqs. (44)–(46), approximating Ω as b^3, and using Eqs. (12) and (23), we have

$$I = \frac{6}{\pi b K'^3 \alpha^2 \theta_c} \frac{\sigma D_v}{lkT} \left(\frac{\sigma}{\mu}\right)^5 \tag{50}$$

for intracrystalline nucleation. Using Eq. (43) instead of Eq. (12), the equation for marginal nucleation is obtained:

$$I = \frac{1}{d}\frac{6}{\pi K'^2 \alpha^2 \theta_c}\frac{\sigma D_v}{lkT}\left(\frac{\sigma}{\mu}\right)^4 \tag{51}$$

6. Scaling relation

Here, we neglect the surface energy term in Eq. (40) and assume

$$F = MF_{strain} \tag{52}$$

Combining Eq. (2) with Eqs. (52), (38), (39), (41), and (42), and using Eq. (36) and either Eq. (50) or Eq. (51), the steady-state grain size in continuous DRX is derived as

$$\frac{d}{b} = B\left(\frac{\sigma}{\mu}\right)^{-p}\left(\frac{wD_{gb}}{bD_v}\right)^{\frac{1}{m}} \tag{53}$$

where

$$p = \frac{5}{4} = 1.25,\ m = 4 \tag{54}$$

for intracrystalline nucleation and

$$p = \frac{4}{3} = 1.33,\ m = 3 \tag{55}$$

for marginal nucleation (Shimizu, 1998b; 2008). B is a non-dimensional constant given by

$$B = \left(\frac{a^4 \pi K'^{m-2}\alpha^2}{4}\frac{\lambda\theta\theta_c l}{b}\right)^{1/m} \tag{56}$$

Using Eq. (48), Eq. (56) can be rewritten as

$$B = \left[\frac{a^4 K'^{m-2}\alpha^2 \lambda\theta\theta_c}{8}\ln\left(\frac{\alpha\mu}{2\sigma}\right)\right]^{1/m} \tag{57}$$

Although σ is included in the right-hand side, the stress dependence of B is negligibly small. Using Eqs. (39) and (49), Eq. (53) can be re-expressed as

$$\frac{d}{b} = K^\circ\left(\frac{\sigma}{\mu}\right)^{-p}\exp\left(-\frac{\Delta Q}{mRT}\right) \tag{58}$$

where

$$K^\circ = B\left(\frac{wD_{gb}^\circ}{bD_v^\circ}\right)^{1/m} \tag{59}$$

and

$$\Delta Q = Q_{gb} - Q_c \tag{60}$$

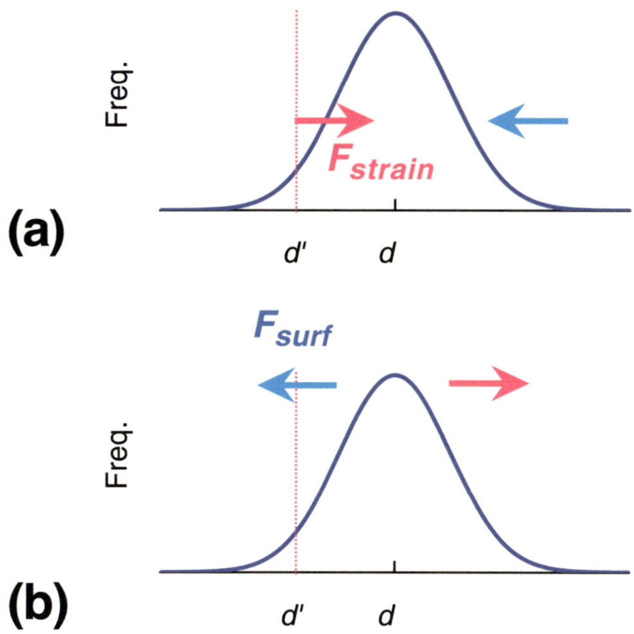

Fig. 5. Schematic representation of grain size evolution due to (a) strain-energy-driven grain growth and (b) surface-energy drag.

As Q_{gb} is generally smaller than Q_c, the recrystallized grain size is predicted to have a weak positive dependence on T. The constant K in Eq. (1) can now be written as a function of T:

$$K = K^{\circ} \exp\left(-\frac{\Delta Q}{mRT}\right) \tag{61}$$

7. Influence of surface energy

We now consider the influence of surface energy (grain boundary energy). In the case of surface-energy-driven grain coarsening in single-phase materials under static conditions (known as normal grain growth), large grains are energetically favorable and grow at the expense of small grains; the evolution of individual grain size has the opposite sense to that considered for DRX in Sec. 3 (Fig. 5). Therefore, when new grains grow by the strain-energy difference, the surface energy acts as a drag force.

In the theory of normal grain growth (Hillert, 1965), grain size evolution is described by

$$\dot{R}_k = Mc\Gamma\left(\frac{1}{R} - \frac{1}{R_k}\right) \tag{62}$$

where R_k and $\dot{R}_k$ are respectively the radius and the growth rate of the k-th grain and $c \sim 1$ is a statistical factor. If R_k is smaller (larger) than the mean radius R, the above expression becomes negative and the k-th grain shrinks (grows). By comparison with Eq. (37), the driving force for the growth of the k-th grain can be written as

$$2c\Gamma\left(\frac{1}{d}-\frac{1}{d_k}\right) \tag{63}$$

where d_k is the diameter of the k-th grain. In the nucleation and growth processes in DRX, the influence of the surface-energy drag is largest for small nuclei. Thus, we introduce a modified factor c' and express the surface-energy-driven force in Eq. (40) as

$$F_{surf}=2c'\Gamma\left(\frac{1}{d}-\frac{1}{d'}\right)\simeq-2c'\Gamma\frac{1}{d'} \tag{64}$$

With this equation and Eq. (42), Eq. (40) can be approximated as

$$\dot{R}\simeq M\left(E_{sub}-\frac{2c'\Gamma}{d'}\right) \tag{65}$$

Using this equation, Eq. (56) can be modified as follows (the parameters p, m, and ΔQ remain the same).

$$B=\left[\frac{a^4K'^{m-2}\alpha^2\theta_c}{8}\left(\frac{3\lambda\theta}{2}-\frac{2c'\Gamma}{b\mu}\right)\ln\left(\frac{\alpha\mu}{2\sigma}\right)\right]^{1/m} \tag{66}$$

8. Comparison of theory with experiments

8.1 Stress dependence of recrystallized grain size

In Fig. 6, p values of rock-forming minerals determined by triaxial or uniaxial or compression tests are plotted against the n-th power of dislocation creep flow laws ($\dot{\varepsilon}\propto\sigma^n$), which reflect the rate-controlling processes of dislocation creep; for climb-controlled creep, n is generally 3–5. The figure also shows the experimental result for a hexagonal Mg alloy (Magnox Al80), which was studied as a quartz analogue (De Bresser et al., 1998). The observed p values are almost independent of the power-law exponents and are well explained by the present model for continuous DRX.

8.2 Application to quartz

The theoretical model for the recrystallized grain size was applied to quartz using the equations presented in Sec. 6 (Shimizu, 2008; 2011). However, the previous model accounted only for strain energy; it neglected the effects of surface energy. Moreover, it turned out that the previous calculation involved a numerical error; when this error is corrected, the theoretical σ–d lines (Fig. 8 of Shimizu (2008)) shift to higher σ. Here, we recalculate the σ–d relation of quartz using the revised equations in Sec. 7.

Because experimentally deformed quartzite samples exhibit intracrystalline SGR at moderate stresses (Hirth & Tullis, 1992; Stipp & Tullis, 2003), we apply the intracrystalline nucleation model (Eq. 54). In addition to the material constants given in the caption of Fig. 4, we use $b=5\times10^{-4}$ μm (Twiss, 1977), $\theta=2^\circ$, $\theta_c=12^\circ$, and D_v and D_{gb} of oxygen in β-quartz (Farver & Yund, 1991b; Giletti & Yund, 1984). For grain boundary energy, we use $\Gamma=0.27$ Jm^{-2}

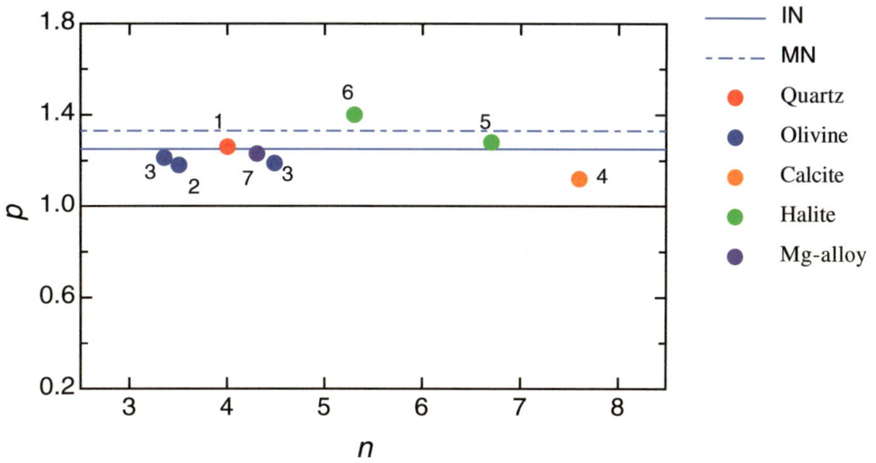

Fig. 6. Stress exponent p of recrystallized grain size plotted against the power-law exponent n of dislocation creep (IN: intracrystalline nucleation model; MN: marginal nucleation model). 1: Stipp & Tullis (2003) for p, Gleason & Tullis (1995) for n, 2: Karato et al. (1980) for p, Karato et al. (1986) for n, 3: van der Wal et al. (1993) for p, Chopra & Paterson (1984) for n, 4: Rutter (1995) for p, Schmid et al. (1980) for n, 5: Guillopé & Poirier (1979) for p and n, 6: Ter Heege et al. (2005) for p, Carter et al. (1993) for n, 7: De Bresser et al. (1998) for p and n. Microstructures of SGR and GBM are reported from all experiments except Ref. 3.

(Hiraga et al., 2007) and assume $c' = 1$. The steady-state grain size [μm] is then expressed a function of σ [MPa] and T [K] as

$$d = 1.82 \times 10^3 \times \sigma^{-1.25} \exp\left(\frac{7.25\ \mathrm{kJ/mol}}{RT}\right) \quad ; \beta-\text{quartz} \tag{67}$$

In this expression, the weak stress dependence of B in Eq. (66) is neglected and $B = 1.01$ at σ=50 MPa is chosen as a representative value. The calculation results (Fig. 7a) agree well with the empirical data for β-quartz (Stipp & Tullis, 2003). For comparison, the σ–d relation based on the marginal nucleation model is also shown.

In Fig. 7(b), the theoretical model is extended to the α-quartz stability field in which D_v of oxygen in α-quartz (Farver & Yund, 1991a) is used and α- and β-quartz are assumed to have the same Q_v/Q_{gb} ratio. The recrystallized grain size of α-quartz is predicted to be

$$d = 9.98 \times 10^2 \times \sigma^{-1.25} \exp\left(\frac{12.4\ \mathrm{kJ/mol}}{RT}\right) \quad ; \alpha-\text{quartz} \tag{68}$$

With decreasing temperature, the steady-state grain size shifts to higher stresses. If the empirical σ–d relation is directly applied to natural rocks that have deformed under low-T ($\leq 400°$C) metamorphic conditions, the stress states will be considerably underestimated.

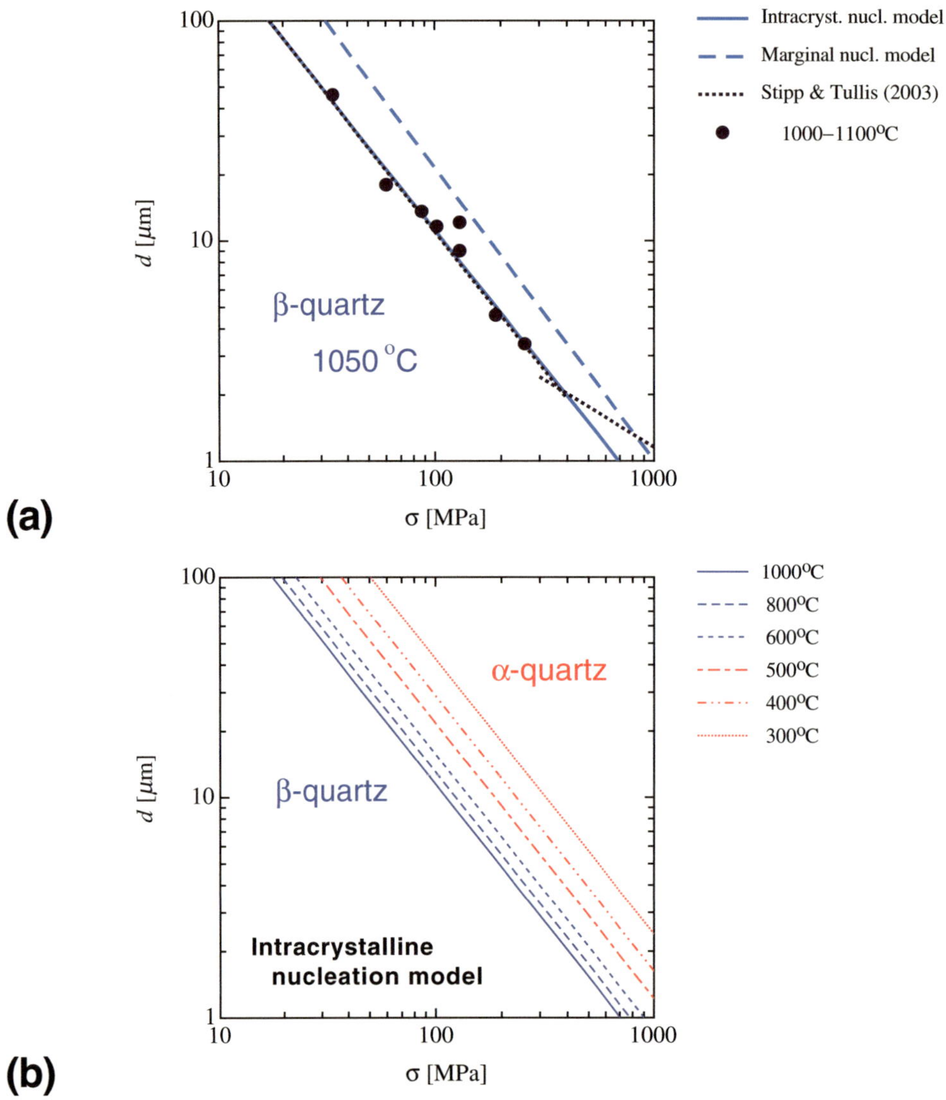

Fig. 7. Recrystallized grain size of quartz. (a) Theoretically calibrated σ–d relations for β-quartz at 1050°C and the experimental results of Stipp & Tullis (2003). Solid line: intracrystalline nucleation model. Dotted line: marginal nucleation model. Solid circles: recrystallized grain size at 1000–1100°C after Stipp & Tullis (2003). Black dotted line: empirical d–σ relation across the temperature range of 700–1100°C after Stipp & Tullis (2003). (b) Theoretically predicted σ–d relations for β-quartz (blue lines, 1000–600°C) and α-quartz (red lines, 500–300°C) using the intracrystalline nucleation model.

9. Summary

High-T dislocation creep of minerals is characterized by the occurrence of continuous DRX. The steady-state grain size is determined by the dynamic balance between SGR nucleation and grain growth by GBM. Surface energy acts as a drag force for strain-energy-driven GBM. The negative dependence of recrystallized grain size on stress is well explained by a theoretical model for continuous DRX. The theory also predicts a weak positive dependence of recrystallized grain size on temperature.

10. References

Carter, N.L.; Horseman, S.T.; Russel, J.E.; Handin, J. (1993). Rheology of rocksalt, *Journal of Structural Geology*, Vol. 15, 1257–1271.

Chopra P.N. & Paterson, M.S. (1984). The role of water in the deformation of dunite, *Journal of Geophysical Research* Vol. 89, 7861–7876.

De Bresser, J.H.P.; Peach, C.J.; Reijs, J.P.J.; Spiers, C.J. (1998). On dynamic recrystallization during solid state flow: effects of stress and temperature, *Geophysical Research Letters*, Vol. 25, 3457–3460.

Derby, B. (1992). Dynamic recrystallization: The steady state grain size, *Scripta Metallurgica and Materialia*, Vol. 27, 1581–1586.

Derby, B. & Ashby, M.F. (1987). On dynamic recrystallization, *Scripta Metallurgica*, Vol. 21, 879–884.

Farver, J. & Yund, R.A. (1991a). Oxygen diffusion in quartz: Dependence on temperature and water fugacity, *Chemical Geology*, Vol. 90, 55–70.

Farver, J. & Yund, R.A. (1991b). Measurement of oxygen grain boundary diffusion in natural, fine-grained, quartz aggregates, *Geochimica et Cosmochimica Acta*, Vol. 55, 1597–1607.

Gleason, G.C. & Tullis, J. (1995). A flow law for dislocation creep of quartz aggregates determined with the molten salt cell, *Tectonophysics*, Vol. 247, 1–23.

Giletti B.J. & Yund, R.A. (1984). Oxygen diffusion in quartz, *Journal of Geophysical Research*, Vol. 89, 4039–4046.

Guillopé, M. & Poirier, J.-P. (1979). Dynamic recrystallization during creep of single-crystalline halite: An experimental study, *Journal of Geophysical Research*, Vol. 84, 5557–5567.

Hillert, M. (1965). On the theory of normal and abnormal grain growth, *Acta Metallurgica*, Vol. 13, 227–238.

Hiraga, T.; Nishikawa, O.; Nagase, T.; Akizuki, M.; Kohlstedt, M. (2007). Interfacial energies for quartz and albite in pelitic schist, *Contributions to Mineralogy and Petrolgy*, Vol. 143, 663–672.

Hirth J.P. & Lothe J. (1982). *Theory of Dislocations*, Second edition, John Wiley & Sons, ISBN 0-471-09125-1, New York.

Hirth, G. & Tullis, J. (1992). Dislocation creep regimes in quartz aggregates, *Journal of Structural Geology*, Vol. 14, 145–159.

Humphreys, F.J. & Hatherly, M., (2004). *Recrystallization and Related Annealing Phenomena*, 2nd ed., Elsevier, ISBN 0-08-044164-5, Amsterdam.

Karato, S.; Paterson, M.S.; Fitz Gerald, J.D. (1986). Rheology of synthetic olivine aggregates: Influence of grain size and water, *Journal of Geophysical Research*, Vol. 91, 8151–8176.

Karato, S.; Toriumi, M.; Fujii, T. (1980). Dynamic recrystallization of olivine single crystals during high-temperature creep, *Geophysical Research Letters*, Vol. 7, 649–652.

Kohlstedt, D.L. & Weathers, M.S. (1980). Deformation-induced microstructures, paleopiezometers, and differential stress in deeply eroded fault zones, *Journal of Geophysical Research*, Vol. 85, 6269–6285.

Nabarro, F.R.N. (1987). *Theory of Crystal Dislocations*, Dover, ISBN 0-486-65488-5, New York.

Poirier, J.-P. (1985). *Creep of Crystals*, Cambridge University Press, ISBN 0-521-26177-5 (hardback) 0-521-27851 (paperback), Cambridge.

Rutter, E.H. (1995). Experimental study of the influence of stress, temperature, and strain on the dynamic recrystallization of Carrara marble, *Journal of Geophysical Research*, Vol. 100, 24651–24663.

Sakai, T. (1989). Dynamic recrystallization of metallic materials, In: *Rheology of solids and of the Earth*, Karato S. & Toriumi M. (Eds.), 284–307, Oxford University Press, ISBN 0-19-854497-9, Oxford.

Sakai, T. & Jonas, J.J. (1984). Dynamic recrystallization: mechanical and microstructural considerations, *Acta Metallurgica*, Vol. 32, 189–209.

Schmid, S.M.; Paterson, M.S.; Boland, J.N. (1980). High temperature flow and dynamic recrystallization in Carrara marble, *Tectonophysics*, Vol. 65, 245–280.

Stipp, M. & Tullis, J. (2003). The recrystallized grain size piezometer for quartz, *Geophysical Research Letters*, Vol. 30, 2088, doi:10.1029/2003GL018444.

Shimizu, I. (1998a). Lognormality of crystal size distribution in dynamic recrystallization, *FORMA*, Vol. 13, 1–11.

Shimizu, I. (1998b). Stress and temperature dependence of recrystallized grain size: A subgrain misorientation model, *Geophysical Research Letters*, Vol. 25, 4237–4240.

Shimizu, I. (1999). A stochastic model of grain size distribution during dynamic recrystallization, *Philosophical Magazine A*, Vol. 79, 1217–1231.

Shimizu, I. (2003). Grain size evolution in dynamic recrystallization. *Mater. Sci. Forum*, Vol. 426–432, Trans Tech Publ., Switzerland, 3587–3592.

Shimizu, I. (2008). Theories and applicability of grain size piezometers: The role of dynamic recrystallization mechanisms, *Journal of Structural Geology*, Vol. 30, 899–917.

Shimizu, I., (2011). Erratum to "Theories and applicability of grain size piezometers: The role of dynamic recrystallization mechanisms" [J Struct Geol 30 (2008) 899–917], *Journal of Structural Geology*, Vol. 33, 1136–1137.

Takeuchi, S. & Argon, A.S. (1976). Steady-state creep of single phase crystalline matter at high temperatures, *Journal of Materials Science*, Vol. 11, 1547–1555.

Ter Heege, J.H.; De Bresser, J.H.P.; Spiers, C.J. (2005a). Dynamic recrystallization of wet synthetic polycrystalline halite: dependence of grain size distribution on flow stress, temperature and strain, *Tectonophysics*, Vol. 396, 35–57.

Twiss, R.J. (1977). Theory and applicability of a recrystallized grain size paleopiezometer, *Pure and Applied Geophysics*, Vol. 115, 227–244.

van der Wal, D.; Chopra, P.; Drury, M.; Fitz Gerald, J. (1993). Relationships between dynamically recrystallized grain size and deformation conditions in experimentally deformed olivine rocks, *Geophysical Research Letters*, Vol. 20, 1479–1482.

16

Recrystallization: A Stage of Rock Formation and Development

R.L. Brodskaya and Yu B. Marin
Saint-Petersburg State Mining University, Saint- Petersburg
Russia

1. Introduction

The goal of the paper is to show the place and mechanism of recrystallization in the complicated and long-term rock evolution. Theoretical preamble to the study, research methods and their results are discussed.

2. Theoretical recrystallization process model

The rock, like any complex system undergoes several significant stages during its development. To each stage of development in time corresponds its own physiographic expression. Let us recall that rock physiography depends on its texture and structure, i.e., relative amount of minerals in the rock, relative and absolute size of mineral grains, their mutual arrangement, orientation and distribution in space. All of these characteristics describe the structure of mineral aggregate (including rock).

We call development stages of mineral individua and aggregates as stages of their ontogenetic development by analogy with the evolution of biological organisms: initiation, growth, and destruction. Inherently, the transition from one stage to another cannot be gradual or smooth. There must be an interval fixed in time between these stages. There are large taxa of rock evolution: effusive, vein, intrusive, orthometamorphic... Rate and duration of crystallization and the formation of magmatic and metamorphic bodies are key evolution factors in this series. Continuing parallels with biological evolution, it is possible to assume that this series corresponds to rocks phylogenesis within one family. Then, for example, basalt - dolerite - gabbro constitute one series of basic rock evolution (phylogenesis). In gabbro mass, there are always mineral aggregate areas that correspond to the processes of late- or post-magmatic alteration, which result in the emergence of new textural and structural relationships in the mineral aggregate. One can observe different development stages of one mineral aggregate. This is our understanding of the difference between ontogenesis and phylogenesis as applied to the rocks.

Rock alterations during its evolution can be recorded at different levels of organization - isotopic, geochemical, mineral. Changes in a mineral aggregate or a real rock that correspond to a certain stage of its development, correspond to the stage of its ontogenesis. Mineral level of investigation taken by the authors assumes that the rock can be polymineral or monomineral, but it is always polycrystalline natural formation, natural mineral aggregate (volcanic glass in this case is not considered). Mineral crystals are formed and exist under conditions of an assembly, collective growth and functioning.

In magmatic rocks, owing to specific character of the crystallization substrate, its dynamic properties and the volume of the crystallization, kinetic characteristics of crystal formation are inconsistent in time. Mineral crystallization takes place under different conditions. This affects the morphology of resulting crystals, their intergrowths and spatial distribution. Usually, mineral crystals in the rock are called "grains". In metamorphic rocks, all transformations proceed in solid state. Dynamic geological conditions associated with new processes cause changes in the structure and composition of mineral aggregates. In magmatic rock, the rate of mineralization reactions and, consequently, crystal growth usually decreases from first portions of the crystallization to last ones. In metamorphic rocks, kinetic inversions in process parameters are possible both towards the increase in the mineral formation rate and towards the decrease. It is reflected in the increase in the mineral grain intergrowth boundaries area, i.e., the roughness of the boundaries increases. Then the process can follow different scenarios. One of them is the granulation of mineral grains, decrease in their size. Another way of system development is the formation of new grains in the area of inequilibrium boundaries intergrowth. This phenomenon is known to material scientists as "mechanical hardening". The formation of new generation individua that absorb "excess" local energy. With another set of circumstances, the totality of energy loading may be beyond the elastic and plastic deformation of the crystal assembly that can result in brittle deformation of solid bodies. One of the thermodynamic process scenarios after the selection of the way of development by the system is its "straightening" in the course of time. The mineral aggregate is adapted to this choice of the system by the flattening of its internal boundaries, i.e., migration of individual sections of the boundaries to a plane parallel to the plane crystal structure grid, which energy corresponds to local potential of the mineral aggregate system in the intergrowth boundary area at the given stage of its development.

Rock physiography in the accepted hierarchy of consideration is a mineral sublevel created by morphology of mineral grains or boundaries of their intergrowth. Mineral individua or grains exist within internal boundaries of a mineral aggregate and differ from one other in different internal structure: some grains are zonal, others contain mineral and/or fluid inclusions, low-angle misorientation of individual blocks of the crystal lattice, etc. Using biological terms, it is possible to say that grains of one mineral in the mineral aggregate can be of different anatomy. It is clear that different anatomy of mineral individua is due to different conditions of their formation, including growth and dissolution in different kinetic regimes.

Changes in the texture and structure of mineral aggregate, as well as the coexistence of mineral grains with different internal structure, are closely related to changing geological conditions of their formation and existence. The "geological conditions" are some external (with respect to the aggregate) physical fields, their energy, forces and orientation such as areas of tectonic stress but occurring within elastic deformation of minerals and rocks, the area of heating from fluid flows located and crystallized near magmatic bodies, etc. If external fields of force are changed, the internal energy of the mineral aggregate must come into compliance with the external energy. In the balancing process, the structure of the mineral aggregate adapts the whole system of mineral grains (mineral aggregate) to new conditions. The adaptation of the assembly of grains is due to the adaptation of the framework of their boundaries, by changing the composition and energy of the boundaries, i.e. by changing the orientation and the area of mineral grain intergrowth boundaries in the aggregate, by changing intergrowth matrix. It is necessary to remind here that the internal energy of the grain assembly consists of the energy of crystal lattices of mineral individua and the energy of mineral grain intergrowth boundaries. Also, we would like to remind that in the massive mineral aggregate, individuum intergrowth boundaries are boundaries of the individua, i.e., as a whole, they comprise the morphology of each mineral individuum and the framework of internal boundaries of mineral individuum aggregate.

One of initial processes of the framework adaptation of aggregate internal boundaries to the changed conditions is its recrystallization, sometimes accumulative recrystallization. In Russian geological literature, it is common practice to call the process of changing the size and shape of mineral grains in the solid state the recrystallization, but there are two types of recrystallization - one with decrease in the size of mineral individua (it is called *recrystallization*) and another with increase in the medium-sized grains (in Russian literature "perecristallizatsia" or *"overcrystallization"*). We would like to repeat that the grain boundaries migration changes the texture of the mineral aggregate. The accumulative recrystallization controls the structure of the mineral aggregate. The mineral individuum boundaries change the orientation relative to the crystal lattice of mineral grain and possibly in space, taking the position that provides them with such an amount of stored energy that can save the grains under new geological, i.e., thermodynamic and kinetic conditions.

However the grain boundaries migration to a stable state under new conditions requires an initial impulse to overcome stable nonequilibrium. A heat flow from approaching or crystallizing intrusion or a fluid flow either an energy flow of tectonic nature can serve as such an impulse. Not only a new compression or stress can be such an impulse, but the decompression as well. In this case, the system of mineral individua adapts to new growth conditions of individuum well-oriented in a new field of force, or the process of accumulative recrystallization. The authors believe that good orientation of the mineral individuum in the field of force is when an individuum occupies a position when the most stable, i.e., the most atomically dense mineral individuum face occurs normally towards the acting stress. Probably, the schistose structure of mica schist forms in such a manner. It is quite possible that this phenomenon is the cause of gneissose structure. It is not inconceivable that that the interaction of external fields of force and aggregate mineral grains hinders the realization of the described scenario. Then the aggregate adaptation will involve the accumulative recrystallization process. Grains of one mineral form glomero-grained clusters, i.e., subaggregates consisting of grains of one mineral. But there were cases when generated monomineral subaggregates formed rather stable distinct boundary between the subaggregate and grain matrix in the aggregate. The monomineral subaggregate attains crystal-like morphology, i.e., a shape when part of its boundaries with mineral aggregate look like simple forms inherent in this mineral. The process of levelling, balancing of intergrowth energy of mineral individua (recrystallization) continues inside the subaggregate. (Fig. 1, 2, 3)

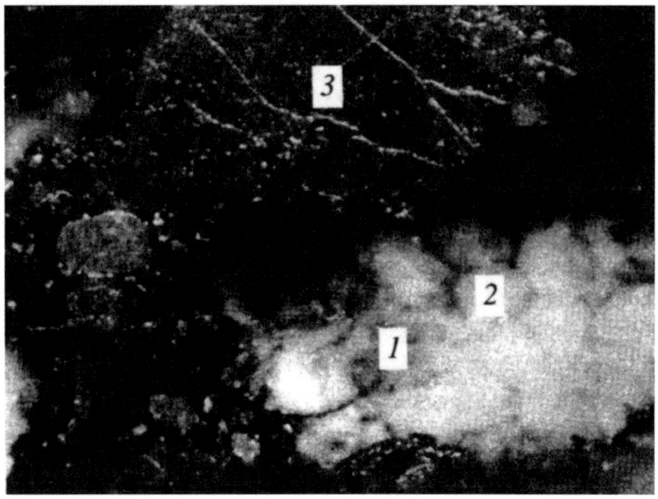

Fig. 1. Fragment of conglomerate from Carbon–Lider ridge. 1- quartz pebble; formation stage of inner boundaries of subaggregate, which is relevant to grain faces - attractor(2); 3 - pebble of pyrite grain. 1.5x.

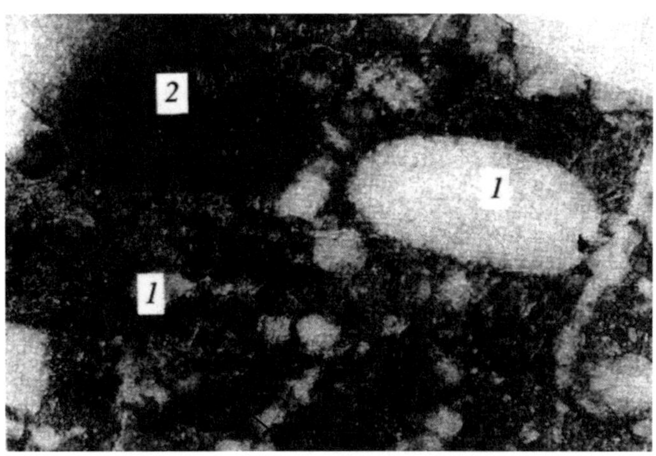

Fig. 2. Fragment of conglomerate from Ventersdop ridge. 1 – pebble of quartz of 1 kind; 2 – quartz pebble of 2 kind (amoebic contours, in the center quartz are free from inclusions, chlorite micrograins paint margin to dark-green color). White points on the right – accumulation of fine grained pyrite. Gray angular segregations are phyllite. 1.2x.

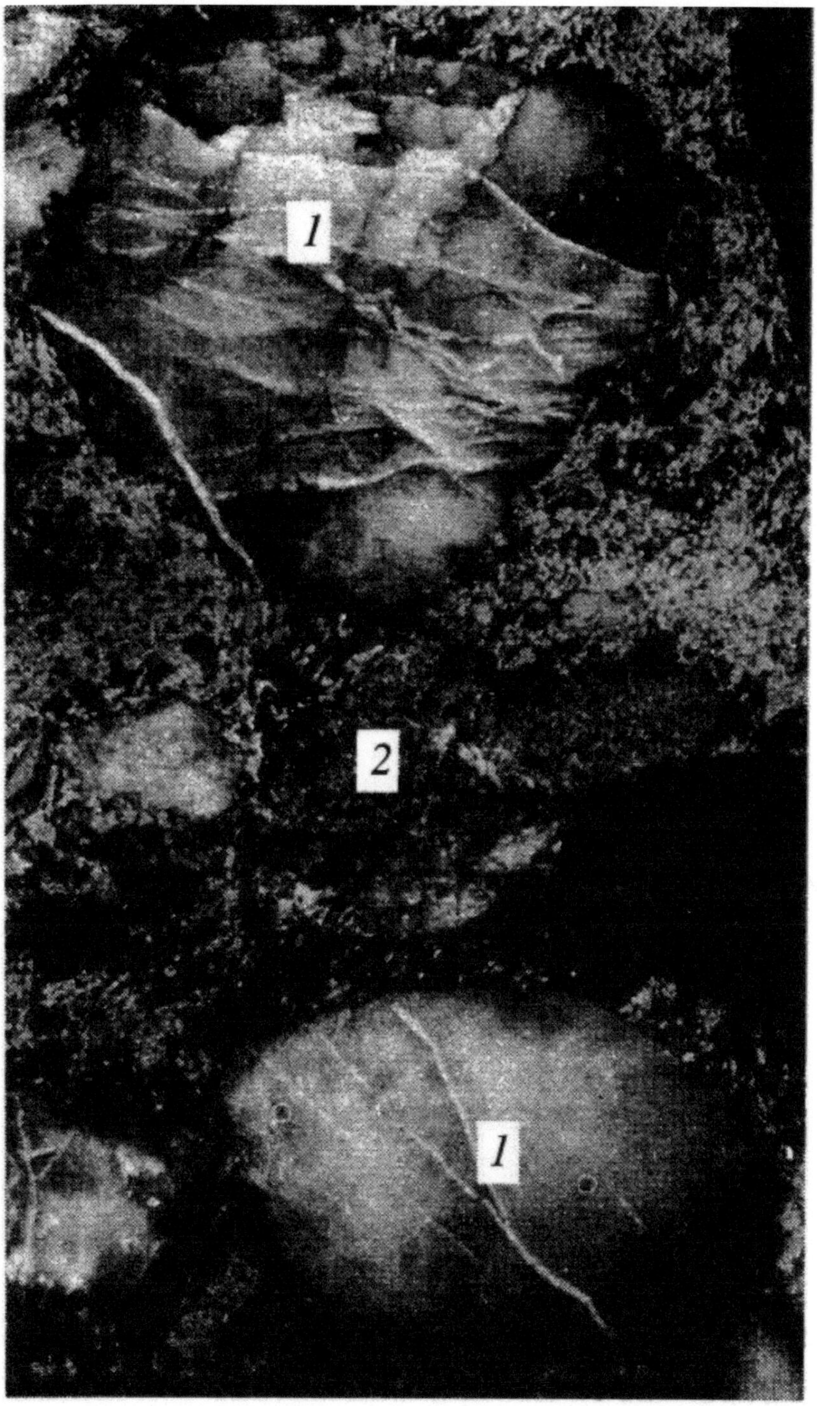

Fig. 3. Fragment of conglomerate from Carbon–Lider ridge. Cataclastic quartz pebble of 1 kind; 2 - pebble of third kind, composition and color is similar to basic matrix. White mass around pebble - aggregate of pyrite. 1,5x.

External flow or the initial impulse induces the energy flow from each mineral grain. This is the energy of edge dislocation of mineral individuum, energy of its boundaries. Energy of dislocations and defects in the crystal lattice of each mineral individuum is involved in the general flow. Thus, the stable equilibrium becomes unstable. Trace elements located in defects and dislocations migrate from their places together with induced energy flow. The flow is directed towards the edge dislocation of mineral grain - its boundary. This energy and its flows provide the grain distillation from trace elements, mineral and fluid inclusions, subboundaries - low-angle boundaries within the crystal lattice (e.g., subboundaries between blocks of cloud extinction in quartz, "loops" and "oblique walls" in olivine). When the impulse energy is sufficient, the process of solid solution disintegration is being formed. Many minerals represent such solid solutions of one mineral in another one. The process of solid solution decomposition results in the appearance of specific, easily recognizable decomposition structure (Fig. 4 (a, b). This is a new instability, which activates migration of subboundaries within the grain. It's possible to indentify by means of displacement character of subboundaries and stimulated movement forces two types of recrystallization - rotational and migrational (Fig. 5,6). This is the way of changing the anatomy of mineral individua; this is the way of replenishing the impulsive force energy for the formation of new mineral grain boundaries. Accumulative overcrystallization within the grain, i.e., the aggregation of micro- and nano-individua clusters takes place simultaneously with the migration of boundaries. Trace elements and newly formed mineral phases are "squeezed-out" to grain boundaries to generate their own mineral form with its own boundaries. This is the mechanism of overcrystallization and mineral formation at grain boundaries in the solid state (Fig.7).

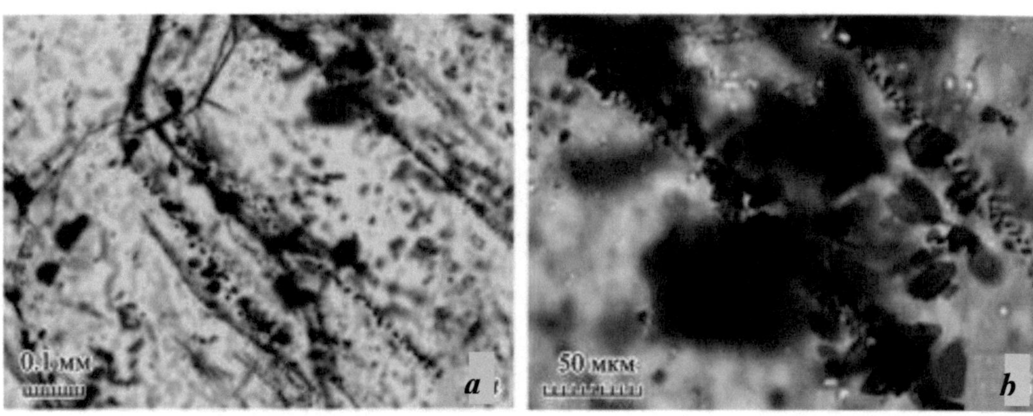

Fig. 4. (a,b) Structures of the solid solutions decomposition in olivine; the newly formed phase is chrome spinelide. Dunite, Gulinsky massif; photographs in transparent light, without analyzer

Fig. 5. Rotational recrystallization of the olivine aggregate from dunite of Gulinsky massif; photograph in transparent light, with analyzer.

Fig. 6. Migrational recrystallization of the olivine aggregate from dunite of Galmoenansky massif; photograph in transparent light, with analyzer.

The surface energy of mineral grain edge dislocations and the boundaries of their intergrowth in the aggregate is, as already mentioned, an instrument in the mechanism of balancing between the internal energy of the aggregate and the external energy of the field of force under changing geological conditions. The amount of the mineral grain surface energy consists of the edge dislocation energy of the crystal lattice of mineral individuum and the presence of some admixtures, i.e., first of all, depends on the orientation of the boundary (edge dislocation) relative to the individuum crystal lattice (Fig. 8 (a, b, c). However, main role in boundary migration is traditionally given to the specific energy of the surface area rather than to the surface energy.

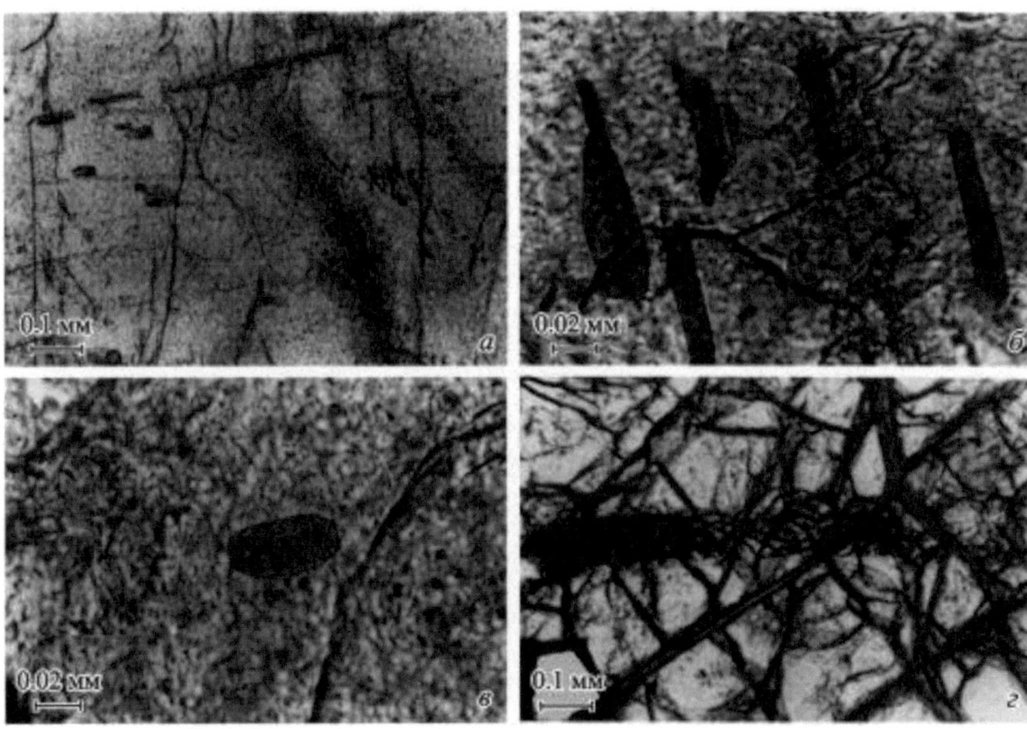

Fig. 7. Forming of new minerals in deformational substructures of olivine (Arai Shoji, 1978). Photographed with different magnifications in transparent light

Fig. 8. (a, b, c). Regularly oriented lamellae of chrome spinelide and the skeleton inclusions of spinel in the olivine grain. Dunite, Gulinsky massif; photographs in transparent light, without analyzer, in different thin sections

Increase in some components of the external field of force (stress, lithostatic pressure, fluid flow pressure, heat flow) results in the increase in the amount and energy of internal boundaries of mineral aggregate. This process inspires the increase in the boundary density in the mineral aggregate space as well as the specific surface energy. The decrease in the external energy flow necessitates the decrease in the internal energy of the mineral aggregate. To this change in external geological conditions the aggregate is adapted due to the decrease in the surface energy of mineral individua. In the first case (increase in the field of force) porphyraceous structures form, in the second - monomineral subaggregates. The selection of the development path system depends on the necessity to decrease the internal energy. This is possible owing to increase in the area of mineral individuum boundaries and decrease in its specific surface energy. This is also possible due to the decrease in the grain intergrowth energy of one mineral. In polymineral aggregate, the least amount of the energy is absorbed by intergrowth boundaries of one mineral.

At grain boundaries of one mineral, the intergrowth energy is lower than in intergrowths of different minerals. Most likely, just the "energy benefit" is the motivation of the accumulative recrystallization that covers vast mineral aggregate spaces. This is the way of formation of glomo-grained subaggregates within the massif aggregate matrix with regular grain distribution of all the minerals. Which mineral in the "struggle for survival" will decrease or increase the density of its boundaries depends on marginal conditions of existence of this or that mineral in the aggregate. It is quite possible that material supply into the crystallization system can result in the generation of such conditions (Fig. 9).

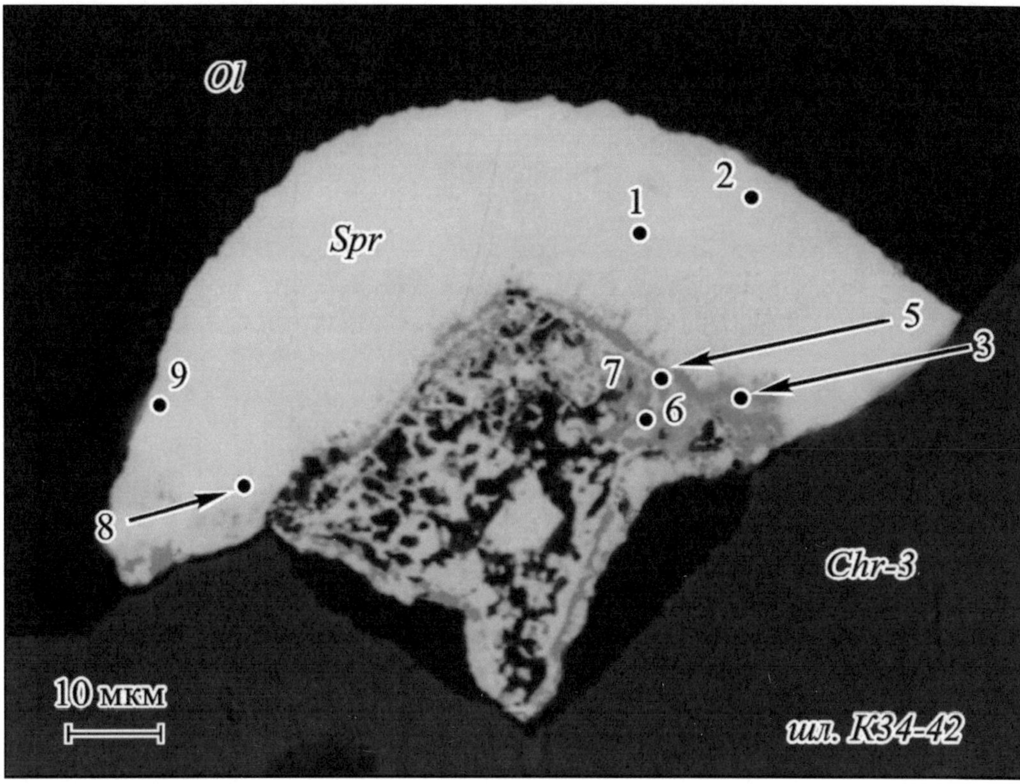

Fig. 9. Subaggregate of grains of platinum and iridium arsenides at the boundary of intergrowth with subaggregates of the third generation chromite (*Chr-3*) and olivine; numbers of points correspond to microprobe analyses: *1, 2, 8, 9*—sperrylite, *5, 6*—raresite (transparent polished section K-34-42); electron microscope.

The character of subaggregate boundaries is an important kinetic feature of the occurring processes. It may be flat or irregular to variable degrees. Under some conditions, the migration of grain boundaries results in their straightening and integration into intergrowths, i.e., increase in the grain volume.

Studying the energy models of internal boundaries relationship in the mineral aggregate shows that changed position of mineral individuum intergrowth boundaries or recrystallization is a response of the mineral grain assembly to new geological conditions under which the mineral aggregate occurs. Main tool of aggregate adaptation to new conditions is the migration of individuum intergrowth boundaries to such a position in space and relative to the crystal lattice of each of the intergrown grains that ensures its stability in the field of changing force and energy. It is quite possible that not all boundaries and not all minerals have to migrate when they change their orientation in space. Thus, atomically dense boundaries, i.e., simple-shape faces are the most stable mineral individuum boundaries. Boundaries of mineral grains, which in crystallographic coordinates have sufficiently high symbols, i.e., have relatively low store energy, are capable of migration in the aggregate space. It is natural that the boundary migration is possible until they reach equilibrium state with one another. But other situations are also possible. For example, if recrystallization accompanies metasomatosis with additional supply of some material. In this case, growth of this material from new crystallization centers is possible either the growth due to the increase in mineral individuum volume if its composition corresponds to that of the supplied material. Porphyroblast growth is possible when supplied material flow rates are rather high.

3. Mineral aggregate ontogenesis and physiography

Earlier it was said that the necessity to decrease the internal energy of mineral aggregate during its adaptation to the external energy (e.g., while decompression) results in the change of the structure of the mineral grain assembly - accumulative recrystallization. Probably the spotty structure of phyllite slate is a result of this process. Quartz grains are accumulated into such glomero-grained subaggregates characterized by irregular distribution and rather variable volume in the rock space. High rate of changes in external conditions can result at some stage or other in the arrangement of a boundary between a monomineral glomero-grained unit and aggregate matrix. In this case, the external boundary is formed by the combination of identical boundaries of mineral grains located at edges of glomero-grained bundles. Individuum boundaries usually correspond to a simply-shaped face of this mineral with relatively high stability factor under given kinetic conditions. The authors came to this conclusion while studying texture of the so-called quartz pebbles in auriferous reefs of Witwatersrand. Macroscopic investigations of samples from conglomerate outcrops of the Carbon Leader and Ventersdorp reefs, suggest the similarity of the morphology of "pebbles" and that of quartz crystals due to the presence of surfaces in the pebble that resemble prism face. Thin sections were made of several quartz "pebbles", which had in their faceting external prism "faces". In the quartz grain subaggregate, microscopic investigations revealed the presence of sections, which boundaries were similar to those of external morphology of the pebbles. The authors interpreted this fact as an existence of a grain-attractor within quartz individuum subaggregate. It is just these boundaries are the most stable for quartz, and therefore most beneficial for the conservation of quartz subaggregate in the regime of unstable parameters of changes in geological conditions, which also imply thermodynamic ones. Most likely, this is the alternation of compression and decompression in the course of compression and extension of host rocks during hydration and dehydration of intergranular space of mineral aggregate. Occurrence of hydrofilms in the mineral aggregate not only increases the plasticity of the rock as a whole, but also helps to increase the resistance of mineral individua to elastic and plastic deformations. Under certain conditions, such quartz monomineral subaggregates can be transformed into blastoporphyric quartz "crystals" if quartz individua will be able to adapt to each other not only by boundaries of appropriate density, but also due to coherent orientation of crystal lattices of porphyroblasts. Such examples are recorded not only in Witwatersrand, but also among porphyraceous dunite aggregates of the Inagli, Galmoenan massifs. It is interesting that in these cases, grains-attractors with orientation and morphology of the subaggregate are also observed inside olivine subaggregates.

4. General procedure of quantitative ontogenetic analysis

Stereometric analysis of rocks and ores made in thin sections and polished sections was the main method of implementing the above mentioned ideas. Quantitative assessment of the parameters of the structure allows the estimation of the recrystallization degree of the aggregate and mineral individua. Structural characteristics become parameters of the texture and structure because of the application of crystallographic and topological methods of analysis. Anatomy, i.e., internal structure of mineral individua also has its quantitative measure. Density of subboundaries, density of fluid and mineral inclusions and other features of refining mineral individua during recrystallization are estimated here.

This research trend (quantitative ontogenetic analysis) allows unbiased assessment of the stage of mineral aggregate evolution by distinguishing individual generations and paragenetic (simultaneous) associations of mineral generations. Not only is the history of mineral aggregate restored, but a place of mineralization in ontogenesis as well.

Thus, the main method of mineral aggregate structure interpretation is the ontogenetic analysis of the mineral aggregate and its individua. Main method of implementation of the expressed ideas is stereometric analysis of rocks and ores. A procedure of studies using polished sections and thin sections was elaborated. A representative area is necessary to obtain metric assessment of the mineral aggregate structure under the microscope. Standard area of thin sections is used, but the design of the integration device MIU-5M allows analysis of thin sections with an area of no more than 40×40 mm. Analysis sensitivity is not worse then 4 μm. It means that using scanning table, it is possible to get grid coordinates of points occurring within the specimen plane spaced at a distance of 8-10 μm from one another. Quantitative measure of fabric parameters allows evaluation of degree of recrystallization of the aggregate and mineral individua. Structural characteristics become texture/structure parameters due to simultaneous employment of well-known crystal-optic, crystallographic, geometric, and topologic analytical methods. Anatomy of mineral individua has quantitative measure. Following features are estimated here: subboundary density within mineral grain, density of fluid and minerals inclusions and other features both residual, primary, and refinement features of mineral individuum anatomy while recrystallization. Roughness of intergrowth boundaries of all mineral grains or grains of one mineral in the aggregate either grains of individual mineral generations can be changed and calculated using several methods.

Such parameters of mineral aggregate fabric as total area of internal boundaries of mineral aggregate, modal portion of individuum boundaries of each mineral, modal and normative granulometric compositions, character of mineral grain distribution in aggregate or frequency index of individuum intergrowth of one mineral with grains of other minerals, etc. are measured simultaneously while scanning a specimen (thin section or polished section). These are the so called integral characteristics of rock structure. Frequency characteristics of mineral grain boundaries can be obtained in the course of their analysis using the device of fractal dimensions or Fourier harmonic decomposition. All quantitative characteristics and fabric parameters necessary for ontogenetic analysis of mineral aggregates are real functional capabilities of the Mineralogical Integration Device (MIU-5M in Russian). The elaborated procedure enables to get and use 22 fabric parameters or part of them in any combination and amount.

This trend of studying rocks and ores can be named quantitative ontogenetic analysis. Its use in investigating thin sections and polished sections allows unbiased assessment of stage of mineral aggregate evolution and identify separate generations and paragenetic (simultaneous) associations of mineral generations. Not only the history of the mineral aggregate, but also a place of mineralization in the ontogeny is reconstructed in the transformations sequence of mineral individua assembly.

5. Conclusions

We discussed the recrystallization process in a mineral aggregate as a migration process of mineral individuum boundaries, mineral grain intergrowth boundaries, as the process of changes in the framework of internal boundaries of the aggregate. Migration reasons were formulated as a mechanism of mineral aggregate adaptation to changed (as compared to initial conditions of formation) geological conditions of rock existence. On the way of aggregate adaptation to changing geological conditions, the recrystallization, similar to accumulative recrystallization, is possible at all levels of mineral matter existence.

6. References

Arai Shoji. Chromian spinel lamellae in olivine from the Iwanai-dake peridotite mass, Hokkaido, Japan /Earth and Planetary Science Letters. 1978. N 39. C. 267—273.

Brodskaya R.L., Shumskaya N.I. // Transactions of the USSR Ac. Sci.. 1998. V. 362. No. 3, pp. 378-381(in Russian)

Brodskaya R.L., Bilskaya I.V., Kobzeva Yu.V., Lyachnitskaya V.D., Rachmanova N.V., Talovina I.V. Formation of surface and properties of the mineral individual borders in aggregate destruction. IGC, 2000. Brazil.sec. of phys. and chem. minerals, publishing in CD.

Brodskaya R.L., Marin Yu.B. Problem of Internal Structure Modeling of Ordered and Equilibrium Mineralogical-Petrographic Systems// ZVMO 2001. P. CXXX. No. 6, pp. 1-14 (in Russian)

Brodskaya R.L., Bilskaya I.V., Kobzeva Yu.V., Lyachnitskaya V.D. Typomorphic Structural Features of Ultramafite Mineral Aggregates and Mechanism of Chrome Spinellide Concentrations in them // ZVMO. 2003.P. CXXXII. No. 4, pp. 18-37 (in Russian)

Brodskaya R.L., Marin Yu.B. Rock Structuring: Adaptation Mechanism of the System to Inequilibrium Thermodynamic Processes /Collection of Articles "Rocks". 2004. Apatity, pp. 19-26(in Russian)

Brodskaya R.L., Bilskaya I.V., Lyachnitskaya V.D., Markovsky B.A., Sidorov E.G. Formation of PGE Mineralization in Ultramafite of the Galmoenan Massif (Koryakia) // Transactions of the 8th International Conference "New Ideas in Geoscience". Moscow. 2007. V. 5, pp. 37-39 (in Russian)

Brodskaya R.L., Marin Yu.B. Formation Model of Mineral Aggregate Internal Boundaries and Examples of Its Application. Transactions of the 8th International Conference "New Ideas in Geoscience". Moscow. 2007. V. 3, pp. 56 - 59 (in Russian)

R.L. Brodskaya, I.V. Bil'skaya, V.D. Lyakhnitskaya, B.A. Markovsky, E.G. Sidorov. Boundaries of Intergrowths between Mineral Individuals: A Zone of Secondary Mineral Formation in Aggregates.// Geology of Ore Deposits. 2007. Vol. 49. No. 8, pp. 669–680. © Pleiades Publishing, Ltd.

R.L. Brodskaya, I.V. Bilskaya, B.A. Markovsky. Ontogenic Analysis of Individual Olivine Grains in Ultramafic Rocks. // Geology of Ore Deposits.2010. Vol.52. No.7, pp. 566-573.

Ponomarev V.S. Energy Saturation of Geological Environment. Transactions of Geological Institute. 2008. Moscow. Nauka. Issue 582. 379 p.(in Russian)

Shcheglov A.D., Shumskaya N.I. // Transactions of the USSR Ac. Sci. 1995. V. 340. No. 5, pp. 667-671(in Russian)

Witwatersrand Gold - 100 Years/ Ed. by E.S. Androbus. Geol.Soc.S.Africa, 1986.P. 298.

Part 3

Recrystallization in Pharmacology

17

Recrystallization of Enantiomers from Conglomerates

Valérie Dupray
Université de Rouen
France

1. Introduction

An object is considered as "chiral" if it is not superimposable on its mirror image. Due to the presence of asymmetric carbon atoms, numerous molecules are chiral but other stereogenic centers provide the asymmetric character of molecular compound (e.g. sulfoxides). Metal-organic complexes can also be asymmetric even if the ligands do not have any stereogenic center. The equimolar mixture of two enantiomers is the racemic mixture. Except for their rotatory powers which are equal in value but of opposite signs, enantiomers present identical chemical and physical properties in achiral environment. Conversely, their behaviors in chiral environment such as biological systems are often different. In particular, for chiral drugs, the pharmacological and the toxicological effects can be drastically dissimilar.

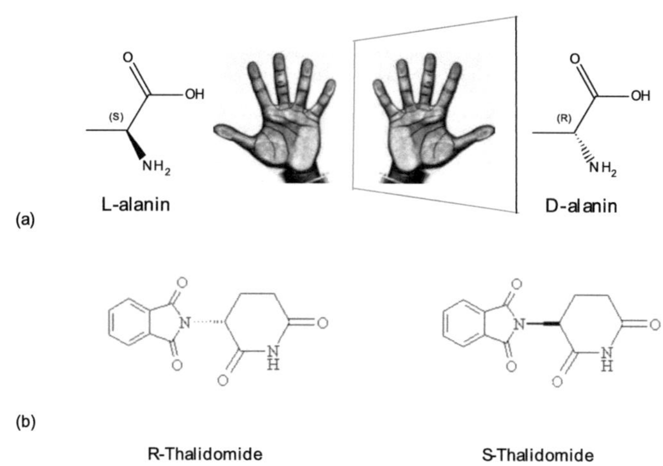

Fig. 1. Enantiomers of (a) Alanin and (b) Thalidomide

A well-known example is the case of thalidomide. Thalidomide was a drug prescribed for nausea during pregnancy in the later 1950s and was marketed as the racemic mixture. The R-enantiomer showed the desired curing effect while the S-enantiomer caused serious teratogenic effects (Kocher-Becker et al, 1982). The health scandal that follows led to the removal of the drug in 1961. Since, the pharmaceutical regulatory authorities have increased the pressure for chiral drugs to be administered in an enantiomerically pure form. Thus, chiral drug industry has been in constant development and has become the most growing segment of the drug market. It represents now more than one third of the drug sales worldwide (Stinson, 2000). At the same time, chemical producers have developed new enantiomeric intermediates for industry and especially enantioselective technologies. Asymmetric synthesis (a chemical reaction of an enantiomeric agent or catalyst with a substrate to produce a single enantiomer of the desired molecule) remains the first mode of production of pure enantiomers. However, the reaction product can appear to be a racemic mixture due to synthesis conditions or in rare occasions to racemization. Commonly, it is also cheaper, easier or cleaner to synthesize directly the racemic mixture. In those cases, a chiral resolution (separation) has to be performed to recover the pure enantiomer.

A direct enantiomeric separation can be carried out via chiral chromatography. To this purpose, the inner surface of a chromatographic column is bonded or coated with a chiral selector or alternatively the chiral selector can be incorporated directly to the stationary phase. As a result, the two enantiomers have different retention times which allow the resolution (Subramanian, 2006). However, this process which leads to high purity products, the productivity (which can be poor due to a high separation time and/or a small rate of injection of the product in the column) is not always compatible with industrial standards. Another solution consists of using crystallization.

The most popular crystallization method certainly remains the Pasteurian resolution (Pasteur, 1853) for which a chiral resolving agent is used to obtain the crystallization of diastereoisomers. Typically, the resolving agent (an enantiomerically pure acid (resp. base)) is added to the racemic base (resp. acid) to form diastereoisomeric salts. Contrary to the enantiomers, diastereoisomers do not exhibit the same symmetry (Coquerel, 2000) and as a consequence do not present the same chemical and physical properties (in particular solubility). Dissolution-recrystallization in an appropriate solvent or mix of solvents permits to obtain a single solid crystalline phase (the pure salt). Then, the pure enantiomer can be recovered by salting out to remove the resolving agent.

In order to improve the yield and to reduce the quantity of resolving agents, several variations have been proposed. Let us just mention the Marckwald's method (Marckwald, 1896), the Pope and Peachey's method (Pope & Peachey, 1899) and more recently the Dutch resolution (Vries et al, 1998), (Kaptein et al, 2000). Note also that the accurate determination of the phase diagrams under the conditions of the experimental process allows an optimization of the resolution (Marchand et al, 2004). Even imperfect (a yield close to 100% is rarely obtained), this method most often fits industrial and laboratories requirements.

Besides the Pasteurian method, the resolution can be performed by "preferential crystallization" (PC) also called "crystallization by entrainment". PC is a stereoselective process in which, alternatively, for a given period of time, only one enantiomer crystallizes although both enantiomers are supersaturated in the mother liquor. In 1866, Gernez (Gernez, 1866) was the first to observe that a saturated solution of one enantiomer, seeded by the same enantiomer, allows the formation of enantiomerically pure crystals. Conversely, he noted that if the seeding was done with the other enantiomer, no crystallization was observed. The entrainment phenomenon itself was described by Jungfleish (Jungfleish, 1882) who underlines the predominant influence of supersaturation.

In the 1990s, the phenomena involved in the preferential crystallization have been explained in details in literature using phase diagrams. For a complete description please refer to Jacques and his co-workers book "Enantiomers, Racemates, and Resolution"(Jacques et al, 1994).

Several improvements of this cyclic process have been proposed such as the auto-seeded variant AS3PC developed by Professor Gerard Coquerel and co-workers at the French University of Rouen (Ndzié et al, 1997; Coquerel, 2007). AS3PC by its robustness, its low cost and reproducibility offers real possibilities for industrial applications of preferential crystallization. As a proof, the AS3PC process has been the subject of several patents for the University of Rouen (Coquerel et al, 1994; Coquerel et al, 1995; Helmreich et al, 2010)

The main advantage of PC and its derivatives is certainly that they do not require any resolving agent. However they suffer from a serious limitation: the compound to be resolved or accessible derivatives such as salts solvated or not, must crystallize as a conglomerate (ie. an equimolar mechanical mixture of crystals, each one containing only a single enantiomer). The requirement of a conglomerate is an important restriction to the application of preferential crystallization due to the low occurrence of conglomerates among the molecular crystallized compounds (5-10% of the racemic species only)(Jacques et al, 1994).

Because the detection of a conglomerate is a key step of the resolution process, we chose to focus this contribution on this subject. The first part constitutes a necessary reminder about crystal packing of chiral molecules and includes the definition of the different types of structures that can be encountered crystallizing racemic mixtures. The second part is dedicated to the enantiopurification by crystallization. The benefits of working with conglomerates are underlined. The next part describes the detection of conglomerates itself. Classical methods are first recalled and then the prescreening of a conglomerate via a high throughput technique involving nonlinear optics is presented. In the last part, we propose a sequential diagram to optimize the detection of conglomerates.

2. Crystal structure and packing of chiral molecules

Speaking about crystallization and recrystallization implies to take into account the way the molecules are packed inside the crystal. It is all the more important as the chiral nature of a molecule imposes some limitations.

2.1 Types of packing of chiral molecules

Three most common types of packing are usually observed when crystallizing a racemic mixture (Coquerel, 2000):

- The racemic compound is the most common (90-95% of racemic species). In the vast majority of the cases it is a <1-1> stoichiometric compound.
- The conglomerate is a mechanical mixture of single crystals containing homochiral molecules only. As mentioned previously, conglomerates represent only 5 to 10% of the racemic species (Jacques et al, 1994).
- The racemic solid solution has a low occurrence. It is a solid solution containing an equal number of molecules of each enantiomer but contrary to the racemic compound, the arrangement is a random distribution (see fig.2)

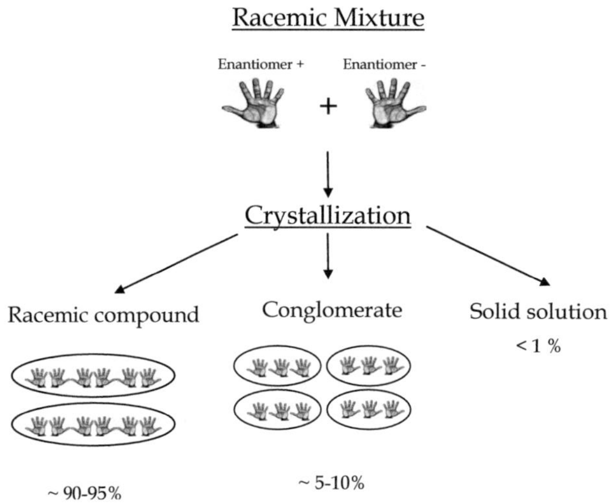

Fig. 2. Types of packing of chiral molecules

2.2 Chirality of crystalline structures

The nature of the crystalline structure formed by the chiral molecules is of major importance when considering the recrystallization processes. At this stage, it is imperative to distinguish between the chirality of the molecule itself and the chirality of the crystalline structure. Indeed, crystalline structures formed from enantiomers (chiral molecules) present in racemic mixture are not necessarily chiral. Let us just first recall that chiral crystals belong to space groups that contain only symmetry operation of the first kind (rotation, translation). It excludes the symmetry operations of the second kind (rotoinversion) which are allowed in non-centrosymmetric structures.

Crystalline structures can then be classified into three categories (Flack, 2003; Galland et al, 2009):

- Centrosymmetric (achiral) structures (type CA) which corresponds to point groups $\bar{1}$, 2/m, mmm, 4/m, 4/mmm, $\bar{3}$, $\bar{3}$m, 6/m, 6/mmm, m3 and m3m.
- Non centrosymmetric achiral structures (type NA) for point groups m, mm2, $\bar{4}$, 4mm, $\bar{4}$2m, 3m, $\bar{6}$, 6mm, $\bar{6}$m2 and $\bar{4}$3m.
- Non centrosymmetric chiral structures (type NC) associated with point groups 1, 2, 222, 4, 422, 3, 32, 6, 622, 23 and 432.

Racemic compounds can theoretically be part of any space groups (CA, NA or NC) but 95 % of the known racemic compounds crystallize in centrosymmetric space groups (CA). The predominant space groups are: $P2_1/c$, C2/c, Pbca and $P\bar{1}$ (95 %) of the centrosymmetric racemic compounds)(Dalhus et al, 2000). Non centrosymmetric racemic compounds (NA) like DL-Allyglycine (Dalhus et al, 2000) represent a proportion of 4.5-5 %, mainly placed in space groups Pna21, Pca21, Cc and Pc. Rare cases of racemic compounds crystallizing in chiral space groups (mainly $P2_12_12_1$ and $P2_1$) have been reported (ortho-thyrosine, α-methylsuccinic acid or camphoroxime (Jacques et al, 1994; Brock et al, 1991; Kostaniovsky, 2008). Their occurrence is estimated to be only 0.02%(Flack et al, 2003).

Regarding the conglomerate, the chiral nature of the molecules imposes restriction in the construction of the crystalline structure so it is impossible to form achiral crystal structures by crystallization of enantiomerically pure chiral molecules (Jacques et al, 1994), (Flack et al, 2003). Consequently, conglomerates crystallize only in one of the 65 chiral space groups (spaces groups $P2_12_12_1$, $P2_1$, C_2 and P_1 represent 95 % of the known conglomerates (Belsky et, 1977).

		Achiral structure		Chiral structure
		CA Centrosymmetric Achiral	**NA Non centrosym. Achiral**	**NC Non centrosym. Chiral**
Racemic compound 90 - 95 %	*Structure*	Permitted	Permitted	Permitted
	Proportion	~ 95 %	4.5-5 %	0.02%
	Predominant space groups	$P2_1/c$, C_2/c, Pbca and $P\bar{1}$	$Pna2_1$, $Pca2_1$, Cc and Pc	$P2_12_12_1$ and $P2_1$
Conglomerate 5-10%	*Structure*	Forbidden	Forbidden	Mandatory
	Proportion	0%	0%	100%
	Predominant space groups			$P2_12_12_1$, $P2_1$, C2 and P1
Solid Solution Less than 1%	*Structure*	Permitted	Permitted	Permitted

Table 1. Formation of crystalline structures from racemic mixtures

Concerning the solid solution, the three crystal types are permitted but no data seems available about the predominant space groups in the case of racemic solid solution. These various possibilities for the crystallization of racemic mixtures are summarized in Table 1.

3. Recrystallization of enantiomers

Enantiopurification by crystallization is presented here for two cases:

- a racemic compound-forming system using the classical selective crystallization processes
- a conglomerate for which a crystallization by entrainment is possible

The two recrystallization procedures require a good knowledge of the behavior of the two enantiomers regarding melting (binary phase diagrams) and solubility in a given solvent (Ternary phase diagrams). For a detail description of the various phase diagrams encountered for these systems of enantiomers, please refer to (Jacques et al, 1994), (Lorenz et al, 2006) and (Coquerel, 2000).

3.1 Enantiopurification from racemic compound-forming system

On figure 3 is depicted the usual ternary phase diagram observed for a racemic compound-forming system (S and R enantiomers, racemic compound RS and solvent V_1) at a given temperature T_1. Six regions are delimited by the solubility curves (see legend of figure 3) and can contain from 1 to 3 phases. The size of theses different regions varies of course with temperature.

Assuming that the mixture of enantiomers to purify (represented by point M) has an enantiomeric excess exceeding that of the invariant liquid I_1, the best process of enantiopurification consists in adding solvent V_1 in order to place the overall composition into the two-phase region (region 3 if enantiomer S is concerned). A new equilibrium is then established between <S> and the mother liquor (point K). Then, a selective crystallization using an appropriate cooling program can be used to obtain the desired enantiomer.

Depending on the location of I_1 and K, the maximum mass of pure enantiomer <S> that can be recovered is given by:

$$m<S> = m_T \times \frac{KI_1}{SI_1} \quad (1)$$

with m_T, the total mass.

Obviously, this mass is always lower than the total mass of enantiomeric excess since e.e.(I_1) ≠ 0.

It is then possible to proceed to a recrystallization of a pure enantiomer in the case of a racemic compound-forming system. However, this is only applicable to a starting system presenting a sufficient enantiomeric excess. Thus, the process requires a first enrichment step that can be performed by chiral chromatography or via diastereoisomer formation. Moreover, it is not possible to recover the totality of the enantiomeric excess.

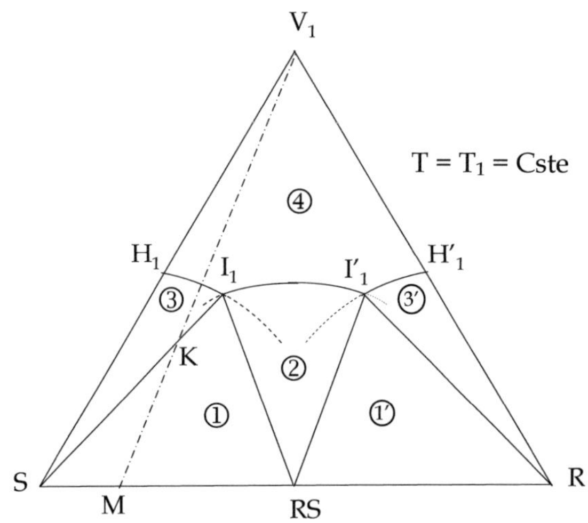

Triphasic domains
Region 1 : Invariant liquid I_1 doubly saturated + <S> + <RS>
Region 1': Invariant liquid I'_1 doubly saturated + <R> + <RS>
Biphasic domains:
Region 2: Saturated solution + <RS>
Region 3: Saturated solution + <S>
Region 3': Saturated solution + <R>
Monophasic domain:
Region 4: Undersaturated solution

Fig. 3. Ternary phase diagram observed for a racemic compound-forming system

3.2 Enantiopurification from a conglomerate-forming system

On figure 4 is depicted the ternary phase diagram observed for a conglomerate forming system (S and R enantiomers, solvent V_1) at temperature T_1. This diagram is simpler that the one corresponding to the racemic compound-forming system since only two solid phases (S and R) can coexist. Four regions are delimited by the solubility curves (see legend of figure 4).

Here, as for the racemic compound, starting from a mixture M of enantiomers, the process consists in adding the exact quantity of solvent (at T_1) so that the overall synthetic mixture is represented by point K. Point K is situated on the tie-line SI separating the 3 phase domain I-S-R (which contains: pure <S>, pure <R> and the doubly saturated solution I) and the biphasic domain S-H-I (where the phases in equilibrium are pure <S> and the saturated solution). Then the selective crystallization can be pursued.

Contrary to the previous case, the whole enantiomeric excess of mixture M can be retrieved (on the condition that the compound crystallizes as a conglomerate without partial solid solution). It constitutes the best situation to obtain an efficient recrystallization in thermodynamic equilibrium.

In practice (see figure 5), it is better to heat the system above T_1 so that the suspension at $T_2 > T_1$ contains only pure <S> and a saturated solution. The smooth cooling program will mainly consist in crystal growth leading to a better filterability and a more efficient washing of the filtration cake.

It is even sometime possible to retrieve more than the initial enantiomeric excess by cooling the system at T_0. Point K is then situated in the 3-phase domain and as long as the crystallization is stereoselective the liquid composition can evolve from I_1 at (T_1) to Z_0 at T_0 (figure 6). Note that Z_0 is located on the metastable solubility curve of S at T_0. This corresponds to an "entrainment" (i.e. a single operation of the so-called AS3PC process).

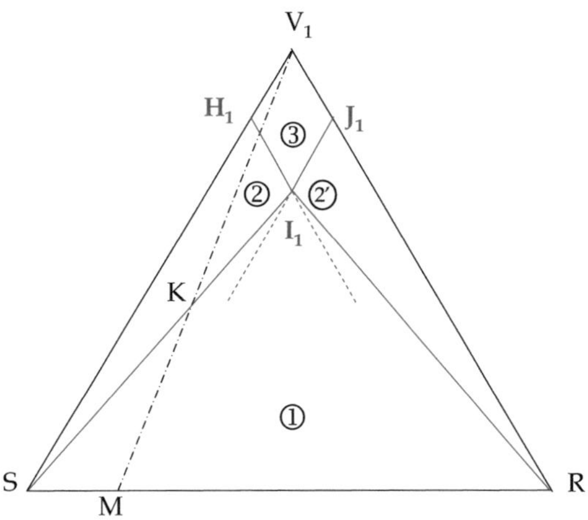

Triphasic domain: 1 - Invariant liquid I doubly saturated + <S> + <R>
Biphasic domains: 2 - Saturated solution + <S>; 2′ - Saturated solution + <R>
Monophasic domain: 3 - Undersaturated solution
Dashed lines stand for metastable solubilities

Fig. 4. Ternary phase diagram observed for a conglomerate forming system

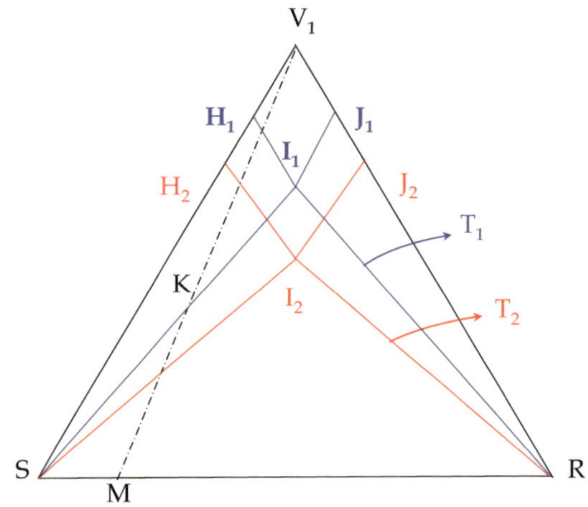

Fig. 5. Ternary phase diagrams observed for a conglomerate-forming system at T_1 (Blue) and T_2 (Red)

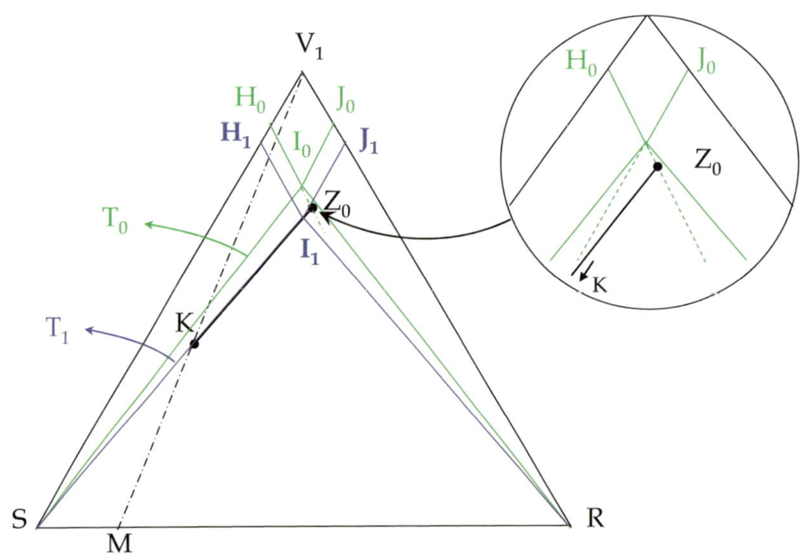

Fig. 6. Principle of crystallization by entrainment

When the initial mixture M corresponds already to a high enantiomeric excess, solvent V_1 at temperature T_1 won't be appropriate because the slurry will be simply much too viscous. Thus, it is necessary to find another solvent V_2 in which the pure enantiomer <S> is poorly soluble (figure 7).

3.3 Enantiopurification from conglomerate-forming derivatives

Figure 8 shows that a conglomerate-forming solvate (here a monohydrate) can also give the full discrimination in the solid state even if a stable racemic compound exists for the anhydrous chiral components. The enantiopurification by means of crystallization can be operated in a similar way as that presented above (point K for a mixture M at T_1). When the initial point M is close to the pure enantiomer (e.g. S), the medium can be changed by mixing an anti-solvent. The second solvent must be miscible with water, should not induce a miscibility gap and should not inhibit the formation of the conglomerate.

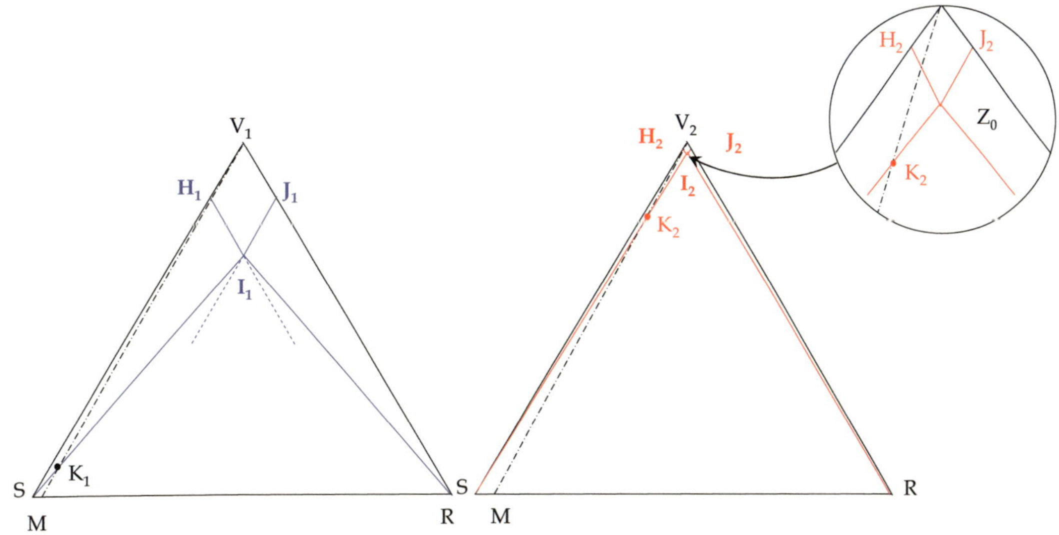

Fig. 7. Enantiomeric purification with a high enantiomeric excess - Choice of another solvent

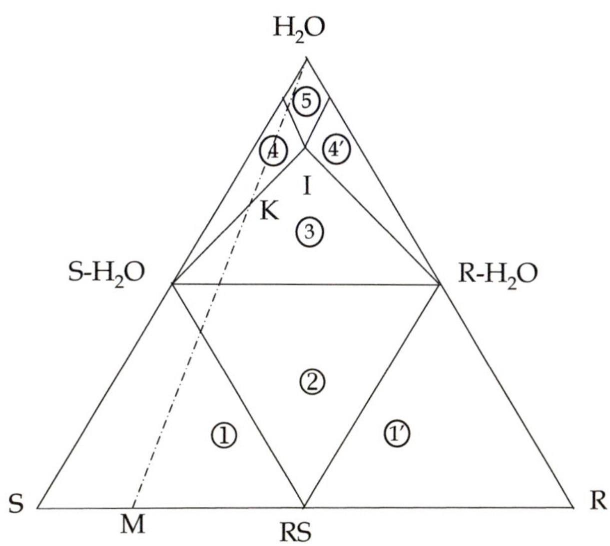

1 : <S-H$_2$O> + <S> + <RS> ; 1′: <R-H$_2$O> + <S> + <RS>
2: <S-H$_2$O> + <R-H$_2$O> + <RS>
3: Invariant liquid I doubly saturated + <S-H$_2$O> + <R-H$_2$O>
4: Saturated solution + <S-H$_2$O>
4′: Saturated solution + <R-H$_2$O>
5: Undersaturated solution

Fig. 8. Enantiomeric purification from a derivative (case of a monohydrate)

3.4 Benefits of conglomerate

It is now obvious that proceeding to the enantiopurification with a starting product crystallizing as a conglomerate is even more beneficial since:

- an optimized recrystallization can 'virtually' be carried out without any loss of enantiomeric excess which is not possible with a racemic compound.
- the choice of appropriate solvent and crystallization conditions usually allows to reach an e.e greater than 99.5 %.
- an optimized recrystallization can be carried out on derivatives such as hydrates (crystallizing as conglomerates) even if the molecule of interest crystallizes as a stable racemic compound.

Note also, that when coupled to racemization, almost pure enantiomer can be obtained by attrition from the racemic mixture (Levillain et al, 2009).

4. Detection of conglomerate

To perform preferential crystallization, the racemic mixture should crystallize as a stable conglomerate. In most cases, the chiral resolution can not be envisaged on the original compound because it does not fulfill this condition. An interesting alternative is to proceed to the resolution on derivatives such as salts, co-crystals, hydrates or solvates that crystallize as conglomerates (at the end of the resolution process, the pure enantiomer can be easily recovered by salting out, dehydration or desolvation). To multiply the chances of spotting a conglomerate, experiments have to be performed on a large number of non chiral acids (for a chiral base) with various stoichiometries and solvents, with different co-crystal formers and under various crystallization conditions.

Consequently, starting from the racemic mixture it is often necessary to synthesize series of derivatives that have to be analyzed in order to isolate at least one conglomerate forming system. Due to the low occurrence of conglomerates, it usually implies to investigate a large number of derivatives.

Conglomerate detection which is an essential step of the resolution method can be pursued by various methods which are described and discussed in the following.

4.1 Classical screening of conglomerates

4.1.1 Comparisons between IR, Raman, solid state NMR or XRPD patterns

Several spectroscopic techniques can be used to differentiate the racemic mixture and the conglomerate via comparison between the racemic pattern and that of the enantiomer. In case of conglomerate without partial solid solution, the patterns of the pure enantiomer obtained from IR, Raman, solid state NMR or XRPD should be superimposable to the ones obtained for the racemic mixture.

Even if Raman spectroscopy and IR spectroscopy can be of a certain help in spotting a conglomerate, they cannot be considered as totally reliable. Indeed, vibrational bands mainly arise from molecular vibrations. Only low frequency vibrations are directly related to the vibrations of the crystal lattice. If the neighborhood of a given molecule can generate variations in the high frequency domain, two close structures can generate similar spectra. Considering these elements, the perfect match of the XRPD patterns is surely the most dependable way to conclude on the conglomerate nature of a sample. XRPD is of particular interest as it also permits to check the crystallinity of the sample and to spot partial or total solid solutions (Wermester et al, 2007; Renou et al, 2007). Moreover, if a single crystal of sufficient size is available, a structure can be resolved by single crystal X-Ray diffraction. The knowledge of the crystal structure definitively confirms the conglomerate nature (or not) of the sample.

4.1.2 Alternatives methods

Other methods consist of isolating a unique particle (single crystal) from the racemic mixture to proceed to the analysis. This particle can be dissolved in a nematic phase (Jacques et al, 1994) or analyzed by chromatography (Chiral CG, HPLC) (Pirzada et al, 2010) and/or polarimetry. However, there are serious limitations to these techniques since the collection of an isolated particle can be difficult and also because the analysis of a single particle necessitates high detection levels. To be exhaustive, let us also mentioned that the detection of a conglomerate can also be performed via thermal analysis (the construction of the binary phase diagram by simple measurements of the melting temperatures of the enantiomers and that of their corresponding racemic mixture can be used to identify the nature of the sample (conglomerate or racemic compound)(Neau et al, 1993) or Solid state circular dichroïsm (SS-CD) and CD microscopy (Kuroda et al, 2000; Claborn et al, 2003). The construction of a ternary isotherm is also a reliable method especially when a conglomerate of solvated phases is suspected.

4.2 High throughput technique for conglomerate prescreening

Unfortunately and despite technical improvements, all the previous methods are time consuming and often require the pure enantiomer to be available at an early stage of the process. Consequently, there is a need for a faster (and cheaper) method compatible with a combinatorial approach of crystallization derivatives (salts, hydrates, solvates, cocrystals, etc.)

Considering the data summarized in table 1, it appears that racemic compounds crystallize not often in non centrosymmetric structures (less than 5% for NA + NC). Consequently, a method that would detect the absence of center of symmetry in crystals obtained from racemic mixtures will be able in most cases to detect conglomerates. To this purpose, nonlinear optics can be very useful. In the next paragraphs are presented the theory of second harmonic generation in crystals and described the principle of conglomerate prescreening.

4.2.1 Nonlinear optics and second harmonic generation (SHG)

In a given medium, the propagation of light is mainly driven by the dielectric properties and the response to electromagnetic fields. The application of an electromagnetic field (light) to a molecule modifies the shape of the electronic cloud and consequently creates an induced electric dipole moment.

The term polarization (**P**, vectorial dipole moment per unit volume) is used to describe this phenomenon on a macroscopic scale:

$$\mathbf{P} = \varepsilon_0\, \chi^{(1)}\mathbf{E} \quad (2)$$

with **E**, the applied electric field, ε_0 is the vacuum permittivity and $\chi^{(1)}$ the linear (first order) susceptibility of the material.

The linear susceptibility is a 2nd rank tensor related to the permittivity ε and the refractive index n of the material:

$$\varepsilon = n^2 = \varepsilon_0\,(1 + \chi^{(1)}) \quad (3)$$

For week electric fields, the polarization varies linearly with **E**. However, for high light intensities (typically greater than 1 MW/cm^2), the polarization becomes a nonlinear function of the applied electric field. It can be expressed as a power series expansion of the macroscopic field:

$$\mathbf{P} = \varepsilon_0\,(\chi^{(1)}\,\mathbf{E} + \chi^{(2)}\,\mathbf{E}^2 + \chi^{(3)}\,\mathbf{E}^3 + \ldots) = \varepsilon_0\,\chi^{(1)}\,\mathbf{E} + \mathbf{P}_{\mathbf{NL}} \quad (4)$$

with $\chi^{(2)}$ and $\chi^{(3)}$ the second and the third order susceptibility tensors respectively and $\mathbf{P}_{\mathbf{NL}}$, the nonlinear polarization. This equation implies that during the propagation of light at a frequency ω, nonlinear components of the polarization at frequencies 2ω and 3ω arise and harmonics of the original optical field at 2ω and 3ω are generated. However, with increasing order, a rapid decrease in the susceptibility coefficients is observed. As a consequence, the nonlinear polarization $\mathbf{P}_{\mathbf{NL}}$ can be approximated by the quadratic term $\varepsilon_0\,\chi^{(2)}\,\mathbf{E}^2$ for moderate energies. Considering the usual expression of the electromagnetic field

$$\mathbf{E} = \frac{1}{2}\mathbf{E_0}\ \exp[j(\omega t - \mathbf{k.r})] + \text{c.c} \quad (5)$$

with $\mathbf{E_0}$, the amplitude and k the wave vector and c.c, the complex conjugate of the formula.

the polarization then becomes equal to:

$$\mathbf{P} = \varepsilon_0\,\chi^{(1)}\,\mathbf{E} + \varepsilon_0\,\chi^{(2)}\,\mathbf{E}^2 \quad (6)$$

$$\mathbf{P} = \frac{1}{2}\left\{\begin{array}{l} \varepsilon_0\,\chi^{(2)}\,\mathbf{E}_0^2 \\ +\varepsilon_0\,\chi^{(1)}\,\mathbf{E_0}\exp[j(\omega t - \mathbf{k.r})] \\ +\frac{1}{2}\varepsilon_0\,\chi^{(2)}\mathbf{E}_0^2\exp[2j(\omega t - \mathbf{k.r})] \end{array}\right\} \quad (7)$$

Finally, the net polarization is composed of three components: a continuous one corresponding to the phenomenon of optical rectification, a component at frequency ω (optical polarizability) and a second harmonic term (frequency 2ω) corresponding to the phenomenon known as SHG (second harmonic generation). In SHG, a fundamental wave of amplitude $\mathbf{E}_\omega$, angular frequency ω (wavelength λ) and wave vector $\mathbf{k}_\omega$ passing through a crystal generates a second harmonic wave of amplitude $\mathbf{E}_{2\omega}$, angular frequency 2ω (wavelength $\lambda/2$) and wave vector $\mathbf{k}_{2\omega}$.

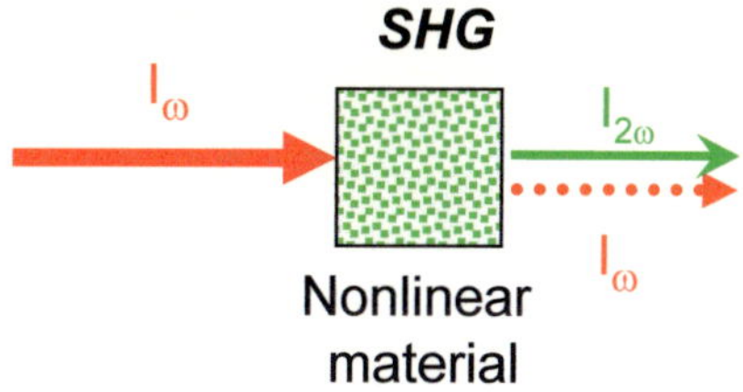

Fig. 9. Second harmonic generation

With distance traveled through the crystal, the second harmonic amplitude varies. The energy generated at frequency 2ω can be obtained by solving the propagation equation:

$$\nabla^2 \mathbf{E} - \mu_0 \varepsilon_0 \frac{\partial^2 \mathbf{E}}{\partial t^2} = \mu_0 \frac{\partial^2 \mathbf{P}}{\partial t^2} \tag{8}$$

with μ_0, the permeability of free space.

The average intensity of an electromagnetic wave of amplitude **E** is:

$$I = \frac{\mathbf{E}.\mathbf{E}^*}{2\varepsilon_0 c} \tag{9}$$

with c, the velocity of light. Assuming that the waves are traveling in the z direction and that the conversion efficiency is low (so amplitude of the fundamental is almost unchanged), the second harmonic intensity after a distance ℓ through the crystal is (Armstrong, 1962):

$$I_{2\omega}(\ell) = \frac{\omega^2 \left(\chi^{(2)}\right)^2 \ell^2}{2\varepsilon_0 c^3 n_\omega^2 n_{2\omega}} \frac{\sin^2\left(\frac{\Delta k \ell}{2}\right)}{\left(\frac{\Delta k \ell}{2}\right)^2} I_\omega^2 \tag{10}$$

with :

n_ω, the refractive index of the crystal at angular frequency ω

$n_{2\omega}$, the refractive index of the crystal at angular frequency 2ω

Δk , the phase mismatch: $\Delta k = k_{2\omega} - 2k_\omega = \frac{2\omega}{c}(n_{2\omega} - n_\omega)$;

$I_{2\omega}$ depends on several parameters and among them on the value of Δk . The case of $\Delta k=0$ corresponds to materials called "phase-matchable materials" for which the longer the distance traveled inside the crystal, the greater the SHG intensity (i.e. large particles will generate a better SHG signal). Phase-matchable materials such as potassium diphosphate (KDP) are used to double high power lasers or as a nonlinear standard. But most materials are "non phase-matchable" (ie. $\Delta k \neq 0$) with as a consequence an intensity value oscillating with ℓ . $I_{2\omega}$ will reach a maximum only for the discrete values of ℓ given by:

$$\ell_c = \frac{\pi}{\Delta k} = \frac{\lambda}{4(n_{2\omega} - n_\omega)} \tag{11}$$

ℓ_c is called the "coherence length".

This implies that the SHG intensity will be optimized for particles of size equal to ℓ_c or equal to an odd multiple of ℓ_c. The consequence of the coherence length will be discussed later. $I_{2\omega}$ depends also on $\chi^{(2)}$, the second order nonlinear susceptibility, and this is the main point for the detection of conglomerates. 27 numbers constitute the electro-optic components of $\chi^{(2)}$ which is a 3rd rank tensor. However, the number of independent coefficients can be reduced to 10 if the absorption of the material is negligible at ω and 2ω. In this case, the tensor is invariant by circular permutation of its three indices.

Moreover, the number of independent coefficients non equal to zero can be determined by tacking into account the symmetry elements of the 32 crystallographic classes. In particular, all the components of $\chi^{(2)}$ are null for centrosymmetric structures. Thus, in crystals with a centre of inversion, all the components of $\chi^{(2)}$ tensor are zero. Consequently, these types of crystals can not exhibit a SHG signal. According to the Kleinman symmetry rules (Kleinman, 1962) three chiral point groups (NC) present also a null $\chi^{(2)}$ (422, 622 and 432). However, Kleinman symmetry is not always applicable. These different possibilities are summarized in figure 10.

As a result, the absence of a center of symmetry can be determined via the detection of a SHG signal. Therefore, this test was chosen as a pre-screening method for spotting conglomerates.

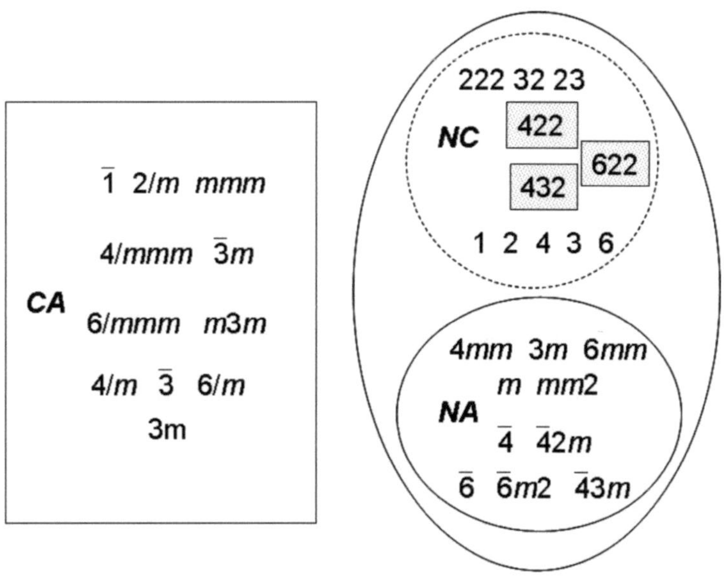

□ : centrosymmetric point groups (no SHG activity)
0 : non centrosymmetric point groups (SHG activity)
░ : no SHG activity if Kleinman symmetry is applicable

Fig. 10. SHG activity among the 32 crystallographic classes

4.2.2 Experimental set-up

The experimental set up proposed by Kurtz and Perry (Kurtz et al, 1968) gives quick information about the SHG activity of powder samples. Because the majority of the pharmaceutical crystalline samples are available in the form of powders, this set-up was chosen for the pre-detection of conglomerates.

Figure 11, shows the experimental set-up used for the SHG test. The laser is a Nd:YAG Q-switched laser operating at 1.06 µm. It delivers 360 mJ pulses of 5 ns duration with a repetition rate of 10 Hz. An energy adjustment device is made up of two polarizers (P) and a half-wave plate (λ/2).

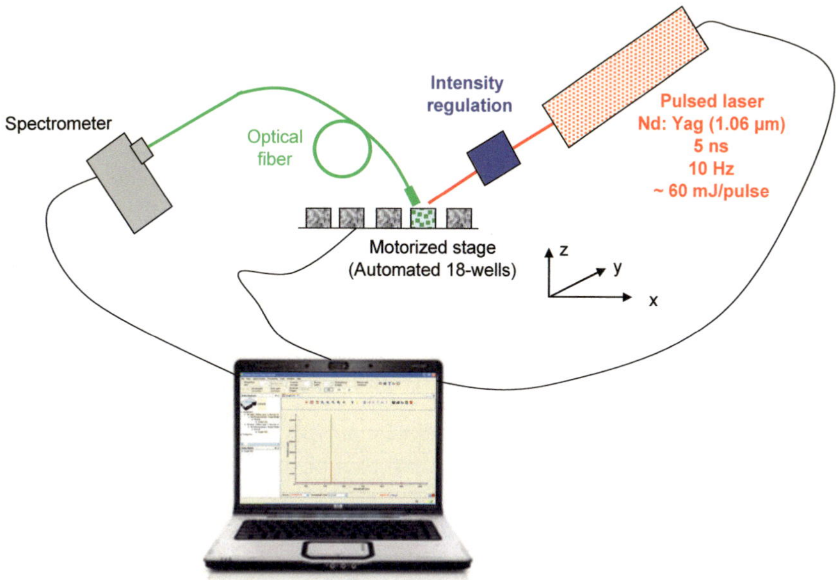

Fig. 11. Experimental Set-up

It allows the incident energy to vary from 0 to circa 200 mJ per pulse. A RG1000 filter is situated after the energy adjustment device to removes light from laser flash lamps. The samples (18-wells plate filled with the various powders) are placed on a motorized stage and irradiated with 60 mJ pulses (beam diameter of 4 mm). The signal generated by the sample (diffused light) is collected into an optical fiber (500 µm of core diameter) and directed onto the entrance slit of a spectrometer. A boxcar integrator allowed an average spectrum (spectral range 250-700 nm with a resolution of ± 0.2 nm) to be recorded over 1 second (10 pulses).

4.2.3 Accuracy of the SHG method

Because of the low occurrence of conglomerates, it would be extremely detrimental to fail to spot one. Conversely, it is not awkward to have a limited number of false positive responses. The diagram presented on figure 12 gathers the various situations that can be encountered during the SHG prescreening. It can be used to evaluate the consistency of the SHG response observed for a given sample.

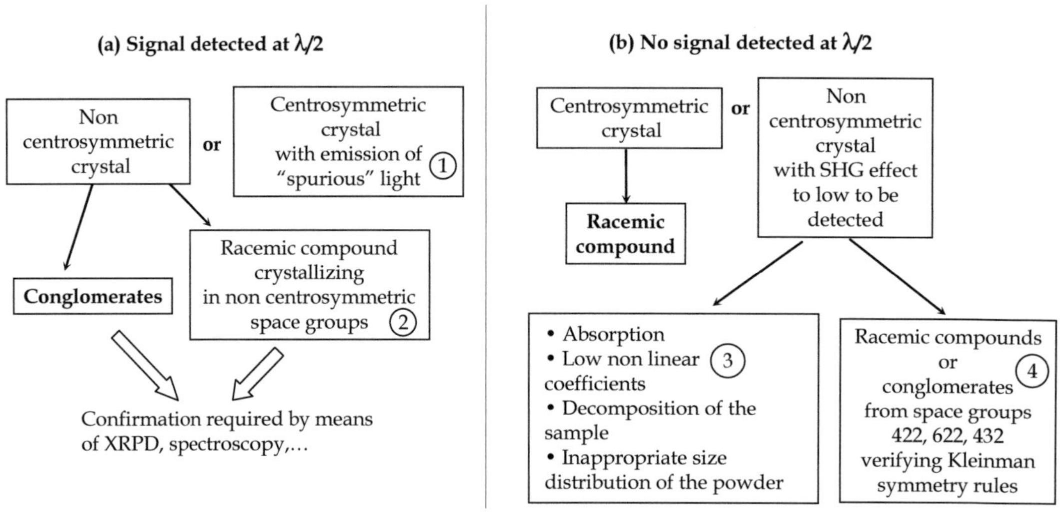

Fig. 12. Decision diagram for SHG test

Panel (a) of the diagram corresponds to the samples for which a signal has been observed. It is necessary to distinguish between two cases: (1) the observation of a "non SHG" signal resulting from optical phenomena such as two-photon fluorescence (TPF) or other photoluminescent processes and (2) the observation of a "real SHG" signal resulting from racemic compounds crystallizing in non-centrosymmetric space groups.

False positive response arising from case 1 can be simply encountered. Indeed, TPF signals present a spectral bandwidth much broader than that of the SHG signal (cf. figure 13).

A base line correction around the SHG wavelength allows the distinction between light resulting from the SHG process itself (signal of interest) and "spurious" light. This simple procedure strongly limits the occurrence of false positive responses.

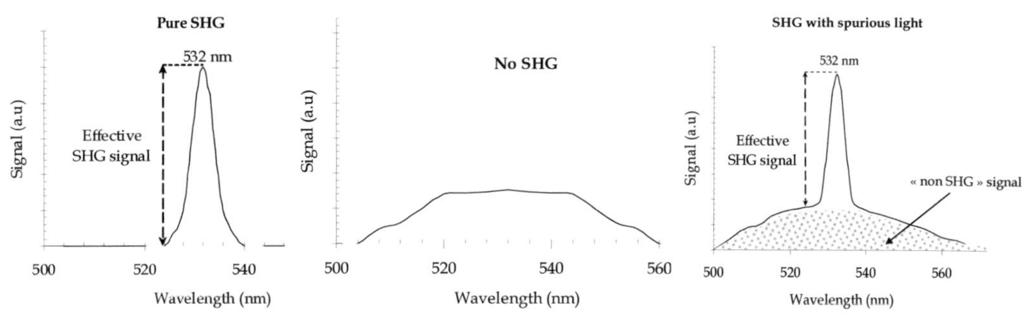

Fig. 13. Removal of false positive responses

As mentioned in table 1, it is possible for a racemic mixture to crystallize in a non-centrosymmetric space group. These types of crystals generate a SHG response and can also be considered as false positive. They constitute case 2. It is also worth mentioning that slightly disordered racemic structures and solid solutions can fall in this category. The influence of the degree of disorder (or pseudo-symmetry) of the structure is difficult to establish so as the consequences on the $\chi^{(2)}$ coefficients. However, the detection of partial solid solutions can be of interest since some of them can be potentially adequate for preferential crystallization (Wermester et al, 2007).

Panel (b) of the diagram concerns samples that do not present any signal at $\lambda/2$. The critical situation of non detected conglomerates is considered in case 3. The main reasons of the "no-detection" are an SHG effect too weak to be detected (due to for example to low non linear coefficients consequence of a low hyperpolarizability of the molecule) or an inappropriate crystal size distribution. The second harmonic signal generated by a crystalline powder is the sum of the contribution of each individual particle (electromagnetic fields are uncorrelated). Because the fundamental beam passes through a large number of particles for which a random orientation is assumed, the intensity of the SHG beam is not easy to optimize and depends on the particle size (especially in non phase matchable materials). To preserve the reliability of the method, too fine particles as those present in submicronic or nanocrystalline powders should be avoided (coherence length is in order of magnitude of several microns for most materials). To limit the number of undetected conglomerates, it is also necessary for the experimental set-up to present a detection level better than 1/100 of the SHG signal generated by the quartz powder (standard non phase-matchable material for SHG measurements - mean diameter 50 μm). This condition is in most cases considered as sufficiently constraining. To be exhaustive, let us mentioned that a non detectable (or a decreasing) signal can also be the consequence of an absorption in the sample at the wavelength of irradiation (λ) or at the wavelength of the re-emitted radiation ($\lambda/2$). This absorption can generate a decomposition of the sample when exposed to the laser beam. The use of an alternative laser source must then be envisaged.

The last case (Case 4) concerns chiral crystals associated to point groups 422, 622 and 432. These should be SHG inactive due to the application of Kleinman permutation rules. However, experience refutes the general applicability of these rules. Recently, SHG activity has been observed in N-acetyl-methylbenzylamine which crystallizes as a conglomerate in space group $P4_12_12$.

Considering all these elements, spotting a SHG active substance can not guaranty the existence of a conglomerate. Only a complementary study of the samples by conventional methods can conclude on the conglomerate nature or not. That is why this method is proposed as a prescreening technique only. However, SHG presents numerous benefits listed below:

- Only a small quantity of the racemic mixture is necessary (ca. 20 mg typically).
- There is at this first stage of pre-screening no need for comparison between results of SHG tests on the racemic mixture and the pure enantiomer. The screening can be undertaken even when the pure enantiomer is not available and thus be carried out at an early stage of the development of the molecule.
- The response is instantaneously delivered; it is thus conceived to be a true high throughput pre-screening method which allows on a short period of time to test numerous derivatives.
- It is a priori a non destructive method.
- It is cheap and can be fully automated: a high throughput device including motorized sample holders and a computer assisted treatment of the spectra should allow in a realistic way to select the "good" candidates with a 50% probability or more.

4.3 Rationalization of conglomerate detection

When a positive SHG signal is obtained, the conglomerate nature of the compound has to be confirmed via XRPD. However, it is worth mentioning that some bias in the conglomerate detection can be introduced at this stage if a strict control of the crystallization conditions is not applied. To avoid misinterpretation of the results, the diagram of figure 14 is proposed for an optimization of conglomerate detection (Gonella, 2011).

Once the SHG prescreening has led to suspect the existence of a conglomerate (step 1 completed), the next step consists in comparing the XRPD patterns of the racemic mixture and the corresponding enantiomer. At this stage it is imperative to apply the same crystallization conditions for the two samples to avoid discrepancies due to polymorphism or solvation / desolvation. It concerns in particular the nature of the counter-ion or cocrystal former, the stoichiometry, the temperature (during the crystallization but also during the SHG detection), the nature of the solvent, etc.

Once both compounds are available (step 2), a precise comparison of the XRPD patterns can be pursued.

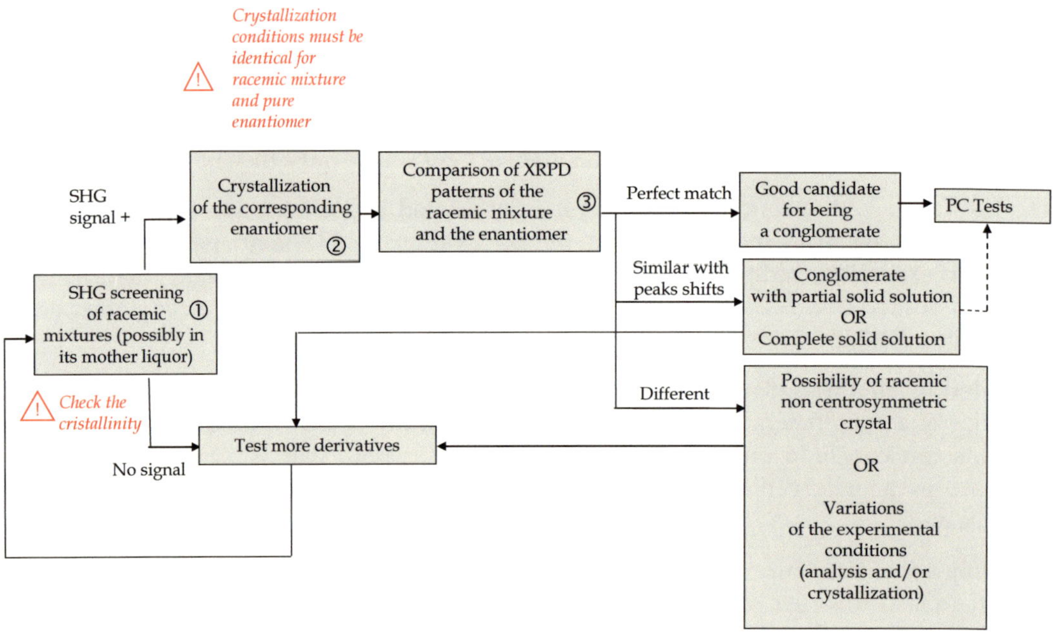

Fig. 14. Workflow for conglomerate detection

Three different cases can be encountered on step 3 : perfect match, similar diffractograms with some peak shifts or different diffractograms). On Figure 15 are presented the XRPD patterns obtained for three different compounds. The upper traces correspond to the racemic mixture (1a, 2a, 3a) and the lower ones to the pure enantiomer (1b, 2b, 3b).

Diffractograms 1a and 1b perfectly match which permits to conclude that compound 1 is with a high level of confidence a conglomerate. PC tests can be undertaken. Diffractograms 2a and 2b present numerous differences. Compound 2 is unlikely to be a conglomerate

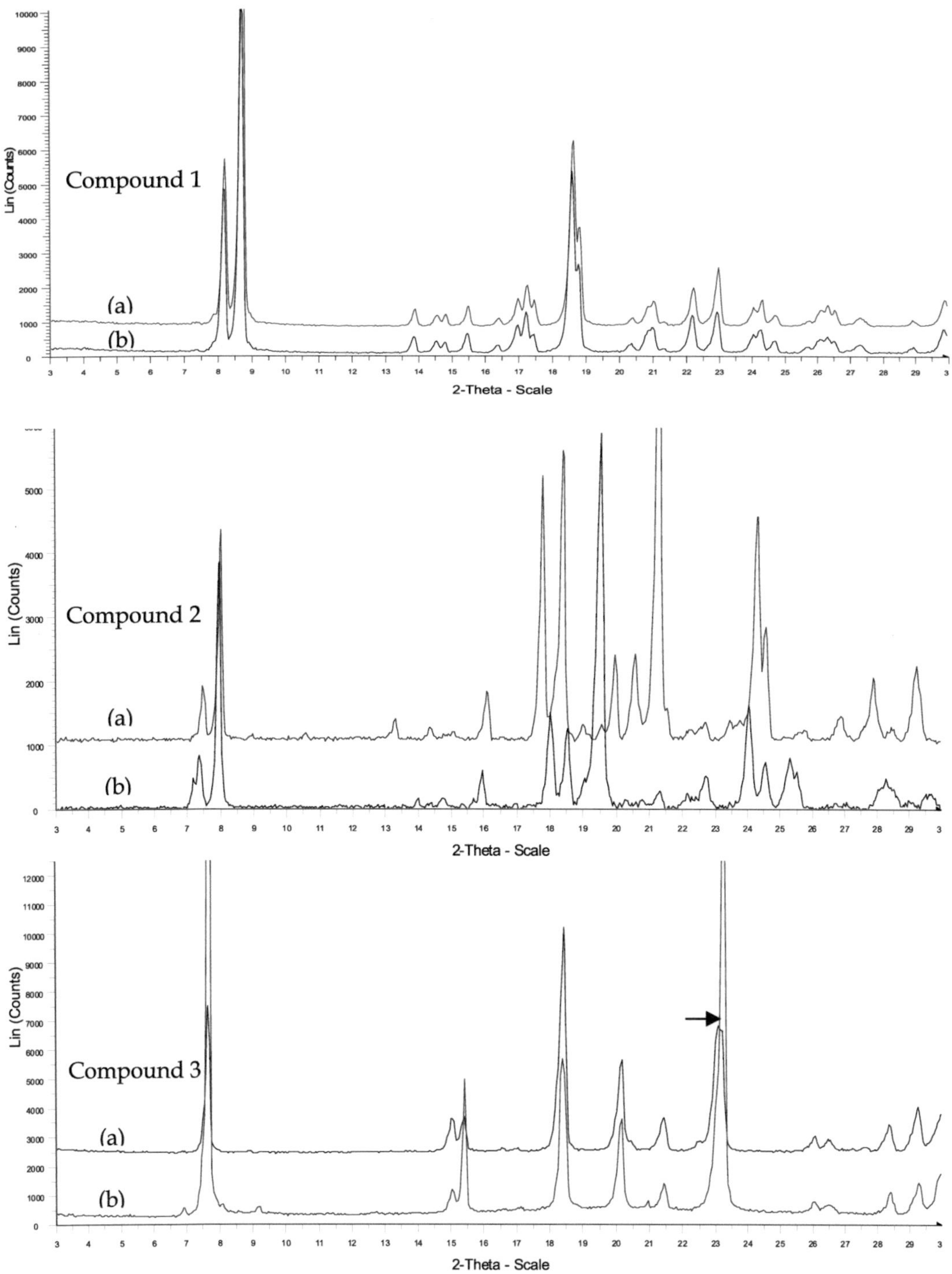

Fig. 15. Comparison of X-Ray patterns for the detection of conglomerate

forming system. However, it is worth mentioning that interesting conglomerates may exist only in equilibrium with their mother liquor (because of one or several of the following reasons: non congruent solubility, solvates with an efflorescent character, solvates which are hydrolyzed by moistures, CO_2 sensitive solid, etc). So it is of primary importance to check if the conditions used during the SHG test are identical to those used for XRPD. Indeed, due to the quickness of the SHG test, an efflorescent solvate could be detected as a conglomerate. Its desolvation prior or concomitantly to the XRPD analysis could then lead to contradictory results. To overcome these problems, SHG and XRPD tests should be run under strictly identical conditions of crystallization.

Diffractograms 3a and 3b are extensively similar but show slight shifts in 2θ positions of some peaks. This couple of derivatives deserves more investigations. Nevertheless, it is likely that this system exhibits at least partial solid solutions.

5. Conclusion

The objective of this chapter was a better understanding of chiral discrimination in the solid state and of the mechanisms involves during recrystallization of enantiomers.

We described two methods of enantiopurification and underlined the benefits of preferential crystallization. Because the detection of a conglomerate is an essential step of this process, we detailed a recently developed technique of prescreening of conglomerates. We finally proposed a workflow to follow in order to optimize the detection of conglomerates and avoid bias that can be induced by different crystallization parameters.

6. References

Armstrong, J. A., Bloembergen, N., Ducuing, J., & Pershan, P. S. (1962). Interactions between light waves in a nonlinear dielectric. *Physical Review*, Vol. 127, No. 6, (September 1962), pp. 1918–1939

Belsky, V. K. & Zorkii, P. M. (1977), Distribution of organic homomolecular crystals by chiral types and structural classes, *Acta Crystallographica Section A*, Vol. 33, No. 6, (November 1977), pp. 1004–1006, ISSN 1600-5724

Brock, C. P., Schweizer, W. B., & Dunitz, J. D. (1991), On the validity of wallach's rule: on the density and stability of racemic crystals compared with their chiral counterparts, *Journal of American Chemical Society*, Vol. 113, No. 26, (December 1991), pp. 9811–9820, ISSN 0002-7863

Claborn, K., Puklin-Faucher, E., Kurimoto, M., Kaminsky, W., & Kahr, B. (2003), Circular dichroism imaging microscopy: application to enantiomorphous twinning in biaxial crystals of 1,8-dihydroxyanthraquinone, *Journal of the American Chemical Society*, Vol. 125, No. 48, (December 2003), pp. 14825–14831, ISSN-0002-7863.

Collet, A. (1999), Separation and purification of enantiomer by crystallization methods, *Enantiomer*, Vol. 4, No. , (1999), ISSN

Coquerel, G., Petit, M.-N., & Bouaziz (1995). Method of resolution of two enantiomers by crystallisation. *PCT Patent WO 95/08522.*

Coquerel, G.; Petit, M.-N.; Bouaziz, R.(1994) Procédé de dédoublement (AS3PC) de deux antipodes optiques par entraînement polythermique programmé et auto-ensemencé. PCT N°94/01.107, 22/09/1994, 1994.

Coquerel, G. (2000), Review on the heterogeneous equilibria between condensed phases in binary systems of enantiomers, *Enantiomer*, Vol. 5, No. 5, pp. 481-498, (Mai 2000), ISNN 1024-2430

Coquerel, G. (2007). Preferential Crystallization, In : *Novel optical resolution technologies - Topics in Current Chemistry*, Sakai, Kenichi et al (Eds), pp. 1–51, Springer, ISBN 978-3-540-46317-7, Berlin.

Dalhus, B. & Görbitz, C. H. (2000), Non-centrosymmetric racemates: space-group frequencies and conformational similarities between crystallographically independent molecules, *Acta Crystallographica Section B*, Vol. 56, No. 4, (August 2000), pp. 715–719, ISSN 1600-5740

Flack, H. (2003), Chiral and achiral crystal structures, Helvetica Chimica Acta, Vol.86, No.4, (January 2003), pp. 905–921, ISSN 1522-2675

Flack, H. D. & Bernardinelli, G. (2003), The mirror of galadriel: looking at chiral and achiral crystal structures, *Crystal Engineering*, Vol. 6, No. 4, (December 2003) pp. 213–223, ISSN 1463-0184

Galland, A., Dupray, V., Berton, B., Morin-Grognet, S., Sanselme, M., Atmani, H., & Coquerel, G. (2009), Spotting conglomerates by second harmonic generation. *Crystal Growth & Design*, Vol. 9, No. 6, (May 2009), pp. 2713–2718, ISSN 1528-7483

Gernez, D. (1866) Séparation des tartrates droits et des tartrates gauches à l'aide de solutions saturées, *Compte-rendus de l'Académie des Sciences*, Vol. 63, (July-December 1866), pp. 843-888

Gonella, S., Mahieux, J., Sanselme, M., & Coquerel, G. (2011). Spotting a conglomerate is just halfway to achieving a preparative resolution by preferential crystallization. *Organic Process Research & Development*, In press, ISSN 1083-6160.

Helmreich, M.; Niesert, C.-P.; Cravo, D.; Coquerel, G.; Levilain, G.; Wacharine-Antar, S.; Cardinaël, P, (2010), Process for enantiomeric separation of racemic dihydro-1, 3, 5 triazines via preferential crystallization. *PCT patent WO2010109015 (A1), 2010-09-30*

Jacques, J., Collet, A., & Wilen, S. (1994). Enantiomers, Racemates and Resolutions (3rd Edition), Kriger Pub. Co., ISBN 0894-648764, Malabar Florida, USA.

Levilain, G., Rougeot, C., Guillen, F., Plaquevent, J-C., Coquerel, G. (2009) Attrition enhanced preferential crystallization combined with racemization leading toredissolution of the antipode nuclei, *Tetrahedron: Asymmetry*, Vol. 20, No. 24, (December 2009), pp. 2769–2771, ISSN 0957-4166

Levilain, G., Coquerel, G. (2010) Pitfalls and rewards of preferential crystallization, *CrystEngComm*, Vol. 12, No. 7, (May 2010), pp. 1983-1992, ISSN 1466-8033

Jungfleish, M. E. (1882) , *Journal of Pharmaceutical Chemistry*, Vol. 5, pp. 346.

Kaptein, B., Elsenberg, H., Grimbergen, R.F.P., Broxterman, Q.B, Hulshof, L.A., Pouwer, K.L. & Vries, T.R. (2000) Dutch resolution of racemic 4-hydroxy- and 4-fluorophenylglycine with mixtures of phenylglycine and (+)-10-camphorsulfonic acid, *Tetrahedron: Asymmetry*, Vol. 11, No. 6, (April 2000), pp. 1343-1351, ISSN 0957-4166

Kleinman, D. A. (1962). Theory of second harmonic generation of light. *Physical Review*, Vol. 128, No. 4, (1962) pp. 1761–1775.

Kostyanovsky, R. G., Kostyanovsky, V. R., & Kadorkina, G. K. (2009) The enigma of a (±)-tartaric acid-urea cocrystal, Mendeleev Communications, Vol. 19, No.1, (2009), pp. 17- 18, ISSN 0959-9436

Kuroda, R. & Honma, T. (2000) Cd spectra of solid-state samples, Chirality, Vol. 12, No. 4, (April 2000), pp. 269–277, ISSN 0899-0042

Kurtz, S. & Perry, T. (1968) A powder technique for the evaluation of nonlinear optical materials, *Journal of Applied physics*, Vol.39, No. 8, (July 1968), pp.3798–3813, ISSN 0021-8979.

Lorenz, H., Perlberg, A., Sapoundjiev, D., Elsner, M. P., & Seidel-Morgenstern, A. (2006), Crystallization of enantiomers, *Chemical Engineering and Processing: Process Intensification*, Vol. 45, No. 10, (April 2006), pp. 863 - 873, ISSN 0255-2701

Marchand, P., Lefebvre, L., Querniard, F., Cardinael, P., Perez, G., Counioux, J-J & Coquerel, G. (2004), Diastereomeric resolution rationalized by phase diagrams under the actual conditions of the experimental process, *Tetrahedron: Asymmetry*, Vol.15, No.16, (August 2004), pp. 2455-2465, ISSN 0957-4166

Marckwald, W., (1896) Ueber ein bequemes Verfahren zur Gewinnung der Linksweinsäure, Berichte der deutschen chemischen Gesellschaft, Vol.29, No.1, (April 1896), pp.42–43

Ndzié, E., Cardinael, P., Schoofs, A. R., & Coquerel, G. (1997), An efficient access to the enantiomers of [alpha]-methyl-4-carboxyphenylglycine via a hydantoin route using a practical variant of preferential crystallization as3pc (auto seeded programmed polythermic preferential crystallization), *Tetrahedron: Asymmetry*, Vol. 8, No. 17, (September 1997), pp. 2913–2920, ISSN 0957-4166

Kocher-Becker, U. , Kocher, W. & Ockenfels, H. (1982), Teratogenic activity of a hydrolytic thalidomide metabolite in mice, *Naturwissenschaften*, Vol. 69, No. 4, (Avril 1982), pp. 191-192, ISSN 0028-1042

Pasteur, L. (1853), Transformation des acides tartriques en acide racémique - Découverte de l'acide tartrique inactif. Nouvelle méthode de séparation de l'acide racémique en acides tartriques droit et gauche. *Compte-rendus de l'Académie des Sciences*, Vol. 37, (Juillet 1853), pp.162-166

Pope, W.J. & Peachey, S.J. (1899), The application of powerful optically active acids to the resolution of externally compensated basic substances. Resolution of tetrahydroquinaldine, *Journal of Chemical Society Transactions*, Vol. 75, (January 1899), pp. 1066-1093, ISNN 0368-1645

Renou, L., Morelli, T., Coste, S., Petit, M.-N., Berton, B., Malandain, J.-J. & Coquerel, G., (2007), Chiral discrimination at the solid state of the methyl 2-(diphenylmethylsulfinil)acetate, *Crystal Growth & Design*, Vol. 7, No. 9, (August 2007), pp.1599-1607, ISSN 1528-7483

Stinson, S.C. (2000). , Chiral Drugs, *Chemical and Engeniering news*, Vol. 78, No. 43, pp. 55-78 (October 2000), ISSN 0009-2347

Subramanian, G., (2006). Chiral Separation Techniques: A Practical Approach (3rd - 18th October), Wiley-VCH Verlag GmbH, ISBN: 978-3-527-31509-3, Weinheim, Germany.

Vries, T., Wynberg, H.,Van Echten, E.A, Koek, J.A, Ten Hoeve, W.A., Kellogg, R.M., Broxterman, Q.B., Minnaard, A.B, Kaptein, B.B, Van Der Sluis, S., Hulshof, L. & Kooistra, J. (1998), The family approach to the resolution of racemates, *Angewandte Chemie - International Edition*, Vol. 37, No 17, (September 1998), pp. 2349-2354, ISSN 1433-7851

Wermester, N., Aubin, E., Pauchet, M., Coste, S., & Coquerel, G. (2007). Preferential crystallization in an unusual case of conglomerate with partial solid solutions. *Tetrahedron: Asymmetry*, Vol. 18, No. 7, (2007), pp. 821–831, ISSN 0957-4166

18

Recrystallization of Drugs: Significance on Pharmaceutical Processing

Yousef Javadzadeh, Sanaz Hamedeyazdan and Solmaz Asnaashari
Biotechnology Research Center and Faculty of Pharmacy, Tabriz University of Medical Sciences
Iran

1. Introduction

Not surprisingly, the wide range of effective medicinal agents available today is one of the greatest scientific achievements. Regardless of the advancements in effectiveness and safety of the medicines embedded in dosage forms, the pharmaceutical concept of the latter is growing to be ever more eminent (Adibkia et al., 2011). Following on from recent advancements, in a time of increased considerations to the level of sophistication in designing pharmaceutical dosage forms keeping pace with advances in drug discovery methods, it seems as important as ever to study the physicochemical properties of active pharmaceutical ingredients, prerequisite for a successful product formulation.

As far as we know, the molecular structure of any drug compound typically defines all of its physical, chemical and biological actions. Owing to the fact that a certain kind of drug might be offered in a variety of solid forms, including polymorphs, solvates, hydrates, salts, co-crystals and amorphous solids, the choice and design of the ideal solid-state chemistry of the pharmaceutical solid form would be critically important to a superior drug development. Accordingly, drug crystals could be modified in different ways including recrystallization, which would affect the physical and physicochemical properties such as melting point, solubility, true density, drug release profile, flowability and tabletability of the pharmaceutical dosage forms (Harbury, 1947; JamaliMitchell, 1973; Jozwiakowski et al., 1996).

Recrystallization is a simple and inexpensive method for scaling up the drug developments to a commercial level. Significant advances in the different pharmaceutical dosage form technologies renders drug recrystallization as a green technique due to the savings of costs, time, energy and less machinery as well as fewer personnel. Recrystallization is one aspect of precipitation obtained through a variation of the solubility conditions and the amount of dissolved solute in an increased temperature. In general, production of another crystalline from of a drug and also purification are the two major sets of applications for drug recrystallization processes. Briefly, in a drug recrystallization process, a hot saturated solution of the drug is prepared with only enough solvent to dissolve it at the boiling point of the solution. Once the solution is cooled the purified drug component or a new crystal form of the drug separates as a result of the lower solubility of that crystalline form of a drug in the respective solvent at lower temperatures.

Since impurities are present in fairly small amounts of drug solutions they do not crystallize in recrystallization and they are ready to separate from the formed drug crystal (Tiwary, 2001). For instance, in the case of natural medicinal compounds obtained from natural sources which almost always contain impurities, in order to obtain a pure drug, usual major steps in the recrystallization process have been schematically demonstrated in figure 1. Purifying a sample drug of compound X which is contaminated by a small amount of compound Y, would be established with an appropriate hot solvent in which all of compound Y is soluble at room temperature and the impurities will stay insoluble in and pass through filter paper, leaving only pure drug crystals behind, as has been shown in figure 1.

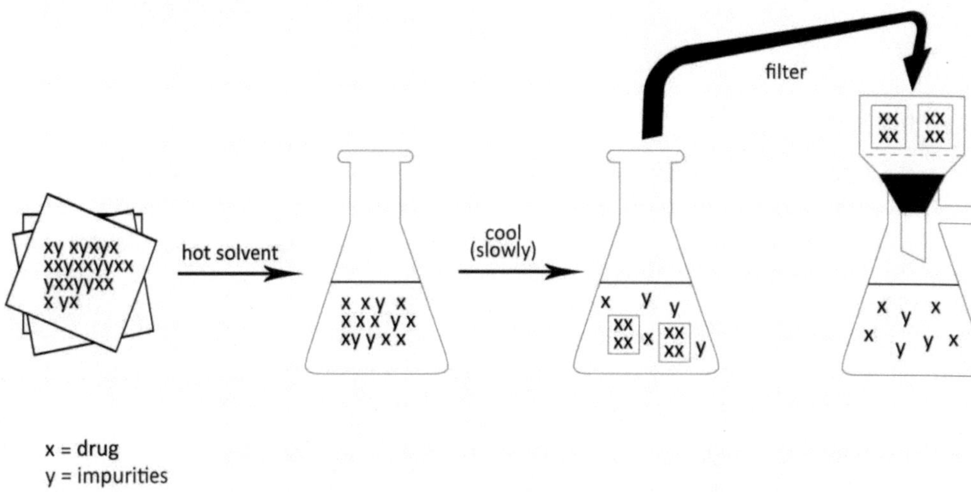

Fig. 1. Schematic protocol of drug recrystallization for further purification.

It is worth mentioning that in any recrystallization technique some drug loss is inevitable and the total recovery would be less than 100%, seeing that even at the lower temperatures the target drug has some finite solubility in the solvent and is lost subsequently when solvent and soluble impurities are removed. Moreover, selection of the right solvent in the recrystallization seems to be one of the crucial features of the process and is made on a case-by-case basis (MirmehrabiRohani, 2005; ChenTrout, 2008). This is because of solubility variation of different drug compounds in different solvents so that a certain drug not only should have the highest solubility in the solvent of choice at its boiling point, but also it should show a markedly diminished solubility at lower temperatures of the same solvent. Although recrystallization is a very common technique used to purify drugs, it has a basic limitation for the compounds that are mostly pure and other techniques of separations are of use for the drug mixtures containing several major components which could not be purified by recrystallization methods.

Ever since, the leading physicochemical properties of a unique form of a drug could seriously influence the bioavailability, manufacturability purification, stability, solubility, and other characteristics, identifying these potential liabilities allows us to predict, control and avoid any complexities that may arise during drug development stages (Krishnaiah, 2010). This would be beneficial in preventing the drug development efforts to costly late stage product failures throughout the manufacture and storage periods too. However, in order to fully control the crystallization process, the link between a particular solid form of a drug structure and its functional physicochemical properties, still challenges to be better established to facilitate the suitable drug production. A large and growing body of literatures has been published on recrystallization techniques and physicochemical properties of a drug. Herein, we gathered some of the related reports of drug recrystallization to have an overlook on the crystal habit of a drug on some basic physical properties of pharmaceutical dosage forms signifying how these factors are interrelated.

2. Impact of crystal habit on pharmaceutical processing

Drug discovery and characterization relies on the nature of the target molecule and the relative physicochemical properties of drugs. Identifying all relevant crystal habit of a drug which is an important variable in pharmaceutical manufacturing at the development phase from research to commercialization is of substantial value. Due to the different crystal form variations of some basic physical properties like, solubility, dissolution rate, melting behavior, and certain micromeritic properties or performance characteristics, e.g. tablet compressibility, mechanical strength, powder flow provide alternatives to select a form that presents the suitable balance of critical properties for development into the drug product. Establishing such modification information at an early stage of drug development process lessens the risk of process alterations given form changes and brings in the opportunity to attain more comprehensive rational property coverage.

The merit of changes in crystal surface form and habit of drug powders by recrystallization method is much more realized when there is an essential to diminish variations in raw material characteristics, to certify reproducibility of results during drug preformulation, and also to judge fairly about the cause of poor performance of a dosage form. Besides, the changes in crystal habit of a drug going together with or without polymorphic transformation at some point in processing storage could account for serious implications of physical stability in dosage forms. Thus, it seems underlying to have a deeper insight to the crystal structures and control the solid-state chemistry of drug substances to design a more systematic and intellectual pharmaceutical dosage forms.

In a survey carried out by Sinclair et al. ibipinabant a potent and highly selective cannabinoid receptor antagonist was evaluated for its solid-state physical stability and recrystallization kinetics in tablet dosage forms using fourier transform raman spectroscopy. The findings of the study showed that exposure to moisture had notable influence on the crystallinity of amorphous ibipinabant. The recrystallization kinetics measurements revealed a two-step process with an induction period (nucleation) followed by rod-like crystal growth by application of the Johnson–Mehl–Avrami kinetic model. On the whole their method provided reliable and highly accurate predictive crystallinity assessments after exposure to a variety of stability storage conditions for ibipinabant (Sinclair et al., 2011).

Recently, Dahlberg et al. analyzed the stability of the amorphous drug, flutamide, by a combination of localized nuclear magnetic resonance (NMR) spectroscopic and NMR imaging techniques. Owing to the fact that, NMR relaxation is sensitive to both the crystalline and amorphous state and the size of the drug substance, it allows for an in situ monitoring of the state of the drug during tablet disintegration and dissolution periods. With regard to the results of the NMR experiments, recrystallization was believed to be related to its enabling factors such as local hydration level and local mobility of the polymer matrix. Eventually, it was verified that the primarily amorphous flutamide may recrystallize either by nanoparticle coalescence or by ripening of crystalline particles (Dahlberg et al., 2011).

The solid-state properties of sulfathiazole and chlorpropamide were modified through recrystallization using supercritical antisolvent process by Yeo et al. They confirmed that the operating conditions of the system such as carbon dioxide injection rate, type of solvent, and temperature significantly had an effect on the physical characteristics of the resulting crystals. Considering the results of the study, drug crystals processed with supercritical system exhibited more ordered appearances with clean surfaces and sharp angles compared with the unprocessed particles where crystal habit changed from tabular to acicular when the carbon dioxide injection rate increased. Photomicrographs of sulfathiazole crystals with methanol as a solvent, confirmed a needle-like acicular and a tabular crystal habits in rapid and slow injections, respectively. Whereas, in the case of chlorpropamide, processed drug particles in the rapid injection experiment exhibited columnar habit in a regular shape, while relatively large crystals with sharp angles were observed in the slow injection mode when acetone was used as the solvent. Overall experimental observations suggested that the supercritical antisolvent process could provide favorable environment for the solid growth of a single type of crystalline drug, minimizing the conditions for growth-related imperfections (Yeo et al., 2003).

According to the fact that thermal analysis has been frequently used to identify crystal forms of drugs and in the course of thermal analysis, crystal transformation is often observed as well as melting and decomposition, Suzuki et al. studied mechanisms of thermal crystal transformation through melting and recrystallization. They characterized two anhydrates (α-from and β-from) and two hydrates (hemihydrate and monohydrate) forms of a novel fluoroquinolone antibiotic, sitafloxacin, in addition to sesquihydrate which is used in the marketed drug products. The results of crystal structural that were characterized by infrared spectroscopy, X-ray powder diffractometry and thermal analysis revealed quinolone rings of sitafloxacin had distorted planar structure and quinolone ring of the drug in α-form and monohydrate hold opposite torsion to those in β-form and sesquihydrate. These kinds of thermal analysis are often recommended as a routine tool for quality control of thermal dehydration and subsequent crystal conversion of drugs (Suzuki et al., 2010).

2.1 Compaction and flowability

Acquiring a clear notion of why certain drug materials are prone to problems during compaction and dominate the relative constraints to offer a successful compaction and tableting strategy of pharmaceutical powders would involve an understanding of the fundamental properties of drug powders. Therefore, it is important to determine the effect of different physicochemical properties such as particle size, shape, surface area, polymorphic form, crystal habit, hydrates, and processing conditions on the compaction of powders. As different crystal habits of a certain drug hold dissimilar planes, they have differing points in their specific surface and free surface energies. Even so, alternative recrystallization solvents could develop crystal with defined crystal habit, size and shape as well as compressibility properties. The nature and amount of these changes count on the recrystallization conditions including the presence of impurities, type of solvents and cooling rates. This view is supported by a variety of papers at molecular level developing the knowledge of solid-state properties such as crystal structure, crystal habit, and polymorphism influence on the mechanical properties of powders in an attempt to identify and modify physical properties of bulk solids of drugs (Liebenberg et al., 1999; Maghsoodi et al., 2007).

Seton et al. evaluated the particle morphology of ibuprofen, an anti-inflammatory drug, by recrystallization from a range of solvents and investigated the following influence on compaction properties. The compaction data achieved from properties of the ibuprofen control and recrystallized samples at different compaction forces and speeds revealed equal or better tablet strength than the control, whilst ibuprofen recrystallized from 2-ethoxyethyl acetate exhibited lower levels of elastic energy during compaction. In addition, the recrystallized ibuprofen samples demonstrated flowablity equivalent quality to the ibuprofen control, excluding the ibuprofen recrystallized from acetone which showed excellent flow properties. Generally, the results displayed ibuprofen recrystallization from various solvents could offer advantages in terms of particle morphology, flowability and compaction properties (Seton et al., 2010).

In another study an anti-epileptic drug, phenytoin, crystals in the form of free acid, having distinct types of habits, was modified via different recrystallization conditions and techniques by Nokhodchi et al. Several sets of experimental conditions for temperature, solvent evaporation and watering-out techniques were applied for evaluation of the drug recrystallization in ethanol and acetone solvents. The solid state characteristics and compaction properties of the crystal habit with factors affecting the resultant crystals were also evaluated. The physical characteristics of the crystals were investigated using scanning electron microscopy, X-ray powder diffractometry, FT-IR spectrometry and differential scanning calorimetry. They confirmed that using watering-out technique as a crystallization method, produced thin plate crystals, while the crystals obtained by other methods were needle shape for alcoholic solutions and rhombic for acetone solutions. Although the crystallization medium had central effect on phenytoin crystal habit modification, altering crystallization temperature had no effect on crystal habits except a change in size of crystals. In the case of compaction, the crystals produced from alcohol or acetone showed high crushing strengths as a result of lower porosity and lower elastic recoveries (Nokhodchi et al., 2003).

As we know, ascorbic acid crystals are unsuitable for direct tableting due to their poorly compactible properties, Kawashima et al. designed spherically agglomerated crystals of ascorbic acid with improved compactibility for direct tableting. They precipitated ascorbic acid crystals by a solvent change method, followed by their agglomerations with the emulsion solvent diffusion or spherical agglomeration mechanism, depending on the solvent combination for crystallization. Considering the results of the study, under static compression, effectively the proper compact with a sufficient strength was produced. After all improved micromeritic properties, such as flowability and packability for the spherically agglomerated crystals were obtained for crystals of ascorbic acid with the spherical crystallization technique (Kawashima, 2003).

Designing a suitable dosage form with an ideal physicochemical and mechanical property is an important basic principle of drug delivery systems. As follows, crystal structure, shape, and size of drug substances have a huge economical and practical effect at all stages of development from research to commercialization. So, there is a necessity to control the critical properties of drugs for their readiness and capacity to form a tablet which are dominating dosage forms in pharmaceutical dosage form manufacturing.

2.2 Solubility, dissolution, and bioavailability of drugs

Nowadays, in pharmaceutical companies drugs with restricted aqueous solubility have become ever more prevalent and challengeable in the research and development stages. Slow drug dissolution in biological fluids, insufficient and inconsistent systemic exposure and subsequent inadequate efficacy in patients, are some of routine challenges to be coped with during the development of poorly water-soluble drug substances especially when they are administrated orally. Notable numbers of drugs especially new drug candidates are in a biopharmaceutical classification of low solubility (BCS Class II and IV) keeping drug dissolution rate as the limiting factor for the drug absorption and attaining suitable blood-levels of the drugs (LobenbergAmidon, 2000). These inadequacies in solubility of clinically established drug substances in water and in the gastric fluids make problems in drug dissolution rate and oral bioavailability of drugs, as well (Blagden et al., 2007). Consequently, there is a basic requisite to deliver such drugs in a way that gives a chance of sufficient dissolution rate, absorption,and demonstrating suitable clinical efficacy.

Numerous scientific and technological advancements have been made in the research and development for improving and maximizing dissolution rates of the mentioned types of drugs. Despite enhancements in solubility and dissolution rate and oral bioavailability of poorly water-soluble drugs with the customary pharmaceutical technologists, still there are concerns about the success of those methods in the complexities arise from the specific physicochemical nature of the drug molecule itself (Krishnaiah S.R., 2010). One of the thriving trends in enhancing the solubility, dissolution rate and subsequent bioavailability of poorly soluble drugs is to deal with crystal forms of materials which could potentially be applicable to a broad range of drugs with different crystalline habits (Yeo, Kim et al., 2003). On account of many factors such as crystal habit, size and even polymorphic forms of a drug, dissolution rates would enhance through habit recrystallization. A number of reports in the literature validated the effects of crystal morphology variation on solubility, in vitro dissolution rate, holding potentials for improving drug bioavailability (Kawashima et al., 1986; Carino et al., 2006).

Several studies in this filed have shown that exposure of diverse crystal faces determines the nature of the wettability and consequent enhancements in dissolution rate of the drugs with different crystalline shapes (Heng et al., 2006). In 2000, Kobayashi et al. published a paper in which they presented different dissolution rates for carbamazepine, where the dihydrate form of the drug in simulated fluids (pH 1.2) had notably slower dissolution rates than the anhydrous forms (forms I and III). Although the metastable polymorph (Form III) possessed greatest rates of dissolution at the initial stages, reductions in dissolution rate at later time points of the profile was achieved due to the rapid conversion of metastable polymorph (Form III) to the dihydrate. Nevertheless, in another study carried out by Tian et al. the behavior of carbamazepine and dihydrate compacts during in vitro dissolution tests various factors were evaluated. Considering the results, presence of excipients such as polyethylene glycol (PEG) and hydroxyl propyl methyl cellulose (HPMC) inhibited the conversion of carbamazepine to the hydrated form following decreased rates of drug dissolution (Tian et al., 2007). Application of different drug habits in pharmaceutical dosage forms could vary the dissolution rates, as the use of metastable polymorphs in enhancing drug dissolution rates. They also performed bioavailability tests in dogs to determine the effects of physicochemical properties of drug form I, form III and dihydrate on the plasma level of carbamazepine. Similar to the findings of other dissolution studies for carbamazepine, drug bioavailability that had been measured was lowest for the dehydrate form. The lower drug bioavailability established with metastable form was in consistent with the probable conversion of the drug habit to the dihydrate (Kobayashi et al., 2000).

Intrinsic solubility of three crystals of diclofenac, a nonsteroidal anti-inflammatory drug, was investigated by Llinas et al. The crystal habits were characterized and detected by thermo gravimetric analysis, differential scanning calorimetry, and X-ray diffraction. They recrystallized the anhydrous sodium salt of commercially available diclofenac with ethanol and precipitated as a hydrated drug that provided consistent results for the intrinsic solubility. Regarding the broad range of values which have been reported for aqueous diclofenac solubility in the literature, they claimed their solubility records were at the smaller end of the range (Llinas et al., 2007).

Perlovich et al. analyzed four new crystal structures of the sulfonamides by X-ray diffraction experiments and comparative analysis of molecular conformational states and hydrogen bonds networks by graph set notations in the crystal lattices. They established temperature dependencies of the solubility in water, *n*-octanol as well as thermodynamic functions of solubility and solvation processes for the compounds. According to Perlovich et al. distinguishing between enthalpy and entropy leads to the insight that the mechanism is different for the different molecules where it may be of importance for further assessment of distribution of drug molecules and provide a better understanding of biopharmaceutical properties of drugs (Perlovich et al., 2008).

Dipyridamole as a critical antiplatelet and peripheral vasodilator drug is known to have properties of water insolubility and poor bioavailability which are the limitations of its effectiveness in clinical usage. Adhiyaman et al. characterized dipyridamole crystals with different types of habits by recrystallization from selected solvents. Physicochemical characteristics of the crystals were assessed via scanning electron microscopy, X-ray powder diffractometry, IR spectrometry and differential scanning calorimetry. The developed crystals of dipyridamole under optimized conditions ensue in different crystalline habits that significantly improved dissolution rate compared to original dipyridamole. Recrystallized dipyridamole with benzene and acetonitrile, produced needle shaped crystals and the ones recrystallizaed with methanol produced rectangular shaped crystals. Whereas smooth needle shaped crystals were obtained with the methanolic solution of the drug in the presence of Tween-80, Povidone K30 and PEG-4000 (AdhiyamanBasu, 2006). Generally, these results were in consistence with the possibility of controlling and enhancing the drug release properties following by a probable improvement in bioavailability of drug particles through characterization of drug crystals.

Recrystallization of phenytoin in ethanol and acetone by Nokhodchi et al. was shown to produce needle-like and rhombic crystal habits which brought about identical dissolution rates of crystals obtained from both solvents. Considering the results, nature of recrystallization solvents in this case had no effect on dissolution profiles. It was suggested that the differences in dissolution rates for phenytoin was related to the surface area of various crystals with different shapes (Nokhodchi, Bolourtchian et al., 2003). Wettability and the changes in intrinsic dissolution rate of doped phenytoin crystals were evaluated by Chow et al. They stated that the differences in dissolution rates of phenytoin crystalline powders with different morphology were mainly because of the changes in surface area rather than the improvements in the wetting of more polar surface moieties. However, they stated that the areas of the relatively polar faces seem to be valuable determinants of the drug release profiles of doped phenytoin crystals along with the correlation of the surface tensions (Chow et al., 1995).

Talinolol is a cardioselective beta blocker agent that is known to have different crystal structures with strongly differing solubilities when pure water, acetate, or phosphate buffers are employed as dissolution media. Wagner et al. have studied the impact of different dissolution media controlling the crystal structures of talinolol influencing the dissolution rate and solubility of the drug. The crystal structures were analyzed by means of light microscopy, differential scanning calorimetry, and X-ray powder diffraction, detecting the variations of talinolol crystal structures being the source of incomplete and unpredictable nature of the drug bioavailability (Wagner et al., 2003).

Carbamazepine, a routinely used drug in the treatment of epilepsy and trigeminal neuralgia, exists in four polymorphic forms and as a hydrate which could modify the physicochemical properties of the drug. In our previously published paper we established enhanced physicomechanical properties of carbamazepine via recrystallization at different pH values. The resultant habits of carbamazepine crystals varied from flaky or thin plate-like to needle shape structures which were ascertained using scanning electron microscopy and X-ray powder diffraction. Considering the results of the in vitro dissolution evaluations of carbamazepine samples, a higher dissolution rate for carbamazepine crystals were obtained from media with pH 11 and 1 compared to the original carbamazepine sample. After all, the carbamazepine particles recrystallized from aqueous solutions with different pH values revealed superior mechanical properties which were generally in consistence with the similar studies of drug recrystallization (Grzesiak et al., 2003; Javadzadeh et al., 2009).

In spite of the absolute potential of drug habit modifications in dissolution rate promotion, far too little attention has been paid to gather detailed documentations of the usage of the approach in enhancement of systemic drug efficiencies following drug bioavailability in human subjects or in suitable animal models. So as to affirm drug recrystallization as an efficient practice in intentionally increasing the bioavailability of poorly soluble drugs, further investigation in this field is mandatory. Ultimately, we could consider the potential management of crystal habits of poorly water soluble drugs as an approach for designing efficient pharmaceutical dosage forms.

3. Polymorphism

Very early on, in 1832 Wöhler and Liebig reported the first observation of polymorphism upon cooling a boiling solution of benzamide where needle-shaped crystals would initially formed followed by a conversion to rhombic crystals upon standing (WöhlerLiebig, 1832). After a longtime history, polymorphism has maintained its innovation for scientists as a curiosity and an urgent challenge of commercial relevance in manufacturing industries, as well.

In general, polymorphism is known to be the ability of a compound to crystallize in more than one distinct crystal structure. Evidently, structures of different crystals would receive various scopes of the possible ranges of intermolecular interactions like, van der Waals, ionic, and hydrogen bonds. It would not be far from expectations that the different polymorphs of the same molecule will have different free energies affecting all the basically physicochemical properties of the compound, in consequence. Therefore, the crystal structure of drugs have leading signature on both physical and chemical properties in a way that the solid-state forms might demonstrate variations in, color, stability, processability, solubility, dissolution and bioavailability, ranging from the subtle to the severe (Rodriguez-Spong et al., 2004). Despite the fact that distinct crystal habits have different processing issues or different chemical stability, these variations usually have direct solutions and the real impact of crystal forms is the difference in solubility and bioavailability. Figure 2 gives a fair picture of the probable consequences of a different crystal form in solubility. Decreased solubility of a certain crystalline form of a drug not only brings about lower bioavailability but also reduced rates of drug clearances are inevitable which are the inferences to the safety and efficacy of any drug product.

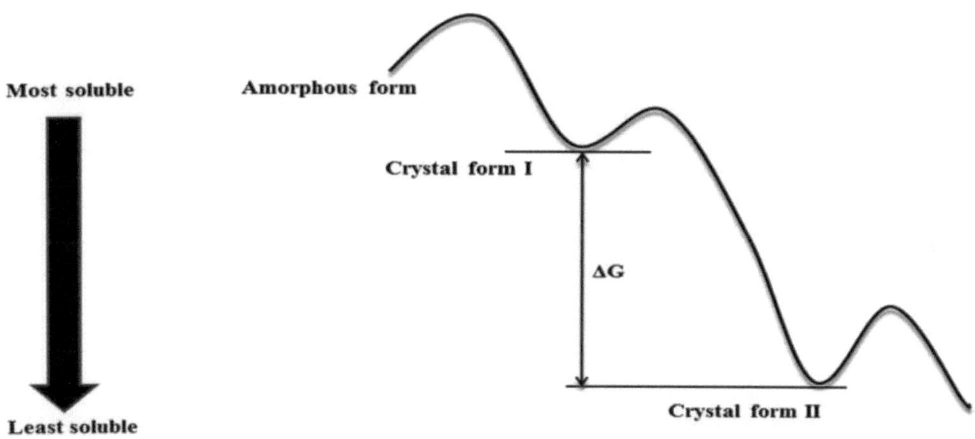

Fig. 2. Relative solubility of different crystal forms as a function of thermodynamic stability

Normally, producing the most thermodynamically stable polymorphic form of a drug in the range of interest is compulsory. The pharmaceutical science has been chiefly in charge for a change in this situation ever since the majority of drugs delivered orally receive rigid approval for a single crystal form or polymorph. Establishing the most thermodynamically stable form of a drug relies on obtaining comprehensive information about the existence and the interrelation of the polymorphs for a given active ingredient. Different polymorphs of a drug could be prepared by recrystallizing the drug with a range of solvents under the optimized conditions which has currently been arisen as an active research province of pharmaceutical science for improving the formulation related problems of drug molecules.

Accordingly, drug polymorphism investigations have an important role in any reformulation study since it has an impact on the development potential of a drug molecule, so as understanding the polymorphic tendencies of a drug molecule thorough characterization of the observed forms is of prime value.

The first important step in characterizing a polymorph of a drug is to distinguish between different structures of a molecule by its class. As it has been shown in figure 3, the most commonly observed forms in drug solids are the polymorphs, amorphous, crystals, solvates, and hydrates forms, that are fairly tractable from a processing notion. More to the point, hydrates or solvates are considered as pseudo-polymorphs that may either be an entirely different crystalline entity or simply incorporated in the parent crystal lattice; however, it is often possible to remove water or solvate by recrystallization.

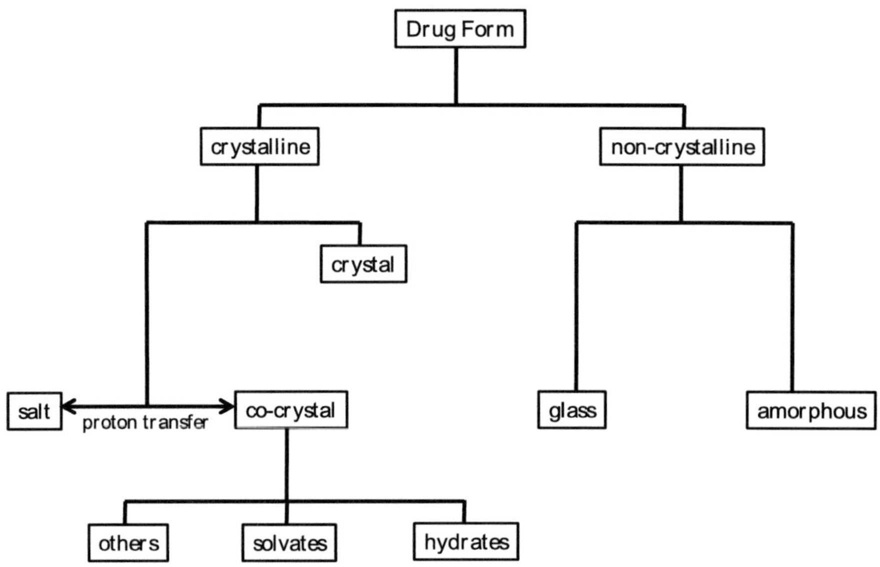

Fig. 3. Pharmaceutical solids in different forms.

While the molecular organization of drugs may differ having impact on the development potential of a drug molecule, the underlying concern for the drug performance is the same. Herein we tried to meet examples and detailed information on how recrystallization was used to improve the relative characteristics of drugs.

Park et al. designed a supercritical anti-solvent process for recrystallization of fluconazole, a triazole antifungal drug which has different polymorphic forms, to modify solid state characteristics of fluconazole by a range of operating conditions such as temperature, pressure, and type of solvent. Fluconazole particles were detected by means of differential scanning calorimetry, thermogravimetry analysis, powder X-ray diffraction, fourier transform infrared spectroscopy and scanning electron microscopy. Regarding the findings of the study, polymorphic forms of fluconazole were remarkably affected by the experimental conditions as the anhydrate form I of drug was obtained at low temperature and the anhydrate form II was obtained at higher temperatures. On the whole they suggested that the solid state characteristics of fluconazole, including the polymorphic form could effectively be managed by altering the operating conditions of the recrystallization process such as temperature, pressure, and solvent (Park et al., 2007). In another study, recrystallization of two anti-cancer drugs, erlotinib hydrochloride and fulvestrant was investigated employing the same method of recrystallization by Tien et al. Polymorph conversion from the original form B to form E or a prior form A for erlotinib was demonstrated under appropriate operational conditions which improved the dissolution rate of the drug. The micronized fulvestrant drug particles showed consistent polymorph like the original drug, but with differences in crystal habits. They also confirmed the positive effect of recrystallization in drug modifications (Tien et al., 2010).

Chen et al. evaluated the effects for the type of solvent, temperature and pressure as well as the solution flow rate on sulfathiazole particle formation using the same supercritical antisolvent technology for recrystallization. In the optimum operating condition of acetone as the solvent, temperature at 308 K with 12 MPa pressure and flow rate of 2mL/min the micronized sulfathiazole including polymorphic forms were obtained. Moreover, it was determined that various solvents resulted in different polymorphisms where the polymorph form III changed to form IV when ethanol was employed as the solvent. As a result, recrystallization of sulfathiazole produced in optimally micronized particles which exhibited a much narrower particle size distribution with an enhanced in vitro dissolution rate by 3.2 times to the original form of the drug (Chen et al., 2010).

4. Micro-crystal formation

Apart from the final quality of the drug crystal product in terms of purity, polymorphism, habit and morphology as well as crystal mechanical strength characterized in recrystallization procedure, size distributions of the drug particle is another crucial aspect of the pharmaceutical dosage forms. With reference to the Noyes–Whitney equation, application of a drug substance in a reduced particle size encourages bioavailability of the poorly water-soluble drug substances owing to the enhanced dissolution rate of micron- or nano-size drug particles (Chaumeil, 1998). An alternative area where small sized drug particles are indispensable is the pulmonary drug administration, inhaler drugs, in which drug powders should have a narrow particle size distribution and a mean particle size of 5 µm with almost no particles larger than 10 µm. Pharmaceutical dosage forms of these kinds are supposed to follow low particle agglomeration tendency, sufficient flow properties, and good batch-to-batch conformity in favor of the relative drug powders (IslamGladki, 2008).

Several techniques could be applied for the preparation of micron sized pharmaceuticals, such as mechanical comminution of the previously formed larger drug particles by crushing, grinding, milling and etc. that are the most common ways in this field. However, these methods not only provide limited opportunity for the control of important drug characteristics, like size, shape, morphology, surface properties but also ascertain distorted drug properties in a principally uncontrolled manner (RasenackMuller, 2004). As the surfaces in mechanically micronized drug powders are not naturally grown as the crystal cleaves at the crystal face and the surface energy changes, processing properties like flowability, agglomeration and stickiness to surfaces are the prevalent phenomenon. Employing milling processes such as jet milling, pearl-ball milling, or high-pressure homogenization other than influencing the preformulation behavior of drug structures, call for high energy and manpower being evidence for insufficient coverage of this method for application in reduced particle size drug production. It seems that these commonly being used techniques do not meet the ideal way for the production of small sized drug particles.

Since the suitable physicochemical and biopharmaceutical properties of a drug substance add to the time and cost of drug development, any postern to resolve these problems and produce small particles of drugs in a controlled process maximizes the opportunity to succeed in drug product manufacturing. Unlike the former techniques, production of small particles using controlled production processes such as spray drying, precipitation from supercritical fluid and recrystallization could be applied for the preparation of properly characterized micron sized pharmaceuticals. Microcrystallization in which the solubility and dissolution rate is improved by forming high specific surface area is used for preparation of drug microcrystals by recrystallization methods to reduce the size of the poorly water-soluble drug particles.

The usual technologies for recrystallization are fulfilled in this framework and it is to use solvent change or precipitation method by immediate mixing two liquids in presence of stabilizing agents. Regardless of the absolute efficiency of the recrystallization approach in production of large drug crystals, producing the small drug particles is still a sort of a challenge due to the high surface area of these particles, exerting tendency of a particle growth. So stabilizing agents would be foremost part in this system preventing particle growth by stabilizing the high specific surface area of small particles (Lechuga-BallesterosRodriguez-Hornedo, 1993). Nevertheless, recrystallization still continues to be one of the important parts of small sized drug production in drug development strategies. Therefore, microcrystals precipitate in presence of stabilizing protective polymers and a large and hydrophilized surface would be formed in a one process step having advantages over traditional milling techniques.

Hence, developing micro-crystallization as an efficient approach that modifies the biopharmaceutical and technological behavior of drug through selection of the process variables to reach an optimal pharmaceutical product has evolved to meet drug development challenges. Exploration on the growing number of publications in domain of micronized drug particle developing techniques apparently to enhance drug dissolution rate, considering the widely increasing number of poorly water-soluble drugs affirms the declaration.

Rasenack et al. prepared microcrystals of a poorly water-soluble drug ECU-01, an anti-inflammatory drug in preclinical state of development by a precipitation practice in the presence of stabilizing agents such as gelatin, chitosan, and different types of cellulose ethers and then spray-drying of the formed dispersion. Considering the low specific surface area of the nearly cuboid-like form of ECU-01, the aim of the survey was to enhance the drug dissolution rate by using microcrystals. Precipitation came off through dissolving the drug in acetone followed by an instant pouring an aqueous solution of the stabilizer into the drug solution. Via the use of cellulose type ethers as a stabilizer employed in this technique the thermodynamically unstable small particles were stabilized forming a protective layer on the crystal surface of the homogeneous microcrystals dispersions. Due to the polymorphic nature of the drug, the newly formed crystals appeared in a needle-shaped habit, highly increasing the specific surface area. Consequently, the dissolution rate rose up to 93% after 20 min compared to the 4% in common drug, indicating large surface of the microcrystals. Recrystallization of the poorly water-soluble drug ECU-01 was considered as a superior method which is easy to handle and only entails ordinary equipment (Rasenack et al., 2003).

A nonsteroidal anti-inflammatory drug, indomethacin, which has a hydrophobic and pH-dependent solubility nature, was developed and studied by Kim et al. through a microcrystallization technique to improve its physicochemical properties. Microcrystals of indomethacin was produced using a pH-shift procedure in which the drug was dissolved in an alkaline water to prepare saturated indomethacin solution thereafter the pH of the solution was decreased by adding 0.5N hydrochloric acid and stored at 20°C for 24 h to form microcrystal. The findings of the study exhibited similar physicochemical properties for the microcrystals produced and the standard crystalline powder in X-ray diffraction, differential scanning calorimetry, and Fourier transform infrared spectroscopy analyses, exclusive of a lower peak height in X-ray diffraction and somewhat lower melting temperature. The plate-like with uniform sized microcrystals of indomethacin dissolved about twice over the standard crystalline powder in the initial phase of dissolution study. Furthermore, the in vitro biological activity of the indomethacin microcrystals was assessed in their capacity to inhibit the proliferation of colon cancer cells that showed 20% greater activity than that of the standard crystalline powder. This view might have implications for improving the efficiency of chemotherapy in treating patients with malignant neoplasms using this technique for production of indomethacine microcrystals (Kim et al., 2003).

More recently, Talari et al. evaluated gliclazide microcrystals; a widely used drug for the treatment of non-insulin-dependent diabetes mellitus which shows a low solubility of 55 mg/L in water and gastric fluids leading to a low dissolution rate and variable bioavailability. The gliclazide microcrystals were prepared by in situ micronization techniques based on solvent and pH-shift and were examined for the drug absorption and pharmacokinetics of GL after oral administration in rats. Compared to the original drug, scanning electron microscopy showed significant changes in the shape and size of the prepared crystals using both methods. Recrystallized samples not only showed enhanced dissolution rates than untreated drug particles but also a reduced particle size of about 30 and 61 times by solvent-change and pH-shift methods were detected for drug crystals, respectively. Regarding results of the in vivo biological assays for hypoglycemic activity, microcrystallization of gliclazide using both methods resulted in an increased pharmacodynamic effect of glucose-lowering in diabetic rats which could be relevant to the improved dissolution rate of the drug (Talari et al., 2010).

Concisely, the microcrystallization of the drug particulates which has an effect on crystal habit, could also improve the drug absorption characteristics and the subsequent drug bioavailability.

5. Lyotropic liquid crystals formation

Liquid crystals (LCs) are a state of matter that has properties between those of a conventional liquid and those of a solid crystal. For instance, an LC may flow like a liquid, but its molecules may be oriented in a crystal-like way. There are many different types of LC phases, which can be distinguished by their different optical properties. When viewed under a microscope using a polarized light source, different liquid crystal phases will appear to have distinct textures. The contrasting areas in the textures correspond to domains where the LC molecules are oriented in different directions. Within a domain, however, the molecules are well ordered. LC materials may not always be in an LC phase (just as water may turn into ice or steam).

Liquid crystals can be divided into thermotropic, lyotropic and metallotropic phases. Thermotropic and lyotropic LCs consist of organic molecules. Thermotropic LCs exhibit a phase transition into the LC phase as temperature is changed. Lyotropic LCs exhibit phase transitions as a function of both temperature and concentration of the LC molecules in a solven . Metallotropic LCs are composed of both organic and inorganic molecules; their LC transition depends not only on temperature and concentration, but also on the inorganic-organic composition ratio.

Examples of liquid crystals can be found both in the natural world and in technological applications. Most modern electronic displays are liquid crystal based. Lyotropic liquid-crystalline phases are abundant in living systems. For example, many proteins and cell membranes are LCs. Other well-known LC examples are solutions of soap and various related detergents, as well as the tobacco mosaic virus.

A lyotropic liquid crystal consists of two or more components that exhibit liquid-crystalline properties in certain concentration ranges. In the lyotropic phases, solvent molecules fill the space around the compounds to provide fluidity to the system. In contrast to thermotropic liquid crystals, these lyotropics have another degree of freedom of concentration that enables them to induce a variety of different phases.

A compound, which has two immiscible hydrophilic and hydrophobic parts within the same molecule, is called an amphiphilic molecule. Many amphiphilic molecules show lyotropic liquid-crystalline phase sequences depending on the volume balances between the hydrophilic part and hydrophobic part. These structures are formed through the micro-phase segregation of two incompatible components on a nanometer scale. Soap is an everyday example of a lyotropic liquid crystal.

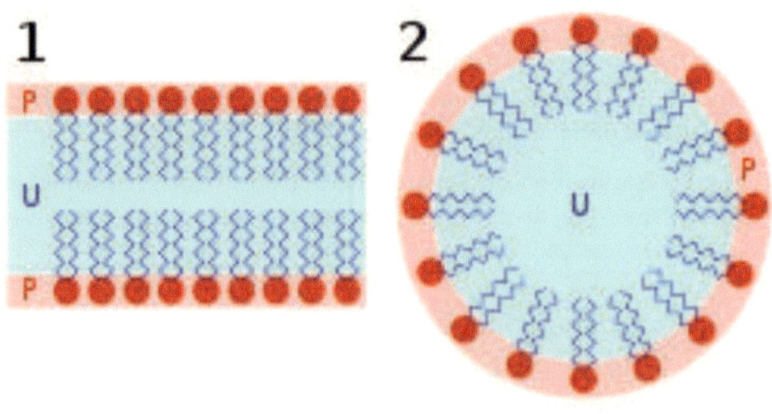

Fig. 4. Structure of lyotropic liquid crystal. The red heads of surfactant molecules are in contact with water, whereas the tails are immersed in oil (blue): bilayer (left) and micelle (right)

The content of water or other solvent molecules changes the self-assembled structures. At very low amphiphile concentration, the molecules will be dispersed randomly without any ordering. At slightly higher (but still low) concentration, amphiphilic molecules will spontaneously assemble into micelles or vesicles. This is done so as to 'hide' the hydrophobic tail of the amphiphile inside the micelle core, exposing a hydrophilic (water-soluble) surface to aqueous solution. These spherical objects do not order themselves in solution, however. At higher concentration, the assemblies will become ordered. A typical phase is a hexagonal columnar phase, where the amphiphiles form long cylinders (again with a hydrophilic surface) that arrange themselves into a roughly hexagonal lattice. This is called the middle soap phase. At still higher concentration, a lamellar phase (neat soap phase) may form, wherein extended sheets of amphiphiles are separated by thin layers of water. For some systems, a cubic (also called viscous isotropic) phase may exist between the hexagonal and lamellar phases, wherein spheres are formed that create a dense cubic lattice. These spheres may also be connected to one another, forming a bicontinuous cubic phase (VroegeLekkerkerker, 1992).

The objects created by amphiphiles are usually spherical (as in the case of micelles), but may also be disc-like (bicelles), rod-like, or biaxial (all three micelle axes are distinct). These anisotropic self-assembled nano-structures can then order themselves in much the same way as thermotropic liquid crystals do, forming large-scale versions of all the thermotropic phases (such as a nematic phase of rod-shaped micelles).

For some systems, at high concentrations, inverse phases are observed. That is, one may generate an inverse hexagonal columnar phase (columns of water encapsulated by amphiphiles) or an inverse micellar phase (a bulk liquid crystal sample with spherical water cavities).

A generic progression of phases, going from low to high amphiphile concentration, is: Discontinuous cubic phase (micellar cubic phase), Hexagonal phase (hexagonal columnar phase) (middle phase), Lamellar phase, Bicontinuous cubic phase, Reverse hexagonal columnar phase, and Inverse cubic phase (Inverse micellar phase)

Even within the same phases, their self-assembled structures are tunable by the concentration: for example, in lamellar phases, the layer distances increase with the solvent volume. Since lyotropic liquid crystals rely on a subtle balance of intermolecular interactions, it is more difficult to analyze their structures and properties than those of thermotropic liquid crystals. Similar phases and characteristics can be observed in immiscible diblock copolymers.

Lyotropic liquid crystals transitions occur with influence of solvents or recrystallization method. Lyotropic liquid crystals occur as a result of solvent induced aggregation of the constituent mesogens into micellar structure. This types of liquid crystalline states were mostly used for designing the sustain release drug delivery system. It is also used for improving the solubility and stability of insoluble drugs by incorporating it into micellar structure of liquid crystals (Lechuga-Ballesteros et al., 2003).

6. Spherical crystallization

Among the particles designed for solid pharmaceutical dosage forms, tablet supports for the half of all oral drug delivery system and 70% of the all pharmaceutical preparations produced. Direct tabletting, simple mixing and compressing drug powders have been widely applied to a large number of drugs on the industrial scale as an excellent technique. A successful tabletting and also compression of any drug material is reliant on the micromeritic properties of the drug crystals. In this context, crystals of needle-shaped or plated-shaped are challengeable to be handled with, due to the poor flowability of these crystals (Kaerger et al., 2004). In 1984 Kawashima et al. introduced crystal agglomeration with controlled properties as spherical crystallization to the pharmaceutical manufacturing and expressed that the spherically dense agglomerates were suitable for direct tabletting. Accordingly, crystallization and agglomeration of the drug substance particles concurrently in one step to transform crystals directly into compacted spherical form during the recrystallization process is defined as spherical crystallization (Kawashima, 1984). Spherical crystals could be established through two different techniques, either by typical spherical crystallization technique or non typical spherical crystallization technique (Nokhodchi et al., 2007). Non typical spherical crystallization technique might be regarded as the traditional crystallization process like salting-out, cooling, precipitation, whereas, the typical spherical crystallization is a three-solvent system employing three solvents; one is the drug dissolution medium known as the good solvent, another is a medium which partially dissolves the drug and has wetting feature that is named bridging liquid, and the last one is immiscible with the drug substance recognized as the bad solvent (Nokhodchi, Maghsoodi et al., 2007; Mahanty et al., 2010).

On average, spherical crystallization techniques are believed to be promising techniques in which the drug crystals are modified using different solvents for direct compressible spherical agglomerates, which can save money and time for tabletting. These approaches not only helps to achieve good flowability, compressibility and micromeritic properties of the drug substances, but also it is known to improve the the wettability, bioavailability, and dissolution rate of some poorly soluble drugs (Kawashima, Handa et al., 1986; Di Martino et al., 1999; NokhodchiMaghsoodi, 2008). Besides, drug materials produced by the spherical crystallization technique result in the economical process in the development of the solid dosage forms for scaling up to a commercial level since it provides reduced time and cost by enabling faster operation, less machinery and fewer personnel.

Seeing that, magnesium aspartate and acetylsalicylic acid crystals in common are tetragonal and prism-shaped with different sizes they show poor flowability and compactibility properties indicating the crystal habit and the electrostatic charge. Szabo-Revesz et al. developed magnesium aspartate and acetylsalicylic acid via spherical crystallization since they are used in direct tablet-making and capsule-filling, the particle size and the spherical form are fundamental in view of their processibility. They prepared drug agglomerates through non-typical (magnesium aspartate) and typical (acetylsalicylic acid) spherical crystallization techniques. Crystal agglomerates of these drugs produced by these different spherical crystallization techniques created opportunity for a comparison between the results obtained. Considering the findings of the study, a higher initial cooling rate and a lower stirring rate were favorable in producing crystal agglomerates. The growth of particle size and the spherical form as well as the associated decreased specific surface of the magnesium aspartate and acetylsalicylic acid crystal agglomerates created better compactibility and cohesivity characteristics than the control samples in a softer flow time, and a higher bulk density. According to the authors both types of spherical crystallization (non-typical and typical) can be effectively used not only for spherical particle forming but also for size growing of drug materials that are pressed directly into tablets or made into filled capsules without excipients (Szabo-Revesz et al., 2002).

Celecoxib, a non-steroidal anti-inflammatory drug which is the first selective cyclooxygenase-2 inhibitor used in the treatment of osteoarthritis and rheumatoid arthritis exhibits poor flow and compression characteristics as well as incomplete and poor oral bioavailability due to its low aqueous solubility. Variety of papers in this case is representative of the suitability of this drug for spherical crystallization process to enhance the flow, compressibility and solubility properties of seems to be a beneficial objective to improve the following therapeutic efficacy of celecoxib (Banga et al., 2007). Paradkar et al. improved the micromeritic and compressional properties of celecoxib by a spherical crystallization process using the solvent change method. Selection of the solvents depended on the miscibility of solvents and also the solubility of drug in the solvents, which candidated acetone as the good solvent, dichloromethane as the bridging liquid, and water as the bad solvent. A solution of celecoxib in acetone was added to a solution of hydroxy propyl methylcellulose in dichloromethane. Drug was crystallized by adding the solution to wall-baffled vessel containing distilled water followed by a continuous mixing in a controlled speed style creating agglomerated spherical crystals. Celecoxib agglomerates exhibited satisfactory micromeritic, mechanical, and compressional properties demonstrating comparable in vitro drug release performance with the marketed capsule formulation (Paradkar et al., 2002). Elsewhere, Gupta et al. prepared spherical crystals of celecoxib using a more hydrophilic polymer, polyvinylpyrrolidone K-30 (PVP) and acetone, water and chloroform as solvent, non-solvent and bridging liquid, respectively. The agglomerates were determined by differential scanning calorimetry, X-ray diffraction, IR spectroscopic and scanning electron microscopy. They showed that the crystals possessed a good spherical shape with smooth and regular surface exhibiting significantly improved micromeritic properties compared to pure the drug. Moreover, the aqueous solubility and dissolution rate of the drug from crystals was notably increased nearly two times, with an increase in PVP concentration. In general, this technique may be applicable for producing oral solid dosage forms of other drugs with improved dissolution rate and oral bioavailability (Gupta et al., 2007). Correspondingly, use of these systems has the potential to facilitate drug development by saving valuable time in selecting the optimal physical or chemical characteristics of a given compound.

On the whole, lessening the risk of drug process modifications and providing the opportunity to gain more comprehensive rational property coverage would be established if such information is established through recrystallization processes at an early stage of drug developments. Eventually, it is of the utmost importance to strictly monitor the processing of drug substances with regards to the different crystal habits of drug materials, as well as to obtain a comprehensive understanding of the physical and chemical stability of these polymorphic states.

7. References

Adhiyaman, R. & Basu, S.K. (2006). Crystal modification of dipyridamole using different solvents and crystallization conditions. *International Journal of Pharmaceutics*, Vol.321, No.1-2, pp. 27-34, ISSN 03785173

Adibkia, K., Hamedeyazdan, S. & Javadzadeh, Y. (2011). Drug release kinetics and physicochemical characteristics of floating drug delivery systems. *Expert Opin Drug Deliv*, Vol.8, No.7, pp. 891-903, ISSN 1744-7593

Banga, S., Chawla, G., Varandani, D., Mehta, B.R. & Bansal, A.K. (2007). Modification of the crystal habit of celecoxib for improved processability. *J Pharm Pharmacol*, Vol.59, No.1, pp. 29-39, ISSN 0022-3573

Blagden, N., de Matas, M., Gavan, P.T. & York, P. (2007). Crystal engineering of active pharmaceutical ingredients to improve solubility and dissolution rates. *Adv Drug Deliv Rev*, Vol.59, No.7, pp. 617-630, ISSN 0169-409X

Carino, S.R., Sperry, D.C. & Hawley, M. (2006). Relative bioavailability estimation of carbamazepine crystal forms using an artificial stomach-duodenum model. *J Pharm Sci*, Vol.95, No.1, pp. 116-125, ISSN 0022-3549

Chaumeil, J.C. (1998). Micronization: a method of improving the bioavailability of poorly soluble drugs. *Methods Find Exp Clin Pharmacol*, Vol.20, No.3, pp. 211-215, ISSN 0379-0355

Chen, J. & Trout, B.L. (2008). Computational study of solvent effects on the molecular self-assembly of tetrolic acid in solution and implications for the polymorph formed from crystallization. *J Phys Chem B*, Vol.112, No.26, pp. 7794-7802, ISSN 1520-6106

Chen, Y.-M., Tang, M. & Chen, Y.-P. (2010). Recrystallization and micronization of sulfathiazole by applying the supercritical antisolvent technology. *Chemical Engineering Journal*, Vol.165, No.1, pp. 358-364, ISSN 13858947

Chow, A.H.L., Hsia, C.K., Gordon, J.D., Young, J.W.M. & Vargha-Butleff, E.I. (1995). Assessment of wettability and its relationship to the intrinsic dissolution rate of doped phenytoin crystals. *International Journal of Pharmaceutics*, Vol.126 pp. 21-28, ISSN 0378-5173

Dahlberg, C., Dvinskikh, S.V., Schuleit, M. & Furo, I. (2011). Polymer Swelling, Drug Mobilization and Drug Recrystallization in Hydrating Solid Dispersion Tablets Studied by Multinuclear NMR Microimaging and Spectroscopy. *Mol Pharm*, Vol.8, No.4, pp. 1247-1256, ISSN 1543-8392

Di Martino, P., Barthelemy, C., Piva, F., Joiris, E., Palmieri, G.F. & Martelli, S. (1999). Improved dissolution behavior of fenbufen by spherical crystallization. *Drug Dev Ind Pharm*, Vol.25, No.10, pp. 1073-1081, ISSN 0363-9045

Grzesiak, A.L., Lang, M., Kim, K. & Matzger, A.J. (2003). Comparison of the four anhydrous polymorphs of carbamazepine and the crystal structure of form I. *J Pharm Sci*, Vol.92, No.11, pp. 2260-2271, ISSN 0022-3549

Gupta, V.R., Mutalik, S., Patel, M.M. & Jani, G.K. (2007). Spherical crystals of celecoxib to improve solubility, dissolution rate and micromeritic properties. *Acta Pharm*, Vol.57, No.2, pp. 173-184, ISSN 1330-0075

Harbury, L. (1947). Solubility and melting point as functions of particle size; the induction period of crystallization. *J Phys Colloid Chem*, Vol.51, No.2, pp. 382-391, ISSN 0092-7023

Heng, J.Y., Bismarck, A. & Williams, D.R. (2006). Anisotropic surface chemistry of crystalline pharmaceutical solids. *AAPS PharmSciTech*, Vol.7, No.4, pp. 84, ISSN 1530-9932

Islam, N. & Gladki, E. (2008). Dry powder inhalers (DPIs)--a review of device reliability and innovation. *Int J Pharm*, Vol.360, No.1-2, pp. 1-11, ISSN 0378-5173

Jamali, F. & Mitchell, A.G. (1973). The recrystallization and dissolution of acetylsalicylic acid. *Acta Pharm Suec*, Vol.10, No.4, pp. 343-352, ISSN 0001-6675

Javadzadeh, Y., Mohammadi, A., Khoei, N.S. & Nokhodchi, A. (2009). Improvement of physicomechanical properties of carbamazepine by recrystallization at different pH values. *Acta Pharm*, Vol.59, No.2, pp. 187-197, ISSN 1330-0075

Jozwiakowski, M.J., Nguyen, N.A., Sisco, J.M. & Spancake, C.W. (1996). Solubility behavior of lamivudine crystal forms in recrystallization solvents. *J Pharm Sci*, Vol.85, No.2, pp. 193-199, ISSN 0022-3549

Kaerger, J.S., Edge, S. & Price, R. (2004). Influence of particle size and shape on flowability and compactibility of binary mixtures of paracetamol and microcrystalline cellulose. *Eur J Pharm Sci*, Vol.22, No.2-3, pp. 173-179, ISSN 0928-0987

Kawashima, Y. (1984). Development of spherical crystallization technique and its application to pharmaceutical systems. *Arch Pharm. Res.*, Vol.7, No.2, pp. 145-151

Kawashima, Y. (2003). Improved flowability and compactibility of spherically agglomerated crystals of ascorbic acid for direct tableting designed by spherical crystallization process. *Powder Technology*, Vol.130, No.1-3, pp. 283-289, ISSN 00325910

Kawashima, Y., Handa, T., Takeuchi, H., Okumura, M., Katou, H. & Nagata, O. (1986). Crystal modification of phenytoin with polyethylene glycol for improving mechanical strength, dissolution rate and bioavailability by a spherical crystallization technique. *Chem Pharm Bull (Tokyo)*, Vol.34, No.8, pp. 3376-3383, ISSN 0009-2363

Kim, S.T., Kwon, J.H., Lee, J.J. & Kim, C.W. (2003). Microcrystallization of indomethacin using a pH-shift method. *Int J Pharm*, Vol.263, No.1-2, pp. 141-150, ISSN 0378-5173

Kobayashi, Y., Ito, S., Itai, S. & Yamamoto, K. (2000). Physicochemical properties and bioavailability of carbamazepine polymorphs and dihydrate. *Int J Pharm*, Vol.193, No.2, pp. 137-146, ISSN 0378-5173

Krishnaiah S.R., Y. (2010). Pharmaceutical Technologies for Enhancing Oral Bioavailability of Poorly Soluble Drugs. *Journal of Bioequivalence & Bioavailability*, Vol.02, No.02, pp. 28-36, ISSN 0975-0851

Krishnaiah, Y.S.R. (2010). Pharmaceutical technologies for enhancing oral bioavailability of poorly soluble drugs. *Journal of Bioequivalence & Bioavailability*, Vol.2, No.2, pp. 028-036, ISSN 0975-0851

Lechuga-Ballesteros, D., Abdul-Fattah, A., Stevenson, C.L. & Bennett, D.B. (2003). Properties and stability of a liquid crystal form of cyclosporine-the first reported naturally occurring peptide that exists as a thermotropic liquid crystal. *J Pharm Sci*, Vol.92, No.9, pp. 1821-1831, ISSN 0022-3549

Lechuga-Ballesteros, D. & Rodriguez-Hornedo, N. (1993). Growth and morphology of L-alanine crystals: influence of additive adsorption. *Pharm Res*, Vol.10, No.7, pp. 1008-1014, ISSN 0724-8741

Liebenberg, W., de Villiers, M.M., Wurster, D.E., Swanepoel, E., Dekker, T.G. & Lotter, A.P. (1999). The effect of polymorphism on powder compaction and dissolution properties of chemically equivalent oxytetracycline hydrochloride powders. *Drug Dev Ind Pharm*, Vol.25, No.9, pp. 1027-1033, ISSN 0363-9045

Llinas, A., Burley, J.C., Box, K.J., Glen, R.C. & Goodman, J.M. (2007). Diclofenac solubility: independent determination of the intrinsic solubility of three crystal forms. *J Med Chem*, Vol.50, No.5, pp. 979-983, ISSN 0022-2623

Lobenberg, R. & Amidon, G.L. (2000). Modern bioavailability, bioequivalence and biopharmaceutics classification system. New scientific approaches to international regulatory standards. *Eur J Pharm Biopharm*, Vol.50, No.1, pp. 3-12, ISSN 0939-6411

Maghsoodi, M., Hassan-Zadeh, D., Barzegar-Jalali, M., Nokhodchi, A. & Martin, G. (2007). Improved compaction and packing properties of naproxen agglomerated crystals obtained by spherical crystallization technique. *Drug Dev Ind Pharm*, Vol.33, No.11, pp. 1216-1224, ISSN 0363-9045

Mahanty, S., Sruti, J., Niranjan Patra, C. & Bhanoji Rao, M.E. (2010). Particle design of drugs by spherical crystallization techniques. *International Journal of Pharmaceutical Sciences and Nanotechnology*, Vol.3, No.2, pp. 912-918, ISSN 0974 - 9446

Mirmehrabi, M. & Rohani, S. (2005). An approach to solvent screening for crystallization of polymorphic pharmaceuticals and fine chemicals. *J Pharm Sci*, Vol.94, No.7, pp. 1560-1576, ISSN 0022-3549

Nokhodchi, A., Bolourtchian, N. & Dinarvand, R. (2003). Crystal modification of phenytoin using different solvents and crystallization conditions. *Int J Pharm*, Vol.250, No.1, pp. 85-97, ISSN 0378-5173

Nokhodchi, A. & Maghsoodi, M. (2008). Preparation of spherical crystal agglomerates of naproxen containing disintegrant for direct tablet making by spherical crystallization technique. *AAPS PharmSciTech*, Vol.9, No.1, pp. 54-59, ISSN 1530-9932

Nokhodchi, A., Maghsoodi, M. & Hassanzadeh, D. (2007). An Improvement of Physicomechanical Properties of Carbamazepine Crystals. *Iranian Journal of Pharmaceutical Research*, Vol.6, No.2, pp. 83-89, ISSN 1735-0328

Paradkar, A.R., Pawar, A.P., Chordiya, J.K., Patil, V.B. & Ketkar, A.R. (2002). Spherical crystallization of celecoxib. *Drug Dev Ind Pharm*, Vol.28, No.10, pp. 1213-1220, ISSN 0363-9045

Park, H.J., Kim, M.S., Lee, S., Kim, J.S., Woo, J.S., Park, J.S. & Hwang, S.J. (2007). Recrystallization of fluconazole using the supercritical antisolvent (SAS) process. *Int J Pharm*, Vol.328, No.2, pp. 152-160, ISSN 0378-5173

Perlovich, G.L., Strakhova, N.N., Kazachenko, V.P., Volkova, T.V., Tkachev, V.V., Schaper, K.J. & Raevsky, O.A. (2008). Sulfonamides as a subject to study molecular interactions in crystals and solutions: sublimation, solubility, solvation, distribution and crystal structure. *Int J Pharm*, Vol.349, No.1-2, pp. 300-313, ISSN 0378-5173

Rasenack, N., Hartenhauer, H. & Muller, B.W. (2003). Microcrystals for dissolution rate enhancement of poorly water-soluble drugs. *Int J Pharm*, Vol.254, No.2, pp. 137-145, ISSN 0378-5173

Rasenack, N. & Muller, B.W. (2004). Micron-size drug particles: common and novel micronization techniques. *Pharm Dev Technol*, Vol.9, No.1, pp. 1-13, ISSN 1083-7450

Rodriguez-Spong, B., Price, C.P., Jayasankar, A., Matzger, A.J. & Rodriguez-Hornedo, N. (2004). General principles of pharmaceutical solid polymorphism: a supramolecular perspective. *Adv Drug Deliv Rev*, Vol.56, No.3, pp. 241-274, ISSN 0169-409X

Seton, L., Roberts, M. & Ur-Rehman, F. (2010). Compaction of recrystallised ibuprofen. *Chemical Engineering Journal*, Vol.164, No.2-3, pp. 449-452, ISSN 1385-8947

Sinclair, W., Leane, M., Clarke, G., Dennis, A., Tobyn, M. & Timmins, P. (2011). Physical stability and recrystallization kinetics of amorphous ibipinabant drug product by fourier transform raman spectroscopy. *J Pharm Sci*, pp., ISSN 1520-6017

Suzuki, T., Araki, T., Kitaoka, H. & Terada, K. (2010). Studies on mechanism of thermal crystal transformation of sitafloxacin hydrates through melting and recrystallization, yielding different anhydrates depending on initial crystalline forms. *Int J Pharm*, Vol.402, No.1-2, pp. 110-116, ISSN 1873-3476

Szabo-Revesz, P., Hasznos-Nezdei, M., Farkas, B., Goczo, H., Pintye-Hodi, K. & Eros, I. (2002). Crystal growth of drug materials by spherical crystallization. *Journal of Crystal Growth*, Vol.237, No.239, pp. 2240–2245, ISSN 0022-0248

Talari, R., Varshosaz, J., Mostafavi, S.A. & Nokhodchi, A. (2010). Gliclazide microcrystals prepared by two methods of in situ micronization: pharmacokinetic studies in diabetic and normal rats. *AAPS PharmSciTech*, Vol.11, No.2, pp. 786-792, ISSN 1530-9932

Tian, F., Sandler, N., Aaltonen, J., Lang, C., Saville, D.J., Gordon, K.C., Strachan, C.J., Rantanen, J. & Rades, T. (2007). Influence of polymorphic form, morphology, and excipient interactions on the dissolution of carbamazepine compacts. *J Pharm Sci*, Vol.96, No.3, pp. 584-594, ISSN 0022-3549

Tien, Y.-C., Su, C.-S., Lien, L.-H. & Chen, Y.-P. (2010). Recrystallization of erlotinib hydrochloride and fulvestrant using supercritical antisolvent process. *The Journal of Supercritical Fluids*, Vol.55, No.1, pp. 292-299, ISSN 0896-8446

Tiwary, A.K. (2001). Modification of crystal habit and its role in dosage form performance. *Drug Dev Ind Pharm*, Vol.27, No.7, pp. 699-709, ISSN 0363-9045

Vroege, G.J. & Lekkerkerker, H.N.W. (1992). Phase transitions in lyotropic colloidal and polymer liquid crystals. *Rep. Fmg. Phys.*, Vol.55, pp. 1241-1309, ISSN 0034-4885

Wagner, D., Glube, N., Berntsen, N., Tremel, W. & Langguth, P. (2003). Different dissolution media lead to different crystal structures of talinolol with impact on its dissolution and solubility. *Drug Dev Ind Pharm*, Vol.29, No.8, pp. 891-902, ISSN 0363-9045

Wöhler, F. & Liebig, J. (1832). Untersuchungen über das Radikal der Benzoesäure. *Annalen der Pharmacie*, Vol.3, No.3, pp. 249-282, ISSN 0075-4617

Yeo, S., Kim, M. & Lee, J. (2003). Recrystallization of sulfathiazole and chlorpropamide using the supercritical fluid antisolvent process. *The Journal of Supercritical Fluids*, Vol.25, No.2, pp. 143-154, ISSN 0896-8446

19

Crystal Forms of Anti-HIV Drugs: Role of Recrystallization

Renu Chadha, Poonam Arora, Anupam Saini and Swati Bhandari
University Institute of Pharmaceutical Sciences, Panjab University, Chandigarh, India

1. Introduction

Understanding and controlling the solid-state chemistry of active pharmaceutical ingredients (APIs) is an important aspect of drug development process. APIs can exist in a variety of distinct solid forms, including polymorphs, solvates, hydrates, salts, cocrystals and amorphous solids. Most APIs are purified and isolated by crystallization from an appropriate solvent during the final step in the synthetic process. A large number of factors can influence the crystal nucleation and growth during this process, including the composition of the crystallization medium and the processes used to generate supersaturation and promote crystallization. For development of a pharmaceutical product, it is generally accepted that the stable form should be identified and chosen for development. However, the stable crystal form of the parent compound may exhibit inadequate solubility or dissolution rate resulting in poor oral absorption, particularly for water insoluble compounds whereas a metastable form might have advantageous properties. Moreover, the metastable polymorphs constitute local minima in the energy landscape (Figure 1) with the thermodynamically stable form being the absolute minimum at a given temperature and pressure. Thus, the search for absolute minimum and energy differences between the local minima of the drug substances is the goal of material and formulation scientists in the pharmaceutical industry. While significant efforts are made by drug development groups to identify and characterize thermodynamically stable crystal forms early in development, there are many instances where new crystal forms have been discovered later in development process. The late emergence of thermodynamically stable crystal form is often explained by Ostwald's law of stages which states that the least stable crystal form is likely to crystallize first. Metastable forms appear first during crystallization process as their crystallization kinetics is faster than those of stable forms but eventually transform into a stable form. It is important to study the transformations, because the sudden appearance or disappearance of a polymorphic form in pharmaceutical products can lead to serious consequences.

Therefore, it is of utmost importance to control the crystal formation and produce a desired form. Discovery and characterization of the diversity of solid forms of a drug substance provide options from which to select a form that exhibits the appropriate balance of the critical properties for development into the drug product. Lately, the crystal engineering approaches utilizing high throughput techniques have been applied to crystalline materials to fruitfully generate various crystal forms of pharmaceutical compounds. The ability to engineer pharmaceutical materials with suitable solubility characteristics whilst maintaining suitable physical and chemical stability provides a driving force to utilize modern analytical tools to generate new crystal modifications.

Drugs with multicomponent crystalline phases such as cocrystals also carry desired drug properties similar to single-component polymorphs. The pharmaceutical cocrystals is defined as multicomponent molecular complex where one of the components is an investigational or marketed drug molecule (the active pharmaceutical ingredient or API) and the second component (the coformer) is a safe chemical for human consumption selected from the GRAS list of the US FDA (generally regarded as safe additive chemicals by the Food and Drugs Administration). The two components are present in a definite stoichiometric ratio and interact through noncovalent interactions, predominantly hydrogen bonds. Cocrystals have been found to offer an attractive platform to improve the solubility and dissolution rate of pharmaceuticals without compromising on the stability of the solid form. Many pharmaceutical companies are working actively on cocrystals and this is reflected by the growing number of publications and patent applications for co-crystals in recent years.

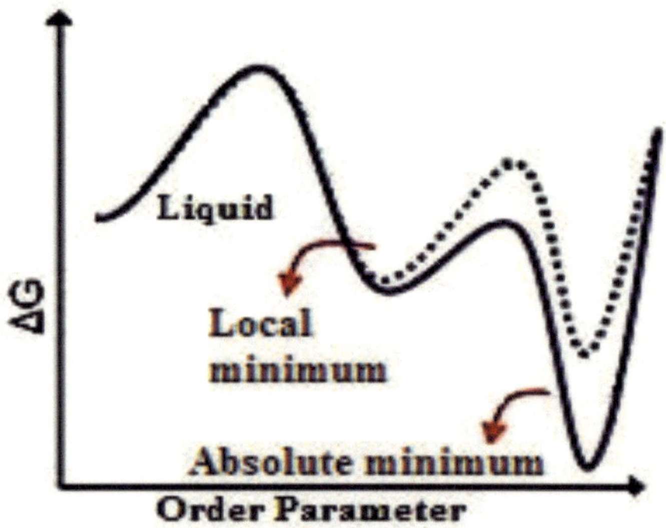

Fig. 1. Energy landscape for polymorphic forms.

The identification and characterization of diverse crystalline forms of the drug substances has become imperative for the solid state chemists in order to select the appropriate form to ensure that the product performance with respect to manufacturability, stability and bioavailability remains unchanged. Therefore, this chapter deals with the effect of crystalline state of pharmaceutical material on its physicochemical properties demonstrated through several case studies describing the phenomenon of polymorphism and a particular attention will also be paid to cocrystallization which is emerging as an important technique to generate crystal forms with improved physicochemical properties (Scheme 1).

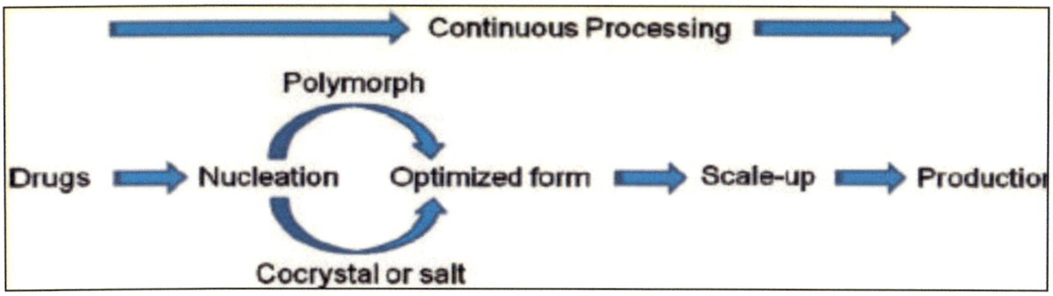

Scheme 1. Scheme representing the path followed by drugs from nucleation to production

1.1 Effect of crystal forms on physicochemical properties

It is well established that different solid forms of drug molecules exist and can affect pharmaceutical drug products with respect to physicochemical properties. It is the variation in the physical and chemical properties of drug molecules that makes polymorphism such a potentially important issue for the pharmaceutical industry. The APIs with abundant hydrogen bond sites and molecular flexibility may be manipulated by proper choice of solvent to form a specific crystal form that has different arrangements or conformations of the molecule in the crystal lattice. Changing the arrangement of the molecules in the crystal lattice changes the solid state properties affecting its solubility, stability, dissolution rate and bioavailability. Different crystal forms can have different rates of uptake in the body, leading to lower or higher biological activity than desired. As a result, significant effort is placed in identifying suitable solid forms of drug substances for use in pharmaceutical drug products.

1.2 Crystallization process

The crystallization process of polymorphous crystals is concerned with evolution of crystalline state from solution or melts and is composed of competitive nucleation and growth. However, crystallization is a complex process which starts with the appearance of most soluble form and ends with transformation to stable form. Alternatively, the nucleation of the stable form can be initiated by the dissolution of this metastable form and growth of the stable form continues until the solubility of stable forms is reached. In many instances the metastable form with desirable properties may precipitate out which is stable for years. To selectively crystallize polymorphs, the mechanism of each elementary step in crystallization process needs to be clear in relation to operational conditions and the key controlling factors. The nucleation process is the most important for control of polymorphous crystallization. Thus control over solid form throughout the drug development process is of paramount importance.

A number of factors (Figure 2) affect polymorphic behavior of a pharmaceutical solid and type of solvent has a major factor in polymorphic selectivity and crystal morphology.

1.3 Effect of solvent on crystal form

Polymorph selectivity is primarily based on the polarity of the solvent. Thus a systematic approach for selecting the right solvent is beneficial for better experimental design in control of crystallization. This effect arises from the solvent-solute interaction at the molecular level and have been explained by few of the researchers. The solvent-solute interactions during cluster formation for nucleation and growth significantly affect the ultimate crystal structure and morphology. If the solvent-solute are strongly bonded at a special crystal face, the rate-limiting step of growth would be the removal of the solvent from that face. In this case, the bonded surface grows slowly or does not grow.

However, Threlfall showed that if crystallization occurs in a region that is supersaturated with respect to one polymorph (the less soluble form) and under-saturated with respect to the other one, the solvent has no influence on the nucleation of the polymorphs and the thermodynamics will lead the process toward the production of the less soluble polymorph (Threlfall T., 2000). Moreover, the thermodynamically stable polymorph is the most stable form irrespective of the type of the solvent used. Since the thermodynamic stability of different polymorphs does not change with type of the solvent, then the solvent effect on polymorphism is attributed to the kinetic parameters.

Fig. 2. Factors which affect the crystallization of a drug substance

1.4 Other factors

Within the same solvent system, many factors are known to influence the crystal habit including supersaturation, cooling rate and agitation. At a constant temperature, supersaturation has a direct effect on the nucleation rate. As the supersaturation is increased, the rate of nuclei formation is greater than crystal growth and growth occurs mainly in one direction, producing elongated crystals. On the other hand, when there is lesser degree of supersaturation, solute solvent interactions are insignificant, producing platy crystals.

Similarly, rate of cooling alters the crystal habit by its influence on degree of supersaturation. Crystallization at a slower cooling rate produces more symmetric crystals compared with faster cooling. During faster cooling, nucleation is faster than crystal growth rate; therefore, many small crystals appear instead of few crystals growing to sufficiently larger size.

Agitation has also an important effect on the process of crystallization. The aspect ratio (ratio of horizontal maximum and vertical maximum distance of particle) is highest for unstirred conditions than during stirring. The crystals obtained under stirring conditions are fine since stirring facilitates the rate of nucleation by an even distribution of the solute molecules in the solvent. Increased nucleation rate is the result of collision of initial crystals with the stirrer and formation of smaller seeds for further crystallization. Additionally, stirring can also break larger crystals to smaller ones. Thus external appearance of a crystal can be altered by changing the growth environment to suit the requirements.

2. Polymorphism in anti-HIV drugs

This chapter throws light on the different crystalline forms reported so far for nevirapine, efavirenz, lamivudine, stavudine and zidovudine. Preparation and isolation methods, structural characterization and properties of polymorphic/solvatomorphic/cocrystal systems as well as phase transformations are illustrated.

2.1 Ritonavir

A number of studies have successfully demonstrated the appearance of different crystalline forms of some anti-HIV drugs upon recrystallization. An early example being that of ritonavir, marketed as Norvir. The late emergence of a thermodynamically more stable form (Form II) which unexpectedly precipitated from the semisolid capsule formulation led to the removal of the product from the market. The new crystal form (form II) appeared after conversion of metastable crystalline form I (Chemburkar et al., 2000; Bauer et al., 2001; Desikan et al., 2005; Miller et al., 2005). Ritonavir polymorphism was investigated using solid state spectroscopy and microscopy techniques including solid state NMR, near infrared spectroscopy, powder X-ray diffraction and single crystal X-ray analysis. Ritonavir was found to exhibit conformational polymorphism with two unique crystal lattices having significantly different solubility properties. An unusual conformation was found for form II that results in a strong hydrogen bonding network. Although the polymorph (form II) corresponding to the "cis" conformation has a more stable packing arrangement, however, nucleation, even in the presence of form II seeds, is energetically unfavored except in highly supersaturated solutions. The coincidence of a highly supersaturated solution and a probable heterogeneous nucleation resulted in the sudden appearance of the more stable form II polymorph.

Form	Melting point, °C	ΔHfus, J/g	Solid state structure
I*	122	78.2	Monoclinic
II*	122	87.8	Orthorhombic
III	78–82	60.3	Monoclinic
IV	116	59.8	Not assigned
V	97	32.0	Monoclinic

*Bauer *et al.*

Table 1. Comparison of physical parameters of ritonavir crystal forms

This polymorphic shift in ritonavir illustrated the need for early and comprehensive identification of solid-form diversity of this API. The polymorphic behavior of ritonavir was explored by Morissette et al (Morissette et al., 2003) in 2003 using CrystalMax, a high-throughput crystallization platform, with the aim of finding known and novel crystal forms of the drug molecule. Three additional crystalline forms of ritonavir were discovered when about 2,000 screening experiments were carried out (Table 1). These forms were found along with both known forms I and II, which were obtained from previously unreported solvent mixtures. Form III is a crystalline formamide solvate that converts to form V, a previously unknown hydrated phase, upon exposure to aqueous medium. Form V which is a trihydrate obtained from exposing the form III to aqueous conditions, in turn converts spontaneously to needle-like form I crystals. The process of preparing form I from III is an unusual route to a "disappearing polymorph" and provides a novel strategy for control of particle size and morphology. Form IV is a true, unsolvated, metastable previously unreported polymorph of ritonavir. Optical imaging (Figure 3) and *in situ* Raman spectroscopy were used to characterize newly formed crystals. Each of the novel forms found by means of high-throughput crystallization was scaled up to multiple milligram and gram levels. Thus the high-throughput crystallization for solid-form discovery and exploration of large numbers of parallel crystallization trials led to identification of more polymorphic forms of ritonavir.

Fig. 3. Video micrograph of crystal Form I (left) and Form II (right).

2.2 Nevirapine

Three polymorphs and many solvatomorphs have been reported for nevirapine, a non-nucleoside reverse transcriptase inhibitor, depending upon the recrystallization method. Specifically, from a crystal engineering viewpoint, the presence of the amide function CO-NH in the nevirapine molecule indicates the possibility of alternative modes of self association, namely via a dimer or a catemer synthon, leading to crystal polymorphism, whereas interaction with the solvent molecules having complementary donor and acceptor functions result in formation of solvates, significantly extending the solid state chemistry of the drug. Form I was prepared by recrystallizing nevirapine from alcohols, ethers, esters or their mixtures while refluxing from toluene, n-butanol or methyl-isobutyl ketone and subsequent cooling to 0-10 °C yielded form II. Form III was prepared by refluxing nevirapine in chloroform and using dichloromethane as antisolvent to the reaction mixture (Reguri and Chakka, 2005, 2006). However, many other experiments to investigate the existence of any other thermodynamically more stable form at room temperature, led to appearance of different solvatomorphs of nevirapine with varying stoichiometries depending upon the solvents selected. The first report on preparation of different solvatomorph by Pereira et al appeared in 2007 (Pereira et al., 2007). Six different solvates of nevirapine with different morphology were obtained by saturating the solvent systems with drug at room temperature and cooling in refrigerator. Despite different morphologies the DSC profiles did not show any relevant differences for raw material and other forms at crystal fusion peaks. However, a thermal event was observed below the melting temperature in NEV 3, NEV4 and NEV6 indicating the loss of some solvent molecules. The presence of solvent is confirmed by TGA and Karl fischer analysis. The authors have reported the crystal structure of two solvate forms of nevirapine, hemihydrate (NEV3) and hemiethylacetate (NEV4) (Figure 4).

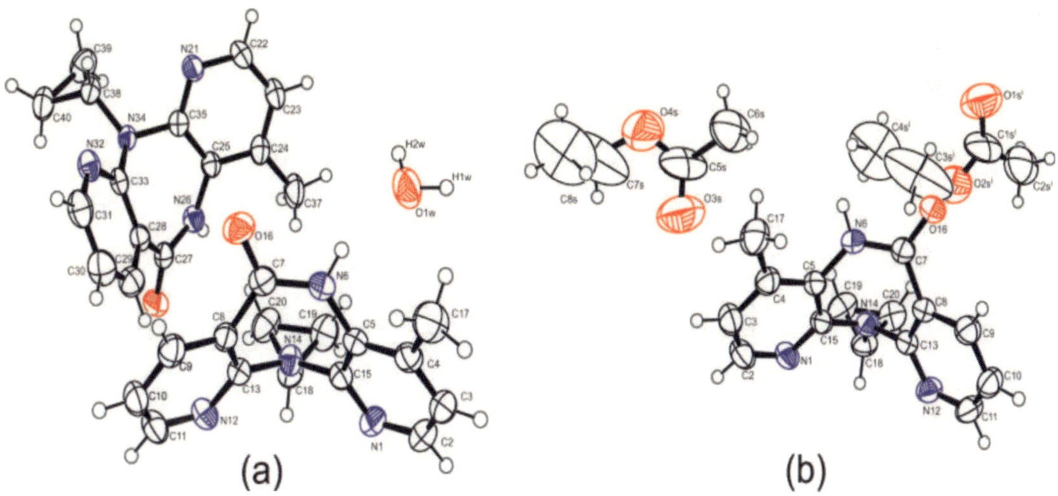

Fig. 4. Ortep representation of structure of (a) nevirapine hemihydrate and (b) nevirapine hemiethyl acetate solvate.

Caira et al have also reported five different solvates of nevirapine (Caira et al., 2008). The initial characterization of all the forms has been performed utilizing DSC and TGA. The results show that the drug molecule displays significant variability in its modes of self assembly while accommodating the different solvent molecules. The DSC results show less thermal stability for toluene solvate while ethylacetate solvate, dichloromethane solvate, hydrate and 1,4-dioxane solvate were stable. The ethylacetate solvate and dichloromethane solvate are isostructural and their dimmers are packed in identical fashion, generating continuous channels parallel to the *b*-axis that accommodate the solvent molecules (Figure 5). However, in the toluene solvate containing a larger guest molecule, the host dimers are packed in a different mode, but the guest molecules are again situated in channels.

The effect of series of alcohols on solvate formation capability of nevirapine molecule has also been illustrated by Caira et al (Caira et al., 2010). The structures of all the solvates were based on a common isostructural framework comprising centrosymmetric hydrogen-bonded nevirapine dimers and contain a common channel parallel to the crystal b-axis in the series which accommodates the various solvent molecules. Thermogravimetric results yielded a guest–host ratio close to 0.5 for the 1-butanol solvate and a steady decrease in this ratio from 0.43 to 0.32 for other solvates. This anomalous stoichiometric variation was resolved following successful X-ray analysis of 1-butanol solvate which revealed that the length of disordered 1-butanol molecule is proportionate with the channel cavity, resulting in a stoichiometric association while in other solvates significant disorder for the solvent molecules was observed which is attributed to their increasing chain lengths being disproportionate with the channel cavity.

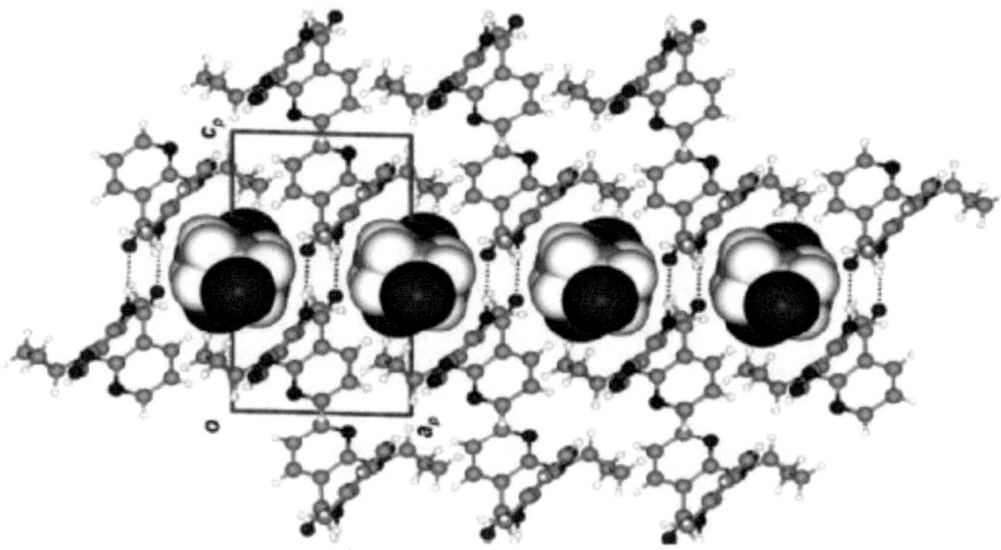

Fig. 5. Inclusion of ethyl acetate molecules within channels in crystals of solvate of nevirapine.

In the view of tendency of solvate formation of nevirapine molecule, other workers have also prepared and characterized the various solvates of nevirapine. In one of the studies, the authors have prepared the solvates by dissolving excess drug in selected solvent systems at 60 °C. However, the choice of solvent system was based on the polarity index of the solvents (Sarkar et al., 2008). Five different crystal forms have been obtained along with an amorphous form. The crystallization of nevirapine under a variety of crystallization conditions resulted in a change in the crystal habit of the drug without change in the internal crystal lattice as suggested by their similar enthalpy of fusion. The mass loss from the TGA was found to be negligible for all the solvates in comparison to the theoretical mass loss indicating that solvents used for crystallization formed weak solvates in this study. However, chadha et al have determined the binding energy of the solvent in the crystal lattice using differential scanning calorimetry which is found to be higher than the enthalpy of vapourization of the corresponding solvent for all the solvates except for toluene solvate indicating that the solvent molecules are tightly bound into the crystal lattice of nevirapine molecule (except in case of toluene solvate) (Chadha et al., 2010). These authors have also calculated the enthalpy of solvation by determining enthalpy of solution of solvate and drug in the solvent which is entrapped in the crystal lattice of the solvate using solution calorimetry technique. The enthalpy of solution when determined in buffer system indicated that out of the six solvates formed ethanol solvate exhibited the maximum ease for molecular release of the solvent molecule from the lattice.

2.3 Efavirenz

The abundant hydrogen bonding sites in efavirenz, another non-nucleoside reverse transcriptase inhibitor, make it a potential candidate to exhibit crystal modifications upon recrystallization. Driven by this aspect, various authors investigated the solid-state structures of recrystallized products of efavirenz. The varying recystallization conditions such as rate of stirring and cooling, antisolvent addition, refluxing/heating, drying under vacuum, and presence of impurities, along with solvents and/or their mixtures with varying polarity have yielded different forms of efavirenz. The patent literature till date reveals 23 different polymorphic forms of efavirenz, one monohydrate and an amorphous form although there is some ambiguity about the actual number of solid forms (polymorphs and solvates) of this API (Radesca et al., 1999, 2004; Sharma et al., 2006; Khanduri et al., 2006; Reddy et al., 2006; Dova, 2008). In these patents, inventors have claimed novel solid forms of efavirenz based on XRPD and DSC analysis. The patent data shows that Form I is the most stable form and all the polymorphs revert to this form under some condition or the other. However, the characterization of these reported forms is not adequate enough to prove them novel.

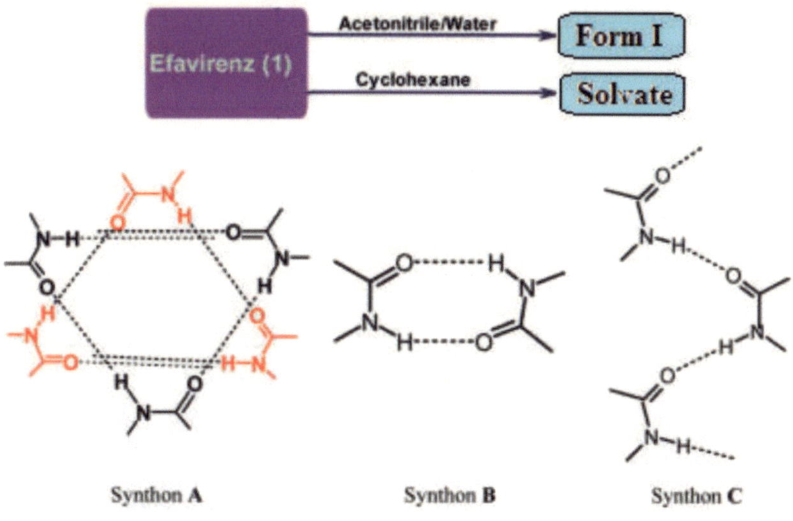

Fig. 6. Solid forms of efavirenz, their transformations and some synthons present in the polymorphs of efavirenz

Recently, two new polymorphc forms structurally characterized by single crystal X-ray diffraction have been reported by Cuffini et al and Ravikumar et al (Cuffini et al., 2009; Ravikumar et al., 2009). These two forms do not correspond to the Form I reported in various patent.

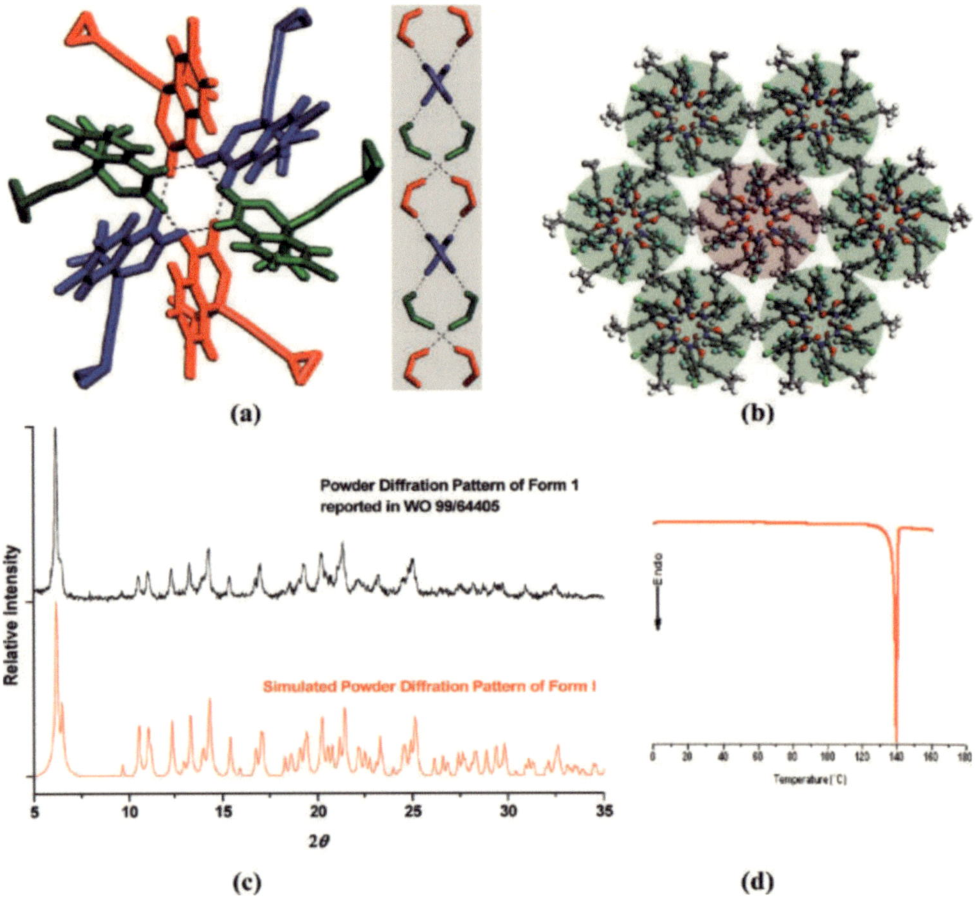

Fig. 7. Efavirenz, form I: (a) A view of the double helical chain viewed down the c-axis with the three symmetry independent molecules represented in red, green, and blue. The inset shows the formation of the double helical chains. (b) Hexagonal close packing of helices. (c) PXRD pattern for form 18g and form I. (d) DSC trace for form I.

The structural information of the stable Form I was first reported by Mahapatra et al (Mahapatra et al., 2010). The authors obtained this form (Form I) at the interface of an acetonitrile-water solvent system as well as from methylcyanide-water mixture (Figure 6). They also reported a solvate of efavirenz which was obtained by dissolving the drug in cyclohexane at 60 °C with stirring and leaving the solution undisturbed for 2 days after filtration to get rectangular crystals. The DSC analysis of these two forms showed that the solvate converts to Form I after desolvation. Both the forms have been further characterized using single crystal X-ray analysis. The efavirenz molecule exhibits a noticeable degree of conformational disorder in the cyclopropyl group. After exhaustive study it was found that Form I crystallized in double stranded helices stabilized by N-H···O molecules (Figure 7). However, in the solvate, two molecules of efavirenz form the amide dimer which are stabilized by C-H···π interactions to form a bilayer stacked in three dimensions to form columns. The solvent molecules are lined in these columns and are stabilized by interactions with CF_3 and cyclopropyl residues (Figure 8).

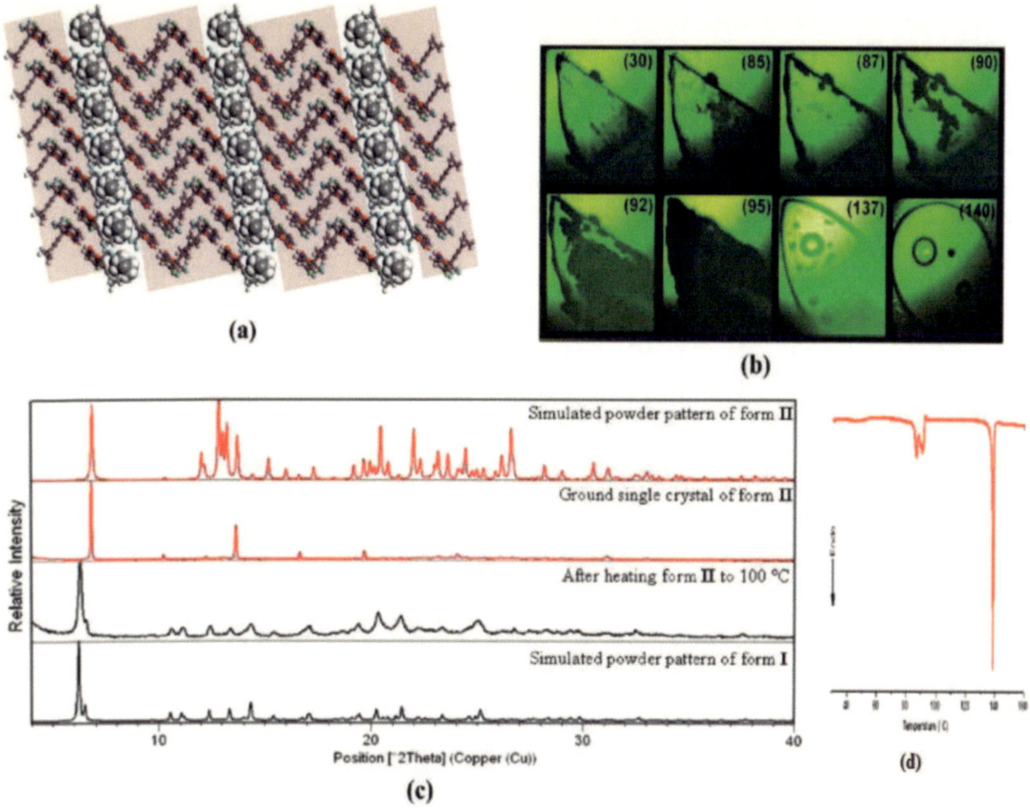

Fig. 8. (a) Crystal packing of efavirenz with the guest cyclohexane molecules represented in the space filling mode (b) HSM images of II (corresponding temperatures are given in parentheses)(c) PXRD patterns of forms I and II and the heated sample of form II (d) DSC plot for form II.

2.4 Lamivudine

Further studies on these categories of drugs have shown the existence of different crystal forms of lamivudine which is a nucleoside analog reverse transcriptase inhibitor. The solid state chemistry of this drug is of significant pharmaceutical interest as the drug is reported to exist in three crystalline forms. The two forms (Form I and II) reported in 1996 were again studied by Harris et al in 1997 (Jozwiakowski et al., 1996; Harris et al., 1997). Later in 2007, a new patent showing the existence of another polymorphic form III appeared (Singh et al., 2007). Michael et al have shown that Form I of lamivudine has been prepared by dissolving Form II in hot water and then adding an equal volume of methanol to reduce the solubility of lamivudine. The Form II has been obtained as a result of a synthetic process. The DSC studies showed heat mediated transformation of Form I to Form II. Besides this, these authors have also calculated the enthalpy of solution from solubility data and compared the results with experimently determined enthalpy of solution by solution calorimetry technique. The enthalpy data revealed that the enthalpy of solution value agrees more closely in systems of low solubility than in system with high solubility.

Harris et al have recrystallized form I of lamivudine as needles from solutions in water, methanol or aqueous alcohols while form II as tetragonal bipyramids on slow recrystallization from dry ethanol and propanol or mixtures of ethanol and less polar organic solvents (Figure 9). Probably, the difference in the polarity of organic solvents led to two different crystalline forms. The authors have used cross-polarization magic angle spinning (CPMAS) NMR to differentiate the two forms. The Form II showed a simple spectrum indicating one molecule in the crystallographic asymmetric unit whereas the spectrum of Form I was found to be extremely complex due to differences in intermolecular packing environment or intramolecular geometry/conformation differences. These results were further confirmed by single crystal X-ray analysis. The Form II was bipyramidal crystals with one molecule in the asymmetric unit and Form I showed five molecules in the asymmetric unit of crystal lattice.

The crystals of Form III are obtained by subjecting the hot saturated solution of lamivudine in water to controlled cooling. The DSC and TGA showed this form to be different from Form I and II. The single crystal X-ray diffraction reveals it to be a hemihydrates with two molecules of water associated with four molecules of lamivudine in a crystal lattice.

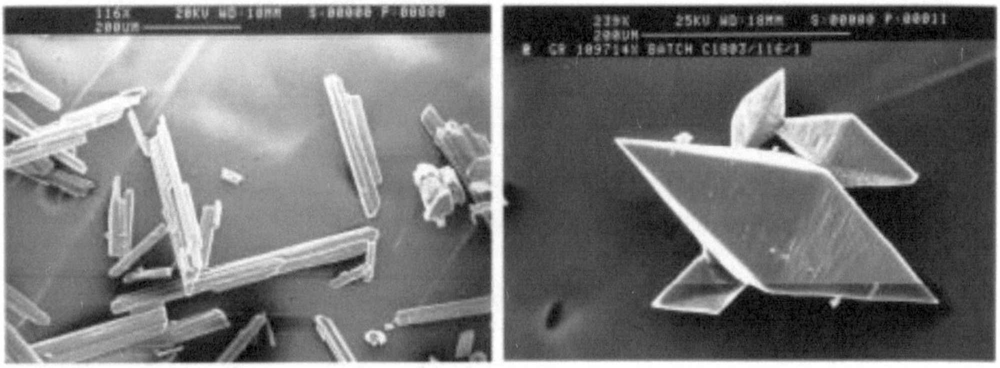

Fig. 9. Scanning electron micrographs of the two forms of Lamivudine

2.5 Stavudine

Stavudine, a thymidine nucleoside reverse transcriptase inhibitor, has been reported to exist in two anhydrous polymorphic forms and one hydrate. Harte et al and Guruskaya et al have reported the single crystal of Form I and Form II respectively (Harte et al., 1991 as sited in Gandhi et al., 2000; Guruskaya et al., 1991 as sited in Gandhi et al., 2000). However, the production of pure Form I i.,s reported by Gandhi et al who have established the conditions of recrystallization governing the formation of this thermodynamically most stable form (Gandhi et al., 2000). The stavudine obtained during the synthesis process is recrystallized from hot organic solvents as the final step in the synthesis. The cooling of hot isopropanol solution from 80 to 70 °C for over an hour and then to 0-5 °C over 1.5 h yielded a mixture of Form I and II. However, Form I was found to be thermodynamically more stable. Therefore, some of the crystallization parameters such as rate of cooling or stirring were studied during recrystallization from isopropanol to selectively obtain Form I. After a lot of experimentation, it was found that slow cooling of hot isopropanol solution reproducibly yielded Form I. Kinetically both forms may be present initially but Form II redissolves and precipitates as Form I with slow cooling.

2.6 Cocrystals

Recently, cocrystallization has emerged as an attractive technique to recrystallize molecular solids that contain two or more distinct chemical components held together by non-covalent interactions. Anti-HIV agents have also been explored by this approach. The cocrystals of efavirenz with 4,4′-bipyridyl and 1,4-cyclohexanedione prepared by recrystallization of their grounded mixture from THF and from a mixture of n-heptane and THF respectively have been reported by Mahapatra et al (Mahapatra et al., 2010). Similarly, the multiple hydrogen bond donor and acceptor groups of lamivudine and zidovudine have been utilized by Bhatt et al for designing their cocrystals (Bhatt et al., 2009). These authors have designed zidovudine cocrystal using retrosynthetic approach where two drug molecules and one molecule of 2,4,6-triaminopyrimidine are held together by a three point synthon which forms basic part of their cocrystal structure (Figure 10).

Besides this, a cocrystal hydrate of lamivudine with zidovudine has been reported. Lamivudine and zidovudine molecules are expected to form synthon II with each other as shown in figure 11. However, during cocrystallization a hydrated 1:1 cocrystal is formed and the synthon formed is the extended IIA synthon rather than II. The observed synthon (IIA) is formed when a molecule of water intervenes in the hydrogen bond pattern of the synthon II. The cocrystallization without use of water resulted in no cocrystal formation in this case, perhaps due to the large repulsions between the carbonyl groups in the two API fragments (Figure 12).

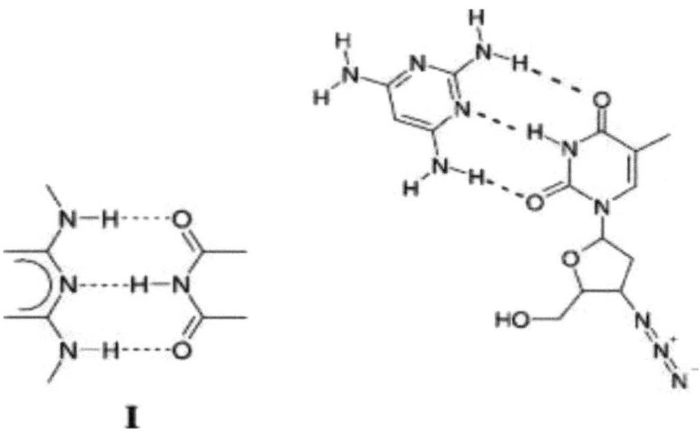

I

Fig. 10. Three-point synthon I and possibility of this synthon between zidovudine and 2,4,6-triaminopyrimidine

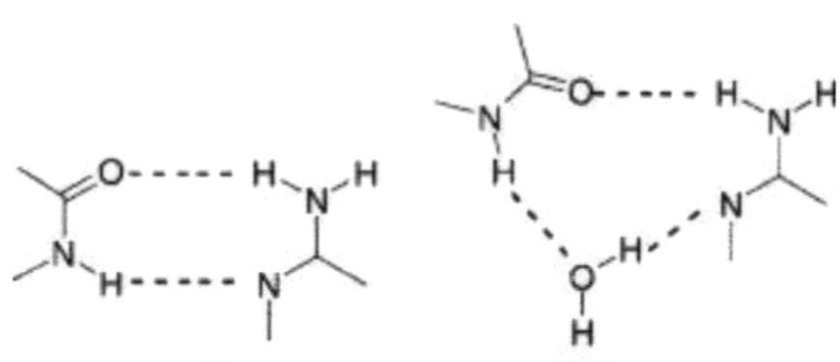

II IIA

Fig. 11. Predicted two-point synthon II between lamivudine and zidovudine and observed synthon IIa in the hydrated co-crystal

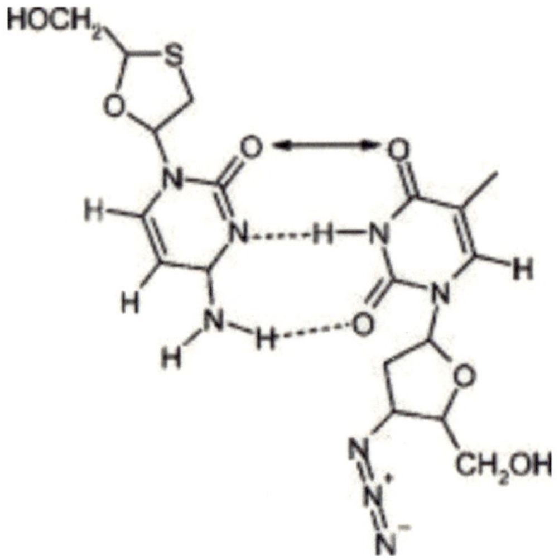

Fig. 12. Possible carbonyl - carbonyl repulsion in the putative lamivudine-zidovudine co-crystal with non-hydrated synthon II

Another cocrystal of lamivudine is designed and prepared with 3,5-dinitrosalicylic acid based upon synthon containing carboxylic acids and 2-aminopyridines. This cocrystal is an example of acid base interaction with a very complex hydrogen bond pattern. These authors also reported cocrystal of lamivudine with 4-quinolinone in a stoichiometry of 1:1. This cocrystal is stabilized by multiple N-H···O and O-H··· O hydrogen bonds. This cocrystal is obtained during the screening process and is not based on any synthon theory. Thus the lamivudine-4-quinolinone cocrystal emphasizes the importance of cocrystal screening to obtain new cocrystals rather than completely depending on the synthon theory.

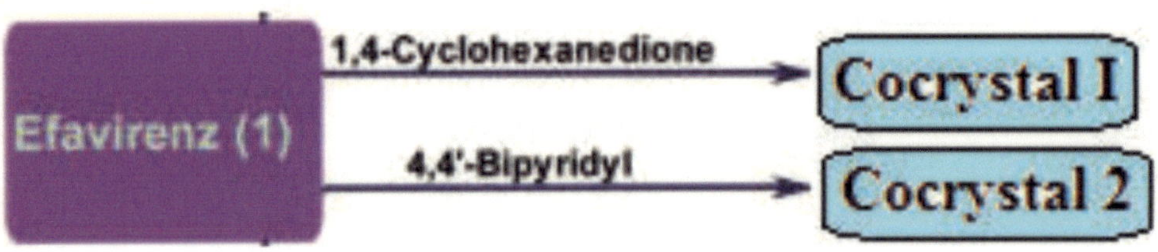

Fig. 13. Cocrystals of efavirenz and their preparation

The possible formation of cocrystals of efavirenz with 1,4-cyclohexanedione and 4,4′-bipyridyl was indicated by X-ray diffraction analysis (Mahapatra et al., 2010). The solids obtained from solvent drop grinding experiment, were subjected to recrystallization in particular solvent systems (Figure 13). Crystals of efavirenz-1,4-cyclohexanedione were obtained from heptane-THF solution. These crystals exhibited two symmetry independent dione molecules in a skewed conformation. One of these two dione molecules makes bifurcated N-H··O/C-H··O hydrogen bonds with two symmetry related molecules of efavirenz, forming a three molecular entity, while the other dione molecule makes linear chains with bifurcated C-H...O hydrogen bonds. These two patterns stacked over one another to form grill-ribbon structure of the cocrystal (Figure 14).

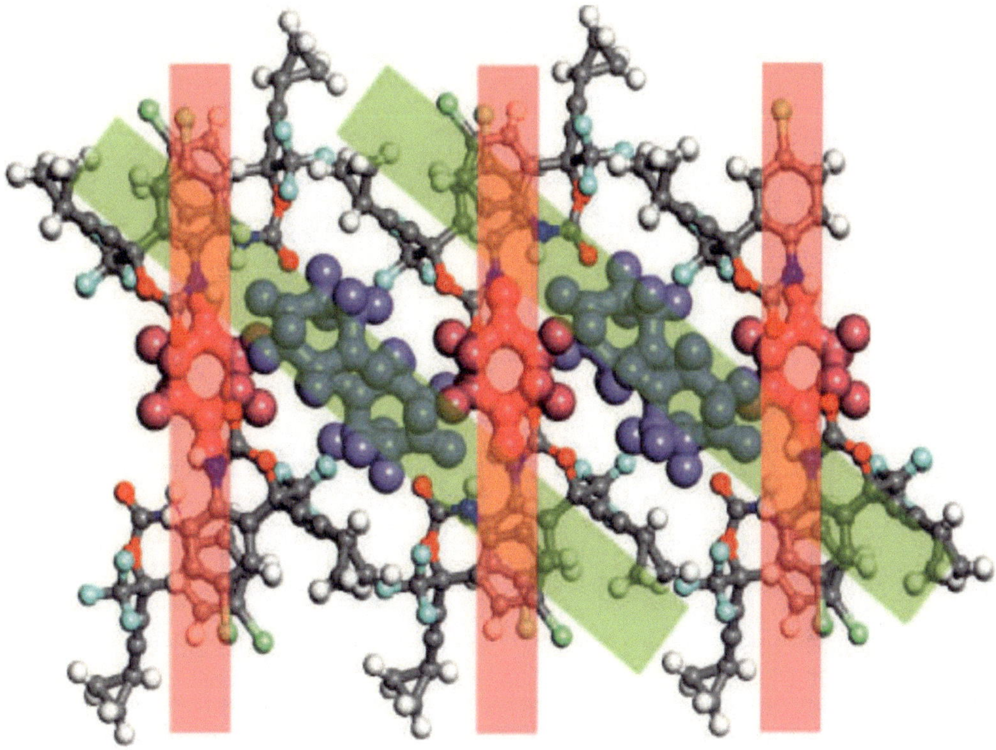

Fig. 14. Cocrystal of efavirenz and 1,4-cyclohexanedione: crystal packing along with the conformation in the native dione crystal (green).

The cocrystal of efavirenz with 4,4′-bipyridyl was obtained by recrystallization from acetonitrile. Unlike the cocrystal with yclohexanedione in which the major synthon couldnot be anticipated, in this cocrystal two distinct heterosynthons were observed. While one of the efavirenz molecules interacted with bipyridyl through cyclic N-H···N/C-H·O hydrogen bonds, the other end of pyridine compound made a single point N-H···N hydrogen bond with other molecule of efavirenz resulting in pillered assemblies separated by hydrophobic cyclopropyl residues (Figure 15).

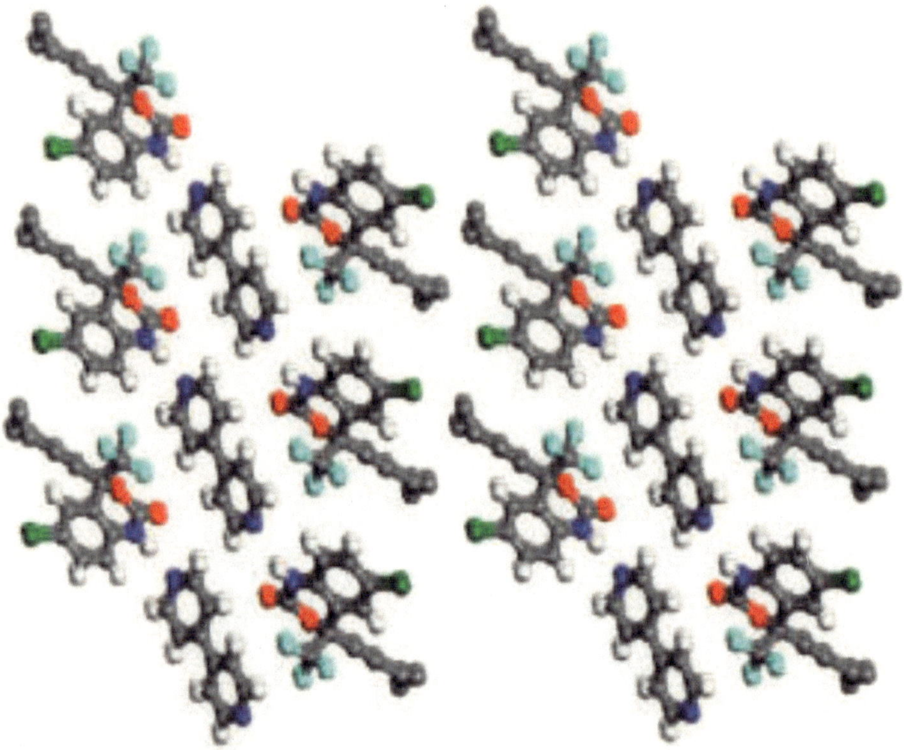

Fig. 15. Crystal packing.of efavirenz- 4,4′-bipyridyl cocrystal

From the above examples it is clear that solvent recrystallization method has been the most common technique to prepare different crystal forms of a compound. However, with increasing demands to accelerate the process of crystal form identification, newer and advanced technologies need to be used as is exemplified by the isolation of three more polymorphic forms of ritonavir by high-throughput crystallization. In addition, cocrystallization technique has also been employing solution based crystallization methods to generate cocrystals of various pharmaceutical compounds. The studies on anti-HIV agents cited here are thus not simply academic exercises but have practical implications in preparation and identification of diverse crystalline forms of pharmaceutical material and can be of considerable benefit in formulation optimization.

3. References

Bauer J., Spanton S., Henry R., Quick J., Dziki W., Porter W., Morris J. Ritonavir: An extraoridinary example of conformational polymorphism. *Pharm. Res.*, 18, 859-866 (2001).

Bhatt, P.M., Azim, Y., Thakur, T.S., Desiraju, G.R. (2009). Co-crystals of the anti-HIV drugs lamivudine and zidovudine. *Cryst. Growth Des.*, *9*, 951–957.

Caira, M.R., Stieger, N., Liebenberg, W., De Villiers, M.M., Samsodien H. (2008). Solvent inclusion by the anti-HIV drug Nevirapine: X-Ray structures and thermal decomposition of representative solvates. *Cryst. Growth Des.*, 8, 17-23.

Chadha, R., Arora, P., Saini, A., Jain, D.V.S. (2010) Solvated Crystalline Forms of Nevirapine: Thermoanalytical and Spectroscopic Studies . *AAPS PharmSciTech.*, 11, 3, 1328-1339.

Chemburkar, S. R., Bauer, J., Deming, K., Spiwek, H., Patel, K., Morris, J., Henry, R., Spanton, S., Dziki, W., Porter, W., Quick, J., Bauer, P., Donobauer, J., Narayanan, B. A., Soldani, M., McFarland, D., McFarland K. (2000). Dealing with the impact of ritonavir polymorphs on the late stages of bulk drug process development. *Org. Process res.* deve., 4, 413-417.

Cuffini, S., Howie, R. A., Tiekink, E. R. T., Wardelld, J. L., Wardelle, S. M. S. V. (2009). (S)-6-Chloro-4-cyclopropylethynyl-4-trifluoromethyl-1H-3,1-benzoxazin-2(4H)-one. *Acta Crystallographica Section E, Cryst.*, E65, o3170–o3171.

Desikan, S., Parsons, Jr. R. L., Davis, W. P., Ward, J. E., Marshall, W. J., Toma, P. H. (2005). Process development challenges to accommodate a late appearing stable polymorph: A case study on the polymorphism and crystallization of a fast-track drug development compound. *Org. Process res. deve.*, 9, 933-942.

Dova, E. (2008). Polymorphic forms of efavirenz. WO 2008/108630 A1.

Gandhi, R.B., Bogardus, D.E., Bugay, D.E., Perrone, R.K., Kalpan, M.A. (2000). Pharmaceutical relationships of three solid forms of stavudine. *Int. J. Pharm.*, 201, 221-237.

Gurskaya, G.V., Bochkarev, A.V., Ahdanov, A.S., Dyatkina, N.B., Kraevskii, A.A. (1991). X-ray crystallographic study of 2′,3′-dideoxy-2′,3′-didehydrothymidine conformationally restricted termination substrate of DNA polymerases. *Mol. Biol.*, 25, 483.

Harris, R. K., Yeung, R. R., Lamont, R. B., Lancaster, R. W., Lynn, S. M., Staniforth, S. E. (1997). 'Polymorphism' in a novel anti-viral agent: Lamivudine. *J. Chem.. Soc., Perkin Trans.*, 2, 2653-2654.

Harte, W.E., Starrett, J.E., Martin, J.C., Mansuri, M.M. (1991). Structure studies of the anti-HIV agent 2′,3′-didehydro-2′,3′-dideoxythymidine (d4T). *Biochem. Biophys. Res. Comm.*, 175, 298.

Jozwiakowski, M. J., Nguyen, N-A T., Sisco, J. M., Spancake, C. W. (1996). Solubility behavior of lamivudine crystal forms in recrystallization solvents. *J. Pharm. Sci.*, 85, 193-199.

Khanduri, H. C., Panda, A. K., Kumar, Y. (2006). Processes for the preparation of polymorphs of efavirenz. WO 2006/030299 A1.

Mahapatra, S., Thakur, T. S., Joseph, S., Varughese, S., Desiraju, G. R. (2010). New Solid State Forms of the Anti-HIV Drug Efavirenz. Conformational Flexibility and High Z′ Issues New Solid State Forms of the Anti-HIV Drug Efavirenz. Conformational Flexibility and High Z′ Issues. *Cryst. Growth Des.*, *10*, 7, 3191–3202.

Miller, J. M., Collman, B. M., Greene, L. R., Grant, D. J., Blackburn, A. C. (2005). Identify the stable polymorph early in drug development process. *Pharm. Dev. Technol.*, 10, 291-297.

Morissette, S.L., Soukasene, S., Levinson, D., Cima, M.J., Almarsson, O. (2003). Elucidation of crystal form diversity of the HIV protease inhibitor ritonavir by high-throughput crystallization. *Applied physical sciences*, 100, 2181-2184.

Pereira, B. G., Fonte-Boa, F. D., Resende, J.A.L.C., Pinheiro, C.B., Fernandes, N.G., Yoshida, M.I., Vianna-Soares, C.D. (2007). Pseudopolymorphs and intrinsic dissolution of Nevirapine. *Cryst. Growth Des.*, 7, 2016-2023.

Radesca, L., Maurin, M., Rabel, S., Moore, J. (1999). Crystalline efavirenz. WO 99/64405.

Radesca, L., Maurin, M., Rabel, S., Moore, J. (2004). Crystalline efavirenz. US 6,673,372 B1.

Ravikumar, K., Sridhar, B. (2009). Molecular and crystal structure of efavirenz, a potent and specific inhibitor ofHIV-1 reverse transcriptase, and its monohydrate. *Mol. Cryst. Liq. Crystal*, 515, 190-198.

Reddy, B.P., Rathnakar, K., Reddy, R.R., Reddy, D.M., Reddy, K.S.C. (2006). Novel polymorphs of efavirenz. US 2006/0235008.

Reguri, B.R., Chakka, R. (2005). Novel crystalline forms of 11-cyclopropyl-5,11-dihydro-4-methyl-6H-dipyrido[3,2-b:2′,3′-e][1,4]diazepin-6-one (Nevirapine). United States Patent 0059653A1.

Reguri, B.R., Chakka, R. (2006). Crystalline forms of Nevirapine. United States Patent 0183738 A1.

Sarkar, M., Perumal, O.P., Panchagnula, R. (2008). Solid-state characterization of nevirapine. *J. Pharm. Sci.*, 70, 619-630.

Sharma, R., Bhushan, H.K., Aryan, R.C., Singh, N., Pandya, B., Kumar, Y. (2006). Polymorphic forms of efavirenz and processes for their preparation. WO 2006/040643 A2.

Singh, G.P., Srivastava, D., Saini, M.B., Upadhyay, P.R. (2007). A novel crystalline form of lamivudine. WO 2007/119248 A1.

Stieger, N., Liebenberg, W., Wessels, J.C., Samsodien, H. Caira, M. R. (2010). Channel inclusion of primary alcohols in isostructural solvates of the antiretroviral nevirapine: an X-ray and thermal analysis study. *Stru. Chem.*, 21, 771-777.